THE CERTIFICATE LIBRARY

* * *

REGIONAL GEOGRAPHY

OCL/GEO 2/A

THE CERTIFICATE LIBRARY

THIS SERIES of books has been designed to cover the various syllabuses of the Ordinary Level Examination for the General Certificate of Education. The books are equally suitable for all students who wish to improve their standards of learning in the subjects of the various volumes, and together they form an invaluable reference library.

The Editor of each book is an experienced Examiner for one or more of the various Examination Boards. The writers are all specialist teachers who have taught in the classroom the subject about which they write; many of them have also had experience as Examiners.

Each volume is comprehensive and self-contained and therefore has far more than an ordinary class textbook: each has a full treatment of the subject; numerous illustrations; questions on the chapters with answers at the end of the book; and a Revision Summary for each section of the book.

These summaries give the information in a concise and compact way so that examination candidates may easily revise the whole contents of the book—they can soon see how much they readily know, and how much they have forgotten and need to study again. The facts in these summaries serve, therefore, as a series of major pegs upon which the whole fabric of the book hangs.

Advice is given on answering the examination paper and there are questions of examination standard together with suggested answers.

Series Executive Editor

B. E. COPPING, B.A.HONS.

THE CERTIFICATE LIBRARY

* * *

REGIONAL GEOGRAPHY

Edited by

K. WALTON, M.A., Ph.D.

Professor and Head of Geography Department, University of Aberdeen

Contributors:

R. E. H. MELLOR, B.A.

Professor of Geography, University of Aberdeen

R. B. SALMON, M.A.

Principal Lecturer in Geography, Moray House College of Education, Edinburgh

AGNES M. LEARMONTH, M.A.

Formerly Geography Mistress, Merchant Taylors' School for Girls, Liverpool

W. J. KING, M.Sc.

Formerly Senior Geography Master, Wembley County School

A. SMALL, M.A.

Senior Lecturer in Geography, University of Dundee

and

K. WALTON

THE GROLIER SOCIETY LIMITED

LONDON

First published 1965
Reprinted 1968
Revised 1971

ISBN: 0 7172 7708 9

Made and printed in Great Britain by
Odhams (Watford) Ltd., Watford, Herts

PREFACE

THE increasing complexity of man's relationship with his environment and the wide variations within even a single national unit are difficulties that must be overcome when producing a short regional geography of the world in one volume. Obviously it would be easy simply to list the features of relief, economic products and basic population data for every country, but the result would not be geography.

Therefore each contributor to this book has selected those aspects which are the most important to the understanding of the geography of the region. The length of the treatment is roughly proportional to the importance of the area in the syllabuses of the various examining boards.

It must be stressed that general and regional geography are complementary. An understanding of man's relationships with his regional environment demands, as a basis, an adequate knowledge of land-forms, soils and vegetation. Both are required to make the study of geography complete and relevant to the world we live in and the life we lead.

This volume has been prepared with the G.C.E. "O" Level examinations in mind, but the reader will find that the many maps and photographs illustrate not only the academic study of geography but its social consequences and meaning also.

K. WALTON

Note to 1971 *edition:* In this revision the chapters describing Britain have been entirely re-written in order to give an up-to-date concept of the situation today.

CONTENTS

THE BRITISH ISLES by K. WALTON

EUROPE AND THE U.S.S.R. by R. E. H. MELLOR

AFRICA by R. B. SALMON

SOUTH ASIA by A. M. LEARMONTH

NORTH AMERICA by W. J. KING

SOUTHERN CONTINENTS by A. SMALL

Coloured Maps and Illustrations

M4 Motorway, Chiswick.

THE CHANGING PATTERN OF BRITAIN

Oil exploration derrick, North Sea.

Dounreay experimental nuclear reactor.

THE BRITISH ISLES

INTRODUCTORY

WHERE better to begin a study of regional geography than with the British Isles where live nearly fifty-seven million people in an area of approximately 120,000 square miles situated, at the closest point in the Straits of Dover, only 22 miles from the continent of Europe? The islands rise from the shallow continental shelf of north-west Europe which has recently proved to be productive in major power resources—natural gas and oil. The indented coastline permits the penetration deep into the islands of warm oceanic influences which raise the winter temperatures to values higher than those experienced on the adjacent continental land masses (50°-60°N).

The deep inlets on both the eastern and western sides of the islands have also allowed shipping to find good natural harbours well inland, while the comparatively large tidal range provides deeper water every twelve hours in the shallower estuaries and firths and allows large vessels to navigate many miles inland without, until recently, the need for dredging to provide deep water channels. Moreover, some of the estuaries in south-east England face the outlets of the major waterways of Western Europe, the Rhine and the Scheldt, which carry now, as in the past, much of the goods traffic of Western Europe. Where these rivers meet the North Sea the major towns of the Low Countries are situated, so that south-east England was especially well placed for trade with the main concentrations of population on the continent of Europe. But those same inlets, the Thames, the Wash, the Humber, the Tyne, the Forth, Tay, Clyde, Solway, Mersey and Severn, are now an embarrassment, in that their wide estuaries obstruct land communications. Many have now been bridged, e.g. the Severn and the Forth, while others such as the Mersey and Clyde have been tunnelled.

Early Immigrants

Into these inlets, and to the other coastlands, came the immigrants of prehistoric and historic times. We would scarcely recognize the islands they encountered; the hills and valleys would seem familiar if we were to return in a time-machine, but we would be much confused by the wild appearance

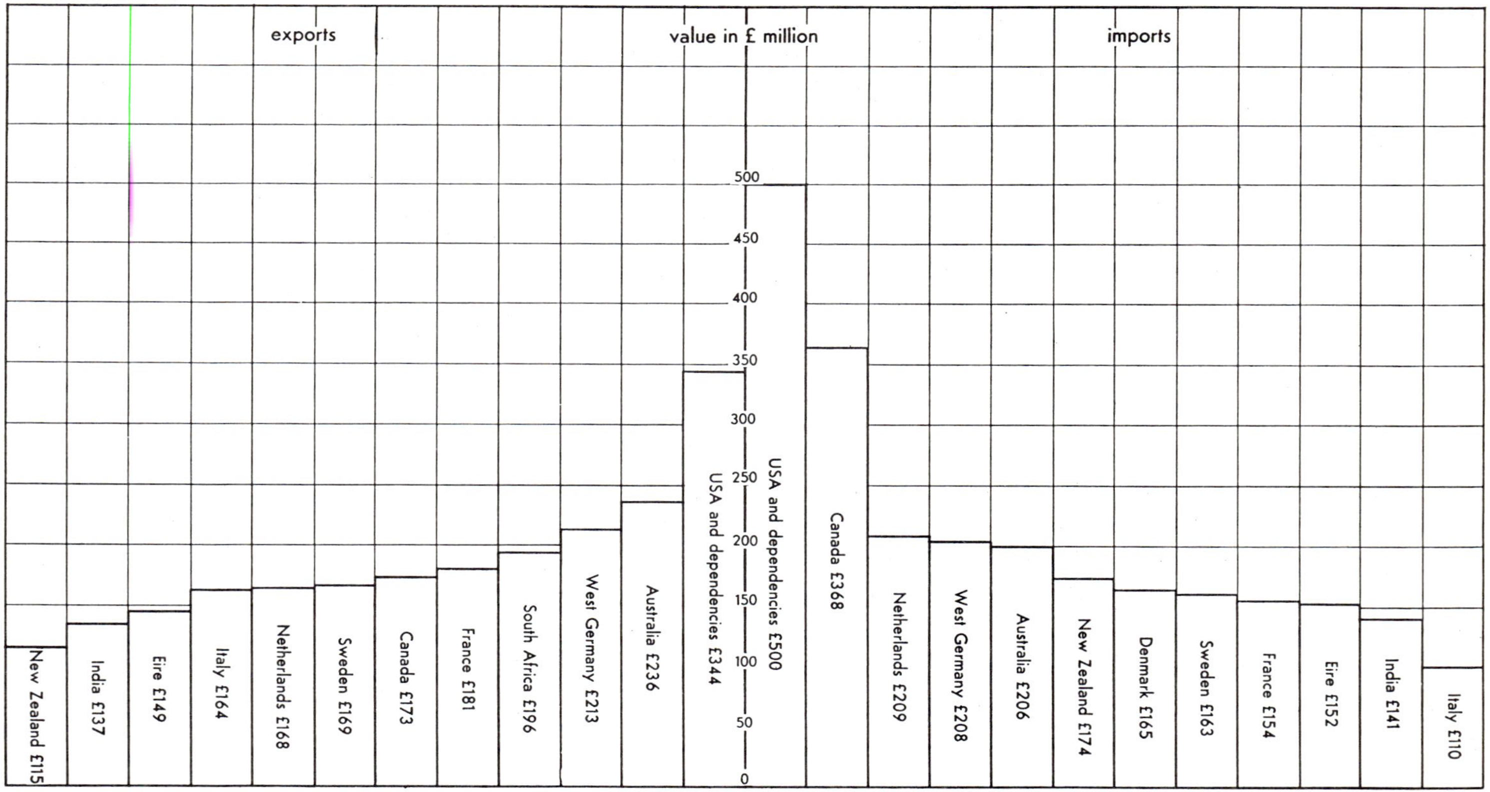

imports

Italy £110
India £141
Eire £152
France £154
Sweden £163
Denmark £165
New Zealand £174
Australia £206
West Germany £208
Netherlands £209
Canada £368
USA and dependencies £500

value in £ million

500 450 400 350 300 250 200 150 100 50 0

exports

USA and dependencies £344
Australia £236
West Germany £213
South Africa £196
France £181
Canada £173
Sweden £169
Netherlands £168
Italy £164
Eire £149
India £137
New Zealand £115

FIG. 1. *Trade: Britain's chief customers and suppliers in a recent typical year.*

of the land with forests, swamps, bogs and moors where now there is an ordered pattern of fields, hedgerows, stone walls, roads, houses and towns. Over the last few thousand years men have worked to transform the land as, according to their skills and inclinations, they have made use of its wide range of natural resources. The imprint of their activities in the past is still with us at the present day. The tumuli, the defensive mounds, the castles, the walled towns whose gates offer problems to modern traffic, the pattern of cultivated and forested land—all these represent earlier attempts to use the natural resources of the land to the best advantage according to the needs of the time.

In some areas, in contrast, it is the coal mines, the canals and the railways, the factories and mills which attract our attention. Here the emphasis in the past was on the use of mineral resources, the transformation of raw materials, such as wool, into clothes and blankets, and the transportation of the minerals and manufactured products to the consumers both within the islands and abroad. Very little, if any, of the landscape of Britain now remains unmodified. The patterns created by the use and modification of the natural resources of earth, climate and water help to differentiate the landscape into areas which are sufficiently distinctive to have received regional names, such as the Black Country, Fens, Lake District.

Exploitation of Resources

So effective has been the exploitation of the natural resources, and so well organized is the system of overseas trade, that the land of Britain can now support its large population at a high standard of living. Without overseas trade, however, this would be impossible, since the amount of land available for food-production in a west maritime climate is relatively limited. There are many areas in Britain where the ground is too high, the climate too severe and the soils too poor for productive farm land; in fact less than half the land area of the country produces food. Fortunately, some areas provide the raw materials for industry—such as coal, iron and limestone—and it is through the sale of the products of its manufacturing industry that Britain can obtain foodstuffs from abroad to supplement home production.

This means that the country depends for the maintenance of the numbers of the people and their standard of living on very intensive use of the agricultural land which is available, the products of its manufacturing industries, and, above all, on the ability of its inhabitants to sell their products abroad. One of the remarkable facts about the geography of Britain is that so many people live at such a high standard of living on such a small area of the earth's surface which is separated from its markets and raw materials by the sea. What is even more remarkable is the importance of Britain in the world where, in spite of the reduction in its overseas possessions, its language is so widely spoken, its systems of law and government so much used, and its services in banking, insurance, scientific and technical knowledge so much in demand.

Population

One aim of studying the regional geography of the British Isles is to discover the reasons for the high average population density (nearly 500 per square mile) and for its variations, which range from uninhabited areas to the vast concentration of people in the Greater London area, where there were nearly eight million inhabitants in 1966: excluding Greater London, 23 per cent of the population of England live on only 4 per cent of the land surface of England. Many, of course, live in those other great concentrations of population made up of towns and cities which merge into one another—the large urban areas called *conurbations*. The conurbation of south-east Lancashire centred on Manchester contained, in 1968, almost 3¼ million people, while in the conurbation of the West Midlands, centred on Birmingham, lived 2,374,100 people in the same year. By comparison, in the north of Scotland, in mid-1968 only 47,600 people occupied the 14,204 square miles of Inverness-shire (an area 400 times larger than that of the south-east Lancashire conurbation). The density of population in the county of Inverness-shire is about 11 per square mile, while in south-east Lancashire it is nearly 6,000 per square mile. Clearly, by world standards some parts of the British Isles are very crowded, whether one considers the average of 500 per square mile or the maximum densities which may be over ten times that number.

If we consider the employment of the population, some further basic points emerge about the geography of the British Isles. A look back into the history of the use of the unevenly distributed natural resources is necessary for the understanding of the uneven pattern of present-day economic activity. Before the Industrial Revolution, the majority of the population were engaged in tilling the soil, especially for cereals, or in animal husbandry. In 1750 there were only 12 million people, and the farms could normally supply sufficient food for them, although there were periodic famines and food shortages when the growing seasons were colder, wetter or shorter than usual. Even in 1801, when the Industrial Revolution had already begun, and the population had increased to 16½ million, the people were eating bread and meat produced by their farming fellow-citizens. The latter were clustered most thickly where conditions of relief, climate and soil were most favourable.

The southern part of England offered the best advantages for growing crops, and the farmers had a ready market in London where the population had already risen to one million. One-sixteenth of the total population of the British Isles was thus concentrated in London in 1801. Nevertheless, new industrial developments on the coalfields of the Midlands and the North and especially in the coastal coalfields of north-east England and South Wales were beginning to draw people from the land to manufacturing towns outside London.

By the year 1851, the population had risen to $27\frac{1}{2}$ million—more than double that of 1751. More than half now lived in the towns of the coalfields, in the ports, and in London. The process of rural depopulation was already well established, as the farmworkers, attracted by the higher wages of the factories and coal mines, left their fields and rural cottages to work at the looms of the cotton and woollen mills, in the iron foundries and in the shipyards. Simultaneously, as the towns of the industrial areas increased in size, the factories and the houses for their workers were built on what had been farming land. The great increase in the population in the century between 1751 and 1851 was in part the result of increased prosperity, and in part the result of improvements in public health, which reduced infant mortality and gave greater expectancy of life to the individual in spite of recurrent epidemics, very hard conditions of work, and poor housing in the towns. The increase in population began to create problems of food supply; no longer were the farmers of Britain able to supply enough grain for its inhabitants, although the Corn Laws still protected them from foreign competition.

By 1911, three years before the outbreak of the First World War, the population had risen to 45 million. The move of people from the country to the town had become even more marked, with three-quarters of the population now living in the towns. The livelihood of the inhabitants was more and more concerned with manufacturing than with farming. The export of manufactured goods had increased to the extent that Britain was called "the Workshop of the World." But the people in the workshops fed less and less on food produced in Britain. The repeal of the Corn Laws in 1869, the surplus food available from North America, Australia, New Zealand and South America, and the ability to pay for imported food by the high volume of exported manufactured goods resulted in the production within the British Isles of only two-fifths of their food, even though the fewer farmers were cultivating the land more intensively and specializing in their products. No longer was there the concentration on cereals, which had been the dominant crop in 1851.

Economic Development

The changes in the numbers and types of employment of the population of the country thus originated with the economic developments of the nineteenth century. They produced an economy based on the importation of foodstuffs and of an increasing range of raw materials for the factories. To pay for the imports, the country exported iron and steel goods, machinery, ships and textiles, and also received payment for services rendered to other countries in banking, insurance and sea transport. Investments by private citizens and by the Government in foreign countries, to provide finance for industry and transport, brought in a high income. The overseas colonies were ready markets for the manufacturers of Britain, and provided the raw materials of both temperate and tropical lands to supplement the natural resources of the homeland.

To meet the demands of customers at home and overseas, the industry of the country was concentrated where power was available—i.e. on the coalfields, or in the ports to which the imported raw materials were brought, or in London where the large population provided the biggest individual internal market. Coalfields situated near the coast, as in South Wales, North-east England, West Cumberland and Scotland, had great advantages over the inland coalfields, since the adjacent industries were in a favoured location for the export of their manufactured goods. Industries on the coastal coalfields concentrated in a very narrow range of industries often with very specialized types of employment; in fact, in one town the whole of the labour force might be employed in one industry or even one firm, as was to happen at Jarrow on Tyneside in north-east England. Moreover, coalfields on the coast were better placed for exporting coal, so that in 1913 they were exporting more than 50 per cent of the coal mined, compared with only 13 per cent from the inland coalfields, such as those of the Midlands.

The dependence of Britain on coal as a fuel and the concentration of "heavy" industries on the coalfields has been stressed. But techniques were bound to change, as they were also in the very countries which had been the markets for British exports. These countries, especially during and after the First World War, developed their own heavy industries, built their own ships, and sought local fuels to replace coal imported from Britain.

Technological Revolution

Since 1918 there has been a revolution in technology in Britain which is as important as that which occurred during the Industrial Revolution. New techniques have led to great changes not only in the types of manufactures but also in the distribution of the manufacturing industries and, with them, the distribution of the population. No longer is the industry of Britain dependent on local supplies of coal. Power can be obtained through the generation of electricity by coal, oil, water power, nuclear reactors, and natural gas, and transmitted long distances at high voltages through the National Grid. Pipelines have been, and are being, constructed to carry oil and natural gas from the tanker terminals and from under the North Sea. Industry, freed from its dependence on the coalfields, has moved to more suitable locations, especially to areas with a large labour force with a wider variety of skills than those possessed by the people of the specialized heavy

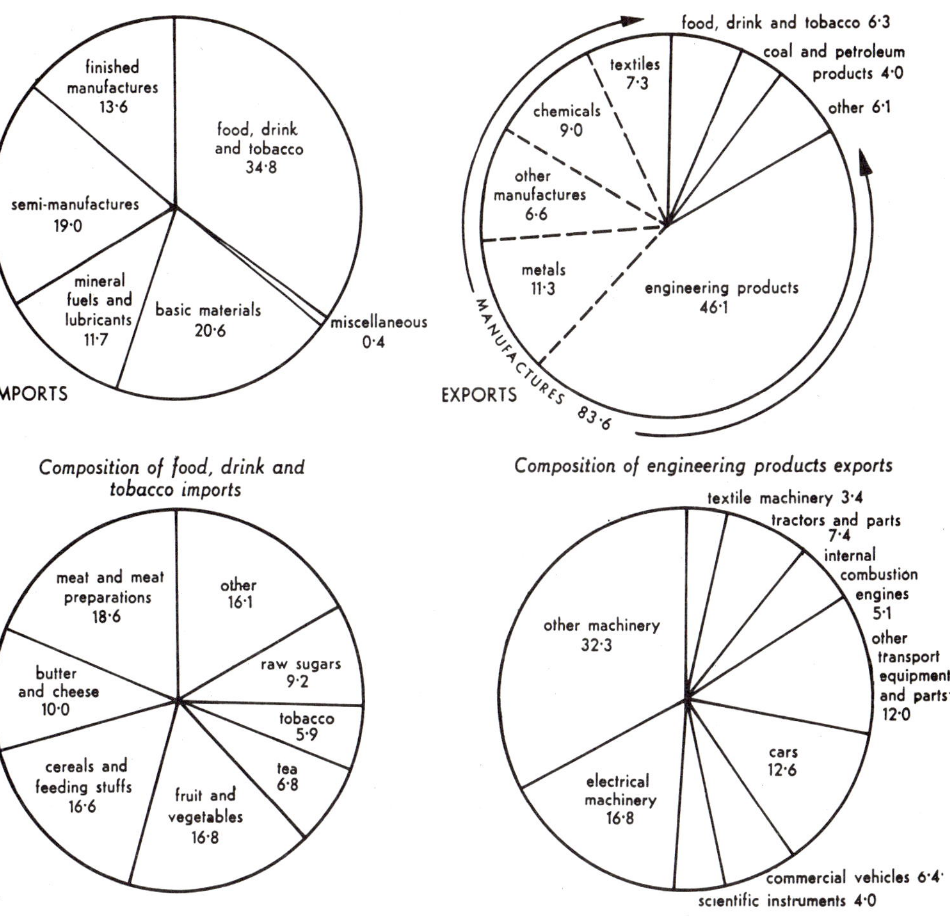

FIG. 2. *Percentage composition of Britain's exports and imports.*

industries of the coastal coalfields. The new locations are often close to the markets, to the concentrations of population and to the points from which their products can be exported and their raw materials easily derived.

New Industries

The changes in the provision of power supplies have been paralleled since 1918 by the introduction of new industries and the rapid growth of science and technology. New industries include the manufacture of road vehicles and aircraft, gramophones and radios—previously produced, if at all, at high cost and on a small scale with few workers in small factories. During the last fifty years the expansion in these industries has been enormous. So also has the manufacture of textiles using man-made fibres such as Nylon and

Terylene as a substitute for, or in conjunction with, the natural fibres of wool and cotton, jute and hemp. Huge petro-chemical factories using imported oil produce the raw materials for plastics which are often a substitute for steel, tin-plate and wood in many domestic articles and are replacing paper for packaging food and other products. The expansion of the chemicals industry to supply new demands is one of the noteworthy changes in industry during the period. To this list can be added cosmetics, ready-made clothing, the components for the motor industry, furniture, electronic components (including transistors) for radio and television, cigarettes and tobacco, electric motors, washing machines and soap powders, and a host of other new products. The scale and range of these new industries is very great.

Compared with the "heavy" industries (iron and steel, shipbuilding, etc.) the new industries do not utilize a great amount of raw material. Their products have a relatively high value compared with their weight, and since some are simply assemblies of components (e.g. television sets) the skills required for their production are not the skills associated with the old industries. Many new manufactured products have to be moved relatively short distances, since the factories have been built near or in the main centres of the population. Many "consumer goods" are moved by road rather than rail, and the fact that the articles can be loaded at the factory and sent direct to local distributors lowers the cost of distribution.

Transport

Thus the railways, which were built to carry the coal, iron-ore, limestone and heavy raw materials and to distribute their products, have now lost some of their former importance for goods traffic. Moreover, passengers whom the railways used to carry in such large numbers now often travel by road except in the vicinity of the large towns and conurbations, where traffic congestion, and the resulting slowness of the journey to work, has resulted in the very heavy "commuter" rail traffic of, say, the Greater London area.

Less traffic on the railways and more traffic on the roads has had a two-fold result. The railway system of Britain has been drastically curtailed and many uneconomic lines closed, while during the same period roads have been widened or completely new "motorway" systems have been built. Attempts to combine the advantages of rail, road and sea traffic are being made through the "container" system, in which the goods are packed at the factory in huge steel boxes which can be carried by road, sea or rail. New rail terminals have been built to handle the "Freightliner" traffic, while some of Britain's port installations have been adapted to speed the flow of imports and exports carried by "container ships". Special oil tanker terminals near, or connected to, refineries by pipelines speed up the flow of oil from the oil-producing areas overseas to the factories and petrol pumps.

FIG. 3. *The new base at Tilbury docks for container traffic.*

The Situation Today

How different is the geography of Britain in the second half of the twentieth century from that of the eighteenth or even the nineteenth century! Now less than 5 per cent of the population of the United Kingdom is directly engaged in farming, forestry or fishing, and they produce only half of the food requirements of England, Scotland, Wales and Northern Ireland (in the Republic of Ireland, however, almost 40 per cent of the population still work on the land). In the United Kingdom, by far the largest proportion of the working population (40 per cent now live in the towns or their rural fringes and work in factories, mines and quarries; another 20 per cent are employed in the distribution and retail industries and in transport and communication; while almost 10 per cent are engaged in building houses, factories, shops and roads (the construction industry). The remaining 25 per cent are in the professional services, i.e. banking, insurance, teaching, medicine, local and national government, and in the defence services.

Such changes in employment, in land-use, and in the location of industry are reflected in the present-day distribution of the population. This still parallels the distribution of mineral resources, the quality of farming land, and, to a lesser extent, the potential of the land for forestry and the sea for fish; in other words, it still reflects the *resource base* of Britain. But the distribution of the inhabitants is not only the result of variations in the quality of the resource base: people tend to accumulate where people already are, so that the more densely populated attract people from the less densely populated areas.

Movement of the population in response to opportunities for employment and a high standard of living leads to a situation in which too many people live in too small an area of the island. The "drift to the South," the increase in the size of the large cities and the extension of their influence over wide rural areas, and the declining population and employment opportunities in remote districts have led to Government intervention. The state now provides inducements to manufacturers to move to areas of industrial decline, creates new towns, provides subsidies to farmers, and designates some areas as in special need of economic assistance. Effectively, this means that the Government, at both national and local levels, is a factor in the changing distribution of the population.

QUESTIONS

1. Why is Britain able to maintain such a density of population at such a high standard of living?
2. Show how the distribution of population in Britain has changed since the Industrial Revolution.

CHAPTER 1

THE LAND

THE resource base of a country is the combination of relief, geology, soils, climate, water and vegetation which constitutes the *physical environment*. Each of the elements may be important in itself, but usually it is the variation in the physical environment as a whole which causes the potential of the resource base to vary from one part of the country to another. People are also part of the resource base, since without them there can be no exploitation of the natural resources. Moreover, men modify the physical environment by accelerating or completely altering the processes and links between its various components, so that they themselves become agents of physical change. The changes men produce, and the parts of the environment which they use, will, of course, vary according to their needs and their technology.

The Romans, for instance, made great changes in the landscape of southern England but did not use the resources of the Highlands of Scotland; in fact, the major resources of the Scottish Highlands—the reserves of water for hydro-electric power and the snow for skiing—have had to wait until the middle of the twentieth century to be exploited. Clearly a resource only becomes a resource when the people at any period in history recognize it to be of value for their economic or leisure activities.

Let us deal first with the relief, geology and soils of Britain since, within the time-scale of human occupation of the island, they have been *relatively* the most stable elements of the resource base. The word *relatively* is stressed because there have been profound changes in the coasts of the island since, say, 250,000 years ago, when the woman whose skull fragment was found at Swanscombe, Kent, was living in the Thames valley. At that time the island was part of the Continent of Europe, and much of the floor of the North Sea was dry land which could be crossed by the elephant and hippopotamus, animals which we associate with a much warmer climate. But Britain was separated from Europe in Middle Stone Age times with a rise in sea level and the creation of the Straits of Dover, some 6,500 years ago.

Even now the Straits are so shallow that the dome of St. Paul's Cathedral would be above sea level if it were sunk in the deepest part.

Relief and Geology

Within the period of human occupation two important elements of the landscape have remained prominent: the blocks of high land of the west and north which contain, and are fringed by, smaller areas of lowland; and the major area of lowland to the south and east. It is sometimes said that the dividing line between these two major components of the relief of Britain joins the mouth of the Tees in Northumberland to the mouth of the Exe in Devon, but this is inaccurate since it omits much of the lowlands of the Midlands of England, the economically important lowland of Lancashire and Cheshire, the lowlands round the Solway Firth, and the economic heart of Scotland, the Central Lowlands.

The blocks of high land (see map on pp. 26-27) in the north and west, where much of the ground lies higher than 1,000 feet above sea-level, are the worn-down roots of ancient mountain chains. The rocks of which they are composed are often very much altered from their original form. By heat, pressure and chemical changes, the original sedimentary rocks have become metamorphic rocks such as gneiss, schist and slate. Large masses of igneous rock, e.g. granite and gabbro, were also intruded into the cores of the ancient mountain chains. Long after these mountain chains had been worn down by the agents of erosion they were dismembered, broken and uplifted to form the high ground of Britain at the present day. Some sedimentary rocks of great age—the Torridonian sandstones of the north-west Highlands of Scotland, which are the oldest sedimentary rocks in Europe—were also uplifted with the adjacent crystalline rocks, and form high mountains providing remarkable scenery.

Generally speaking, the sedimentary rocks, other than those of the Pennines, form the lowlands of Britain. These sandstones, limestones, clays and chalk become increasingly younger in age as one travels from the north-west to the south-east. Each outcrop has reacted differently to erosion, so that the lowlands consist of vales of less resistant rocks overlooked by the abrupt slopes of escarpments whose summits descend gently as dip slopes beneath the younger sedimentary rocks of the next vale to the south-east. Even within the lowland zone there is thus, as a result of the changes in rock type, much variation both in the height and the shape of the land—which is, in turn, paralleled by variations in the use of the land. The most recent rocks occur in the Thames and Hampshire basins, which have extensions geologically on the European Continent in Flanders and the Paris basin and, of course, in the intervening North Sea and English Channel. It is in the sedimentary rocks of the lowlands of Britain that the most important mineral resources are found—coal, iron, limestone, salt and, more recently, oil and natural gas.

Both the highlands and the lowlands were uplifted relatively recently in geological time, but the uplift was greatest in the north and west, and there was even relative subsidence in the south and east. The direction followed by the main rivers of England, Wales and Scotland indicates that the country has been tilted towards the east, while the North Sea basin has subsided.

These movements of the crust were accompanied by fracturing and by the outpouring of lavas, as in Western Scotland. Similar volcanic rocks are also found in north-east Ireland.

Glaciation

The scenery of Britain owes much of its variety to the different rock types, under the agents of erosion. But some of the most important contrasts and some of the reasons for the diversity of the resource base, especially for agriculture, result from a further complication in history when, about a million years ago, the climate became cooler, and the precipitation fell as snow rather than rain and accumulated to become the ice sheets which spread from the Scottish Highlands as far south as a line joining, roughly, the Thames estuary to the Bristol Channel. The Quaternary Ice Age was, however, not a completely cold era; in the interglacial periods it was warmer in Britain than at the present day. The limits of the ice sheets in the cold periods varied; in the later stages of the last glaciation, only parts of Northern Scotland and some of the higher ground to the south were covered by ice.

To the north of the maximum ice-limit Britain has, therefore, a glaciated landscape, while to the south the land was unglaciated. To the south of each limit of the ice, the ground was cold, permanently or seasonally frozen,

FIG. 4. *Classic English scenery in Gloucestershire.*

and supported a vegetation similar to that of the tundra lands of the present day. The rocks north of the ice-limit were covered with ground-moraine (boulder clay) or by sands and gravels deposited by meltwater, while in some areas huge lakes were impounded in which vast spreads of clays accumulated. These were to form the surface deposits from which our present day soils have been created, while to the south of the ice-limit in Southern England the soil parent material is the weathered fragments of the local rock.

On both high and low ground, ice erosion produced many characteristic features as old river valleys were altered by the scouring action of the ice, and corries were formed on the edges of the upland plateaux and their summits. Here, in the areas of intense glacial erosion, the weathered rock fragments were removed by the ice sheets (but it is possible to find, even in the Scottish Highlands, many areas which show comparatively little evidence of ice action). Meltwater from the ice sheets produced deep valleys in some areas and left the mounds of sand and gravel which are now an important source of aggregate for concrete used in building and civil engineering works. Glacial erosion thus accentuated the variety of the scenery of Britain, while glacial deposition, especially in the highland valleys and the lowlands, produced a range of soil parent material different in its distribution from what would have been available had glaciation never occurred. These different parent materials were to provide a very varied resource base for the growing of crops, and help to explain the rapid changes over small areas in the quality of the soil.

The glaciations had another very important side-effect. During the glacial periods sea level was lowered as water was evaporated from the oceans to become temporarily locked up as ice on the lands of the Northern hemisphere. Sea level fell by as much as 300 feet, revealing the shallow former sea floors as dry land. When the ice melted the sea-level rose more rapidly than the land masses recovered from the weight of the ice sheets. The sea flooded into the low ground of the coastal strips, especially in northern Britain, leaving evidence of its limits in deposits which now stand well above sea level. The net effect of these changes in sea level, associated with the glaciations, has been the drowning of the lower ends of river valleys in the south and the glaciated valleys of the north. Drowned valleys are an important asset to the British Isles, since they provide shelter and access inland for shipping.

The Coastline

The coastline of Britain is of great significance to the economy of the country. In some areas it consists of bold high cliffs whose shape reflects the action of the waves on rocks showing varying resistance to erosion. Fortunately many of the cliff coasts are broken by bays and estuaries. In other areas, especially where lowlands front the sea, the low coast is composed of sand, shingle or mud, sometimes backed by fringing sand-dunes. Here, especially in South-east England and East Anglia, drowned river valleys favoured the development of sea transport. The varied coastline of Britain is thus one of its major resources, and it is increasingly recognized to be a major asset, especially for purposes of recreation.

It is perhaps a measure of the ease with which goods and people may enter and leave Britain by sea that only in time of war are the people aware that they live on a group of islands—unless, of course, they live in the remoter isles of Shetland and Orkney in the far north or the Scilly Isles in the south-west. The sea is clearly a part of the daily life of the Shetland islander. although aircraft and modern ships have brought the "mainland" much nearer in time, the cost of transport to his island is reflected in the higher prices of many commodities compared with those on the mainland.

QUESTIONS

1. What are the major differences in the physical environment between Upland and Lowland Britain?
2. How has glaciation contributed to the variety of scenery and landforms in Britain?

CHAPTER 2

CLIMATE, SOILS AND VEGETATION

In addition to the restrictions imposed on living space in Britain by the high ground of the north and west, nearly one-third of the land is virtually unusable because of the climate. Adjectives such as oceanic, mild, equable, temperate, are habitually used to describe the climate of the British Isles, and climate maps and graphs show the average January temperatures as about 3·3° C. (38° F.) and the average July temperatures as 14·4° C. (58° F.). The frequency of change in the weather may be stimulating, but an oceanic climate is not always pleasant when westerly gales, sweeping the land, in a single night devastate thousands of trees, blow down chimneys, and destroy roofs.

The cost of keeping warm when cold air with frost spreads in winter from the Continent, and the cost of the disruptions to the flow of road traffic when snow-bearing winds stream down from the Arctic are considerable. The western Highland areas are depressingly wet, while East Anglia may be so dry that the yields of crops are low and water shortages not uncommon, and, in extreme cases, the arid soil may be blown away by strong spring winds. The level of the treeline in the British Isles is remarkably low; much of the land above 1,500 feet is bare of trees, while in the Shetland Isles and the Outer Hebrides a single tree at sea level is very conspicuous.

Climate

Climatic statistics are thus unreliable as indicators of the true climate of Britain, especially for the needs of the farmer, the motorist and the heating engineer. They are also misleading, since they do not demonstrate the differences between the town and the country. The climate of a conurbation such as London, Manchester, Birmingham or Glasgow is different from the regional climate of the surrounding area, and 80 per cent of the population live in the towns and cities. The temperature in the city centres can be as much as 9° C. (16° F.) warmer than the surrounding rural areas. In spite of legislation, pollution of the city air makes the towns foggier so that they receive fewer hours of sunshine than the country districts.

Clearly there is no such thing as *the* climate of Britain; there are many climates, with marked differences between the south and the north, between the east and the west coasts, between the hills and the lowlands, between the sunny and the shady side of ridges, between the coast and inland, between the towns and the countryside. The variety of climates ranges from

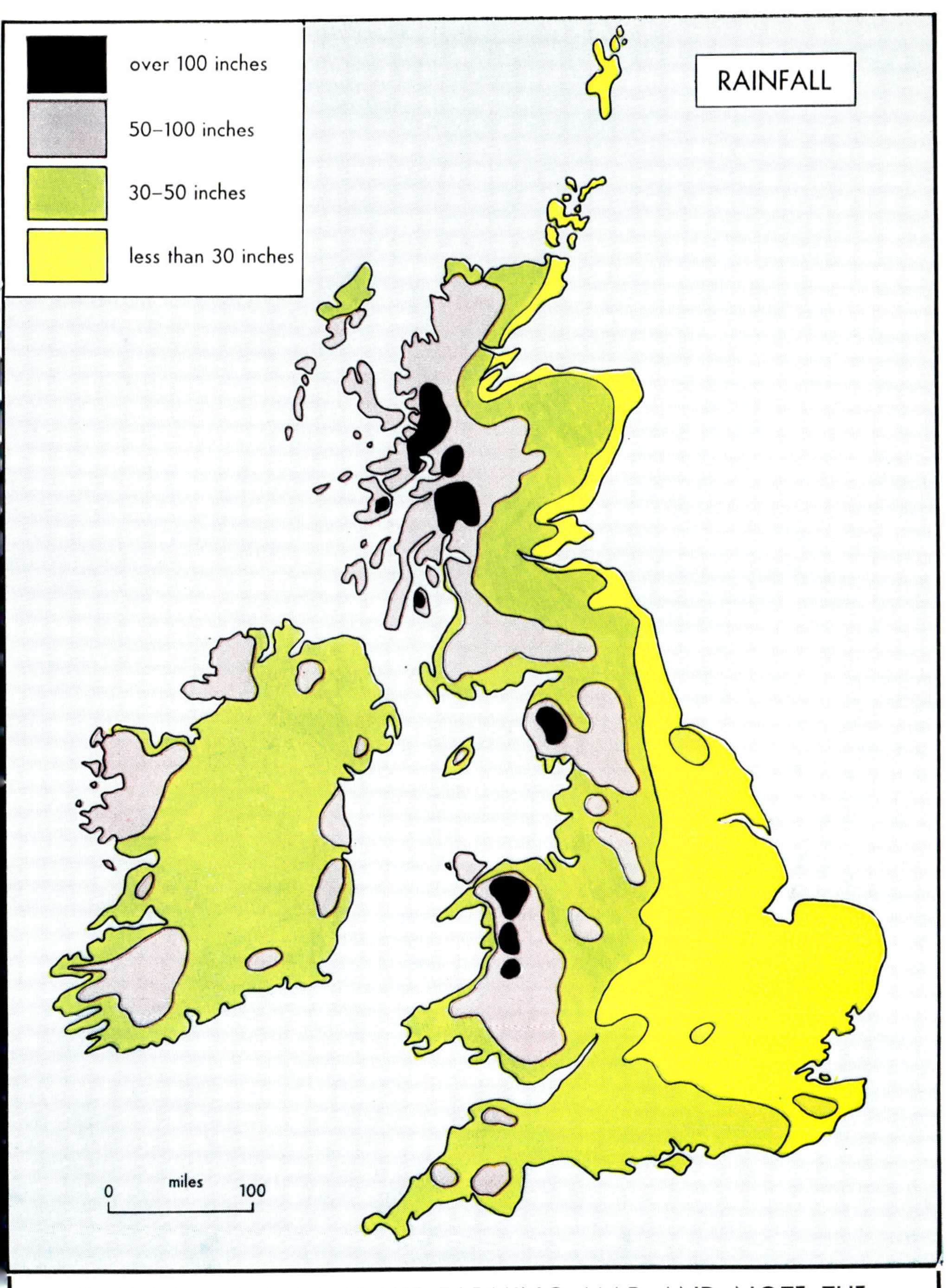

COMPARE THIS WITH THE FARMING MAP AND NOTE THE CONNEXION BETWEEN RAINFALL AND FARMING TYPES

BRITISH ISLES
conurbations
land over 600 feet
ATLANTIC OCEAN
NORTH SEA
SCOTLAND
Inverness
Aberdeen
Spey
Don
Dee
Forfar
Dundee
Arbroath
Perth
Tay
Forth
Kirkcaldy
Stirling
GLASGOW
Falkirk
Edinburgh
Paisley
Berwick-upon-Tweed

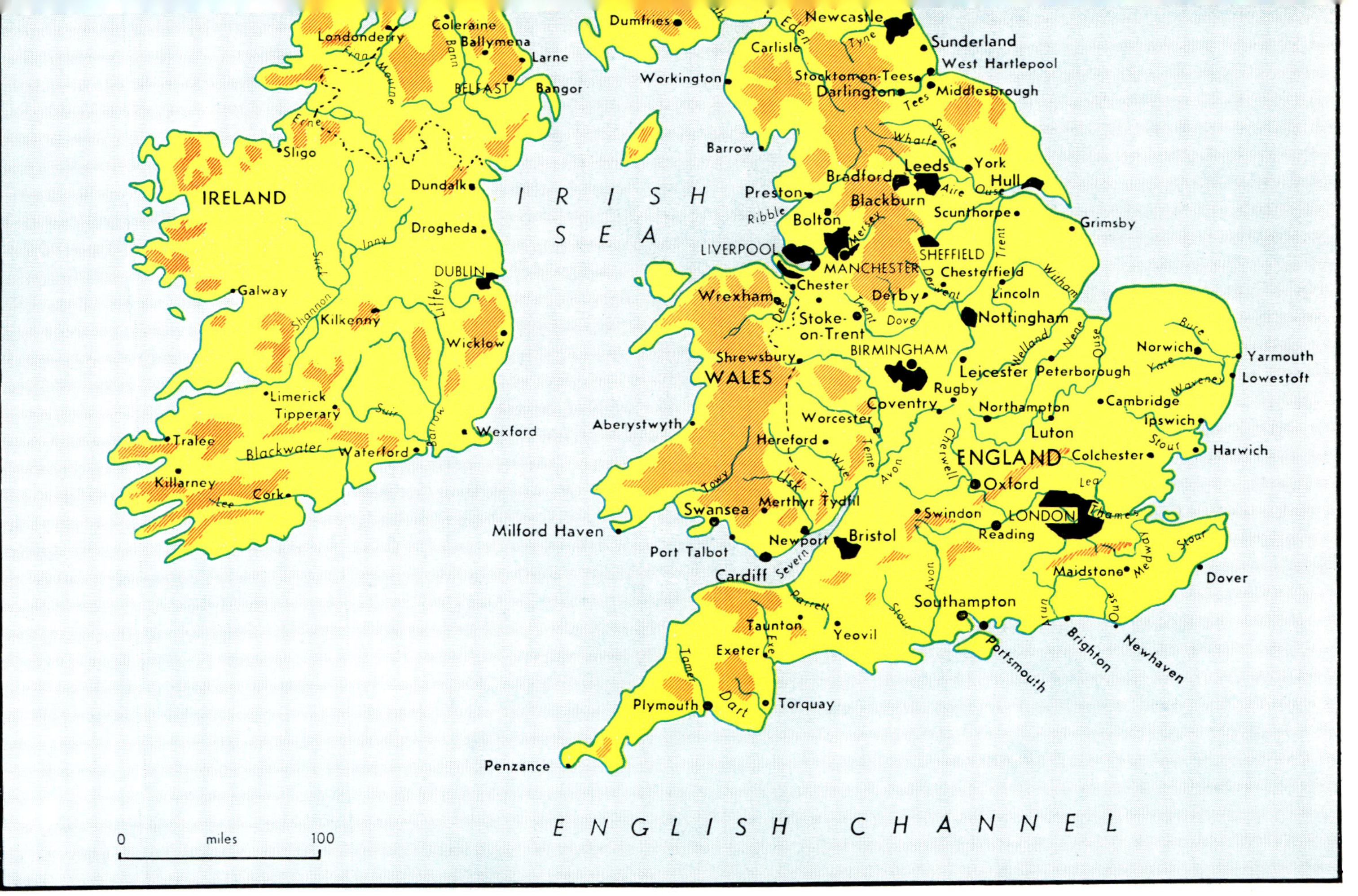
IRELAND
WALES
ENGLAND
IRISH SEA
ENGLISH CHANNEL
Londonderry
Coleraine
Ballymena
Larne
BELFAST
Bangor
Sligo
Dundalk
Drogheda
DUBLIN
Galway
Kilkenny
Wicklow
Limerick
Tipperary
Wexford
Tralee
Waterford
Killarney
Cork
Foyle
Mourne
Bann
Erne
Inny
Suck
Shannon
Liffey
Suir
Barrow
Blackwater
Lee
Dumfries
Workington
Barrow
Carlisle
Newcastle
Sunderland
West Hartlepool
Stockton-on-Tees
Darlington
Middlesbrough
York
Leeds
Bradford
Hull
Preston
Blackburn
Bolton
Scunthorpe
Grimsby
LIVERPOOL
MANCHESTER
SHEFFIELD
Chesterfield
Chester
Lincoln
Wrexham
Derby
Stoke-on-Trent
Nottingham
Shrewsbury
BIRMINGHAM
Norwich
Yarmouth
Lowestoft
Leicester
Peterborough
Rugby
Coventry
Cambridge
Northampton
Ipswich
Aberystwyth
Worcester
Luton
Harwich
Hereford
Colchester
Oxford
Swansea
Merthyr Tydfil
LONDON
Milford Haven
Swindon
Newport
Bristol
Reading
Port Talbot
Cardiff
Maidstone
Dover
Southampton
Taunton
Yeovil
Brighton
Portsmouth
Newhaven
Exeter
Plymouth
Torquay
Penzance
Eden
Tyne
Tees
Swale
Wharfe
Aire
Ouse
Ribble
Mersey
Trent
Witham
Derwent
Dee
Dove
Bure
Yare
Waveney
Welland
Nene
Stour
Cherwell
Teme
Wye
Usk
Towy
Avon
Lea
Thames
Medway
Severn
Parrett
Exe
Tamar
Dart
Arun
0
miles
100

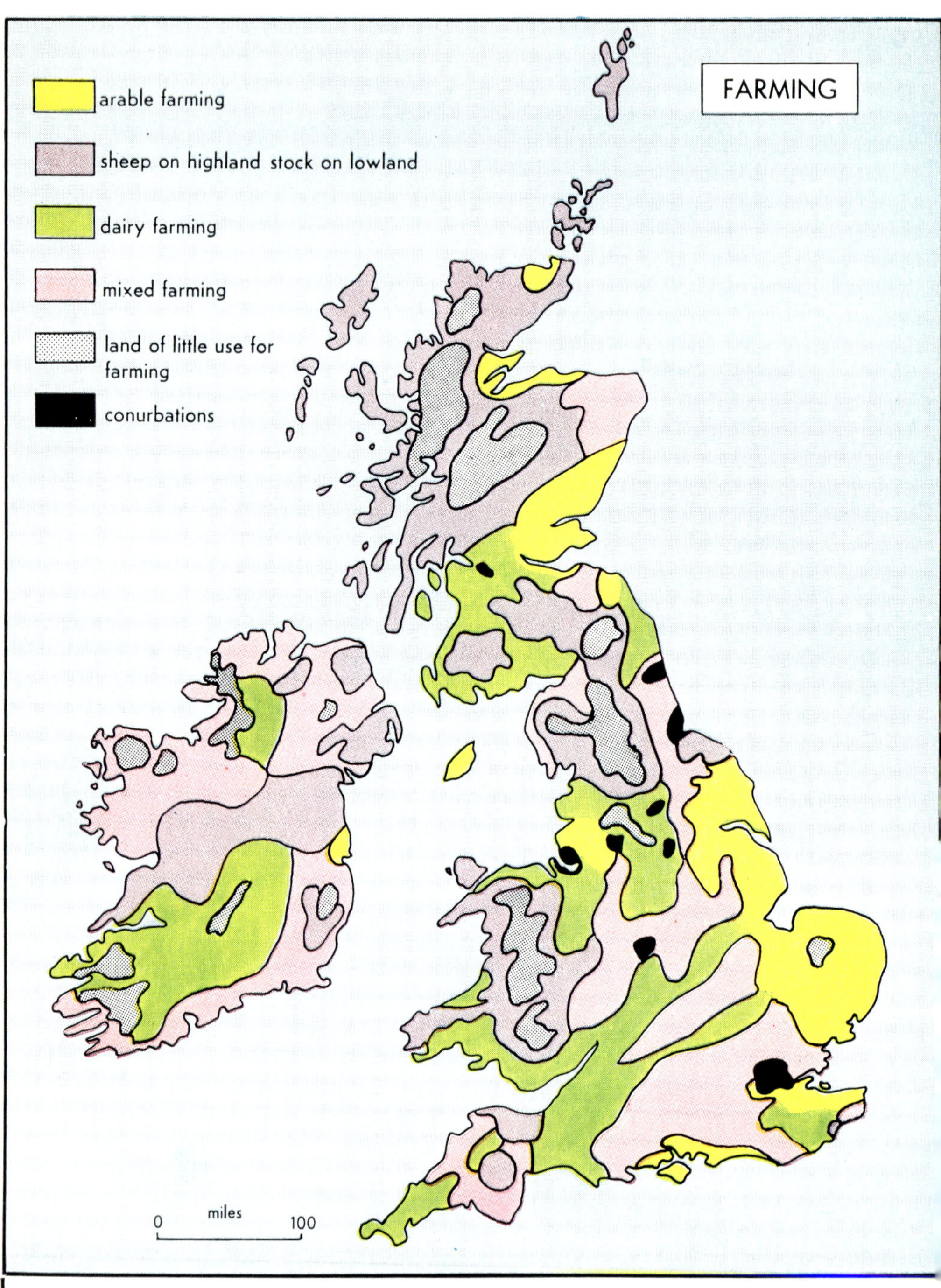

NOTE HOW ARABLE FARMING TENDS TO BE CONCENTRATED ON THE DRIER EASTERN SIDE OF BRITAIN

the sub-Arctic of the summits of the Scottish Highlands to the almost sub-tropical of the Scilly Isles and southern Cornwall. Local climates add to that variety in the resource base already indicated in the relief, geology and coastlines.

Heat waves and blizzards, wide differences in rainfall amounts, frost, hill fog and sea mist result from Britain's location between the Atlantic waters and the adjacent western extremities of the Continental land mass of Europe and Asia. Moreover, the range of latitude (50°-60° N.) is itself great enough to produce temperature differences, since latitude governs the length of day-light and insolation. The story is made more complicated by the nature of the atmospheric circulation in the area of the North Atlantic in which the islands are situated.

Weather

The British Isles lie within the track of the frontal depressions which move along the zone of separation between air masses. Air masses, each with different properties of temperature and humidity according to the source and direction of movement, mix over north-west Europe. The resulting weather is highly variable, not only in time but also in character. It is determined by the types of airflow at any particular time, and the pattern of these airflows may cause several air masses to affect a particular area, as the depressions (pro-duced at the contact of the air masses) move within the general airflow, which is normally from the west towards the east.

The main air masses affecting the British Isles are Polar Maritime and Polar Continental, Tropical Maritime and Tropical Continental, but there is a clear predominance of Polar Maritime air which, when the wind blows from the north-west after the passage of a cold front, produces cool showery weather both in winter and summer, especially on the western coasts where winds often blow with great strength. Similar unsettled conditions are pro-duced under Polar Maritime air mass conditions when the wind blows from the west, although the weather is then often much milder. Tropical Maritime air, often found in the warm sectors of depressions, frequently occurs as south or west winds, giving mild spells and fog in winter or warm, cloudy, and even thundery conditions in summer.

Polar Continental air arrives over Britain mainly in December, January and February. It is very dry, but, in its passage from the east over the North Sea, may take up moisture which later produces the snowfalls of the eastern Pennines and other parts of eastern England and Scotland. Tropical Continental air arrives in the summer as south or south-east winds. It also is very dry and produces the longed-for hot settled weather; unfortunately it is comparatively rare, and, even when it occurs, not all of Britain benefits, since the passage of warm dry air over the colder waters of the North Sea produces sea mist, which affects shipping movements and brings a cool clamminess to the east coasts.

The origins of the weather pattern over Britain are not quite as simple as this account might lead one to believe. Minor complications in the system of depressions may have major effects on the weather. Rarely is Britain covered in its entirety by the airflow of one air mass; this is one of the reasons for the division of the island and the surrounding seas into relatively small forecast regions.

Effect of Relief

The weather is also strongly influenced by the relief which affects not only the upland areas of the north and west but also the adjacent lowlands. The maps which show the distribution of rainfall reflect quite clearly the pattern of high and low ground. Much of eastern Britain lies in the rain shadow of the higher west. In the west, many areas of high ground have an average annual rainfall in excess of 100 inches and are some of the wettest areas in the world, while less than 100 miles away the rainfall may be as low as 25 inches a year. Snow falls, as one would expect, more frequently on the high ground, and also lies much longer there than on the low ground. The length of snow-lie in the Cairngorms has led to an expansion of winter sports there—some patches of snow persist almost the whole year round, and it would require only a very slight fall in the average annual temperature (about 0·3° C., 0·5° F.) to produce permanent snow cover for the highest summits. Put in another way, if Ben Nevis was 5,300 feet above sea level rather than its actual altitude of 4,406 feet, it would probably be permanently capped with snow.

The upland areas of northern Britain—i.e. the Lake District, the Southern Uplands and the Scottish Highlands above 2,000 feet—have frequent severe frosts and snowfalls and relatively long snow-lie in the winter. In summer they may experience very high temperatures. These high uplands thus have the greatest seasonal extremes of climate experienced within the British Isles. The low islands off the west coasts of both Scotland and Ireland experience the lowest ranges of temperature, and are remarkably mild for their latitude. The flowers on the Scilly Isles are well known; not so well known is the bulb-growing area of the Isles of Tiree and of North Uist in the Hebrides.

Length of Growing Season

To the crop farmer the important elements of the climate are temperature and rainfall—not the rather artificial climate recorded in the Stevenson screens of the meteorological stations, but the temperature of the soil and the amount of moisture in it. Data regarding soil temperatures is not generally available for the whole country, but the length of the growing season may be used as a partial substitute. It may be defined as the number of days when the mean daily temperature exceeds 6° C. (43° F.). The south-west of England, especially the islands, has on this basis almost a year-round growing season, but in eastern Scotland there are only about 250 days with such temperatures. There is, of course, a rapid decrease with elevation in every part of the country (about 9 days for every additional 100 feet of elevation in south-west England). The summits of Dartmoor, with a maximum elevation of nearly 2,000 feet above sea-level, thus have a growing season of only about 189 days, compared with almost 365 days on the coast. In Scotland, the decrease is less marked (about 5 days per 100 feet), but the overall effect is great since the growing season on the low ground is shorter. At 2,000 feet above sea level in Aberdeenshire, the growing season is only about 150 days (less than half the year).

Frost

The distribution and intensity of frost often depend on local factors of relief. When cold air drains down-slope into a hollow or basin in the southern parts of Britain the winter temperatures may be as low as on the summits of the Scottish Highlands, especially in cold anticyclonic spells with still air. On the marginal slopes, down which the air is moving, the night temperatures are higher and often above freezing point. Clear air and fog, alternating on hills and valleys, create serious traffic problems on the motorways.

More significant on the national scale are the dates on which the first air frost occurs. In this respect, the coastal areas of southern and western Britain are particularly fortunate. There the first air frost occurs usually after December 1, and the frost period ends before the beginning of March, for example, in the islands of North and South Uist of the Outer Hebrides, and the Isles of Scilly. The midland and south-east lowland areas can usually expect the first air frost by the middle of October, and the frost-free period is unlikely to recur before the middle of May or even the beginning of June. To summarise a rather complex situation, the last air frosts of winter occur later in the year from west to east, from the coasts inland, and from lower to higher altitudes.

City Climates

Each new house, road, or factory modifies the local climate to provide a climate generated by the towns and cities themselves. These modifications result from what is called "the heat island" effect, as buildings and roads radiate heat to the atmosphere. In broad terms, temperatures do not fall so low in the winter in the cities as they do in the country; there are fewer days with frost, and the snow of the surrounding country often turns to sleet or rain within the built-up area.

Over the cities the air is often polluted by the smoke from factories, power stations and domestic chimneys, or from the exhausts of motor vehicles. Smoke particles, highly corrosive to metal and stonework, and damaging to human lungs, are most abundant on still nights in winter, when the smoke hangs as a pall over the built-up areas. Over many British cities this atmospheric pollution reduces solar radiation by as much as 50 per cent in the winter months. In London there are marked differences in the length of mean daily sunshine between the suburbs and the city centre, but since the blanket of "smog" also reduces outgoing radiation at night, higher night temperatures are found in the cities than in the suburbs and rural areas.

Dust particles act as nuclei round which water droplets condense, so that the polluted city air is often prone to fog. Much work had been undertaken by Government through legislation for smoke-free zones to reduce atmospheric pollution, but the city-dweller in Britain, especially in the large conurbations, is clearly living in an unhealthy artificial climate of his own making.

Soils

An important element in the resource base of any country is its soil, which we define here as the "medium in which plants grow." Wheat, oats, turnips, potatoes, grass and fruit trees do not grow on solid rock. The soils of Britain are exceedingly complex in their types and distribution, with many local varieties and changes of soil characteristics even within the same field. The parent materials from which the soil is formed, and the soil-forming processes themselves, vary widely.

The British Isles have only two *zonal* types of soil, i.e. those in which climate is the principal soil-forming factor. These are the *podsols* and the *brown forest soils.* Podsols are typically the soils of the wetter north and west; heavy rainfall and lower rates of evaporation lead to leaching of the soluble salts, which are carried down to the "B" horizon. The raw humus, or the needles of the coniferous forest trees, cover the greyish leached "A" horizon, and the soils are usually acid. In contrast, the lowland zone soils, typically in the south and east (the drier parts of the islands), are less leached and less acid, and would normally bear a deciduous forest cover. Both of these soil types lose their fertility under cultivation, and it must be restored by the application of plant foods—especially nitrates, phosphates and potash and farmyard manure—and by a scientific system of crop rotation.

Complications to this general picture of acid podsols in the north and west, and higher base-status brown forest soils in the south and east, are introduced by the nature of the parent materials—either the solid rock which outcrops relatively infrequently in the lowlands, or the surface materials which were transported by ice for some distance in northern areas. Two points emerge: (1) the soils on the limestones will differ from the soils of the neighbouring clay or sandstone outcrops, and the soils derived from rocks rich

FIG. 5. *Smoking chimneys contribute to atmospheric pollution in Britain's towns and cities.*

in silica, such as granite, quartzites and sandstones, will be more acid than soils derived from calcareous rocks within the same climatic zone; and (2) the soils are not always related to the solid rocks which they cover. This latter fact derives from glaciations, which moved large quantities of rock and crushed them to fragments many miles from their original location so that they are "foreign" to the areas where they now occur. Soils derived from such transported parent materials will often have quite different properties from soils derived from the local solid rock. There are also areas of land where the soil patterns are disrupted by skeletal soils, by accumulations of organic matter (e.g. the peat of the uplands, especially in the wet areas of the west and north), and by high water tables.

One of the most important agents in determining the pattern of the soils has been 7,000 years of forest destruction, and increasing use of the lowland areas for cultivation. Natural soils are now found only where the ground is uncultivated because it is too high, too wet, or too uneconomic to work. Given sufficient time—and the disappearance of man from the British Isles—the soils would tend more and more towards the zonal soil types, the *podsols* and the *brown earths*, each with its characteristic vegetation. But as long as the British Isles are populated, the soils will remain diverse and contribute further to the variety of the islands' resources.

Vegetation

A visitor to lowland Britain might well be impressed by the apparently well-wooded landscapes. On closer inspection, the appearance is seen to be made up of hedgerow trees, tree-lined roads, ornamental woodland in the neighbourhood of large mansions, and shelter belts grown to protect crops and livestock from wind. In the upland areas the open moorlands would stand out in contrast with their wide expanses of heather, rough grassland, peat moor and bare rock surfaces. In the valley floors, and in sheltered situations with a good southern aspect, oak, Scots pine and much scrub-birch would be visible. The altitude of the tree line might be as high as 2,000 feet in the eastern Highlands of Scotland, but only at sea level in the wetter, windier west. In upland or lowland Britain, the only great expanses of woodland to be seen are the plantations of the Forestry Commission or the woodlands managed by private owners. They amount to only about one-twentieth of the total land area.

Vegetation has been modified by centuries of human interference. There is no natural vegetation left in Britain—everything is managed, even the heather moors of the uplands, which are regularly burnt over to provide young shoots for food for the grouse. The upland grasses, which have replaced the forests cut down for fuel, for charcoal, for iron-smelting or for shipbuilding materials, are grazed by sheep or cattle. On the lowlands, fields have replaced forests, and even hedgerow trees are few in number compared with three hundred years ago: in the days of wooden ships, the oak of the hedgerow with its gnarled form and many branches, provided the natural crooks which gave strength to the timbers of ships. To a large extent the natural woodland has disappeared to be used in the economy and the defence of the country.

Earlier Vegetation

Man has not, however, been solely responsible for the changes in vegetation. Analysis of the pollen in the peat bogs and the evidence from the old tree stumps and roots at altitudes well above the present treeline, shows that there have been many "vegetations" of the British Isles since the last disappearance of the glaciers and ice-caps about 10,000 years ago. The climate was sometimes warmer, sometimes cooler, sometimes wetter, sometimes drier, than at the present day. The original tundra was gradually replaced by birch until, about 9,000 years ago, in the warmer and drier climate of the Middle Stone Age (Mesolithic), the birch of the Boreal period lost its dominance to be replaced by hazel and pine, with some elm and oak. In so-called *Atlantic* times, from about 5000 B.C. to 3000 B.C. (the late part of the Middle Stone Age and the New Stone or Neolithic Age), the Atlantic Ocean rather than the Continent became the dominant influence in the climate, which became warmer and wetter with summer temperatures higher than the present day's.

This period is sometimes called the "climatic optimum". Alder and mixed oak forests became the most important and widespread trees in the Bronze Age, but heavy rainfall in the west also encouraged the accumulation of deep peat in blanket bogs. Thereafter, the climate deteriorated, as the summers became cooler and the winters colder and wetter. By about 900 B.C. the Iron Age people lived under much less favourable conditions of temperature and rainfall, while the vegetation was mainly birch, alder and oak on the uplands and peat mosses in the upland areas.

Vegetation Today

The vegetation of Britain presents a varied face. At low altitudes, the remnants of forests are dominated by the Scots pine in northern Britain, and by the deciduous oak woodland in southern Britain. With increase in altitude, the Scots pine gives way to heather moss on the dry eastern side and to wet grass moors on the wetter western side. Above about 2,500 feet the heather is often replaced by grass and heath moors or the deer forests of the Scottish Highlands, and above about 3,000 feet by dwarf juniper and eventually by the sub-Alpine vegetation of dwarf plants. Exactly what kind of vegetation is to be found at these higher altitudes depends on exposure of north- or south-facing slopes and the amount of rainfall and soil drainage. The non-wooded areas, below the sub-Alpine zone, are usually called "rough grazing."

FIG. 6. *A view of Snowdonia: the lowland is covered with trees, but there are none on the mountain tops.*

Regional variations in vegetation are also due to different outcrops of rock and surface materials. The limestone and chalk lands are particularly important; beach and hornbeam are characteristic trees on these calcareous soils. On the clay vales the oakwoods have been mostly replaced by the fields of farmers. Differences in glacial deposits also affect the distribution of trees: sandy areas are often dominated by conifers, while the boulder clays tends to support deciduous trees.

In relation to the climate, vegetation and soils of Britain two important points should be remembered. First, as with the relief and geology, the assets of the resource base are not spread evenly over the whole of the country, the southern and eastern lowlands having an advantage; secondly, there is clear evidence of modification of these assets by centuries of human occupation. This modification has now reached the stage when the effects of pollution of the atmosphere and of the rivers, and the use of the wilder areas of the country for economic purposes, are recognized to be dangerous not only to health and economic resources but also to the quality of life in the densely populated island.

QUESTIONS

1. In what ways does the climate of a city in England differ from that of its surrounding region?
2. What kinds of climate and vegetation have been experienced in the British Isles since the end of the Ice Ages?
3. Describe the problems that the climate poses to the farmer in lowland Britain.

CHAPTER 3

POWER, MINERAL AND WATER RESOURCES

DURING the last two hundred years the economy of the British Isles, with the exception of the Republic of Ireland, has changed from an agricultural to an industrial base. This has been made possible by the quantity and variety of the mineral resources and power supplies made available through the gift of Nature or the inventiveness of man. The variety of the mineral resources—from gold (in very small quantities in Wales and the Scottish Highlands) to iron—is a reflection of the geology which, in only a small area of the earth's surface, includes rocks of almost all known ages. The wet steep slopes of the uplands provided the earliest sources of power from running water, but this was eclipsed with the development of the coalfields. Later, new competitors have appeared as sources of energy—electricity, oil imported from overseas, hydro-electric power, nuclear energy, and supplies of natural gas from the floor of the North Sea.

Coal

Coal is derived from the Carboniferous rocks found mainly in the lowlands and basins on the margins of the upland areas. It has been used for domestic heating for centuries; there are records of its first being carried by sea from Northumberland to London in medieval times. The great exploitation of coal came, however, with the Industrial Revolution, when the invention of the steam engine, fuelled by coal, permitted the exploitation of the coal seams deep below the surface. It was the steam engine that pumped out the water which would otherwise have flooded the mines, powered the winding machinery, and, with the invention of the steam locomotive, permitted the coal to be transported more rapidly from the mines to the furnaces.

Ireland is unfortunate in that its coal seams are few and unprofitable to work. Peat forms the principal fuel available, especially in the Republic, but its low calorific value makes it much less valuable than coal for industrial purposes. Elsewhere in the British Isles there are four major groups of coalfields. The first lies in the Central Lowlands of Scotland where the sedimentary rocks of Carboniferous age are found in the rift valley between the ancient crystalline and metamorphic rocks of the Highlands and the metamorphosed sedimentary rocks of the Southern Uplands (Fig. 26). Within the Scottish coalfield there are four separate fields, Midlothian and Fife (separated by the Firth of Forth), the Lanarkshire, and the Ayrshire coalfields.

In England the chief coalfield groups lie on either side of the Pennines; the Lancashire-Cheshire Field occurs to the west together, with the coalfield of West Cumberland; to the east are the Northumberland-Durham field and, further south, the Yorkshire-Derbyshire-Nottingham fields. Each outcrops on the upland slopes of the Pennines and dips away under younger sandstone clays and limestones towards the sea.

The third major coalfield zone extends from the edge of the Welsh uplands through the Midlands of England. The workable outcrops are discontinuous and occur as separate fields—the Flintshire-Denbighshire, the Shropshire, the Staffordshire, the Warwickshire, and the Leicestershire-South Derbyshire fields. Lying at the economic heart of the country, they are of great importance.

To the south the fourth belt lies from west to east. In the west occurs the South Wales coalfield which, in the past, was one of the most important fields in Britain; further east and of less size and importance are the Forest of Dean, Bristol and Somerset field, and the quite isolated field in Kent (see the map on p. 78).

All the coalfields contain seams of good workable coal, but the mineral varies in its characteristics and its usefulness for different purposes. Anthracite comes from the west of the South Wales field, but the bulk of the coal produced can be divided into two main types—coking and non-coking coal. Coking coal has always been of great importance since it was first used in blast furnaces with iron ore and limestone for the production of pig-iron. Nearly 30 million tons of coking coal are used in British blast-furnaces and foundries each year; it comes mainly from the southern part of Durham, eastern South Wales, South Yorkshire and North Staffordshire. About 20 million tons of coal a year is used in gas-works, but the greater part of the 150 million tons of coal produced annually is used in the production of electricity, in industry, and in domestic fires. Very little coal is used in ships' boilers or by the railways or exported. Electric power stations, industry, gas-works, coke ovens, and domestic consumption are the dominant users of coal in Britain.

Working the Coal

Changes in use have been paralleled by changes in the methods of getting coal. There has also been a concentration of industry on the coalfields with the best quality and biggest reserves. Since 1946 the National Coal Board has closed down many uneconomic mines—as in Lanarkshire, Fife and South Staffordshire. The most important fields are in the Midlands, especially the South Yorkshire, Derby and Nottingham fields, which produce about half the coal mined in Britain. Machines rather than men now cut the coal from the seams and transport it underground and to the surface. Bevercotes, a new pit in Nottinghamshire with reserves of 100 million tons, is the world's first wholly remotely controlled colliery. At Longannet in Scotland, a super-mine complex is linked by computer to a power station.

Heavy machinery is also used in open-cast mines, where the coal seams lie close to the surface. In open-cast mining the overburden of soil and valueless rock is stripped away to reveal the coal beneath. When the coal is exhausted, the ground can be restored for agricultural use. Mechanization, both above and below ground, has been accompanied by a reduction in the number of mines and miners, which has caused important social problems in the mining areas.

Electricity

About 10 per cent of the coal output of Britain is used in the production of electricity, which has now become the major source of power for industry, but oil has increasingly come into use in place of coal in the thermal generating stations. Power is fed into the National Grid transmission lines at 110 kilo-volts and into the new super grid at much higher voltages (over 275 kilovolts), and is carried to the transformer stations where the voltage is reduced to a lower level (about 240 volts) for domestic purposes. Whether powered by coal or oil, the thermal generating stations must be located where fuel is readily available or can easily be transported, especially by water. Water is also required for cooling. Huge cooling towers are especially plentiful along the river Trent in the English Midlands, which has already been described as the area with the greatest output of coal at the lowest cost.

Water also figures prominently in the two other methods of electricity generation. Since the Second World War, hydro-electricity generating stations have multiplied in the Highlands of Scotland and the Uplands of Wales. The physical conditions are not ideal—not because of lack of gradients in the river courses or of high rainfall, but because the catchment areas of the rivers are small. Much civil engineering construction has been necessary to link the catchment areas by tunnels and to store the water in enlarged natural lakes or purely artificial reservoirs. In the Highlands of Scotland some 3,500 million units are produced each year from more than 50 stations. (The Shannon is the only river used directly for hydro-electricity production in the Republic of Ireland.)

Nuclear energy stations are also usually situated beside water—partly to satisfy the demands of cooling, and partly because a situation on the coast can be found remote from the main centres of population, e.g. on the coast of Caithness at Dounreay, at Sizewell and Bradwell on the coast of East Anglia, at Dungeness, Kent, at Hinkley Point in the Severn estuary, at Wylfa in Anglesey, and at Hunterston on the Firth of Clyde. The station at

Trawsfynydd in North Wales is situated inland, using a reservoir into which the water is pumped back for re-use.

Oil and Gas

As well as electricity, other forms of power can be transmitted over long distances. Pipelines carry crude oil from the tanker terminals to the refineries —e.g. from Finnart on the west coast of Scotland to Grangemouth on the east, and from Milford Haven to Llandarcy in South Wales. Other pipelines carry refined petroleum from refineries such as those at Fawley on Southampton Water, Shellhaven on the Thames, and Stanlow on the Mersey. Oil production within Britain (one per cent of the country's requirements) is insufficient to warrant pipelines from the small wells in Nottinghamshire and Devon but discovery of oil under the North Sea may alter the pattern of Britain's oil requirements.

What the United Kingdom lacks in oil is, in part, compensated by the enormous reserves of natural gas under the North Sea, which are equivalent to three times the annual requirements of the British gas industry for thirty years. In 1967 and 1968, gas from two of the biggest deposits began to flow into the national gas grid, to supplement liquid gas (methane) brought by tankers from Algeria to the Canvey Island terminal in the Thames estuary. In 1968 about 30 per cent of all gas sold was based on coal, 55 per cent was based on oil, and 15 per cent was Algerian and North Sea gas.

Apart from its use in domestic heating and cooking, natural gas is valuable for industry, in which its use will increase; sufficiently high temperatures

FIG. 7. *The Llandarcy oil-refinery, one of the biggest in the country.*

(up to 4,000° C.) can be obtained for smelting metals using suitable techniques. Electricity power stations are already powered by naphtha from petroleum; clearly many will soon be producing electricity from natural gas. The 1960s will certainly figure prominently in the history books as the decade when new and previously undreamed-of assets to the resource base were discovered.

Water

Water is converted to steam to drive turbines in thermal generating stations, and is used as a coolant there and in nuclear power stations. Formerly it provided the power for grain and textile mills, and it is now used for hydro-electricity generation. As a power resource water figures prominently in the economic life of modern Britain, apart from its domestic and recreational uses.

Water resources are a problem in modern Britain. There is no lack of it; the difficulty is in making it available to the consumers. Scotland has much more water than England, but the total population is less than that of Greater London and only twice that of the Birmingham or the Manchester conurbation.

The water resources of Britain exist in the rivers and lakes, especially those of the wetter north and west where population is sparse. The rivers which originate in western Britain and flow to the east over the lowlands are the great natural water pipelines. But only half the rainfall reaches the streams and rivers; the other half is evaporated, or is transpired by trees, grasslands and growing crops. Some goes underground to form great subterranean storage reservoirs, especially where the rocks are permeable, as in the outcrops of sandstone, limestone and chalk in the lowlands.

Except in the very remote parts of Britain, the population cannot rely on regular supplies of water from local sources. It is usually available at the wrong time (in winter, when evaporation is least), in irregular amounts (due to variations in rainfall), and in the wrong places. Ground water supplies, which in natural circumstances are the most dependable sources, have been used faster than they can be replenished, as in the London basin, where some of the capital's water formerly came from artesian wells; or in the industrial areas, where ground water has been drawn on for various manufacturing processes. Rivers, formerly a substantial source of domestic supplies, have become polluted by domestic and industrial effluent and can be used as drinking water only after treatment.

Control of Water Resources

In 1963 the Water Resources Board was set up for England and Wales to make an inventory of the water resources and the demands made upon them, to control floods, to reduce pollution, and, by the creation of storage reservoirs, to augment those already in existence. Its aim is to create 200 days' reserve of water. This can only be done by using the high rainfall impermeable uplands of the west and north as storage areas—which immediately raises problems of conflicts in land use, since the upland areas are increasingly used for forestry and recreational purposes. Long-distance water pipelines, like that which has long connected Thirlmere in the Lake District with Manchester, will have to be built, and more water is likely to be diverted to underground storage by assisting it to infiltrate quickly into the water-bearing rocks of the lowlands. Much thought has also been given to the desalination of sea water: a 500,000-gallons-a-day plant was built in Guernsey in 1962, and a 1½-million-gallons-a-day plant has been completed to supplement the natural water supply of Jersey. But desalination is too expensive for general use. Rainwater collection costs about 11p per thousand gallons, while, under the most favourable circumstances (e.g. using the exhaust steam from power station turbines as the energy resource) desalination water costs 18p per thousand gallons.

Water shortages over the whole of Britain at the same time are unknown, but locally there often have to be temporary restrictions for inessential purposes, especially in the eastern coastlands of Britain. That there is no acute water shortage is the result of good administration and civil engineering, but it is only achieved at high cost.

Mineral Resources

Lead, tin, copper, gold and silver have been important elements of the resource base in many parts of the upland area since prehistoric times. Tin and copper, together with iron, were especially significant, since these metals provided the tools by which in Bronze Age and Iron Age times the physical environment was modified. Flint tools from the chalklands of southern and eastern England played a notable part in the evolution of domestic economy before the smelting of metals. Tin, copper, lead, gold and silver occur in relatively small quantities, and have been exploited for so long (over 2,000 years) that they are now almost exhausted, although periodical attempts are made to re-open some of the old tin mines. Britain is, however, fortunate in having large deposits of ferrous metals (i.e. iron-ore), of rarer non-metalliferous minerals such as salt, gypsum and china-clay, and of limestones, granites, clays, sandstones, sands and gravels.

Ferrous Metals. Iron-ore, in association with coal (especially the coking coals) and limestone, was the "king" of metals in Britain, since without its existence in quantity and wide distribution the country could not have led the world in the Industrial Revolution. At the present day, however, over half the iron-ore needs of the country are imported, e.g. from Sweden, North Africa and Canada, and the ore is of higher quality.

By far the greatest quantity of home-produced iron-ore (about 17 million tons per annum) comes from the sedimentary iron deposits of the Jurassic scarplands on the south-east fringe of the Midlands, extending from Oxfordshire to the Humber. These deposits are worked by open-cast methods, or by mining where the beds dip to deeper levels in the east. (The iron-ore of the Cleveland Hills which gave rise to the iron industry at Middlesbrough, at the mouth of the Tees, is now exhausted.) The most northerly of the three main producing areas is North Lincolnshire, where a thick bed of ore, with about 20 per cent iron content, has given rise to the great foundries and steel-making plants at Scunthorpe. Further to the south-west occurs the most important area, extending roughly from Grantham to near Northampton.

FIG. 8. *Opencast iron mining, Corby. The Jurassic ores have a low iron content but are plentiful, easy to mine and lie near to coal deposits needed for smelting.*

These "Northampton sands" yield about 32 per cent iron. Ore from this district, centred on the steel town of Corby, is sent to North Lincolnshire for mixing with ore from that area. The Oxfordshire ore area is located in the Banbury district.

Formerly mining of iron-ore in the coal measures was of great importance since the ore and the coking coals occurred together. The iron from this source did not occur in great quantity, but it was on these iron-ores in particular that the Industrial Revolution in Britain was based, for prior to the development of transport, to have coal and iron in close proximity was essential. Typical of those sources are the clayband and blackband iron stones of the Central Lowlands of Scotland, on which early heavy industry was based.

The coal-measure iron-ores have not been exhausted, but they cannot compete with the lower-quality iron from the Jurassic escarpment in terms of cost of exploitation and competition from imports. Hence they are now chiefly of historical interest, as are the iron-ores of the Weald, where charcoal from the forests was used as fuel in the medieval iron-smelting industry of Kent and Sussex.

Non-Metallic Minerals. Of these the most important is *salt*, obtained either as rock salt or as brine (i.e. rock salt in solution). Both forms occur in the Triassic rocks which underly the Cheshire lowland to the west of the Pennines, but the salt is now mainly exploited as brine. The evaporation of brine in coastal salt pans gave rise to an early chemical industry in North-east England, which has developed into one of the major producers of basic chemicals (acids, alkalis and salts), using local brine pumped from the Triassic deposits, and *anhydrite*, from which ammonium sulphate and sulphuric acid are obtained. *Gypsum*, which is used for plaster of paris and plasterboard and is chemically related to anhydrite, is obtained at only a few places in the Midlands, in the Triassic rocks of East Staffordshire and Nottinghamshire, and in the Keuper marls to the south of the Tees estuary.

Other minerals of importance are *fluorspar*, used principally as a flux for the steel industry, and *barytes*, used in paper and paint manufacture. Both are mined in the northern Pennines and, together with the *china clay* of Cornwall, represent the most valuable mineral products of the upland massifs.

Other Rocks and Minerals

Limestone is a common rock in lowland Britain, and on the upland fringes of England and Wales. The *Carboniferous limestone*, especially from the western edge of the southern Pennines at Buxton, is used in great quantities for the chemicals industry of the Cheshire plain nearby, as a flux in the iron and steel industries, for lime and cement, and as a road metal. *Magnesian limestone* from north-east England, where it occurs as *dolomite*, is also used in the iron and steel industry. Further large reserves of dolomite are known in the north-west Highlands of Scotland, but they do not have the same favourable location in relation to sources of coal and iron. Both the Carboniferous and the Magnesian limestones have been used as building materials, but the Jurassic limestones, which include the Bath stone of the Cotswolds and the Portland stone of Dorset, are better known in this respect. Portland stone is used as a facing in many famous buildings in London.

Chalk, a softer form of limestone, was unimportant until the great demands for cement made it one of the valuable resources of the country. The main chalk quarries for cement manufacture are found on the flanks of the Thames basin, including the North Downs and the Chilterns, and on the north bank of the Humber.

Other building materials. The *granite* of the uplands of Scotland has been used both locally and further afield, but the metamorphic schists and gneisses are of only local significance. The *sandstones* of the lowland areas provided a distinct building material in the past, but natural building stone is now very little used because of the high cost of the skilled labour of the masons and the increasing use of newer building materials based on concrete.

Concrete. For concrete, sand and gravel are required in large quantities. They are available from certain types of meltwater deposits in the northern areas, but are in relatively short supply where they are most needed, i.e. in the areas of the greatest concentration of population. In the well-populated areas of the lowlands of England the most readily available supplies come from the gravels of river terraces, especially those of the Thames and the Trent valley. Flooded gravel workings are being increasingly used as recreational centres for boating and angling.

Sand is also of great importance for making moulds in the iron and glass-making industries. It is obtained from the Triassic series of rocks in the Midlands and the lowlands of Yorkshire and Cheshire, and also from the Lower Greensand series of the Cretaceous rocks in Bedfordshire and Norfolk.

Slates. Roofing slates are available from the metamorphosed shales of North Wales, the Lake District and the Scottish Highlands, but they are infrequently used in modern buildings. They have been replaced by the less costly clay roofing tiles, such as those made in Staffordshire, from the clays and crushed shales of the coal measures, from which fire clays are also manufactured.

Bricks. The major brick-producing industry is focused on the Oxford clay, especially in the zone which extends from Peterborough through Bedford to Bletchley. Thanks to mass production and efficient transport this area now produces over half the country's requirements. This concentration has led to the closing-down of many of the small brick-pits in other parts of the country although, in the urban areas of the coalfields, coal-measure clays are still used for brick-making, since they have the advantage of a short haul to building sites.

The effective mineral resource base of Britain is located in the lowland areas, including the coalfields, iron-ore deposits, limestones for the iron and steel and cement industries, salt, brick clays, roofing tile clays, moulding and glass sands. The upland areas offer relatively few mineral resources—some limestone, the fluorspar and barytes, china-clay and roofing slates. The upland zones certainly provide the basic resource of water and the hydro-electric stations, but both of these can, by modern technology, be transported to the lowlands, as can the natural gas from the bed of the North Sea. Thus the advantages of relief, soils and climate of the lowlands are reinforced by the mineral resources they provided for the Industrial Revolution, especially in coastal situations. Increasingly the overwhelming advantage of the lowlands over the uplands is reflected in their high population densities.

Moreover, some areas of the lowlands, especially the Midlands of England, either have all the basic raw materials readily available or are able, as a result of the increasingly favourable road and rail system, to import raw materials from overseas. The predominance of the Midlands in the total economy of Britain is increasingly apparent. Here occurs the maximum concentration of coal, iron-ore, limestone, brick clays. We shall find in the next chapter that the lowlands are also the main agricultural base of the country.

FIG. 9. *A brick kiln baking bricks of Oxford clay.*

QUESTIONS

1. What are the energy supplies available in Britain and where are they located?
2. Why is there a water problem in Britain?

CHAPTER 4

LAND-USE

OVER the centuries there has been a decline in the number of people employed in agriculture in the British Isles, more particularly in England, Wales and Scotland: that Britain is still able to produce about half the food supplies for its people with only 5 per cent of the working population directly engaged in agriculture points to the fact that Britain's agriculture is probably the most efficient in the world, making the fullest use of the potential provided by the natural resources of relief, climate and soils. Agriculture in Britain is one of the most highly mechanized in the world. Farmers and farmworkers, comparatively few, obtain high yields from crops, assisted by the use of imported artificial fertilisers and the application of weed-killers and pesticides. The quality of farming depends on regional specialization; different areas can concentrate on the crops best suited to the locality because a good transport system can carry the products to the major centres of population cheaply and quickly.

There are areas in Britain where agriculture is marginal and where a poor return is obtained in relation to the amount of work and capital invested. Such areas are normally the most remote from the major centres of population, e.g. in the north-west of the Scottish Highlands and the Islands, where the crofts bear little resemblance to the large farms in the lowland areas of Britain. Into this category also fall the farms near the upper limit of cultivation—about 800 feet above sea level for arable farming and about 1,500 feet for cattle farming. These farms, together with many small farms even in the more favoured areas, depend increasingly on assistance from the Government in the form of subsidies, so that, superimposed on the environmental factors which help to explain the character and distribution of farming in Britain, there is an agricultural pattern that has been created by Government intervention.

The elements of physical geography cannot, in themselves, offer a satisfactory explanation of the patterns shown in the map on page 28. Farmers grow crops to obtain a livelihood; the individual farmer decides which crop will give him the greatest profit and, although he is undoubtedly influenced by the potential of his land in relation to altitude, climate and soils, he is also influenced by the success, or lack of success, of his neighbours and by Government policy. A farmer in any area will produce those crops and livestock which bring the greatest cash return, in view of the factors of advantage or restraint imposed by location and environment, and in the light of the economic and political framework of the time.

Cattle

Britain is essentially a land of mixed farming, in which both crops and livestock are commonly found in any area. Nevertheless, the distribution of the uplands and lowlands, the variations in rainfall between the west and east, the differences in soil quality, and the length of the growing season produce certain basic patterns. The high rainfall of the west and its milder climate, especially near the coasts, favour the growth of grass; and livestock farming, with cattle on the lower ground (below about 1,500 feet) and sheep on the higher ground, is characteristic. Grass, however, is cultivated, part of the "improved" land on which under a scientific system of crop rotation crops, root crops, sown grasses and clover are all grown. There is a clear relationship between the improved stock-rearing grassland and the wetter western parts of Britain.

Upland and lowland animal farmers work together. Young cattle and sheep bred on upland farms in the north and west are sent to the lowlands for fattening for beef production, or for milk production. Many cattle are dual-purpose (reared for both beef and milk), but there is a preponderance of dairy cattle in the west and in the vicinity of large centres of population. Beef cattle are widely distributed, and are not necessarily confined, as is sometime said, to the eastern portion of the country; there are many areas in the west lowlands (and the Republic of Ireland) where beef and dual-purpose cattle predominate. Nevertheless, there is a strong concentration of beef and dual-purpose cattle on the improved pastures of the Midlands of England, while in southern and south-west England they are fattened on the cereals and rotation grasses. In the north-east lowlands of Scotland, they are part of a mixed farming system. The importance of livestock, especially for dairy produce, is demonstrated by the fact that they represent 80 per cent of the value of British agricultural output.

Sheep

Sheep show a more definite relationship with the physical geography of the island. They are typically the animals of the uplands—the Highlands of Scotland, the Southern Uplands, Pennines and Lake District, the plateaux of Wales, and the uplands of south-west England. But they are also reared in lowland Britain on the limestone and chalk escarpments and dip slopes and on the downlands of southern England. They are even found in large numbers in East Anglia, but here, as in other lowland areas, sheep farming tends to be located where the soils are so freely drained or so poor that other types of stock, such as dairy cows, would give poor a financial return. The distribution of pigs and poultry tends to reflect the distribution of the population, with concentrations near the main urban centres.

Crops

Grass. In the west, grass, in the form of permanent or rotation grassland, covers the largest area, but in the eastern lowlands much of the ground is under cereals and root crops. The proportion of the area under crops increases from about a quarter of the total land area in the west Midlands to more than half in the East, while locally, as in the Fens, crops may take up more than threequarters of the land area. The further one travels eastwards in Britain the greater is the proportion of land under crops. Grassland is also often found on heavy clay soils, as in the clay vales between the limestone and chalk outcrops. The arable land in Britain thus varies inversely with rainfall. But it is wrong to believe that there is no arable in the west, and no grassland in the east. Arable land is related to relief as well as to climate and soils, and there is much tillage in the lowland fringes of the uplands of the north and west.

Cereals. Individual crops reflect the predominance of crop land in the east and the lowlands. This is especially true of wheat and barley. Wheat has a more restricted distribution than barley; it is not grown in quantity north of the Forth, and the major concentration is, in England, the zone bounded on the west by the Jurassic limestone escarpment. Barley extends even to the lowlands of the Moray Firth, but in lowland England the major producing area coincides roughly with that of wheat. (The wheat and barley of south-east Ireland are grown in response to a favourable bounty by the Republic's government.)

Oats show a much wider distribution than either wheat or barley. In Scotland and Wales this is the dominant cereal, since it can ripen in the cooler and wetter summers of the extreme north and west which prevent good yields from wheat and barley. Oats are the mainstay of mixed farming in the north-west of Scotland and the Islands, and were grown extensively in preference to other cereals in eastern Scotland before the technique of fattening beef cattle on barley was introduced.

Roots. Of the root crops—potatoes, turnips, swedes and sugar-beet—only sugar-beet tends to be restricted to certain areas, especially East Anglia and Lincolnshire in England, and Fife in Scotland. These locations are mainly determined by the distribution of the sugar-beet factories. Potatoes have a widespread distribution, with seed potatoes in the eastern areas. Turnips, which revolutionized the practice of agriculture in the British Isles in the 18th century, are an essential part of mixed farming, and their importance as winter feed for both cattle and sheep, and their value in the rotation system of cropping, give them a widespread distribution.

Farming Efficiency. Agriculture in Britain has undergone numerous revolutions but none perhaps as important as that which has occurred since the Second World War. Since 1945, yields of nearly all the main crops have risen steeply; the yields of both cereals and potatoes are now almost 25 per cent higher than before the War. Grassland is more intensively managed by controlled grazing, while the extension of grass silage is reflected in a doubling of the production of milk, beef and veal, lamb and mutton. Pork-production has increased by more than three times; that of eggs has almost doubled. In all, the output of British agriculture is three times as great as in 1939, and this increase has been achieved with a reduced labour force.

Farming has thus become more efficient, aided by Government subsidies and by advances in the application of science and technology. Mechanization, including combine-harvesters for cereals and crop-drying plant to overcome the rigours of the British weather, have coincided with the use of artificial fertilisers and selective weed-killers and new systems of marketing and transport. In some cases it has proved more efficient to breed and rear cattle and pigs indoors; but the chicken broiler industry and battery egg-production offer the best example of "factory" farming. Farming in Britain is now a science-based industry, demanding a high level of technical and administrative skill.

In turn this has caused a re-appraisal of the size of British farms. In England and Wales alone, it is estimated that over 40 per cent of the farms are too small to be effective, contributing only 7 per cent of the nation's agricultural produce. Many such farms are run by the farmer and his family with occasional hired labour, and their income, even assisted by Government subsidies, is often below what they could earn as workers in industry. Greater production from the remaining 60 per cent of the farms in England and Wales could make up the low production of the small farms. As more small farmers move away from the countryside, a further increase in the size of farms is seen, and already the enlargement of fields by the removal of hedges and walls is having an effect on the landscape. It marks another stage in the intensification of agricultural production.

The best farming area, as reflected in crop yields and diversity and in intensity of livestock production, is in the area bounded by Liverpool in the north-west, Hull in the north-east, Bristol in the south-west, and London in the south-east. This area coincides with what are broadly termed the Midlands, where we have already observed the major industrial resource base exists, in a comparatively small area, less than 200 miles across in any direction. This area produces, both in industry and agriculture, about 90 per cent of the total economic output of Britain. It is thus the economic heart of the country, with the greatest concentration of population. All the remaining parts of Britain are secondary to it, even the industrial belts of South Wales and north-east England.

Forestry

In spite of the efforts of the Forestry Commission, Britain is still one of the most sparsely wooded areas of Western Europe, although within lowland England the hedgerow timber gives the impression of a wooded landscape. Hedgerow and park trees represent about one-third of the timber stock of the British Isles, and it has been estimated that the broad-leaved trees—oak, elm, ash, beech and sycamore—yield about 10 per cent of the annual increment of the forest resources. These wooded landscapes are confined to lowland England and Ireland; the upland zones of Wales, Northern England and Scotland are relatively bare of timber.

Great inroads have been made into the natural forests for agriculture, fuel and constructional timbers. Especially destructive were the charcoal-burners, who provided fuel for the smelting of iron in the Weald, the Furness area of the Lake District, in Yorkshire and in the Highlands of Scotland. More recently, much forest, planted by estate-owners in the Scottish Highlands in the 18th and 19th centuries, was felled in the two World Wars when imports of timber from the Baltic and North America were reduced.

These depredations, especially during the First World War, necessitated a large-scale planting programme organized by the Forestry Commission set up by the Government in 1919. The aim was to re-plant areas which had been felled, to plant areas which had not recently been forested but were unsuitable for agriculture, and to improve existing woodland.

New Forests. Rapid-growing, high-yielding species were selected for the new forests, especially the coniferous spruce and the deciduous larch, but other trees, were planted especially on the margins of the forests, to reduce the fire hazard and to improve the scenic appearance. Although the Forestry Commission have planted and are responsible for many acres of hardwoods, it is the soft woods which have been planted over a wide range of soils and glacial deposits. On the coast the Foresty Commission have contributed to the stabilizing of mobile sand dunes, as in the Culbin Forest to the east of Inverness. The Commission also maintain the amenity and conserves the wild life of the New Forest. In the remoter areas of the United Kingdom it has attracted many visitors to "forest parks." British forests are thus available for recreation as well as for their timber—in other words, the land is being used for more than one purpose. Planting has also been completed on low ground with poor soils, as in the Breckland of East Anglia.

The problem which now arises in the upland areas is a conflict in land-use, since there is already evidence that forestry is beginning to compete with agriculture, especially stock-raising (sheep) and the recreational use of upland for grouse-shooting and deer-stalking below the economic altitude of afforestation. This altitude, under the best circumstances of climate, exposure and soil, is about 1,700 feet above sea-level.

With careful management forest products provide a sustained yield, since the thinnings can be used before the crop matures. The demand for pit-props has declined as collieries have closed and steel has replaced wood, but thinnings can be used in pulp mills, such as that at Fort William, provided no great transport cost is involved.

Fishing

The sea is another element of the resource base, although remote from the main centres of population. In spite of difficult conditions imposed by wind and sea in the strong tidal waters off north-west Europe, the richness of the fishing grounds within reach of the British Isles has given rise to one of the world's most important fishing industries. The coastal waters, less than 600 feet deep on the Continental shelf, and the waters above the slope to the depths of the Atlantic Ocean, offer the coastal communities a rich sea harvest. The bordering shallow waters receive from the rivers, such as the Thames, Rhine and Forth, mineral nutrients, including nitrates and phosphates, which are distributed by the tidal currents and the vertical movements of the sea water. Minute plants and animals—plankton—thrive on these nutrients in the shallow waters through which the sunlight can penetrate, forming part of the vital food-chain of the sea—in which some fish feed on plankton while others prey on the smaller fish.

The coastal waters—the *near-water* fishing of the North Sea, the west coast of Scotland, the Irish Sea and the English Channel—contain two principal types of fish. The bottom feeders (*demersal* fish) include the important species of turbot and plaice; other demersal fish, such as cod and haddock, feed near the bottom and in the middle depths. *Pelagic* fish live near the surface, feeding on the floating plant and animal life; this type includes the herring and mackerel, of which the herring is by far the more important. *Distant-water* trawling to the Faroes, Iceland, Spitzbergen and the Barents Sea is undertaken only for demersal fish.

Methods of Fishing. Different types of fish require different catching methods. Pelagic fish, such as the herring, were formerly taken wholly by *drift* nets, which trapped the fish by their gills as they swam into the meshes suspended from the sea surface by floats controlled by the movements of the *drifter* in relation to the tide. Nowadays, herring are taken principally by seine nets, especially the *purse-seine* net, in which the vessel manoeuvres round the shoal of fish (detected by echo-sounding apparatus) before drawing the net tight by winches. The *seine-net* works in a similar way, but is used also for the capture of demersal or "white" fish (cod, halibut, haddock, etc.). For the bottom- and middle-water feeders the *trawl* is also used: here the vessel drags a net along the bottom or at a depth controlled by the trawl boards. Trawling is carried on in both near and distant waters. Fishing by lines—

FIG. 10. *Catching fish by means of seine nets off north-east Scotland.*

long-line fishing—is not now important, but it was once the chief method of catching white fish that, like the cod, can be taken by a baited hook.

Near-water trawling and seine-netting for white fish involves only short voyages, but produces high-quality fish—landed at the home port, especially Aberdeen and the many small ports of north-east Scotland. Distant-water trawling involves much larger vessels which spend weeks away from their home ports, such as Hull, Grimsby, Aberdeen and Fleetwood. Half the fish caught by trawlers is landed at Hull and Grimsby. This fishing takes place all the year round.

Herring. In contrast, herring fishing is distinctly seasonal, and the catching areas in any month of the year depend on the movement of the herring. The spring fishing begins off the west coast of Scotland and moves north, reaching the Shetland Islands by the early summer. Here vessels of many nationalities with seine and purse nets take huge hauls (a purse-seine net may take 700 tons of herring in one haul), landing them at Lerwick for sale as fresh fish, for kippering, or for use as fish-meal. In the summer and autumn the shoals move south in the North Sea and are followed by the vessels, which land their catches at more southerly ports; at that time of year Yarmouth and Lowestoft become the principal centres of the herring fishery.

Transport. Increasing specialization is taking place in the fishing industry. Ports such as Hull, Grimsby and Fleetwood concentrate on distant-water trawling, with vessels up to 250 feet long specially equipped to freeze the fish at sea. Many smaller fishing towns in England and Scotland operate smaller trawlers and wooden seine-netters for the inshore and near-water fisheries; they depend on the speedy delivery of the freshly caught fish to the port and from the port to the consumers, who often live hundreds of miles away. As in agriculture, an efficient transport system is an essential part of the industry.

Speedy transport is even more important for the highly perishable shell-fish, especially crabs and lobsters, which are taken in baited creels by many small boats operating all round the rocky coasts of Britain from the Shetland Islands to Land's End. Such fish are sometimes kept alive in special ponds or storage tanks before despatch by air to London and other conurbations.

The most recent development in the industry is fish-farming, carried on experimentally with plaice in some of the lochs of western Scotland. Fish-farming at the edge of the sea will increase as stocks in the traditional fishing grounds decline through over-exploitation, not only by the fishermen of Britain but by all the countries which border the North Sea and by Russian trawlers and factory ships.

QUESTIONS

1. Show how the variations in types of agriculture in Britain are, in part, related to climate, relief and soils.
2. Why are fishing and forestry important in the economy of Upland Britain?
3. What changes in agriculture have occurred since 1945?

CHAPTER 5

THE CORE AREA: (i) THE MIDLANDS

It has long been customary to divide Britain into a number of regions whose boundaries are based on natural features such as the junction between high and low ground, ridges, and river basins. Thus we may read of the Pennines or the Scottish Highlands, the Scarplands and the Weald. Many of these regions have well-known names, but they do not always bear a clear relationship to the reality of the geography of Britain at the present day. Many are based on agricultural regions which, although obviously of value when the majority of the population was engaged in agriculture, are now less important than the industrial areas. Moreover, the movement of population from the countryside to the cities has changed the pattern of population distribution and the relationship of the cities to the countryside around them. Each small town or large conurbation exerts a very strong influence on its surrounding area; modern road transport gives easy access between town and country, allowing the town to act as a centre to which people come from a wide area, for shopping, banking, insurance, and education.

Today the country is considered as divided into a series of regions of varying sizes, each tributary to the small towns which, in turn, are related to the great urban centres. Called *city regions*, they differ from the older regions based on physical features. They are regions or areas which have a unity imposed on them by their dependence on the town or conurbation.

Other regions result from attempts made by the Government to administer the various services provided for its citizens. Some of these regions are specially defined for individual purposes, such as the division of responsibilities between the police forces, fire services, or hospital services, or for planning reasons. Others are areas defined on the basis of their economic needs and problems. These are *administrative regions*, the boundaries of which coincide with the boundaries of the counties included within them. The Government's planning powers can produce great changes in economic geography—by making inducements to industry to move to certain areas, or by refusing to grant permission for houses or offices to be built in others. It tries, in effect, to create conditions throughout the country which will give each citizen a high standard of living in good surroundings. Very often, the physical regions, the city regions and the administrative regions do not coincide. The geographer has to devise divisions of the country sufficiently broad to cover all types of regions.

One method is to consider the areas of Britain in relation to their contribu-

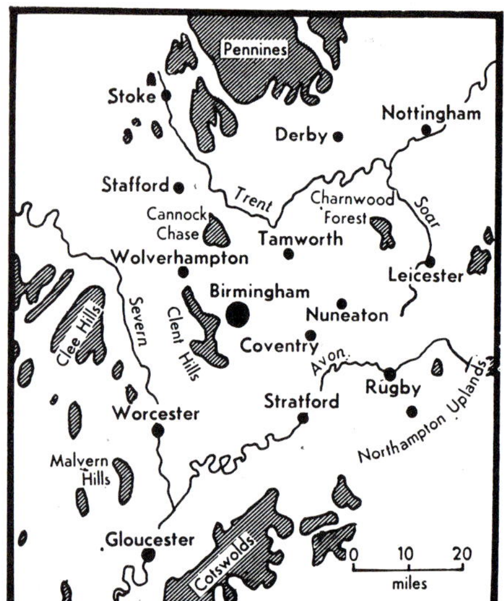

FIG. 11. *Simplified map of the Midlands.*

tion to the economic life of the country, which will be reflected in the density of the population. These considerations reveal that there is a central or "core" area where the population is increasing, where industrial and agricultural output is high, and where the system of roads and railways is highly developed.

This most important region of Britain is sometimes called the "Midlands", which, we have already seen is a comparatively small area, extending from Liverpool to London and from Bristol to Hull. It consists mostly of low ground, although parts of the southern Pennines are included; it contains the modern industries, the most productive coalfields and iron-ore fields, some of the best agricultural land, and most of the large towns and conurbations. In this area were built the first motorways; it has a tight network of roads and a high density of railway lines; although mainly inland, no part is more than 70 miles from the sea, and at its extremities are some of the country's major ports. It thus has good communications with other parts of the country and with lands beyond the seas.

Peripheral areas of Britain surround the "core" area. Some, because of their remoteness, relief and climate, are sparsely populated areas such as the Lake District or the Highlands of Scotland. Some are areas where the ground is given over to intensive agriculture, but with a number of towns in which there is some industry, as in East Anglia. Some are industrial areas whose development depended on the coalfields and local iron-ore but whose economic prosperity has declined as the old heavy industries have declined—such areas are north-east England, centred on the Northumberland and Durham coalfield, and the coalfields of the Central Lowlands of Scotland. All these regions are marginal to the Midlands, to which they are connected by new motorway systems and improved trunk roads. The products of the Midlands can be moved rapidly to these areas on the outer fringe, and to the ports for exports overseas. Raw material imported through the ports, some of which lie in the outer fringe areas, can also be quickly carried into the Midlands.

The contrast in the geography of Britain is now between the central and the peripheral areas. The central area of the Midlands is the growth zone, but much effort is being made by the Government to assist economic growth in the marginal regions.

The Midlands

The Midlands are dominated by the great Birmingham conurbation in the west, the towns of the potteries in the north, and the large cities of Derby, Nottingham and Leicester in the north and east. Here are the great factories producing a large proportion of the modern industrial products of Britain, some of the major electricity generation stations, the most modern and productive coalfields, and important parts of the motorway, railway and internal waterway systems. Between the cities the ground is intensively farmed, the poorer quality ground being used for the recreational activities of the very large populations.

The English Midlands extend from the slopes of the ancient rocks of the Welsh upland in the west to the limestone escarpment in the south and east, forming a rough triangle. The area is mainly underlain by outcrops of sedimentary rocks which are progressively younger in age towards the North Sea and the Thames valley. These gently dipping sedimentary rocks, inclined from north-west to south-east, have been dissected by rivers to produce, by the differential erosion of the resistant and weak outcrops, that series of plateaux, vales and scarplands which is characteristic of the relief of lowland England. Such relief offers few obstacles to settlement and communication, compared with the upland areas.

Rocks. The oldest sedimentary rocks are economically the most important. In the Carboniferous sediments are the coal measures which have long been exploited in the Shropshire, North and South Staffordshire, Derby, Nottingham and Leicester coalfields. A large part of the Midlands is underlain by the sandstones and marls of Triassic age with, to the south-east, the lias clay vale overlooked by the westward facing escarpment of the Jurassic limestone. Further variety in rock type and scenery is provided by the ancient and resistant pre-Cambrian and Cambrian rocks which "peep" through the sedimentary rocks to form the higher ground of the Licky Hills and Charnwood Forest. To the north, the southern margin of the Carboniferous rocks of the Pennines offers a special type of scenery, with moorlands and small escarpments of millstone grit surrounding the mountain limestone with its deep gorges and disappearing streams.

Since the Midlands lie within the area covered by the ice sheets there is much glacial drift lying on top of the solid rocks. Boulder clay, sands and gravels and old lake deposits provide variety not only in relief but also in soils, which are generally of the brown forest type. Modified by cultivation over many centuries, the soil now supports some of the best arable and graz-

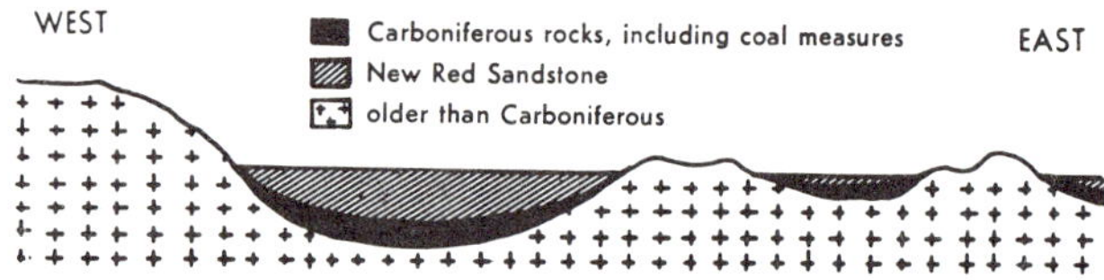

FIG. 12. *Simplified section across the Midlands.*

ing in the country. In addition, there is much good meadow land in the lias clay vale and along the rivers.

Rivers. The major rivers of the region, all flowing near its margins, are important as sources of water supply for industry and cooling water for electricity generating stations. The Trent, from its source in the southern Pennines, flows first south-east before turning north-east near Burton-on-Trent towards Nottingham and the Humber. In the south-west the Severn, Stour and Avon lead towards the Bristol Channel and the Atlantic. Although no major river flows through the heart of this area, the rivers are inter-connected by systems of canals.

The Midlands are not a plain but a series of low plateaux, with much rolling country dominated in the south-east by the steeper slopes of the limestone escarpments. The major river valleys are broad, but do not dominate the relief. Indeed, it is not the relief that provides the major contrasts in the area but the economic activities. On these grounds it is possible to divide the Midlands into a western and an eastern section.

The West Midlands

In the West Midlands are to be found some of the most densely populated industrial areas of Britain, interspersed with more sparsely populated good farming lands. The upland fringes in the west have scarcely any population. It was the presence of coal which provided the stimulus to industrial development; especially following the discovery, at Coalbrookdale in 1709, that it was possible to smelt the iron-ore in the coal measures—later from the Forest of Dean, using coke. Limestone was obtained from the early Palaeozoic outcrops near Dudley and Sedgley Hill. To transport the coal, iron-ore and limestone to the foundries, canal systems were built to overcome the inadequacy of the river system, and Birmingham became famous for its wrought-iron works, its nails, swords, horseshoes and agricultural implements, while the heavy pig-iron was produced in the valley of the Stour. Later the South Staffordshire coalfield, centred on Dudley, with coal, ironstone and limestone all locally available, was to become the iron workshop of Britain—the "Black Country"—where, in the 19th century the blast furnaces and foundries grew up which covered former farming land with industrial waste heaps and workers' houses and polluted rivers and meadows.

Birmingham. Today the Birmingham conurbation, with over 3 million people, dominates the West Midlands, which house about 10 per cent of the population of Britain. Birmingham alone has about $1\frac{1}{2}$ million people, but so close are the towns of Solihull, Sutton Coldfield, Halesowen and Stourbridge that the whole forms one large built-up area with very little farming land in between. To it come workers from Bromsgrove, Kidderminster and even from Stafford to the north, and "new towns" have been created at Dawley and Redditch.

FIG. 13. *A famous example of the new-style urban landscape replacing the sprawl of Britain's nineteenth-century cities: the Bull Ring Centre, Birmingham.*

The older industries based on heavy iron and steel have been supplemented by the "light" manufacturing industries, although in the Black Country of the South Staffordshire coalfield there is still much traditional heavy engineering, using iron-ore brought from Corby in the East Midlands. Since the transport cost of raw materials is high, the area increasingly concentrates on goods which require a great deal of skill—chains, springs, nails, screws, locks and keys. These have been supplemented by manufacture of cars and car components industry (nearly one-fifth of the working population of the Birmingham conurbation is employed in car assembly), machinery and machine tools, electrical goods (Wolverhampton), cocoa and chocolate (Bournville), jewellery, rubber (especially the manufacture of tyres), and many types of consumer goods. The area is one of the most industrially diversified in the country, and its workers receive high wages compared with those in many other parts of the country. To it have been attracted many Commonwealth immigrants, whose presence creates social problems in housing and education.

On Birmingham was focused the first motorway in Britain, the M1, which gives a direct quick road link with London; it was followed by the M5 to the Severn estuary, and the M6 to the industrial areas of Lancashire and beyond to Carlisle. The Birmingham conurbation is thus ideally equipped for the supply of consumer goods by road and railway to all parts of the country. The chief problem is water supply, since its situation on the watershed area of the Midlands means that there are few streams and water has to be brought from reservoirs in the Welsh uplands.

The Potteries. The North Staffordshire coalfield on the borders of the Midlands and the lowlands of Lancaster to the north-west, has a distinctive title—the "Potteries"—but like the "Black Country" of the South Staffordshire coalfield, now bears little resemblance to its former appearance. Here the rivers, which unite to form the Upper Trent, have cut a series of valleys trending roughly north and south in a plateau of coal-bearing rocks with an overburden of glacial deposits to provide a relief unlike the the remaining areas of the Midlands. On the floors of these valleys are situated the towns of the Potteries, connected by roads, canals and railways which lead southwards towards Birmingham and London and northwards to Manchester and Liverpool.

The coalfield has rich seams which can be worked for long-flame coals for firing pottery and tiles, for coking, and for gas coals. Some seams outcrop near the surface and can be worked by adit, or drift, mines, while others demand deep shafts. Along with the coal there are deposits of clay band and black band ironstone, which, with the limestone from the adjacent Pennines, provided the basis for an early iron industry. This still continues, using the iron-ore from the East Midlands at Shelton. Goldendale survives as an iron

FIG. 14. *Old-style industrial landscape: the Potteries.*

foundry, using local coal, because of its specialization in high quality products.

It was, however, the clays suitable for coarse earthenware, firebricks and building bricks that provided the main basis of industrial development. Pottery-making began as a cottage industry, but improved both in quantity and quality with the import of clay and flint from Dorset and Devon by the river Severn and by packhorse. The most important changes took place in the late 18th century when Wedgewood introduced china clay from Cornwall, brought by sea to the Mersey and thence by canal. On the banks of the Trent and Mersey Canal at Etruria was established the first large factory for high-quality china ware. Tunstall, Burslem and Hanley continue to make coarse earthenware with local clays, glazed with salt from Cheshire. The clay marls are also used for engineering, bricks, roofing and flooring tiles.

The smoke haze which hung over the "five towns" of the Potteries has been reduced as electricity has superseded coal for firing the kilns. New industries—the manufacture of tyres, wire and cables—have been introduced to widen the range of employment, but the landscape is still dominated by the buildings of the older industries, with patches of farmland between them or in a ring round the margins.

Farming. The many types of farm in the West Midlands depend in part on the quality of the soil, but principally on the demands of the dense population, which requires great quantities of milk and vegetables. Dairy farming is very important, but there is much arable land to provide cattle feed, and both beef cattle and sheep are brought to the pastures of the low plateaux and the rich river meadowlands for fattening after rearing in the poorer uplands of Wales. The most famous farming area of the West Midlands is, of course, the Vale of Evesham. Here the light soils and early spring provide early crops of all kinds of vegetables, but the vale situation can lead to cold air drainage down the slopes from the limestone escarpment to the south, producing early spring frosts that damage the cherry, plum and damson orchards. Fruit and vegetable canning is an important industry.

By contrast, the hillier lands of Shropshire and the lowland round Hereford are important cattle-rearing areas, but the land-use of the Hereford area is varied by the apple orchards grown for the making of cider. As in the East Midlands, the trend towards "factory farming" is well developed. The emphasis throughout the Midlands is on intensive, highly specialized farming methods dictated by the demands of the conurbation of Birmingham.

The East Midlands

Intensive mixed agriculture is also characteristic of the East Midlands, where the Triassic sandstones and marls, the lias clays and the Jurassic limestone escarpment support cattle and sheep, brought to such market towns as Melton Mowbray and Market Harborough. This is famous hunting country. A traveller along the M1 could be forgiven for imagining that agriculture is the most important activity in the East Midlands.

In fact, the area is dominated by the three large towns of Nottingham, Leicester and Derby, with other urban centres such as Burton-on-Trent, Chesterfield, Lincoln and Grantham. Industry naturally tends to be concentrated on the Nottingham and Derbyshire coalfield, with engineering, coal mining and the production of iron. This coalfield is relatively easy to work since the seams, although not very thick, extend over long distances and are not seriously disturbed by faults and folds. Lack of good coking coal is the greatest drawback, although some is available in the Chesterfield-Mansfield district.

The disadvantage is offset by the great reserves of other coals and by the mechanization and automation of collieries.

The coal is also used in the thermal generating stations aligned along the river Trent, so that the East Midlands already described as the greatest coal-producing area in the country is also the greatest producer of electricity.

Along with the coal were found bands of iron-ore which, with limestone from Derbyshire to the north, gave rise to an iron industry in numerous small centres. Derbyshire limestone for chemicals, the gypsum near Newark for plaster board, and the gravels of the terraces in the Trent valley are now of great importance. There are also producing oil wells at Eakring and Egmanton, although their output is slight.

Textile Towns. The towns of Leicester, Nottingham and Derby are the centre of a textile industry concentrated on knitwear, hosiery and rayon. Nottingham lace is a speciality. But each of these towns has other important industries. Leicester has added a wide variety of engineering products to its textile and footwear industries, including machine tools and electronic equipment. Nottingham is now better known for its cigarettes, chemicals and pharmaceutical products than for its textiles and lace. Derby is world famous as the main centre of Rolls-Royce as well as known for synthetic textiles made by Courtaulds.

The population while not as great as that of the West Midlands, nor so highly concentrated, is just over 3 million which, with the 5 million of the West Midlands, accounts for nearly one-seventh of the population of the entire country.

Like the West Midlands, the East Midlands enjoys an excellent communication system of canals, railways and roads. No other part of the country has the advantages provided by the M1, M5 and M6 motorways. The Midlands are poorly served by civil air services, but the electrification of the railway system offers speedy transport to London: a London-bound train leaves Birmingham every 20 minutes.

QUESTIONS

1. Distinguish between physical regions, city regions and administrative regions, using examples from *either* Scotland *or* England.
2. Why have the Midlands become such an important centre for industry and population in Britain?

CHAPTER 6

THE CORE AREA: (ii) SOUTHERN ENGLAND

To define Southern England we must, as in considering the Midlands, think in terms of its people and their livelihoods rather than purely in terms of relief. Certainly it is a part of the lowlands of Britain, and there is much variety in its scenery, geology and soils, but the dominant and underlying theme of life and work from Northampton in the north-west to Kent, Sussex and Hampshire in the south is the presence of the Greater London conurbation, the metropolis and capital city of the United Kingdom, covering an area of 616 square miles and containing about one-seventh of the entire population of the country.

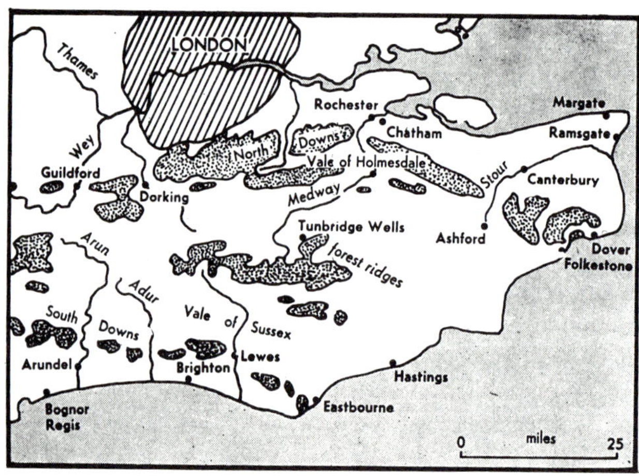

FIG. 15. *Simplified map of the Thames Basin.*

London

While London's built-up area can be outlined by a circle with its centre in Trafalgar Square and radius of roughly 16 miles, it is better to define London by its range of influence. Within the built-up area of Greater London sleep 8 million people, more than in any other area of comparable size in the British Isles and, with the exception of Tokyo and New York, more than in any other city in the world. To the resident population must be added the

1 million "commuters" who travel into the centre every day for work or come to visit the shops, cinemas and theatres. Effectively London is nearer 60 miles in radius than 16 miles.

The capital dominates more than its immediate area of the Thames basin and Southern England. On it are focused the roads, railways and internal air-line services, the telephone and postal services, of the whole country. Its financial centres serve every part of Britain and countries overseas. To it come visitors, businessmen, world statesmen and scholars; to many countries London *is* the United Kingdom, since their embassies and trade missions are all located there.

London's River. Up the great estuary of the Thames, which London straddles at its lowest bridging point, come ocean-going vessels to the 4,000 acres of docks and quays. They bring food and the raw materials for London's industries (it is the largest manufacturing town in Britain) and for the industrial towns of the Midlands. From its docks are exported much of the country's manufactured products, so that the Port of London plays a very important part in the economy of the United Kingdom. London depends on the Thames now as in the past, not only because it is one of the world's great waterways, but also because it provides a source of water for drinking and industry (after suitable purification); the Thames also carries away the refuse, human and industrial, of over 8 million people.

London has risen to its pre-eminence through its situation. Its site is made up of the river terraces and mounds of gravel which offered dry points for settlement close to the limits of the ebb and flow of the tide of the Thames estuary. At London the Thames could be forded in early times and later bridged, and the settlement became not only a land route centre but a seaport as well. To it flowed the country's agricultural produce and "sea-coal" from Northumberland. Moreover, across the narrow seas lie the great estuaries of Europe—the Rhine and the Scheldt.

Employment. Greater London offers an extremely diverse range of employment. Administration, banking, commerce, advertising and insurance are linked with its importance as the political and commercial capital, but also its manufacturing industries have grown immensely since electric power freed factories from their former dependence on coalfields. London offers an unlimited supply of labour attracted by high wages; in reverse, its population provides many purchasers for the products of its own industries, and its communications with the rest of the country have been so well developed that the whole of England, Wales and Scotland is its market. Its industries are of the modern type—electrical engineering, paper, printing and publishing, food and drink, automobile accessories and vehicle components, electronic equipment and precision instruments. These are often assembly industries for which there is a large demand for semi-skilled, quickly trained

FIG. 16. *Part of the Royal Docks, London.*

workers, and like similar industries in the Midlands they attract people from the older industrial areas where there may be unemployment.

Population. People attract more people, especially when wages are high—to sell and distribute the products, to serve in the national airports at Heathrow and Gatwick, to work on the Underground or suburban train services and the buses, to serve in the restaurants. London's problem is its increasing population, which causes congestion on the roads and traffic hold-ups that delay the lorries distributing the goods within London and through London to the docks.

Arrangements have been made for towns some distance from London to take its "overspill" population. They include Ashford, Basingstoke, Bletchley, Bury St. Edmunds, Swindon and Thetford. "New towns"—Stevenage, Crawley, Harlow, Hemel Hempstead, Bracknell and Basildon—have been created since the Second World War, with new houses and factories to draw people away from London itself. But new industry and offices round the fringes of London continue to draw more people to the area.

The traffic problem thus spreads out from the metropolis; there are more and more people to be supplied with water, more and more problems of disposal of refuse. Increasingly, the farming land of the Thames basin is being built over. As well as relieving unemployment in the older industrial areas of the north and west, Government incentives to the development of new industries are intended to create a redistribution of the population by drawing off a proportion from the overcrowded South-east. Living and working space within Greater London is so scarce, and the price of land so high, that the tendency is to build high blocks of flats and offices, and even to discourage the building of offices within inner London at all. London is not just a city; it is a region in its own right which extends from south-west Sussex to Bedfordshire and through Northamptonshire along the line of the M1 motorway. From the west it extends from Oxfordshire to the North Sea coastlands of Essex. With the Midlands, it forms the major part of the economic heart of Britain.

The Southern Landscape

Southern England, dominated by the London region, spreads south-east over the dip-slope of the Jurassic limestone escarpment, across the Oxford clay vale, the small escarpment of the Corallian limestone, the gault clay of the Vale of Aylesbury into the Thames basin. In the Thames basin, the rocks

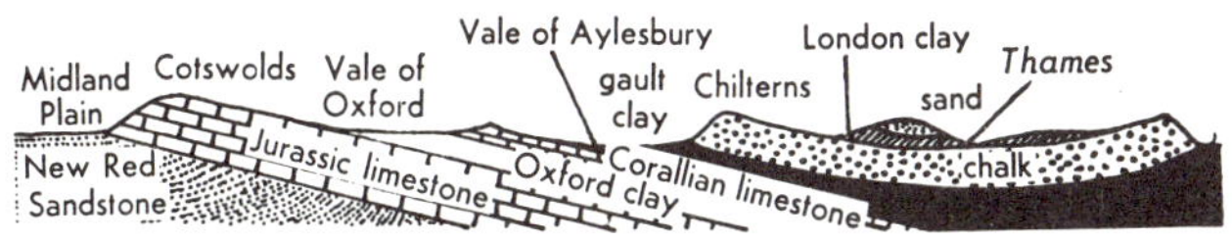

FIG. 17. *Simplified section across the Thames Basin.*

are arranged in a syncline with the youngest rock, the London clay, in the centre. Differential erosion of the Cretaceous and Tertiary rocks has produced outward facing escarpments in the Chilterns to the north and the North Downs to the south. Through the gap in the Chilterns flows the Thames to be joined by north bank tributaries such as the Brent, the Lea and the Roding; from the south come the Wey, the Mole and the Medway. The south bank tributaries have cut gaps in the southern chalk rim through which come the roads and the railways into London from the Weald. They include the Wey gap at Guildford, and the Mole gap at Dorking.

The Weald. Lying to the south of the Thames, the Weald is a dissected anticline in which the relief consists of inward-facing escarpments of chalk (including the North and South Downs). The South Downs have been partly eroded in the south where the waters of the English Channel beat against the white cliffs at Dover and Beachy Head. These downs are of porous rock with a thin, limy soil and patches of clay-with-flints. Towards the centre of the anticline, the overlying chalk has been eroded to reveal low ridges of greensand which give way to clay vales created by rivers tributary to the consequent streams which flowed down the flanks of the anticline to the Thames basin and the English Channel. The highest part of the Weald (the name means "woodland") consists of the Ashdown Forest sandstone ridges.

To the west of the Weald, the North and South Downs converge to form the chalk tableland of Salisbury Plain with its prehistoric remains. This tableland fringes the north side of the Hampshire basin which, like the

London Basin, is a syncline in which the youngest rocks appear in the centre. The southern edge is represented by the chalk ridges of the Isle of Wight, where it forms the bold headland of the Needles in the west and Culver Cliff in the east. The Solent cuts off the southern section of the syncline, which contains, together with the northern part of the Isle of Wight, the more recent Tertiary rocks consisting of sands and clays. The rivers of the Hampshire basin drain towards the great inlet of Southampton Water, where the drowned valleys of the Test and the Itchen converge.

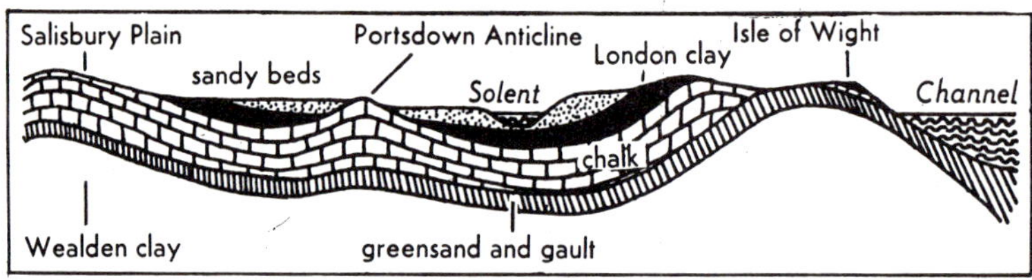

FIG. 18. *Simplified section across the Hampshire Basin.*

Rocks. Limestone, chalk, clays and sands—these are the wide variety of rocks in Southern England with, in addition, very recent accumulations of muds which form the saltmarshes of the Thames estuary and Romney Marsh. There are also huge deposits of shingle on parts of the coast, as at Dungeness. To the north of London much of the land is covered by glacial drift deposited at the maximum extent of the ice sheets, but the Weald and the Hampshire basin were not glaciated. The greatest variety in soils is thus found to the south of the Thames, but even to the north of the Thames the underlying limestones and chalks affect the soils far more than in the Midlands and the north.

Agriculture

Compared with the northern parts of Britain, Southern England suffers few climatic restrictions on agriculture. The growing season is long, and summer temperatures are high. In winter, cold air may, however, spread from the Continent, and heavy snowfalls are common, particularly in Kent. Agricultural land-use is thus more influenced by the nature of the soil than by climate, but it is the presence of the London conurbation that affects it most, particularly with the spread of suburbs into Kent, Surrey and Sussex, with commuters travelling daily to London from as far as Folkestone, Brighton and Eastbourne.

The dip-slopes of the Jurassic limestone, e.g. the Cotswolds, or the Northampton Heights, are often devoted to mixed arable farming, with root crops and grain. In the Middle Ages the sheep on the dip-slope of the Cotswolds, in particular, gave rise to a woollen industry which is still important in small towns such as Stroud and Chipping Norton. There the houses, built of the distinctive Oolitic limestone, were roofed with slabs of thin-bedded local limestone.

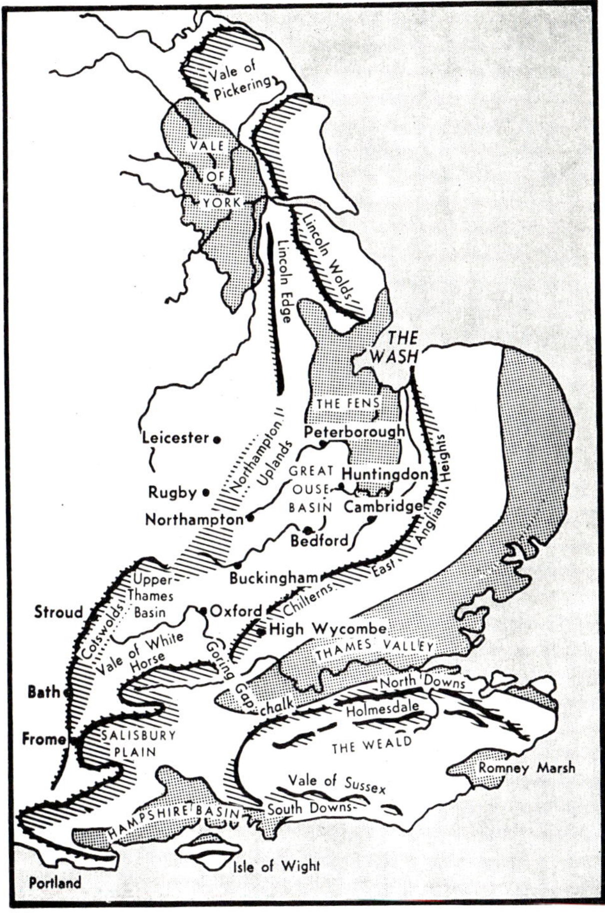

FIG. 19. *Simplified map of South-East England, showing main physical features.*

Parts of the chalk lands have rather heavy soils where patches of clay-with-flints occur. Here are the major woodland areas with beech and oak predominant, but much of the clay soils of the chalklands are under arable with grain, especially wheat and barley, and root crops. Moreover, these chalk lands offer a major recreational area for the London conurbation, especially where the chalk outcrops on the coasts and forms the North and South Downs.

In contrast, the clay vales, such as the Oxford clay vale between the limestone and the chalk, or the London clays, are arable or meadowland, with much market gardening and dairy farming. Particularly distinctive is the poultry industry of the Aylesbury district of the Vale of Oxford, although with the increasing use of "factory farming" and freezing, London is now receiving poultry from areas much farther away. The vales of the Kent and Sussex Weald are especially important for dairy produce, eggs, vegetables and fruit for the London market. Hops from Kent and Sussex are sent to breweries all over Britain. The keynote of all the agricultural land-use is very intensive agriculture with specialities wherever the climate and soils are favourable. The Lea Valley is a good example, with its concentration on vegetables and, especially, its acres of greenhouses.

Industry in Rural Areas

Even the thin dry soils of the sandy rocks of the Hampshire and London basins have some value. They are used for military training grounds, as near Aldershot, and areas considered sterile for other purposes are used increasingly for recreation. Everywhere land is in short supply, and particular attention is paid to the maintenance of "Green Belts" between the built-up areas of the towns. Surprisingly, for an area with such a high urban population, there still remain many districts which have not been touched by the ever-increasing tempo of economic life in Southern England. The old village life still exists in rural Kent, Hampshire and Sussex, offering a striking contrast to the dormitory towns of Sevenoaks, Tunbridge Wells, Maidstone, Guildford, Dorking, etc., whose inhabitants have for long commuted to London for work. Offices and light industries have recently tended to move into the dormitory towns, where land and rents are cheaper than in the centre of London or its more densely built-up fringes.

There has, of course, always been some industry in the rural areas of Southern England. The beech forests of the Chilterns provided the timber for the furniture industry still active at High Wycombe, while, even earlier, the oak forests of the high Weald were burnt for charcoal used in the smelting of iron. Iron now seems remote from the industries of Southern England, but the greatest reserves of iron in Britain are worked on its northern margins: the great steel town of Corby (a "new town") represents the most southerly intrusion of heavy industry, using the ore from the Northamptonshire uplands. Steel-using industries are, however, not uncommon in Southern England: at the western end of the Thames valley is the motor industry of Cowley near Oxford (B.L.M.) and on the north shore of the Thames Estuary the great Ford motor works at Dagenham has its own foundry adjacent to the assembly plant, and an advantageous site for importing fuel. To the north is the motor industry of Luton (Vauxhall), while there are many accessory and component firms in the London area.

Other iron- and steel-using industries are relics of the need for agricultural implements in earlier times. Newbury in the Vale of Kennet produces flour-milling and agricultural machinery, as does Reading at the confluence of the Kennet and the Thames. But both Reading and Newbury also offer employment in modern light industries, including aircraft manufacture, while the research establishment at Farnborough contributes to the aircraft industry of the whole of Great Britain. Harwell, the centre of the United Kingdom Atomic Energy Authority, is another example of a research station which has national, and international, rather than purely local significance. The increase in recreational activities is reflected in factories which mass-produce caravans and fibreglass boats; the latter are gradually replacing the traditional small wooden boat-building yards of the south and east coast. They are often as far inland as Tonbridge in Kent.

Ports and Airports

Southern England includes the great port of Southampton, as well as London. Situated on Southampton Water, with the advantage of a prolonged stand of high water, and within easy reach of London, this port is a major trans-Atlantic passenger terminal, which also handles a large volume of imports and exports. The trend towards the use of containers has led to the complete re-equipment of the port of Tilbury on the Thames Estuary, which will, in the course of time, probably divert an increased volume of trade away from Southampton and from the Port of London. In spite of the great increase in air traffic the Channel ports, especially Dover, maintain their importance by catering for the holiday-maker and for goods vehicles, with frequent services, including roll-on roll-off ship ferries, as well as the new hovercraft. Heathrow airport, to the west of London, handles so much traffic, especially goods of high value, that it is ranked as the third most important "port" in the United Kingdom. The search for a suitable site for another airport for Southern England, additional to Heathrow and Gatwick (south of London), has occupied several years.

The Channel coast and the Thames estuary are important in the oil industry of Britain: to Fawley on Southampton Water and to Shellhaven on the Thames come huge oil tankers bringing crude oil for the refineries round which important petro-chemical industries have been established. The methane-gas tankers from Algeria discharge into vast storage tanks on Canvey Island, Essex, whence a pipeline leads across the Midlands into south Yorkshire. On the desolate shingle foreland of Dungeness have been built two of Britain's coastal nuclear power stations.

The Future

Southern England, dominated by the London conurbation, contains the second concentration of population and economic resources in Britain. It is difficult to foresee any decline in its importance relative to the rest of the island, especially if the proposed Channel tunnel and the closer trading link with the countries of the European Economic Community become a reality. There will then be a major extension of the economic life of Europe through Southern England into the Midlands, based on the rail and motorway systems and container traffic. Other areas of dense population will also be affected, including the industrial areas which lie to the west and the north of the Midlands, especially Lancastria and Southern Yorkshire.

QUESTIONS

1. Write an essay on the London conurbation.
2. Describe the main features of the relief and soils of Southern England.

CHAPTER 7

THE CORE AREA: (iii) LANCASHIRE AND YORKSHIRE CONURBATIONS

THERE are no barriers of relief to separate the Midlands from the lowlands which flank the southern Pennines. The energy, railway and motorway network of Southern Britain extends north to serve the large concentrations of population and industry in the lowlands of Lancastria and Western Yorkshire, and in the valleys which are cut deeply into the Millstone Grits of the Pennines on both the western and the eastern sides. The area south of the river Ribble, drained by the Mersey and its tributaries, must be considered as a northern extension of the lowlands of Midland Britain; similarly the valleys of the eastern Pennines, which contain large cities, such as Leeds and Sheffield, represent another extension of the economic heart of England towards the north. The Pennines themselves are bridged by railways, motorways and waterways in spite of the steep slopes and altitude, so that there is

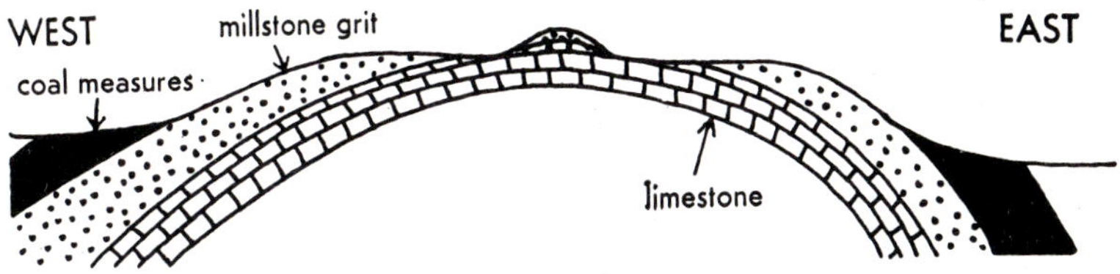

FIG. 20. *Simplified section across the southern Pennines.*

now, although less so in the past, a complete road, railway, gas and electricity link with the industries of the Midlands and with Southern England.

Lancashire and Yorkshire Industries compared

Nevertheless, these areas of Northern England support distinctive manufacturing industries. Lancashire is associated with the cotton, and the West Riding of Yorkshire with the woollen, industry. The industrial history of the whole area has led to a marked concentration on textiles. The common factors are the availability of soft water and water power from the rivers flowing off the impermeable millstone grit, and the coalfields of Lancashire on the west and Yorkshire on the east. Superimposed on these elements of the resource base were individual manufacturers' decisions to develop cotton textiles to the west and woollen goods to the east, although it is difficult on the border of Lancashire and Yorkshire to see a clearly defined break.

These two areas also have different heavy industries. Lancashire and Cheshire are associated with chemicals, and Yorkshire with iron and steel. Common to both is dependence on a relatively small number of industries of traditional type derived from their advantages in the Industrial Revolution, modified since 1945 by a trend towards the diversification of industry. Light modern industry has not reached the level of the East and West Midlands and of Southern England, although the very large car assembly plant at Hailwood near Liverpool and the new industries of Manchester and Merseyside offer much employment to workers made jobless by the reduction in the number of cotton spinning and weaving mills. Both areas retain much of their individuality, expressed in the characteristic industrial landscapes of the textile towns; they also differ in their social attitudes not only from the Midlands and Southern England but also from each other.

Lancastria

The Land. The western lowland is dominated by the Manchester and Liverpool conurbations, which contain over 5 million people. It is floored by Triassic sandstones and marls, similar to those of the West Midlands, and by the coal measures which form the Lancashire-Cheshire coalfield. Within the Triassic rocks are abundant deposits of salt, which have been worked since the Middle Ages. Overlying the solid rock, sometimes in great thickness, are glacial deposits of boulder clay and outwash sands which provide much diversity in relief. The most prominent hills are the sandstone ridges, while large and small river valleys are cut down deeply into the relatively weak rocks. Where drainage is poor there are great areas of peat mosses, such as Chat Moss between Manchester and Liverpool, but some areas are so freely drained they are covered with heath or poor woodland. The coast is low and in its original form was backed by sandhills, lagoons and marsh.

The clayey soils are usually under grass rather than crops. The relatively mild winters and, for the lowlands of Britain, relatively high rainfall, gave Lancashire and Cheshire a well-founded reputation for dairy produce, even before the demand increased from the populations of the textile and mining towns. The milk was, and still is, processed into cheese and butter, using the salt from the Triassic rocks as a preservative, but increasingly the demands for milk in the towns controls the output of the dairy farms. The towns also make heavy demands for potatoes and vegetables, so that much farming, especially in Western Lancashire and in Cheshire, is given over to market gardening, to pigs and to poulty farming.

Textiles. Most of the towns of the eastern edge of the lowland grew from small market towns to large industrial centres through the cotton industry

FIG. 21. *The Lancashire coalfield.*

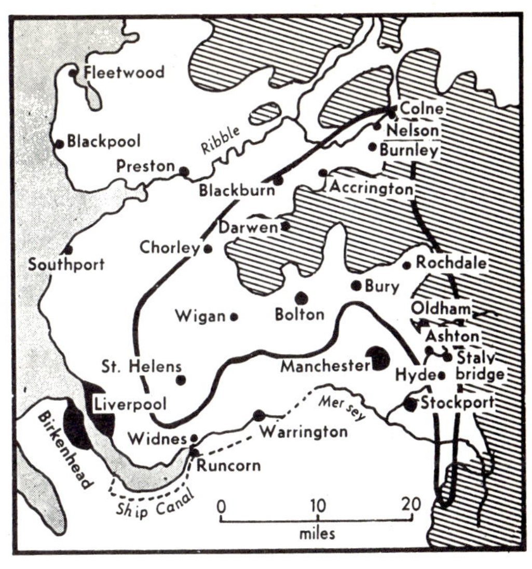

and through the coal-mining. The development of the cotton industry here was due to the combination of certain natural advantages in the resource base, its location, and the decisions, as elsewhere, of its population. The primary factor was the water power available where the tributaries of the Mersey and the Ribble came swiftly down from millstone grit moorlands of the Pennines in deeply incised valleys with falls and rapids. The softness of the water was an undoubted asset, as was the trading connection with the West Indies and North America built up by the merchants of Liverpool. Their ships took full advantage of the deep water provided by the natural scouring of the Mersey estuary, while the much older port of Chester on the Dee estuary became silted up and lost the importance it had had in Roman times. Through the port of Liverpool, connected by the Leeds and Liverpool canal and the Bridgewater canal to the cotton area, the raw cotton could be imported and, just as important, the finished textiles could be exported. The humid climate was an asset, as was the migration of weavers from Flanders, but it was water power and location near a great estuary which provided the favourable conditions for the early textile industry. Later the coalfield provided power for the mill-engines. The cotton mills moved towards the collieries, or to the banks of the canals by which the coal could be brought by barge. Later the railway gave more freedom to the mill owners in deciding where to locate their factories.

In the 19th century cotton textiles were the principal exports of the country. It was this over-concentration on cotton which, after 1918, led to the decline of the cotton towns and turned Lancashire into a depressed area, as competition from India and Hong Kong led to loss of markets overseas. The spinning towns on the southern fringe of the coalfield from Wigan to Stockport, and the weaving towns from Preston to Colne to the north of the high ground of the Rossendale Forest, all suffered severely—as did Manchester, which had grown into the commercial and business centre of the industry, connected to the port of Liverpool by the opening of the Manchester Ship Canal in 1894. This canal, which brings ocean-going ships into the heart of the industrial area, was one of the most far-sighted civil engineering achievements in Britain.

Although rayon, nylon and terylene have now been added to the fibres used in the textile industry, and 90 per cent of the British cotton industry is still located in the Lancashire towns, there has been since 1945 a great decrease in the number of workers in the textile industry, which is predominantly an employer of female labour. The Lancashire coalfield has now a very much reduced output and many of the western pits have been closed. Textile machinery is, however, an important industry and is exported to many of the developing countries.

New Industries. Many attempts have been made since 1945 to introduce new industries, including light engineering, but the old cotton mills are not easily converted to modern needs and the congestion on the valley floors makes the building of new factories difficult and expensive. Outside the coalfield there has been major expansion of industry, especially in Manchester, where Trafford Park by the Ship Canal has over 200 factories engaged in such industries as food processing, corn milling and the making of electrical machinery, while rubber, dyestuffs and aircraft works can all be found within 15 miles of the city centre.

Manchester has grown to dominate the whole of the metropolitan area known as SELNEC (South East Lancashire, North East and Central Cheshire), covering an area of over 1,000 square miles and a population (1968) of 3,232,000. Here there is intense population pressure, great commuter traffic, and "overspill" into the rural areas of Cheshire. New towns have already been built at Runcorn, Cheshire, and Skelmersdale, West Lancashire.

To the west of SELNEC lies a developing industrial area based on the salt and chemical industry of Northwich, Winsford, St. Helens, Runcorn, Warrington and Widnes. To these towns have also come glass and electric goods industries, and the whole area has been given a great locational advantage through its position astride the Ship Canal, the M6 Birmingham-Carlisle motorway, and the M62 from Liverpool to Yorkshire. The extraction of salt, however, has created many problems of ground subsidence which affect the suitability of the land for the building of houses and factories.

Liverpool. Beyond the coalfield and the saltfield the estuary of the Mersey opens up towards the conurbation of Merseyside based on the port of Liverpool with in all (1968) 2,063,000 people. Here are situated some of the major industrial units of the western lowland, based on imported raw materials. At Stanlow is a big oil refinery; soap and margarine are major

DRAINAGE: NOTE CONCENTRATION OF HIGHER LAND IN THE WEST

FUEL and POWER
conventional power stations (over 200 MW), in operation and
being built
oil refinery (over 2 million tons annual capacity),
nuclear power station, in operation, and
being built
tanker discharge points (over 80,000 tons)
hydro-electric power stations (over 45 MW output)
imported natural gas storage
coal-bearing areas
gas terminal
gas well
natural gas pipeline
being laid
submarine pipeline
being laid
Fasnakyle
Rannoch
Clunie
Errochty
Lochay
Kincardine
Sloy
Cockenzie
Hunterston
Grangemouth
Chapelcross
Blyth 'B'
Teesport
Calder Hall
Locton
Heysham
Rough Field
Ferrybridge
Drax
West Sole Field
Fiddler's Ferry
Eggborough
Viking Field
Thorpe Marsh
Easington
Wylfa
Cottam
West Burton
Leman Bank Field
Stanlow
Theddlethorpe
Trawsfynydd
Willington
Ratcliffe on-Soar
Hewett Field
Ironbridge
Drakelow
Bacton
Rheidol
Rugely
Indefatigable
Accumulation
Milford Haven
Sizewell
West Pennar
Berkeley
Pembroke
Llandarcy
Oldbury
Bradwell
Aberthaw
Didcot
Hinkley Point
Dungeness
Fawley
1 Kingsnorth
2 West Thurrock
3 Tilbury 'B'
4 Shellhaven
5 Coryton
6 Canvey Terminal
7 Isle of Grain

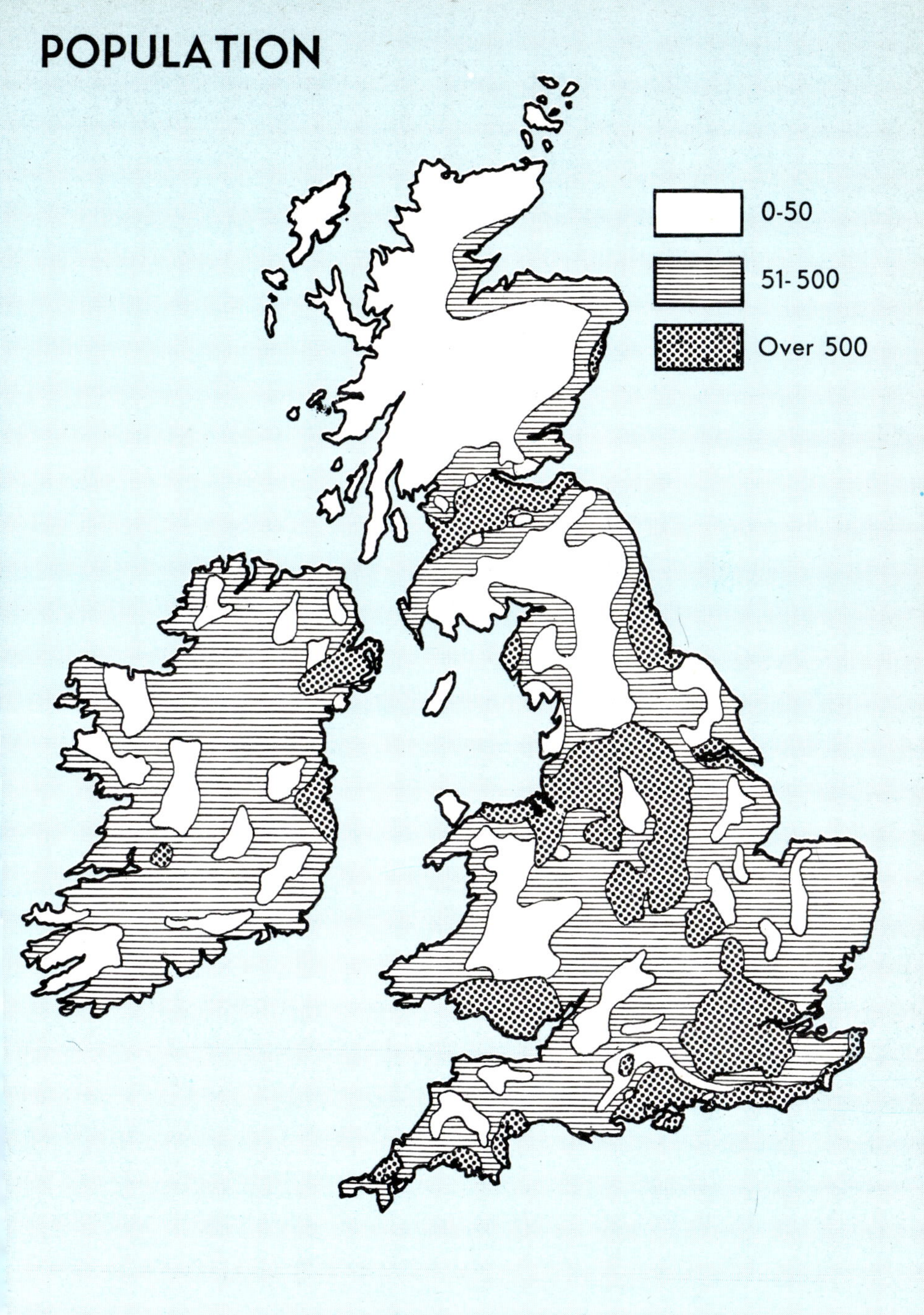
POPULATION
0-50
51-500
Over 500

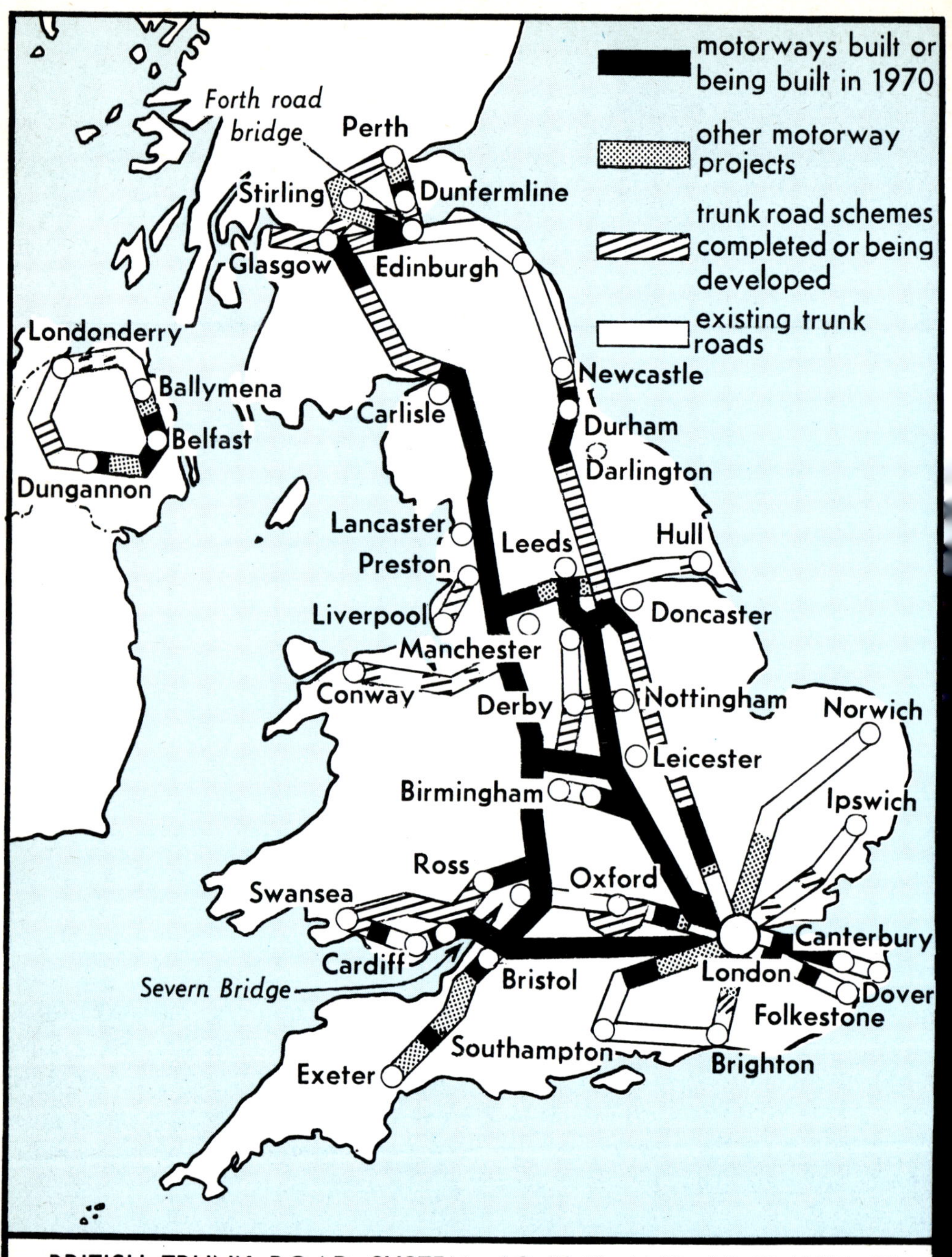

BRITISH TRUNK ROAD SYSTEM, AS IT IS AND AS IT WILL BE. SHOWN HERE ARE MOTORWAYS THAT HAVE ALREADY BEEN COMPLETED, OR ARE EITHER UNDER CONSTRUCTION OR PLANNED FOR THE FUTURE

industries, together with food-processing, meat and meat products, shipbuilding, and at Hailwood the very large Ford car assembly plant. Merseyside, which includes both sides of the estuary, has problems of communication, since there is no bridge downstream from Runcorn; but these have been partly resolved by the Mersey tunnels.

SELNEC and Merseyside face serious problems of water supply, for which the uplands of Wales, the Lake District and the Pennines must provide the necessary resources. Its seaboard, as at Southport and Blackpool, is increasingly used for purposes of recreation, as are the uplands on its margins. Like Birmingham and London, the conurbations attract population from other parts of the British Isles, although the old mining and cotton towns have not advanced as rapidly as the remainder of the area.

Yorkshire

Extending from the western flanks of the Pennines to the mouth of the Humber, this area includes the textile industry of the West Riding, and the coal and steel industries based on Barnsley and Doncaster, Sheffield and Rotherham. Farther to the east, and separated from the western industrial area by the agricultural region of the Vale of York, is Humberside and the steel town of Scunthorpe. In its combination of textiles, coalfield, and industrialised estuary Yorkshire thus resembles the western lowland, with the salt and chemical industries of Cheshire replaced by the iron and steel industry of South Yorkshire.

The Land. The watershed of the Pennine upland lies nearer to the west than the east coast of England, and from it drain, across the Carboniferous limestone, millstone grit and coal measures, rivers such as the Wharfe, Aire, Calder and Don, which unite in the Vale of York before entering the Humber. Beneath the Vale of York lie Triassic rocks which give way towards the east to the Jurassic limestone escarpment and the chalk of the Lincolnshire Wolds, separated from the Yorkshire Wolds by the estuary of the Humber. As in the area west of the Pennines, these rocks are covered by glacial deposits, including the York-Escrick moraine which marks the southern limit of one of the extensions of the ice sheets in the Quaternary period.

There is thus a much greater variety of relief than in the area to the west of the Pennines. The upper reaches of the eastward-flowing rivers have cut deeply into the heathy moors of the Pennine upland, creating broad valleys (the Dales), which give way to the gently undulating country of the vale that was once the site of a large glacial lake, Lake Humber. The escarpments and dip slopes of the limestone and chalk offer drier, limy soils, with much arable farming and sheep-rearing, but it is the Vale of York that forms the main agricultural area, with crops of wheat, barley and oats, dairy-farming and market-gardening catering for the townspeople of the West Riding, the coalfield and Humberside.

Population. The dense population concentrations are, as in Lancashire and Cheshire, in the industrial areas located on the floors of the deep valleys of the Pennines and its eastern flanks. They are the West Riding conurbation of Leeds, Bradford and Huddersfield, and the steel area of Sheffield and Rotherham. There is little agricultural land on these valley floors, as the towns merge into one another between Leeds and Keighley in the Aire valley, and in the valleys of the Calder and Colne between Wakefield and Halifax and Huddersfield. In no other part of Britain, other than the South Wales coalfield and East Lancashire, are there such dense populations served by a road, canal and railway network concentrated on valley floors hemmed in by steep slopes and bare moorlands.

Wool. The West Riding conurbation is based on the woollen textile industry (about 80 per cent of Britain's woollen industry is concentrated here), but there is an overflow of cotton weaving from Lancashire into the western section at Todmorden and even in Huddersfield, Halifax and Skipton. From early beginnings as a cottage industry using wool from the sheep of the Pennine moorlands and dales, the woollen industry expanded greatly in the Industrial Revolution with the use of water power for driving the machinery. The large quantities of soft lime-free water from the millstone grit was then, as now, an important factor in the location and growth of the woollen industry; it was required for washing the raw wool, for fulling and for dyeing. Without the plentiful supplies of water the urban centres could not have grown, and it would have been impossible to create the elaborate canal system that preceded the railways. But it was the coalfield that was mainly responsible for the expansion of woollen textiles. The textile towns on the coalfield grew more rapidly than those dependent on canal transport for fuel supplies, so that Leeds and Bradford, Halifax, Dewsbury and Keighley at the northern edge of the coalfield now form the centre of the West Riding conurbation, with a combined population (1968) of over 1½ million people.

Within the woollen textile area there is specialization, as in the cotton industry of Lancashire. The worsted centres are mainly in the north-west, and the woollens in the south-east. They come together in the Huddersfield-Halifax group of towns. Shoddy—made from woollen waste—is concentrated in the heavy woollen area of Dewsbury and Batley, while Halifax specializes in carpet-making. Bradford, the chief wool market, has important combing and dyeing industries, as have Leeds and Huddersfield. Leeds (840,000 in 1968), the largest city, situated at the eastern end of the Aire gap through the Pennines and connected by the M62 motorway to Manchester and by the M1 to the Midlands and Southern England, is the centre of the clothing trade, and has also important chemical industries for the dyeing process. Like Manchester to the west, it also contains many of the service

FIG. 22. *Tapping an electric arc furnace in a Sheffield steelworks.*

industries, such as banking, commerce and insurance, for the West Riding conurbation.

Iron and Steel. The industrial development of West Yorkshire has also hinged on the iron and steel industry of the south-eastern sections of the coalfield, where ironstones from the coal measures provided the basis for a medieval iron industry. With imported ores there was much manufacture of wrought-iron, later superseded by steel. But by 1932 the iron and steel industry of Bradford and Leeds was almost dead, although not before it had given rise to textile-machinery-making and other engineering industries in Leeds, Wakefield, Huddersfield and Halifax.

The southern part of the Yorkshire coalfield is essentially devoted to mining and iron and steel. It is based on the towns of Barnsley, Rotherham and Sheffield, which contained, in 1968, just over 1 million people. The coal measures contain ironstone nodules which have been smelted since the 12th century. By the Middle Ages Sheffield was already famous for its cutlery. The skills of the Sheffield iron and steel workers in the production of sharp-edged tools, knives and axes, led to early imports of high grade iron-ore to what is a difficult site for industrial development in the deep valley of the Don. The water power which drove the textile mills of the West Riding also powered the early factories. The millstone grit for grinding wheels, and the presence of the coal and gannister (used for the lining of furnaces) and dolomite from the Magnesian limestone, ensured, with the presence of the skilled smiths and foundry workers, continuity in the industry. Today Sheffield is famous for its steels, especially stainless steel (steel and chromium, or other special alloys such as manganese and tungsten), for both domestic and industrial use. Rods, bars, strip, springs and tool steels from Sheffield are used in engineering industries throughout the British Isles.

The Sheffield conurbation, including Barnsley and Rotherham, has spread from its former valley location up the valley side slopes. Up here the residents suffer, as they do in the higher parts of the West Riding, from decreased temperature with altitude, greater snow-fall and longer snow-lie than the valley floors and the lowland of the Vale of York to the east. It is surprising, perhaps, that there should be such a concentration of population in these valleys of the Pennines, since communications are difficult because of the steep slopes, especially in winter. Fortunately the high slopes and plateaux of the Pennines also provide opportunities for recreation, recognized by the creation of the Peak District National Park, as well as ample water supplies.

Humberside. Separated by the agricultural lands of the Vale of York from these western industrial areas is Humberside, on the north and south margins of the Humber estuary where the Trent joins the Yorkshire rivers. The northern side is dominated by Kingston-upon-Hull (350,000 in 1968) which imports a wide range of raw materials, including petroleum, foodstuffs and timber, and exports much of the iron and steel goods and machinery from Western Yorkshire. It has diversified industries—food processing, soap, magarine, paint, shipbuilding and aircraft manufacture. It is, in addition, the most important distant-water trawling port in Britain, and through its docks comes about 50 per cent of the fish landed in Britain.

The Humber is not yet crossed by any bridge, and no tunnel has yet been driven to connect the north and south banks. Nevertheless, like Merseyside, it has large towns, including the ports of Immingham and Grimsby on its south side, together with the town of Scunthorpe (70,000) where almost 50 per cent of the population are employed in the steel industry. Steel plates, girders for construction work, and rolled steel for the car industry are produced, using local Jurassic iron-ore. The population of south Humberside amounts to just over 300,000 people, who also find employment in the production of cement, chemicals and fertilizers. Grimsby is well known as a fishing port for cod, haddock and plaice; it also imports timber and iron-ore from the Baltic countries and dairy produce from the Netherlands.

There is in southern Yorkshire a very high concentration of population employed in a wide variety of industries and in agriculture, of which only the textile area of the West Riding has not acquired the new light industries or those based on advanced industrial technology. Like southern Lancashire it has good communications with the Midlands and with the north, and has access to the world through both Merseyside and Humberside.

QUESTIONS

1. Account for the types of industry in the Lancastrian lowland.
2. Write an essay on the textile industry of the West Riding of Yorkshire.

CHAPTER 8

PERIPHERAL INDUSTRIAL AREAS

THE four remaining major industrial areas of the United Kingdom are all situated on the coastal coalfields, which had many advantages in the 19th century but were to become in the 1930s the "distressed" areas to which special Government action had to be directed. These areas, all separated from the main economic heartland of the Midlands, have a poorer resource base than industrial Lancashire, Yorkshire, the East and West Midlands and southern England; they lack a good system of energy supplies and traffic links and, although efforts have been made to introduce new industries, they suffer from their location away from the major concentrations of people and industry. All in or near upland Britain, they comprise the South Wales coalfield, the north-east of England, west Cumberland, and the central lowlands of Scotland.

South Wales

South Wales forms a very distinctive industrial area where the north-to-south flowing rivers have dissected the southern edge of the Welsh upland to expose the carboniferous rocks and within them the coal measures. In the deep river valleys are situated the collieries, the mining villages, the roads, and the railways, and on the plateau slopes are the great spoil tips where the refuse of the coal-mining industry has been heaped since the beginning of the Industrial Revolution.

To the west of the coalfield the moorlands give way to the low plateau of Pembrokeshire, deeply indented by Milford Haven, which is now a major tanker terminal and oil-refining centre with a pipeline to Llandarcy 60 miles away, near Swansea. On the coast at Port Talbot there is the largest steel works in Britain, which feeds sheet metal to car body-building plant. Here in South Wales is to be seen most strikingly the contrast between the traditional coal industry with a declining population and the growing new industries (oil-refining and car components). Physically, the contrast continues between the high bleak moorlands, 1,000 feet above sea-level, and the good soils of the Vale of Glamorgan, where mixed farming with dairying and horticulture are characteristic. The moorlands also contrast with the cliffs and sandy bays of the Gower peninsula, which attract many summer visitors.

Rocks. South Wales is dominated by Carboniferous rocks preserved in an east-west syncline with gentle dips in the north and steep dips in the south. To the north, the Palaeozoic rocks of the Brecon Beacons (2,900 feet) confine the syncline, while to the south the Triassic sandstones of the Vale of Glamorgan rest against the Carboniferous limestones. The coal measures occupy a large area in the centre of the basin, which is separated into two parts by high gritstone moorlands. In the west, Swansea Bay and Carmarthen Bay break the southern rim of the syncline.

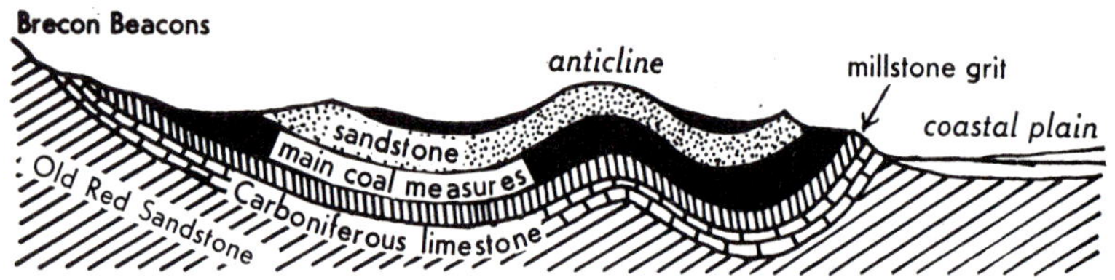

FIG. 23. *Simplified section across the South Wales coalfield.*

Coal. There is a wide variety of coals found in the South Wales coalfield. At the western end (the Pembrokeshire field is no longer worked), anthracite, which contains very little ash or volatile matter, is found west of the Neath valley. It contains less than 10 per cent of volatile matter, provides great heat with practically no smoke, and is thus in great demand as a smokeless fuel. To the east the volatile matter increases (but is still less than 15 per cent) to provide the steam coals of the Maesteg district and the Rhondda valley. These give way to the more volatile but best coking and manufacturing coals (19-30 per cent volatile matter) in the mid-Taff and the valleys of western Monmouthshire.

This coalfield once provided the bulk of the fuel for ships bunkering in British ports and was also exported to bunkering stations overseas. It yielded the best coal for raising steam in locomotives before the change to diesel power on the railways. In consequence, the ports of Swansea and Port Talbot were developed and linked by railway with the coalfield, as were the docks at Barry and Cardiff. The anthracite, 90 per cent pure carbon, was originally less important since it must be burnt in a closed furnace with a good draught of air, where it produces great heat. Today it is highly valued for central heating systems based on solid fuel. The modern mechanized collieries, however, mainly produce coking and manufacturing coals, since oil, nuclear energy, electricity from coal-fired stations and natural gas are serious competitors with household coal.

Iron and Steel. Like the South Staffordshire and the Yorkshire fields, the South Wales coalfield has iron-ores available from the coal measures. The combination led to a great iron industry based on Ebbw Vale and Merthyr Tydfil, and also in other towns in the valleys of the north where flux from the Carboniferous limestone rim was available. Iron-ore is heated in blast furnaces with coke and limestone (the importance of the coking type of coal is that it is strong enough to support the iron-ore and the flux), with blasts of air, which usually led to the siting of the furnaces on high ground. Much of the carbon which the ore contains is oxidized by this process; other impurities rise to the surface, leaving the molten iron which is then drawn off at a lower level. The product of the blast furnaces is pig-iron.

With the change-over between 1860 and 1900 from iron to steel for rails, strip and rolls, a further stage in production became necessary. The making of steel involves removing the small percentage of carbon which exists in the pig-iron, by oxidation in heated furnaces and by adding various alloys, e.g. chromium, molybdenum, manganese or tungsten, to produce the various kinds of steel required for individual needs. In a modern iron and steel works the pig-iron from the blast furnace is run direct to the steel converter or furnace. The change from pig-iron to steel affected all the iron-producing areas of Britain, but was particularly important in South Wales. As higher-grade ores were brought by sea from both home and overseas, there was a shift from the former iron-smelting towns of the north to the coast. The Margam steel works near Port Talbot and the Llanwern Works at Newport are good examples of the trend. But Ebbw Vale still retains an important steel industry which supplies sheet, bar and strip to many factories in the Midlands and southern England.

Thin sheets of iron will rapidly rust in contact with atmospheric moisture, and are thus neither long-lasting nor suitable for use as food containers. If, however, the sheets are dipped in palm-oil to assist an even coating with tin (coming at first from Cornwall but later from overseas), then the protective coating produces tinplate, for which large works are situated near Llanelly, at Trostre, Ebbw Vale and Velindre. The Llanwern steel works has been equipped as an integrated plant to produce tinplate directly from iron sheets. The tinplate, used both at home and overseas for canning food products, suffers today from compctition from plastics and other packaging methods

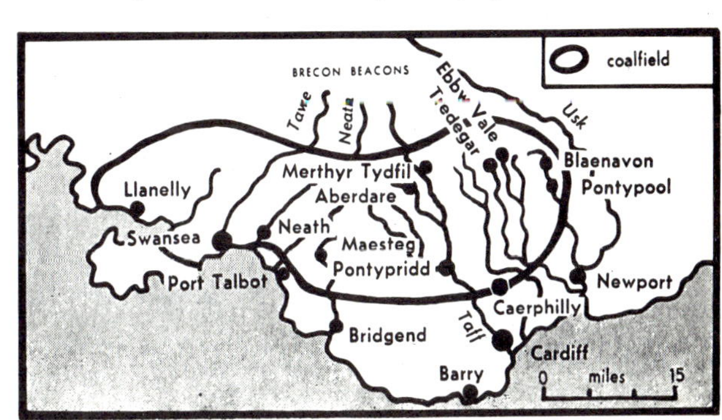

FIG. 24. *Towns of the South Wales coalfield.*

and materials, which have also affected to some degree, the galvanizing industry of South Wales.

Other Metals. Non-ferrous metals produced at Swansea and Newport include copper and aluminium (smelted at Fort William in Scotland), while nickel is also smelted at Swansea, which also produces components for the car industry. Cardiff, the capital of the Principality of Wales and its largest city, is small (about 300,000) compared with, say, Birmingham or Manchester. Although there is much built-up land in the coal-mining valleys, there is no conurbation compared with that of the West Riding or the West Midlands, in spite of the new factories and population which have been induced to move to South Wales since 1945. The bridging of the Severn estuary in 1966 has provided a closer link with the Midlands and southern England than with central and North Wales, to which the uplands offer difficulties in communications.

North-east England

Northumberland and Durham coalfield. This coalfield lies on the eastern flank of the northern Pennines, drained by the Tyne and the Wear, but the industrial area extends south to the northern flanks of the North Yorkshire moors and the valley of the Tees. The population is concentrated in the coastlands, where the people are engaged in the few basic heavy industries—shipbuilding, marine engineering, heavy engineering, coal-mining, steel-

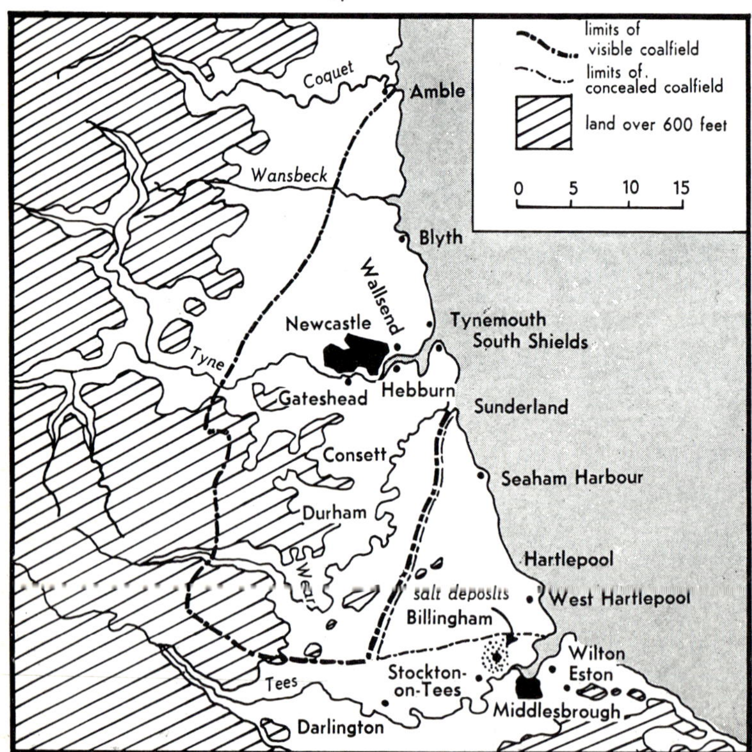

FIG. 25. *The Northumberland and Durham coalfield.*

making, and the manufacture of heavy chemicals. Further inland the industrial countryside gives way to the sandstone escarpments and peat mosses of the high plateaux, before rising finally to Cross Fell (2,930 ft.) on the Carboniferous rocks. This separation from the rest of the country by the Pennines and Cheviot Hills and by the sparsely populated areas of the agricultural lowlands of North Yorkshire makes the region an industrial "island", like South Wales. It is dominated by the Tyneside conurbation (1,026,000 in 1968).

Rocks. Here a new element in the structure of the island is to be seen. North-east England is part and parcel of upland Britain, dominated by the granites of the igneous intrusion of the Cheviots and by the tilted block of the Pennines in the west with millstone grit moorlands which pass eastwards beneath the coal measures, consisting mainly of sandstones, shales and coal. To the south-east the coal measures are in turn concealed by the dolomitic rocks of the Magnesian limestone whose westward facing escarpment is a well-marked feature in East Durham. The dip slope of the Magnesian limestone is covered to the east by the Permian marls, which contain rock-salt and anhydrite. These marls, and the adjacent outcrop of Triassic sandstones, are overlooked from the east by the Jurassic limestone escarpment of the Cleveland Hills. Into these sedimentary rocks, especially in the north, sills and dykes are intruded—especially the quartz-dolerite Whin Sill, the north-facing escarpment of which carries, for long sections, Hadrian's Wall, which marked the effective limit of the Roman occupation of Britain. Until the union of the crowns of England and Scotland in 1603 this was a frontier zone, where many castles took advantage of the abrupt slopes for defensive purposes.

There are many relics of glaciation. The Cheviot Hills had a local ice cap, and its margins are seamed with glacial meltwater channels. On the lowlands towards the coast there are deposits of boulder clay sometimes so thick as to bury completely the underlying relief of the solid rock. Elsewhere the boulder clay is very thin, and it is sometimes entirely missing, as on the rocky escarpments which fringe the main Cheviot dome; in some places there are large areas of outwash sands and gravels. The whole produces a great variety of soils, best suited to agriculture in the lowlands, especially in the lower Tees valley.

Rivers. The Tees has dissected the dip slope of the Carboniferous rocks to produce a broad open valley inland, and it reaches the sea in a broad estuary. Other rivers enter the sea in deeply incised valleys. The Wear, for instance, flows in a deep meander at Durham, where the cathedral is sited on the neck of the meander. The deep valley of the Tyne near the sea acts as a barrier to north-south movement, and has to be bridged by road and rail high above the water. To the west, however, it provides, through the

Tyne gap, an easy link, rising no more than 500 ft. above sea level, to Carlisle and the south-west of Scotland. The lower Tyne is navigable, and its position in the exposed part of the coalfield provided an early stimulus to coal-mining.

Coal has been mined near Newcastle since the Middle Ages, and used to be transported to London as "sea-coal". All types of coal except anthracite are represented, but the most important is the coking coal of the western part of the Durham coalfield, which also has the greatest reserves. The most productive pits, including those of the "concealed" coalfield, are being modernized, and mining takes place even under the sea. The less productive parts of the coalfield have been closed down, and there are many derelict mining villages in need of alternative employment.

The iron and steel industry is based principally in the lower Tees at Middlesbrough, with Consett an example of a large modern plant away from the coast, to the west of Chester-le-Street. The early iron industry was founded on iron-ore from the coal measures and the deposits of haematite from the ore bodies in the Carboniferous limestone of the Cleveland Hills. All these deposits have now been worked out, and all the ore is now imported. Iron and steel was the dominant industry of the north-east, concentrating on the building of ships and railway locomotives and the making of steel plate and colliery machinery. Each town specialized mainly in one product. To provide labour there was much immigration from Scotland and Ireland, from the worked-out tin mines of Cornwall, and even from East Anglia.

Shipbuilding. In the 19th century the coastal location of the north-east was a great asset, and in spite of the competition of the industrial Midlands, it flourished. By the middle of the century, the Tyne and the Tees had been improved for navigation to cope with the rapidly increasing export trade and with the increased size of iron ships. On the Tees and the Wear, more and more land by the river was taken up for shipbuilding, graving docks and marine engineering, as Middlesbrough and Sunderland grew. But the most marked expansion occurred between Newcastle and the sea, where Gateshead, Jarrow, Wallsend, Tynemouth and South Shields became famous for both merchant ships and naval vessels. By 1913, when nearly 2 million people were living in the coastal industrial area, 750,000 inhabited the urban areas of Tyneside, another 275,000 lower Tees-side, and about 170,000 Wearside. The economy was dominated by mining, iron and steel, shipbuilding and heavy engineering.

Industries. One other industry, centre on Tees-side, was and is very important. The salt industry, which had begun with the evaporation of sea water in the coastal salt pans, as at South Shields, became concentrated after 1860, using the salt from the Permian rocks south of the coalfield at the mouth of the Tees. Anhydrite was discovered in the same area, and with the salt (both pumped up as brine) forms the basis of Middlesbrough's chemicals industry. The anhydrite, a variety of calcium sulphate, provides sulphuric acid and the fertilizers ammonium sulphate and nitro-chalk.

During the First World War the perfection of the technique of fixing atmospheric nitrogen led to the creation of a new chemicals industry at Billingham to the north of the Wear. This town grew with the expansion of Imperial Chemical Industries, which continues to provide a major source of employment.

Concentration on heavy industries made this area very vulnerable after the First World War. Not only was there the danger of specialization within each town on a single product, but within one town, Jarrow, 85 per cent of the male working population was employed by one shipbuilding company. With the economic depression of the 1930s, such towns were especially affected; in Jarrow, out of a total population of 35,000, 23,000 were on public relief. Finally the Government made most of county Durham and Tyneside—with the exception of Newcastle and Tees-side—a "special area", like the South Wales, the West Cumberland and the Lanarkshire coalfields and the lower Clyde. The aim was to tempt new industries to move to these areas by providing financial aid and by the building of new industrial estates, e.g. the Team Valley trading estate. The results were not very successful, but the problem was temporarily solved during the Second World War when the renewed demand for ships and munitions created work for the heavy industries.

Since 1945 there have been two main developments. The continued expansion of the chemicals industry makes Tees-side one of the world's major concentrations of that industry, producing pharmaceutical as well as commercial chemicals, paint and soap. The trading estates have also increased in number, the largest being at Team Valley, Tynemouth, Jarrow, South Shields, Sunderland, the Hartlepools, Tees-side, Aycliffe, West Auckland and Spennymoor. Together they have provided at least 70,000 new jobs in light engineering and the consumer industries, and new towns have been built to house the workers, as at Newton Aycliffe, Peterlee and Washington.

The employment structure of North-east England, with its population of about 2¾ million people, reflects the diversification of its industries. Agriculture, usually of the mixed type but with market-gardening in many places, accounts for only 2 per cent of the employed population, shipbuilding and marine engineering only 4 per cent, the iron and steel industry 4½ per cent, the chemical industries 4½ per cent, and coal-mining and other extractive industries 8½ per cent; that is to say, only about one fifth of the population are now employed in the industries for which the North-east became famous. Another fifth work in other manufacturing industries, including electrical engineering on Tyneside, glass making at Gateshead, food, clothing and other consumer goods on the trading estates, and at Newcastle. The remainder are employed in the construction industry (8 per cent) and, especially, in the service industries (about 45 per cent). Even greater diversification of employment is needed before the economic wellbeing of the North-east can compare with that of the Midlands and Southern England. Moreover, the relative isolation of the area remains a drawback: although the A1 trunk road has been much improved by dual carriageways and by-passes, the area is still not connected by motorway to the south.

West Cumberland

On the opposite side of Northern England, fronting on to the Irish Sea, and separated by the uplands of the Pennines and the Lake District from the North-east industrial area, is the West Cumberland coalfield, centred on Maryport, Workington, Whitehaven and the industrial town of Barrow-in-Furness at the south-western tip of the Lake District. The deep indentation of the shallow waters of Morecambe Bay increases the isolation of this area of 500,000 people from the industrial lowlands of Merseyside and SELNEC. But there are plans to reclaim the bay and to build a causeway to shorten the distance for road transport to the south.

Like the other coastal coalfields, West Cumberland was a "special area" between 1934 and 1945, and is now a "development area". Its dependence on the export of coal and on heavy industry following the Industrial Revolution has created many problems, yet it was here that one of the major advances in the technique of steel production took place with the establishment of the Bessemer steel plant and rolling mill at Workington in 1867. The Bessemer converter removes carbon from molten iron by forced blast, which causes the oxygen of the air to combine with the carbon so that the latter is removed with the waste gases. The process is quick and cheap, but demands high-grade ores; if low-grade ores are used the furnaces must be lined to prevent corrosion by the acid impurities (such as sulphur and phosphorus) of the molten ore. The Gilchrist-Thomas process, perfected in 1878, consisted of lining the furnaces with basic materials to allow the use of the acid low-grade ores.

The West Cumberland coalfield extends from the flanks of the Lake District to the west beneath the Irish Sea. Because of faulting, the thinness of the seams and their proximity to the sea floor, it is one of the several British coalfields with above-average production costs. Nevertheless, mining and the export of coal from Whitehaven and Workington has continued since the 16th century. Although the bulk of the coal is used for coking, it is mixed with coking coals from South Wales and Durham before being used at Workington.

The high-grade haematite ores in the Carboniferous limestone fringe of the western Lake District, which gave rise, with the coal, to the iron and steel industry, are now worked out, except in the Furness area near Millom, and iron-ore is imported through Workington from North Africa, Sweden and Spain. Much limestone however is still quarried in Furness.

The three main iron and steel centres centre on Workington, Millom and Barrow-in-Furness. Workington has for a long time fabricated steel rails, and Distington, a few miles to the south, produces steelworks' plant and paper-making machinery. At Millom the concentration is on large-scale castings. Barrow-in-Furness is the main centre of heavy industry, with blast furnaces for iron castings, general engineering, and a steelworks producing ingots. More important are Barrow's shipbuilding yards and marine engineering works, producing naval vessels, including nuclear submarines, as well as merchant ships, dredgers, and floating docks.

During the economic depression of the 1930s, in some districts 80 per cent of the working population were unemployed. The area was designated a "special area" and, as in North-east England, trading estates were created; the Solway Estate to the south of Maryport was the first example, followed since 1945 by new industries varying from textiles, hosiery, boots and shoes, to instruments, paper, cosmetics and detergents. Here, too, are the nuclear reactors of Windscale and Calder Hall; from the latter, the world's first atomic power station, power was fed into the grid as long ago as 1958. Despite these attempts to provide a wider range of employment, the problem of isolation from the other industrial areas of the United Kingdom still remains to be solved.

Central Scotland

Central Scotland contains the last major concentration of population and industry that we shall consider. Here Scotland narrows between the Firth of Forth and the Firth of Clyde. The land has been downfaulted between the Highlands to the north and the Southern Uplands to the south to form a major rift valley. The floor of this lowland is by no means flat. It is broken

FIG. 26. *Simplified section across the Scottish Lowlands.*

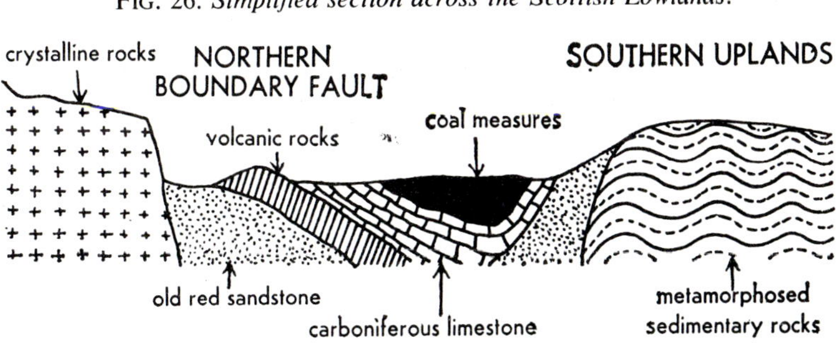

by intrusions of volcanic rock which pierce through the Old Red Sandstone and Carboniferous rocks, mainly sandstones and shales, to form islands of high ground, while the lower ground between is mantled by thick deposits of boulder clay, sometimes in the form of drumlins, and by sands and gravels which form ridges and mounds. The firths are fringed with raised beaches which, in the case of the Firth of Forth, overlook the wide areas of the Carse. These are marine clays which form flat land, but require draining before they can be put to use. In the Carboniferous rocks occur the main coalfields of Scotland, and it was on these and their associated iron that the industrial prosperity of Scotland was formerly based. At either end of this lowland are the two principal cities of Scotland—Edinburgh, the capital, and Glasgow which, with Clydeside and other towns, forms the chief Scottish conurbation.

Fringing the hill-foot of the Highlands in the north lies the agricultural vale of Strathmore, a rich farming district of large farms, with concentration on the production of seed potatoes and soft fruits within a general framework of mixed farming. It is separated from the North Sea and the Firth of Tay by the lavas of the Sidlaws and the Ochil Hills, through which the river Tay breaks in a gap dominated by the route centre and market town of Perth, sometimes called the gateway to the Highlands. Towards the west, the Ochil Hills, with their characteristic hill-foot towns such as Dollar and Tillicoultry, are separated from the volcanic Campsie Fells by the gap through which the meandering upper Forth emerges. Here, where volcanic sills with a steep western and gently sloping eastern face provided a good defensive site at the river crossing, the town of Stirling developed, controlling another route system leading into the Highlands by the Trossachs. The latter form an area of great scenic beauty which also contains Loch Katrine, one of the main sources of water for the Glasgow conurbation.

The main industrial wealth of Scotland lies to the south of the Sidlaws, the Ochils and the Campsie fells. The principal coalfields, with their interbedded iron-ores are in this area, the whole much disturbed by faulting and by the volcanic intrusions. Here, too, the Carboniferous rocks contain the oil shales of West Lothian, which laid the foundation for the great petrochemicals industry, now based on imported oil, at Grangemouth on the Firth of Forth.

Lanarkshire Coalfield. Extending from the Clyde valley to Falkirk, the Lanarkshire coalfield was one of the earliest to be worked. The coal was readily accessible and, just as important, the blackband ironstone provided the basis for an early pig-iron industry. Out of this grew the heavy iron and steel industries, particularly that of nearby Clydeside, with its famous shipbuilding yards. Motherwell, Coatbridge and Airdrie to the south-east of Glasgow shared in the prosperity of heavy iron and steel, and a new steel

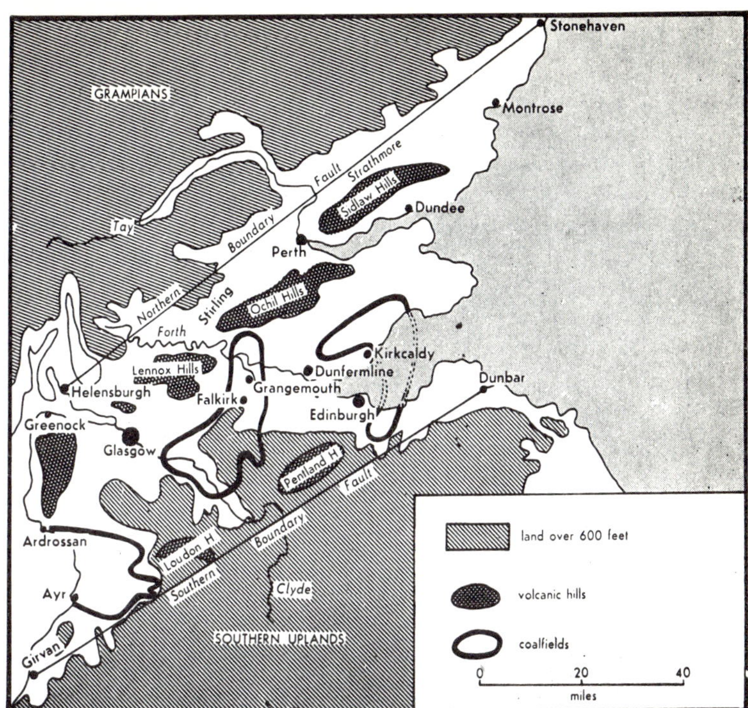

FIG. 27. *The rift valley of the Central Lowlands.*

works has recently been built at Ravenscraig. Coatbridge has certain specialities, such as chains and cables, while Hamilton has extended its industry to include brass founding, radio and electrical components, hosiery, knitwear and carpets. This coalfield is now an area of abandoned workings, and the emphasis on mining has shifted to the more productive and more easily worked fields in the east.

Ayrshire Coalfield. Industrial development here has been mainly in the form of textiles and engineering. The woollen industry, using the wool from the sheep of the Southern Uplands, was established at Kilmarnock on the Irvine. As in Lanarkshire, heavy engineering became the dominant industry with the manufacture of diesel locomotives and machinery, together with steel rails and prefabricated ship sections. The coastal ports are small, mainly exporting coal to Ireland and acting as terminals for the tourist trade to Arran and the Isle of Man. But Ardrossan has an important oil refinery.

Eastern Coalfields. These are now the most important coal-mining areas. The Fife and Clackmannan field, based on Alloa, is very productive, but the most noteworthy developments have been in Fife, where the coalfields extend under the Forth. The East Lothian and Midlothian coalfields similarly offer better opportunities than those in the west of Scotland, and there has been a distinct shift in the emphasis on coal-mining to this eastern area.

Like the other long-established heavy industrial areas of Britain, the central valley of Scotland is in the process of change. Increasing diversification of industry has brought about the introduction of the new light industries, particularly electronics in Edinburgh, added to its traditional printing, publishing and brewing. The jute industry of Dundee, a town now connected to the south side of the Firth of Tay by a new road bridge, has been augmented by the manufacture of typewriters and cash registers, while the towns of southern Fife, once heavily dependent on linoleum and the export of coal, have gained new light industries, aided by the building of the Forth road bridge. As in England, diversification of industry has been accompanied by the building of new towns, notably Cumbernauld to the east of Glasgow, East Kilbride south of Glasgow, the extension of Haddington in East Lothian, and Glenrothes in Fife. In East Kilbride, with its jet engine and electrical factories, and in Linwood and Bathgate with their car factories one sees great changes in the traditional heavy industries of Central Scotland. Major developments are also planned for the Clyde estuary, with new port and manufacturing facilities.

Glasgow. The great conurbation of Glasgow still dominates the industrial and economic life of the area. It is the major west coast port of Scotland, lying deep inland up the river Clyde with outports at Greenock and Port Glasgow. The Clydeside conurbation, with a population of about 2 million people, began as a market town at the lowest bridging point on the Clyde, built on a series of drumlins whose whaleback ridges give variety to the urban site. It acts as a centre for education (two universities), commerce, banking and insurance, not only for the west but for much of the remainder of Scotland—especially the Western Highlands, with which it has long-established sea links. The problem here, as elsewhere in the old industrial towns of Britain, is one of urban reconstruction and the housing of "overspill" population, but it is aggravated by the Clydeside's dependence on shipbuilding, which is now suffering intense competition from other shipbuilding areas both at home and overseas. Together with the rest of Scotland, it forms a "development area", to which the Government is making special efforts to introduce new industries.

Edinburgh. The capital of Scotland, on a hilly site dominated by the Castle Rock and the ancient volcano of Arthur's Seat, with an excellent defensive site, is now an important industrial town, although its population is less than one quarter of that of the Clydeside conurbation. Its port, Leith, the second in Scotland, with much trade with Europe, is really a part of Edinburgh, since there is no break in the built-up area between them. Together they offer employment to over 50,000 people in industries as diverse as electronics, brewing, the manufacture of tyres, distilling, and the manufacture of foodstuffs. Edinburgh is also a major cultural centre with an ancient

FIG. 28. *The new road bridge across the Forth.*

university, the seat of the principal Scottish law courts, and the government administrative centre. These factors, together with its importance as a regional shopping centre for eastern central Scotland, increasingly raise its status as a major urban centre. An adjacent "new town" has been created of Livingstone, and housing developments have taken place at, for example, Haddington.

QUESTIONS

1. Describe the characteristics of industry in *one* of the coastal coalfields of Britain.
2. Why has it been necessary for Government to assist the former heavy industrial areas?
3. Why is Central Scotland the most important economic area of Scotland?

CHAPTER 9

PERIPHERAL RURAL AREAS: (i) SCOTLAND

THE contrast in the geography of Britain lies between the areas of economic growth, with expanding urban populations, improved communications and good distribution of energy supplies, and those areas which have lagged behind. The areas of slower economic growth, or economic stagnation or decline, are the remoter parts, often with high and difficult relief, harsh climates, and short growing seasons. They may include areas which in the past were important agricultural districts, and some that are still agriculturally important but give employment to fewer people as a result of the increased mechanisation and amalgamation of farms. The upland areas of Britain have the lowest density of population; but they offer certain resources which the people of the lowlands increasingly demand: ground for water storage, for forestry and in particular for recreational purposes in areas which are relatively unspoilt.

The Highlands

To the north of the Highland boundary fault, which separates the ancient metamorphic and igneous rocks from the sedimentary Old Red Sandstones and lavas and the Carboniferous rocks of the Central Lowlands of Scotland, occur the high glaciated plateaux of the Scottish Highlands, culminating in the highest mountain in Great Britain, Ben Nevis (4,406 ft.). On the eastern margin of the high plateaux, especially in the Old Red Sandstone districts round the great indentation of the Moray Firth, and in Eastern Aberdeenshire, there are extensive tracts of lowland where the soil is derived from the glacial deposits of the highland ice. In the extreme north the lowlands of Caithness are continued beyond the stormy Pentland Firth into the Orkney Islands, where the soils, derived from the Old Red Sandstone, offer considerable potential for agriculture. In contrast, the Western Isles and the Shetland Islands are composed of the same ancient metamorphic and igneous rocks as the Highlands, and here the acid soils are leached by the heavy rainfall, while strong winds inhibit the growth of trees except in the most sheltered situations.

It is convenient to divide the Highlands proper into two sections separated by the Great Glen. The Glen comprises Loch Ness and Loch Lochy, connec-

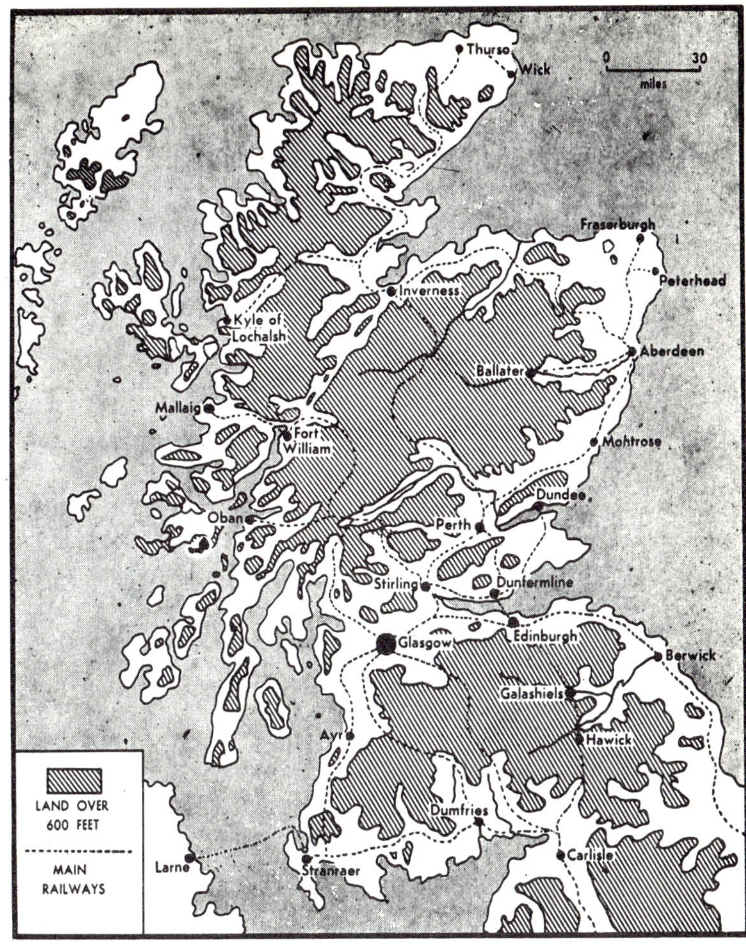

FIG. 29. *Map of Scotland.*

ted by the Caledonian Canal which provides a passage for small, mainly fishing, vessels between Inverness at the head of the Moray Firth and Fort William at the head of Loch Linnhe. To the north-west of the Great Glen a distinctive region of high mountains and monotonous moorlands shows much evidence of glaciation. The western margin is deeply indented by the great sea lochs—drowned glaciated valleys—along the margins of which, and on the coastal fringe, lives a sparse population of crofters. Crofters, who also predominate in the Western Isles, are small farmers cultivating a few acres of hardy oats and potatoes, and keeping a few cattle which are sent for sale to the markets in the east for fattening. More numerous are the sheep which graze on the wet grass moors; like the cattle, they are sent to markets in the east and south. Opportunities for alternative employment are very few.

This is an area of declining population from which much emigration has taken place over the last 200 years. Fewer and fewer people speak the distinctive Gaelic language. But on the eastern fringes of the north-west Highlands, particularly at the heads of the Dornoch, Cromarty and Beauly Firths, there are extensive tracts of good land devoted to mixed farming. Cereals are favoured by the very warm climate of the summer months on the shores of the Moray Firth. There is indeed a great contrast between Easter and Wester Ross, the former typically lowland in its economy and the latter distinctively Highland.

To the south-east of the Great Glen, and particularly to the east of the A9—the trunk road, paralleled by the railway, between Inverness and Perth—there is a remarkable change in scenery. The underlying rocks are similar to those which lie immediately to the west of the Great Glen, but the plateaux are less dissected and extend for miles as high moorlands topped by higher plateaux, culminating in Ben Macdhui (4,296 ft.) in the Cairngorms. Here there are few lochs compared with the north-west; the valleys are deep, but often consist of broad straths broken by narrower sections, as in Strath Spey and Deeside. The imprint of glaciation is strong. Many corries are sculptured into the edges of the plateaux; and the valleys show clear evidence of over-deepening by ice; there are also huge deposits of sands and gravels left by the melt-waters which flowed beneath the margins of the ice. There are, however, few textbook examples of upland glaciation, and often the scenery consists of relatively unmodified survivals of landforms which were in existence before the Ice Ages. As in the north-west Highlands, there is a clear change in vegetation and land-use from low to high ground.

The Land. The Highlands and Islands offer few opportunities for work other than on the land, in forestry, or in fishing. The income per head is low compared with that in the industrial areas of the Central Lowlands of Scotland or of England. The widely scattered communities, although now served by electricity from the hydro-electric generating stations developed since the Second World War, have few amenities other than the superb scenery. The land comprises large estates, in which the ground is managed according to its potential—the highest ground for deer stalking (with some attempts at deer farming for venison); the moorlands of grass in the west for sheep and grouse; and the heather-covered moors in the east for grouse with some sheep. Especially in the east, where there is some shelter from the strong winds, the ground below about 1,500 ft. above sea level is increasingly used for forestry, both private and planted by the Forestry Commission. At lower levels—particularly on the valley floors—there is some arable land, but favourable sites are few and often separated by sterile glacial deposits of sands and gravels. There is competition for the best land between forestry, sheep and cattle grazing, and grouse shooting—a problem which is not yet resolved, although most experts would agree that forestry and cattle offer the best opportunities for conserving the assets of the area.

Each estate manages its land according to its experience of what will give the greatest return on capital invested. The salmon rivers, such as the Spey, Dee and Don, and the smaller but good fishing rivers of the north-west Highlands, are an important source of revenue. Many large estates have been broken into smaller units, which nevertheless cover thousands of acres. These contrast strongly with the crofts of the islands (apart from Orkney) and of the western coastal fringe. On both large estates and small crofts, scientific methods of agriculture are being introduced: poor grasslands and heather moor are reseeded, and lime applied to counteract soil acidity. Subsidies from the Government assist the financing of these improvements, and the Crofters Commission and the advisory services of the Colleges of Agriculture in Glasgow and Aberdeen give advice and conduct experiments. But something more is needed to provide employment and reduce emigration.

Forestry is of long standing in the Highlands. Before the planting and replanting schemes of the Forestry Commission, exploitation was for charcoal or for timber for shipbuilding, with little thought of obtaining a sustained yield except on a few estates. There are a few areas where the old Scots Pine forest survives, such as the Forest of Rothiemurchus near Aviemore on Speyside, but the main forests, apart from those of birch scrub, are the great stands of quick-growing conifers planted by the Forestry Commission. These new forests of spruce and larch employ far more people per acre than other forms of highland land-use—one man to about 25 acres compared with one man to 1,000 acres in sheep-farming, grouse-shooting, or deer-stalking.

Industry. It has long been realized that the introduction of industry to the Highlands was necessary to provide more employment. The creation of the hydro-electric plant for aluminium at Foyers on the shores of Loch Ness, and the later Kinlochleven and Fort William aluminium works, were the first step, followed after 1945 by an immense development of hydro-electric schemes. The establishment of the nuclear reactor at Dounreay near Thurso in Caithness provided a new technology for the Highlands. Smaller scale industries are the Harris tweed of the Outer Hebrides and the cottage knitwear industries of the Shetland Islands. But all these together produce goods of less value than the old-established whisky distilleries. These are concentrated principally in the glens of the Banffshire hills, but isolated distilleries are found from Orkney to the far south-west. Nearly 50 distilleries, using barley from the Moray Firth lowlands and imported from abroad, contribute to Britain's exports more than any other industry in the Scottish Highlands.

With the creation of the Highlands and Islands Development Board, responsible for the development of the seven "crofting counties" from Argyll to Zetland, a concerted attempt was made to re-invigorate the Highland economy. The major plan is for development along the shores of the inner Moray Firth, using, in particular, the deep-water advantages for industrial development on the Cromarty Firth at Invergordon. Here, in 1970, a new aluminium smelter was already under construction, and a major petrochemical industry is planned. The Board has powers to stimulate all types of economic activity in the Highlands, and has assisted many firms and individuals to provide new employment both on the land and in the fishing industry. The largest single bulb-growing unit in Europe is being developed in Uist in the Outer Hebrides, taking advantage of the relatively mild climate of the north-west seaboard.

Recreation. A major value of the Highlands is its potential as a recreational area. At Aviemore there are hotels and a large tourist centre to take advantage of the snow for winter ski-ing in the Cairngorms, and the opportunities in summer for hill walking, pony trekking, sailing and canoeing. The tourist industry has already made major changes in employment opportunities, and great efforts are being made, with the assistance of the Highlands and Islands Development Board, to publicise the holiday attractions of the Highlands. As a result caravan and camping sites, even in the remotest areas, are crowded in summer, and the ski-slopes with chairlifts at Cairngorm, Glen Coe, and the Devil's Elbow have several thousand skiers every winter weekend and at Easter.

The North-East

The Moray Firth lowlands between Inverness and Buckie are underlain by Old Red Sandstone rocks mantled by glacial deposits of boulder clay, sands and gravels. They have a rather dry, sunny climate (less than 25 in. of rainfall a year) and are relatively free from the coastal fog which affects other parts of the east coast of Scotland in the spring and summer months. The glacial deposits provide good farm land with mixed farming, barley and oats forming the main crop. The coast, backed alternately by dunes and by low cliffs, is increasingly of importance to holidaymakers, while the towns of Nairn, Forres and Elgin act as service centres and markets. On the coast, Lossiemouth is the most important fishing town and Burghead is a small port.

Aberdeen. Between the Cairngorms and the Moray Firth and the North Sea to the east of the river Spey, extend the low plateaux of north-east Scotland floored by igneous and metamorphic rocks mantled with glacial drift. The principal rivers, the Spey, Don and Dee, have only small estuaries, but that of the Dee is sufficiently large to serve the most northerly large port in the British Isles, Aberdeen. It is situated where the cliff coast of Kincardineshire, to the north of the Highland boundary fault, gives way to the long sandy bays backed by dunes which are characteristic of the eastern coast of the north-east lowlands. The encircling hills of the lowland, approaching close to the sea, focus routeways towards Aberdeen. Its site on glacial mounds and hollows between the mouths of the Dee and the Don has been developed into a city

FIG. 30. *The Highlands are being developed for recreation.*

and port of nearly 200,000 people, with houses characteristically built of the grey granite. It is the fourth largest city in Scotland, a major centre for services and the distributive centre for much of Northern Scotland, with an ancient university, research centres for agriculture and fishing, and a range of manufacturers, including textiles, engineering, fertilisers, and, especially, paper for which esparto grass from North Africa and pulp from Scandinavia are imported.

The hinterland of Aberdeen, which includes the farming areas of Buchan and Donside, and the more scenic but agriculturally less valuable valley of the Dee with the royal residence at Balmoral, is essentially devoted to agriculture and, increasingly, to forestry. It is an area of small to moderate-size farms where the range of crops is restricted by the relatively short growing season, so that oats and barley are of great importance. The barley, formerly grown for the distilleries, is increasingly used for feeding the beef cattle for which, through the Aberdeen-Angus breed, this area has become famous. Potatoes, turnips and clover and grass are used in rotation with oats, barley and the livestock (including sheep). Beef carcases are exported to the more densely populated areas of Britain. Like other agricultural areas this is a zone of rural depopulation, and much effort is being made to attract light industries to the smaller market towns such as Inverurie (which formerly had an important railway repair works), Turriff and Huntly. Tourism in "Royal Deeside" and in Aberdeen itself is of long standing and increasing importance.

Granite-quarrying, using the grey granite of Aberdeen, the white granite of Kemnay and the salmon pink granite of Peterhead, was formerly a leading industry, and its products are found as buildings, bridges and sea defences in many parts of Britain. Increasing use is made of polished granite cut into thin sheets for the facing of buildings, but the demand for large blocks of the stone has decreased because of its cost in comparison with concrete. The number of people employed in the industry has declined, but its products, including ornamental stones, still play a part in the economy of the north-east.

Fishing. The fishing industry is of particular importance. It is based on the catching of prime quality white fish from near-water grounds and from the mid-water fishing-grounds of Shetland, Faroe and Iceland. The nearer waters are fished by seine-netters, wooden vessels up to about 75 ft. long, which use bag-like nets for the catching of cod and haddock. They land their fish at the numerous small ports, including Buckie, Macduff, Peterhead, Fraserburgh and Stonehaven, as well as Aberdeen. These are dual-purpose vessels which also seek the herring in the summer and early autumn. From the same small ports operate the inshore yawls for shellfish, including crabs and lobsters. All these ports, but in particular Peterhead and Fraserburgh, suffer from over-dependence on the fishing industry, but new factories making, for example, tools and pneumatic machinery have been introduced.

Aberdeen is pre-eminently a trawling port, with a great diversity of employment connected with trawling, including ice-production, curing, kippering, filleting and freezing, frozen food processing, box making, chandlery, marine engineering and repair work. With the fishing industry has developed a ship-building industry which builds not only trawlers but also coastal vessels, vehicle ferries and small bulk carriers.

It is clear that in spite of its industries, the north-east lowlands of Scotland are essentially an area of primary production—agriculture, fishing and forestry—and that, because of its relative remoteness from the "core "area of Britain, it will, like the other parts of northern Scotland, especially the islands of Orkney and Shetland, require special consideration by the Government for development if the existing population is to be retained.

Southern Scotland

To the south of the industrialized and densely populated central lowlands of Scotland, the southern uplands extend from the North to the Irish Seas as a plateau of ancient rocks with subdued relief consisting mainly of unpopulated moorlands with a maximum altitude of about 2,000 ft. The moorlands acted in the past as a barrier, recognized by the Romans in their building of the Wall between Solway and Tyne; and the Borders were for long an area of dispute between Scotland and England. Now, however, the barrier is less obstructive and the dual-carriageways of the Carlisle-Glasgow-Stirling trunk road and the A1 (between Newcastle and Edinburgh) carry much lorry traffic. Nevertheless, road and rail routes keep to the low ground of the east coast, in spite of some steep gradients, or follow the valleys which in the west penetrate deeply into the hill mass from north and south.

In the upland areas, and the flanks of the valleys, where boulder clay and sands and gravels cover the solid rocks, the land-use is either for sheep or sport (grouse-shooting). But in the valley floors, where the main settlements occur, there is some arable ground with hay and root crops which fit into the stock-rearing pattern.

Two areas of southern Scotland more productive than the upland zones are the Tweed valley and the Solway Firth lowlands. The Tweed valley, which reaches the sea at Berwick-on-Tweed, extends in an east-west direction between the Lammermuir Hills and the granite massif of the Cheviots. In its upper reaches, the valley is narrow and its economy is restricted to upland farming, but to the east it opens out into the Merse, a broader lowland floored by a wide variety of glacial deposits with much infertile sand and gravel as well as more productive soils derived from the boulder clay. It is an area of the low-rainfall east-coast type, but much water drains down the Tweed tributaries and provided the basis, with wool from the sheep of the upland rim, for a textile industry producing tweeds, factory knitwear and blankets. Coal was later imported from Northumberland by rail to provide power. Hawick (16,000) is the knitwear-hosiery centre, and Galashiels (12,000), Peebles (5,500) and Kelso (4,000) are dominant in other textiles. This Border area has its own problems of depopulation. arising chicfly from the dominance of agriculture, which is devoted to grass, oats, wheat and turnips in a mixed economy of cattle and sheep. Increasing mechanization, reduction in farm labour forces, and the special problems of the upland or hill farmer all contribute to rural depopulation, but the lower Tweed valley is still an important agricultural area.

So also are the higher-rainfall lowlands of the north side of the Solway Firth, where grassland is the main land-use and the economy is based primarily on dairy cattle. Milk from this area serves both Northern England and Central Scotland, and butter and cheese are made. The lowland, floored by sedimentary rocks with some intrusions of granite which form higher ground (e.g. Criffel near Dumfries), is covered with glacial deposits from which the relatively productive soils are derived. The comparatively small area of arable ground is mostly devoted to fodder crops, including hay and roots. Compared with the eastern section, towns are small, apart from Dumfries (28,000), and have few industries. Most are situated on the formerly navigable inlets, but only Stranraer is an important passenger port—for the short sea-route to Larne.The Chapelcross nuclear power station is the most prominent modern industrial development. But both the towns and their rural hinterlands are attractive to tourists.

Southern Scotland is thus, like the area to the north of the Central Lowlands predominantly rural, with emphasis on farming and forestry, and the inevitable problem of depopulation. But compared with the north of Scotland it has the advantage of lying between two major industrial areas and of being crossed by main road, railway and energy links which could be of use in industrial development.

QUESTIONS

1. Describe the relief soils and land-use in the Highlands of Scotland.
2. Give an account of the fishing industry of north-east Scotland.

CHAPTER 10

PERIPHERAL RURAL AREAS: (ii) ENGLAND AND WALES

The Lake District

To the south of the Solway Firth and north of Morecambe Bay lie the high uplands of the Lake District, sometimes called the Cumbrian Mountains. Although separated from the northern Pennines by the Vale of Eden in the north, the limestone rocks so characteristic of the Pennines fringe the northern and southern edges of the central mountain mass, and link the Lake District to the Pennines in the vicinity of Shap Fell. There is a through valley in this intervening ridge at Tebay, the Lune Gap, which carries the railway and the M6 motorway between Carlisle and the industrial conurbations of Lancastria. For reasons of altitude, high rainfall and severe winter climate, with much of the valley floors occupied by water, the Lake District proper has few natural resources, apart from slate quarries and some grazing for sheep. But its scenery and opportunities for recreation by climbing, fell walking, camping and boating have made it into a major tourist area. With the construction of the M6 motorway, it can quickly be reached by the millions of the conurbations of SELNEC, Merseyside and the West Midlands.

The Welsh Massif

To the north of the industrial area of South Wales, where the Carboniferous rocks provided the basis for coal mining and industrial prosperity, lie the older pre-Carboniferous rocks of the high, eastward sloping plateaux of Central and North Wales. The highest mountains, of igneous rocks, such as Snowdon and Cader Idris, lie in the north-west and owe much of their detailed relief, including the *cwms* or corries, to glaciation, which has also left relics in the form of glacial deposits at both high and low levels throughout the area. The whole area is essentially an upland, with relatively small tracts of lower ground, as in Anglesey which is separated from the mainland by marine and glacial erosion along a zone of weakness, produced by faulting—the Menai Straits. Elsewhere, the Vale of Clywd, floored by downfaulted Triassic rocks, and the lowlands of Pembrokeshire, eroded on ancient rocks of pre-Cambrian age (as in Anglesey), form the principal low ground, apart from the valleys whose flanks and floors bear evidence of glaciation. These

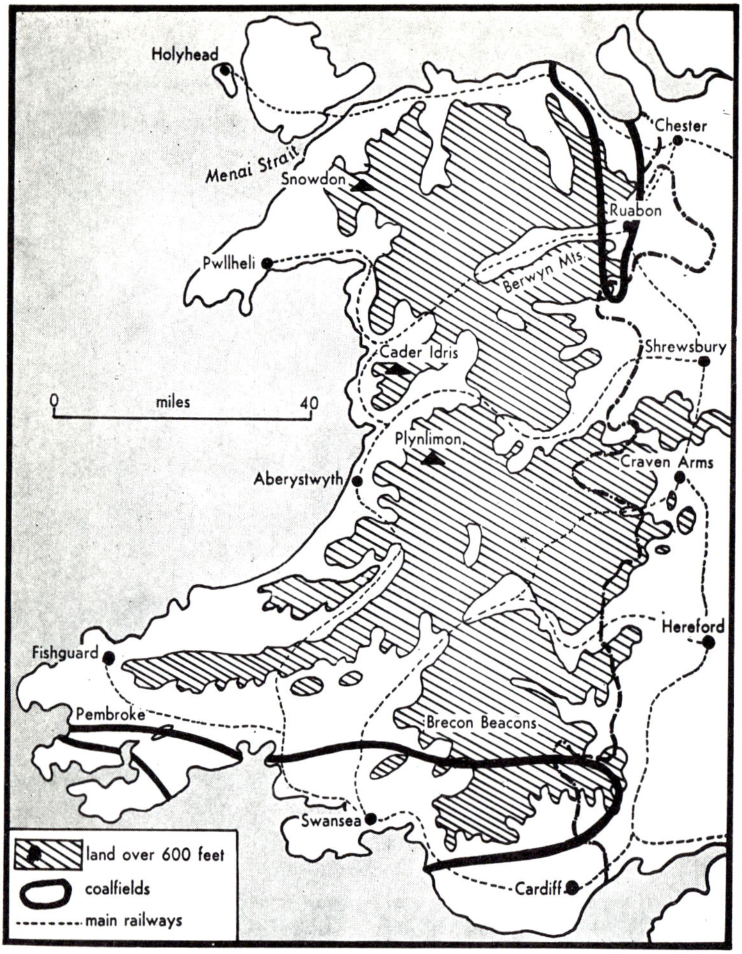

Fig. 31. *Map of Wales.*

valleys usually drain towards the east. The watershed runs parallel to the coast of Cardigan Bay before swinging in an easterly direction in the vicinity of Plynlimon, a prominent residual mountain which overlooks the plateaux of central Wales. Much of this high ground is between 1,500 and 2,000 ft. above sea level; receiving a high rainfall, its soils are often leached and acid, and it is principally moorland. Only in the narrow, lower, coastal plateaux, on the lowlands and in the valleys are there opportunities for farming rather than stock-rearing. The Welsh massif thus resembles the Lake District, the Highlands of Scotland, and the Southern Uplands in that it offers a relatively poor resource base compared with the Midlands or the coalfield of South Wales.

Mining. In the past, minerals such as copper (in Anglesey), manganese and gold gave rise to small settlements, but the largest extractive industry was based on the slates of Ffestiniog, Llanberis and Bethesda. These mineral workings have all declined. Slate, produced by the metamorphism of mudstones into rocks which can be split into thin sheets, formerly provided much local employment and was exported from ports such as Portmadoc to roof the houses built during the Industrial Revolution. The industry, which had exhausted its supplies of good slate and was suffering competition from other forms of roofing material, was already in decline before 1945. Now the slate-quarrying villages are problem villages, and planners are seeking to bring in new light industries.

The Welsh massif plays an important part in the life of the conurbations on its flanks. It has long been important for tourism, both to the high mountains of Snowdonia (now a National Park) and to the coastlands—especially the resorts in the north between Abersoch and Rhyl which cater for the industrial towns of Merseyside, SELNEC and the West Midlands. The river valleys, much altered by glaciation, have offered suitable sites for reservoirs which, like Lake Vyrnwy and those in the Elan valley, provide water for Merseyside and Birmingham. Reservoirs, like hydro-electricity works, do not provide much local employment once the initial engineering has been completed. The most important developments are the nuclear power stations at Trawsfynydd, for which a large reservoir was created, and at Wylfa on the north coast of Anglesey. Otherwise North and Central Wales must look to scientific methods

FIG. 32. *Caban Coch reservoir, Elan Valley.*

of upland land-use which permit the introduction of cattle in place of hill sheep.

South-west England

Compared with the other upland areas of Britain, the peninsula of Devon and Cornwall has several unique characteristics. It is far more densely populated, especially in the summer when tourists bring millions of pounds into the area; it has, on the coasts, a very mild climate with a long growing season and an early spring; and it was not glaciated, since it lay almost entirely south of the maximum extension of the ice-sheets in the British Isles. For these reasons alone there is a different economy and a different landscape from those of the other upland massifs.

As in all British uplands, the landscape consists of a series of plateaux, but the south-west plateaux do not exhibit corries on their margins; they are not dissected by glaciated valleys, nor are they mantled with glacial deposits. Instead, the hills have a mantle of debris produced under the conditions of intense cold of periglacial climates, with much peat on the poorly drained moorlands. Much of the area consists of a pleateau about 400 ft. above sea level, above which rise the higher plateaux of Dartmoor, Bodmin Moor and Exmoor. The river valleys, such as those of the Tame and Torridge (which drain to the Atlantic) or the Exe, Dart, Tamar and Fal (which flow towards the English Channel) are deeply incised into the lower plateau, and their mouths have been drowned to form the rias which provide excellent ports.

All the highest moorlands are composed of resistant rocks. Exmoor and the Quantock Hills are uplifted masses of Old Red Sandstone, but Dartmoor and Bodmin Moor are igneous intrusions of which granite is the most

FIG. 33. *The South-West Peninsula.*

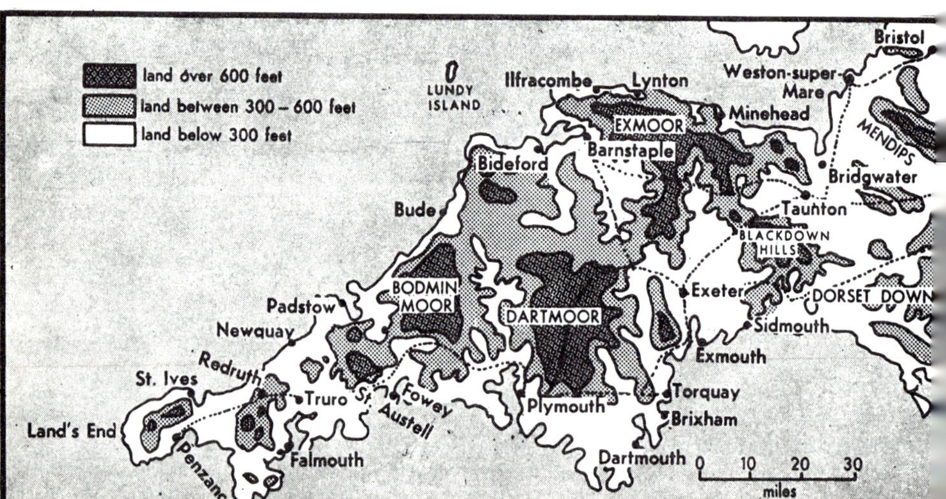

FIG. 34. *The "lunar" landscape of a Cornish kaolin quarry.*

prominent rock. Exmoor, which reaches 1,700 ft. in Dunkery Beacon, is lower than Dartmoor, where the maximum altitude is attained in High Willhays and Yes Tor at over 2,000 ft. The highest point on Bodmin Moor is Brown Willy (1,370 ft.). All possess a landscape of rounded moorlands, with valleys broad and shallowly incised in the upper sections and deeply incised in the lower reaches. The granite moorlands can be distinguished by their "tors," but the principal differences between the granite and the sandstone moorlands are seen in their economic activities.

China Clay. The felspar in the granite has in places decomposed to kaolin or china clay. This occurs in pockets surrounded by relatively unaltered rock; when washed out by hydraulic jets it is separated from the other principal minerals of granite, quartz and mica. Kaolin is used for numerous industrial purposes; the most obvious is in high-grade pottery, but it is also of value in paper making, in the pharmaceutical and cosmetic industries, and in paint and rubber manufacture. Much goes from Fowey and Par by sea to Merseyside and thence to the potteries and Lancashire, but some is also exported abroad. The gullied white hills and quarries of the china clay district form an extraction industry landscape unique to the granite areas of Cornwall.

Mining. Elsewhere in Cornwall mineralisation of the granite intrusions and the surrounding rocks provided the tin and copper ores which brought the Phoenicians and later the Romans to the region. Minerals have been extracted here since pre-historic times. Alluvial mining of the ore eroded from the mother lode and deposited in river beds was followed by deep-mining of the veins themselves. Some of the earliest examples of mining engines, both for pumping out water and for transporting the miners below the surface, were installed in this area. The tin-plate industry of the 19th century in South Wales used much Cornish tin. Supplies are now almost exhausted, but some attempts have been made to re-open one or two of the mines since modern processes can utilize ore formerly considered uneconomical. The only other extractive industry of significance is the working of the Delabole slates.

Farming. The high moorlands, with their generally poor, sandy soils covered by heather, bracken and bog, are bleak and sparsely populated but support a few cattle and sheep. The margins, especially where better soils from sedimentary rocks are present, and the river valleys contain the major farms, where dairy herds yield cream, butter, cheese and condensed milk, as well as the fresh milk needed in the summer tourist season. The Plain of Somerset, between the Quantock Hills and the Mendips in the north and the limestone escarpment in the south, is the most prominent farming area; lush grassland, reclaimed from the original marshy fenland, as shown by the drainage canals and the pumping stations, provides the basis for the major dairying area of the south-west peninsula. Cheddar cheese is the best known product. On the higher, drier areas apple orchards produce cider fruit and dessert apples. Cider is especially linked with Devon, but the industry is now factory- rather than farm-centred.

The coastal strip, narrow and generally fringed with prominent cliffs, is characterised by its mild climate, with relative freedom from late and early frosts, and a long growing season. Here are produced the spring flowers and early vegetables that are sent, because of the high prices they can command early in the season, to shops all over Britain. The Scilly Isles, with an even earlier season, produce the first spring flowers.

A major part of the regional income comes from the seasonal tourist economy. Torquay, Falmouth, Penzance, Newquay and other coastal resorts attract holiday-makers in increasing numbers, and with the proliferation of the motor car and the closure of some of the railways, and the increase in caravanning and camping, even the remoter areas are frequented by tourists. Many are attracted to the small fishing villages to which tourism brings in a much greater income than fishing, although shellfish are still caught in quantity from small boats. Other visitors frequent the larger fishing towns, such as St. Ives, Newlyn and Brixham, whose fishing boats seek pilchards.

Plymouth and Exeter, the regional centres, overshadow the inland market towns such as Truro, Camborne, and Okehampton. Plymouth, with about 200,000 people, is a long-established major seaport and naval base; the drowned valley of the Tamar forms a port commanding the entrance to the English Channel and was the starting point for some of the early voyages of discovery and colonisation, especially of the New World. The Devonport naval dockyard is gradually being replaced as the principal employer by new industries based on imported raw materials.

Exeter, with a population of 80,000, is an ancient regional capital on the lowest bridging point of the Exe. Although formerly a port, its estuary did not have the navigational advantages of the Tamar or the Fal. It is a university town with a wide range of services which are not provided by the smaller market towns such as Barnstaple and Newton Abbot.

Eastern England

The great lowland of Eastern England extends between the Thames estuary and the Humber, as far inland as the dip-slope of the Jurassic limestone escarpment. Predominantly agricultural, it was for centuries the most intensively farmed land in the British Isles, in spite of the difficulties of drainage in some areas and comparatively poor soils in others. Here the first attempts at modern scientific farming began, with a system of crop rotation and the introduction of root crops. Although many other areas in Britain are now farmed intensively, these lowlands still remain in the forefront of agricultural production, with high yields of wheat, barley, rye and sugar beet. Market gardening, especially carrots, peas and celery, is also a highly intensive form of land-use; the products are sent both to the major market in London and to the local food-processing factories. It is an area of rural hamlets, villages and market towns, whose prosperity was once built on textiles (wool) and grain; many towns now possess light industries of some importance on the national scale (e.g. the Pye electronics factory at Cambridge). Nevertheless, this area lies wholly to the east of the main road and railway lines in Britain and must be classed as a "peripheral" zone.

Gas and Power. Although it lacks minerals, it lies adjacent to the main reserves of natural gas discovered in the North Sea, for which the long-established fishing port of Great Yarmouth has acted as a service base. From the gas-producing fields pipelines come ashore at Bacton, Norfolk, and cross the agricultural lowlands to join the main methane gas grid. At Sizewell and Bradwell nuclear power stations feed electricity into the national grid. Rural depopulation will probably continue in this area, although there is evidence that the spread of the London "commuting" belt will extend at least as far as the ancient university town of Cambridge.

East Anglia. Much of Eastern England is known as East Anglia, the ancient kingdom of the East Angles. It mainly comprises the counties of Norfolk and Suffolk, but the north-eastern section of Essex is also included. Forests in the south and fens in the north-west, now changed into more productive

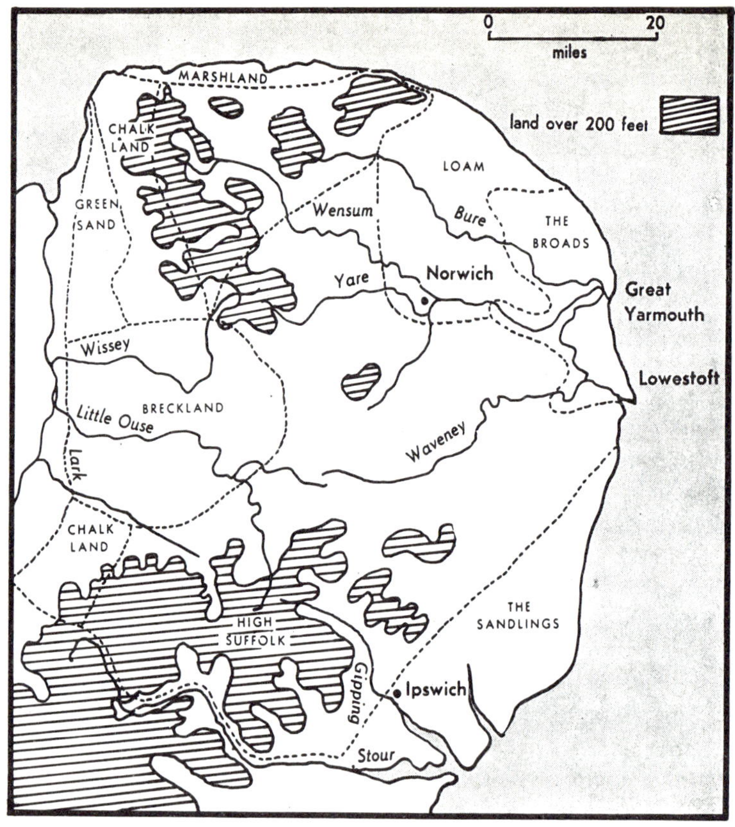

FIG. 35. *East Anglia.*

forms of land-use, formerly served to isolate this region from the rest of England by land. It was by river and sea that the agricultural produce of the area was formerly transported, which gave prominence to the navigable drowned river valleys of Essex and Suffolk and the artificial waterways of the Broads in Norfolk.

The mantle of glacial deposits, ground moraine and outwash sands provides more contrasts in soils than in relief. There is no high ground other than the chalk escarpment (the East Anglian Heights and Norfolk Edge), and there are only minor differences in elevation; the area is predominantly one of gentle slopes. The coastline is composed in places of low cliffs of chalk or clay; in other sections, long shingle spits such as Orford Ness front the sea, and these are often the only natural protection against flooding of the coastlands during severe sea storms. The soils range from clays and sandy loams to sands; the latter are often still heathland, or have been planted by the Forestry Commission with conifers, as in the Breckland. Elsewhere, the broad but shallow river valleys, such as those of the Bure, Wensum and Yare, contain much alluvium and peat. Excavation of the peat in Norfolk in medieval times produced long depressions now flooded to form the Norfolk Broads; the Broads are best known for the recreational facilities they provide for inland cruising and sailing, but they are also of great scientific interest to ornithologists and botanists.

Agriculture remains the dominant land-use in East Anglia. Arable farming, especially for grain, occupies more than half the land under cultivation. Wheat and barley growers take advantage of the favourable low rainfall and summer sunshine, and increase yield by mechanisation and the application of fertilisers. Higher productivity per employed man is the result. But, although East Anglia has a reputation as a cereal growing area, cattle are becoming increasingly important. The practice of feeding cattle and poultry with barley has brought about a great growth in mixed farming, with dairy herds as well as beef cattle. Barley is particularly significant on the chalk soils, but is also spreading to replace wheat on the clayey soils. Sugar beet, both the tops and the pulp, silage made from peas, and the traditional East Anglian root crops are all fed to the stock.

In earlier times sheep were the most common livestock, and the working horses bred from Suffolk stock were used in many parts of Britain. Today there is much greater variety. Although the sandy soils of Breckland still carry many sheep, pigs and poultry are now more important to the farmer's income. Pig farming, as in the rest of Britain, is most commonly found where there are bacon and pork processing factories near by, as near Ipswich. It is organized on the basis of intensive specialization, with scientific feeding and housing. Poultry farming has, however, seen the most notable application of scientific and industrial methods.

Its growth has been phenomenal, especially in Norfolk which had an early start in the production of cheap broiler chickens. Production quickened with the increased demand by supermarkets for "oven-ready" frozen poultry. Controlled environments of temperature and light are linked with carefully assessed feeding to produce broiler chickens of the correct weight and quality in the shortest possible time. Factory farming of poultry began with "battery" hens for egg production; the poultry population is now as high as 100,000 per farm.

Market gardening in East Anglia is widely distributed, but, as with pig farming and sugar beet, it tends to be concentrated near the factories that can and freeze soft fruits and vegetables, such as peas and carrots. Great Yarmouth and Lowestoft, like other fishing towns on the east coast, are important canning and freezing centres. Nevertheless, much of the market garden produce, especially cauliflowers and brussels sprouts, is sent to London vegetable markets.

Such industry, as there is, is chiefly based on agriculture—freezing and canning at, e.g., Thetford and Harleston; sugar beet factories at, e.g., Bury

St. Edmunds and Ipswich. East Anglian industry has always grown from agriculture, particularly woollen cloth and worsteds from the sheep on the border between Suffolk and Essex. There the small villages and market towns have clearly had a "wealthier" past, as is shown by the architecture of the churches and houses. Norwich, once the chief centre of the woollen industry in England, still has clothing factories, but has become one of the major centres of boots and shoes in the country.

Originating with the making of ploughs at Ipswich, the manufacture of farm implements has long been prominent in East Anglia. It grew into engineering on a bigger scale, including steam traction engines, and from these beginnings engineering of a wide variety is now well established in many towns. The factories produce food processing machinery, marine engines, refrigerating equipment and electronic components. Great Yarmouth, Lowestoft and Norwich are the main centres; trading estates have been established at Thetford and Haverhill.

The tourist industry is important to the economy. The coastal resorts and the Broads cater for a large seasonal trade, and redevelopment of the old established estuarine facilities is taking place as more and more people take to the water for recreation. In its participation in the tourist industry, East Anglia thus conforms with the other peripheral areas of Britain; but unlike the Highlands, Lake District, North Wales and South-west England, which are upland areas, its attractions depend almost entirely on water.

The Fens. This district, dominated by the Great Ouse basin and the coastlands at the head of the Wash, forms an area almost unique in the British Isles. Reclaimed, like parts of the Low Countries, from the North Sea, much of the area was marsh and peat, which isolated the drier parts of East Anglia to the east from the lowlands of central England. Lying at the seaward end of the vale between the limestone and the chalk escarpments, it consists of the area drained by the lower Witham, Welland, Nene and Ouse, collectively called the Great Ouse basin. Much of it represents the infilling, by river and marine sediments and peat, of a much larger indentation than the present-day Wash. There is an important distinction between the marine silt-fen and the peat-fen. There are occasional "islands" of glacial or silt deposits which, because they stood as dry land above the fen and marsh, became the major settlement sites. The Isle of Ely, with its great cathedral, is the best known of these, but almost every farm and village site takes advantage of slightly higher ground.

The rich agriculture of the modern Fenland, even more firmly based than East Anglia's, depends on centuries of reclamation and control of the level of the ground water table by sluices and drainage ditches. Mixing of the peats with underlying limey clays has helped to improve the fertility of these polderlands. The first attempts at drainage were made by the Romans, but it was in the 17th century that Dutch engineers, especially Vermuyden, did the most important work. As in the Low Countries, the process of drainage was accelerated when steam pumps (later diesel-engined and electric pumps) were employed. Most of the land has now been drained, but some fens, e.g. Wickham Fen, are retained as nature reserves. Paradoxically, the Fenland, which suffered so severely in the 1953 sea floods, and which has offered great

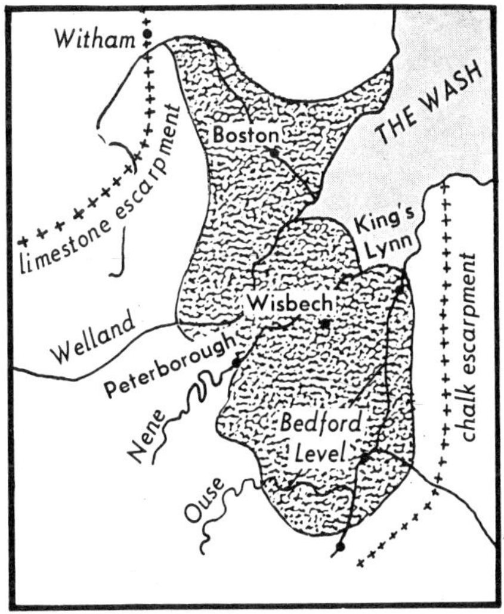

FIG. 36. *Simplified map of the Fens.*

problems to the drainage engineers, is one of the parts of Britain most subject to soil erosion. Much top soil is blown into the drainage channels in dry springs before the crops can provide protection to the soil.

Draining results in shrinkage of the peat and a lowering of the ground surface. This may reveal the underlying clays, as did the earlier methods of stripping and burning the peat prior to cultivation. As a result, the cultivated land is below the level of the rivers and the drainage channels. Constant maintenance of river banks and channels is necessary; its cost could not be justified if the land were not so productive. Although grassland was formerly the principal land-use, offering rich grazing for cattle, arable crops now give the highest profits. Very little of the land is now under grass, and there are very few cattle and even fewer sheep.

Fenland is ploughland. The arable, with no bare fallow, is sown with wheat, barley, potatoes, sugar beet and vegetables. In both Holland (mainly silt-fen) and the Isle of Ely (of which about half is peat-fen) wheat is the dominant crop, with barley and oats of lesser importance; as elsewhere, barley is gaining at the expense of oats since it is of high value in cattle-feeding. Root crops, including potatoes and sugar beet (factories at Spalding and Peterborough), are grown both on peat and silt land; vegetables are concentrated mainly on the silt lands of Ely (near Wisbech) and of Holland. Peas for quick-freezing occupying the largest area. The light soils of the Spalding area bear the distinctive bulb and flower fields, and there is some specialization in fruit at Wisbech.

Although there are factories associated with freezing, canning and sugar beet, agriculture is the biggest employer of labour in the Fens. The income is derived from farming of a most specialized and intensive kind. There has been an increase in the activities of the ports of Boston and King's Lynn, especially in the importation of refined petroleum by small coastal tankers; but there is no industrial hinterland which might permit these ports to overcome the difficulties of navigating the Wash and the shallow channels off it.

The Fenland, although one of the most difficult environments in lowland Britain, is agriculturally the most productive. Although undoubtedly a "peripheral" area, it is relatively close to the industrial "heartland" of the Midlands, Lancashire, Yorkshire and the London region. This location, coupled with the intensity of its agricultural production, give it, and neighbouring East Anglia, an advantage over other marginal areas in spite of relatively poor systems of roads and railways.

QUESTIONS

1. How does the Welsh massif contribute to the life of the lowlands on its margins?
2. Either:
   (*a*) Describe the agriculture of East Anglia, or
   (*b*) State why the Fenlands are such a distinctive area of Britain.

CHAPTER 11

IRELAND

IRELAND became politically divided in 1921 when the twenty-six southern counties were separated from the six northern counties, which remained part of the United Kingdom. In 1922 these southern counties became organized as the Irish Free State; in 1937 they were named Eire; and in 1949 they became the independent Republic of Ireland.

Northern Ireland

Northern Ireland consists of the counties of Antrim, Down, Fermanagh, Londonderry, Tyrone and Armagh. Geologically it is the south-west continuation of Scotland. It is crossed by the Highland Boundary and the Southern Boundary faults which mark the northern and southern limits of the Scottish Central Lowlands, thus forming a rift valley in which newer Carboniferous and Old Red Sandstone rocks have been let down between ancient rocks. North of the Highland Boundary fault lie the ancient Pre-Cambrian Sperrin Mountains and the Antrim Hills in the north-east. These uplands are composed mainly of schists. To the south of the Southern Boundary fault, the Silurian and Ordovician rocks, which outcrop in County Down and Armagh, are similar to those which form the Southern Uplands of Scotland.

Northern Ireland was the centre of great volcanic activity in Tertiary times, and its effects were greater than in Scotland. Large areas of Antrim are covered by thick sheets of basaltic lava: where it reaches the sea it forms the peculiar columnar structure of the Giant's Causeway on the north coast. A large area of subsidence in the basalts formed a hollow which filled with water, the remnants of this large lake now forming Lough Neagh. Before the basalts were extruded, Triassic, Jurassic and Cretaceous rocks were deposited in the north-east. They are mostly covered by the basalts, but outcrop near the north-east coast. As a result, on the cliffs of Antrim white chalk is capped by almost black basalt. The chief outcrop of Triassic rocks forms a narrow strip of lowland in the Lagan valley near Belfast.

The Rift Valley of Northern Ireland differs from the Central Lowlands of Scotland in that it contains only a small unimportant coalfield near Dungannon in south-east Tyrone. The rocks of the western portion of the

Rift Valley are of Carboniferous limestone, and much of the surface has a thick cover of glacial deposits.

Except for bauxite (aluminium ore) in Antrim, which was exploited during the last war, and the small coalfield north of Dungannon, Northern Ireland is almost devoid of minerals of economic value; yet it contains the most important industrial regions in Ireland. These are located in the Lagan and Bann Valleys, the Lough Neagh lowlands, and near Londonderry.

The textile industry has an outstanding position in the economy. In the early eighteenth century the industry was helped by the settlement of Huguenot linen weavers skilled in making fine linen. Northern Ireland is famous for its fine linens; it also makes thread and coarser linen goods, and is easily the most important centre of the linen industry in Britain. A large percentage of its goods are sold abroad, particularly to the U.S.A.

The demand for pure linen goods has declined in recent years, and the industry has been hit by severe unemployment. However, the increasing use of cotton and man-made fibres in the Ulster textile industry, and the development of new industries, has partially relieved the depression.

Some other reasons for the location and development of the textile industry in Northern Ireland are as follows:

1. Locally grown flax formed the basis of an early domestic industry which produced generations of people skilled in spinning and weaving.
2. Abundant supply of soft water needed for retting the flax (removing the fibre from the stem) and for bleaching.
3. The port of Belfast and the position of Northern Ireland opposite to the Ayrshire and Cumberland coalfields, and to the Clydeside engineering area, made the importation of fuel and machinery comparatively easy when textile machines were introduced.
4. Now that the local supply of flax is quite inadequate, it can easily be imported from Western Europe via Belfast.

The textile industry is carried on in many towns in the Belfast region and in the Lough Neagh lowlands. The most important towns are Belfast itself, Lurgan, Portadown, Dungannon, Banbridge, Ballymena and Larne. Linen and cotton goods are also made in the lower Foyle Valley at Londonderry, and small towns nearby. Ballymena is also the site of a large new tobacco factory, which employs 3,000 people.

Belfast (pop. 400,000) is the centre of government and of industry in Northern Ireland. It is well situated on the broad mouth of the Lagan, which affords many miles of sheltered tideway. The textile industry has encouraged clothing manufacture and engineering.

Shipbuilding is the other great industry of Belfast, and like textiles it has fallen on bad times. The industry is entirely dependent on imported raw materials, particularly from Clydeside. There appears to be no satisfactory geographical reason why the mouth of the Lagan became so important for shipbuilding except, perhaps, its sheltered tideway. It now contains the largest shipbuilding dock in the whole world.

A relatively new industry is the making of aircraft, and a new £8-million oil refinery was opened on the north side of Belfast harbour. Other industries include grain-milling, electronic computers, jam, confectionery, tobacco, brewing, distilling and soft drinks. Belfast is the port for all Northern Ireland. The port also has steamer services to Clydeside, Heysham (Lancs.) and Merseyside.

Londonderry (pop. 56,000), a market town on the River Foyle, specializes in making cotton and linen goods. Synthetic rubber is manufactured nearby.

Some of the industries recently started in Northern Ireland are: the making of synthetic fibres at Coleraine; nylon stockings at Newtownards; optical lenses at Lurgan; and light engineering at Larne.

Agriculture. Most of the farms of Northern Ireland are small, and are situated in the lowlands. Mixed farming is usual, with sheep on the hills, and cattle and arable farming elsewhere.

Oats is the most important cereal produced (over 90 per cent of total cereal production), with lesser amounts of barley, wheat and rye. They are entirely used for home consumption.

The chief root crop is potatoes. Seed potatoes have become an important export to Britain and other countries. Many farms still produce flax, but since the climate is unsuitable for the other product of flax, linseed, it is grown almost entirely for its fibre (see p. 106). In this area the strong south-westerly winds and the heavy rainfall (over 60 inches a year) preclude both the growth of trees and all cereal crops.

Most farms keep some dairy cattle. Beef cattle are reared south-east of Belfast. The other principal cattle area is the Lough Neagh Lowlands, dairying being the main occupation in the Lagan Valley area, and beef production on the upland farms of counties Down and Antrim. There is a considerable trade with Britain in live cattle, most passing through Birkenhead to be fattened on the Cheshire and Shropshire plains. However, there is an increasing trend to fatten and slaughter the beef before export.

Fruit and Market Gardening. Strawberries, raspberries and rhubarb are grown on the reclaimed fenland which fringes the southern end of Lough Neagh. They help to supply the jam factories around Portadown. Apples are also produced round the south-eastern side of the lake, mainly on hummocky glacial deposits known as drumlins. Large quantities of cooking and dessert apples are exported to Britain.

Market gardening is important in County Down and near Londonderry, where greens and salad crops are produced. Early potatoes are cultivated in County Down, where the soil is particularly fertile.

THE REPUBLIC OF IRELAND (EIRE)

The greater part of Ireland constitutes the Republic of Ireland. It will be realized from its relief that it cannot readily be divided into clear-cut structural regions. Therefore it will be considered under three somewhat arbitrary headings:

North and North-West Eire

The mountains of Donegal, Sligo, Mayo and Galway, which occur in

FIG. 37. *Ireland, physical.*

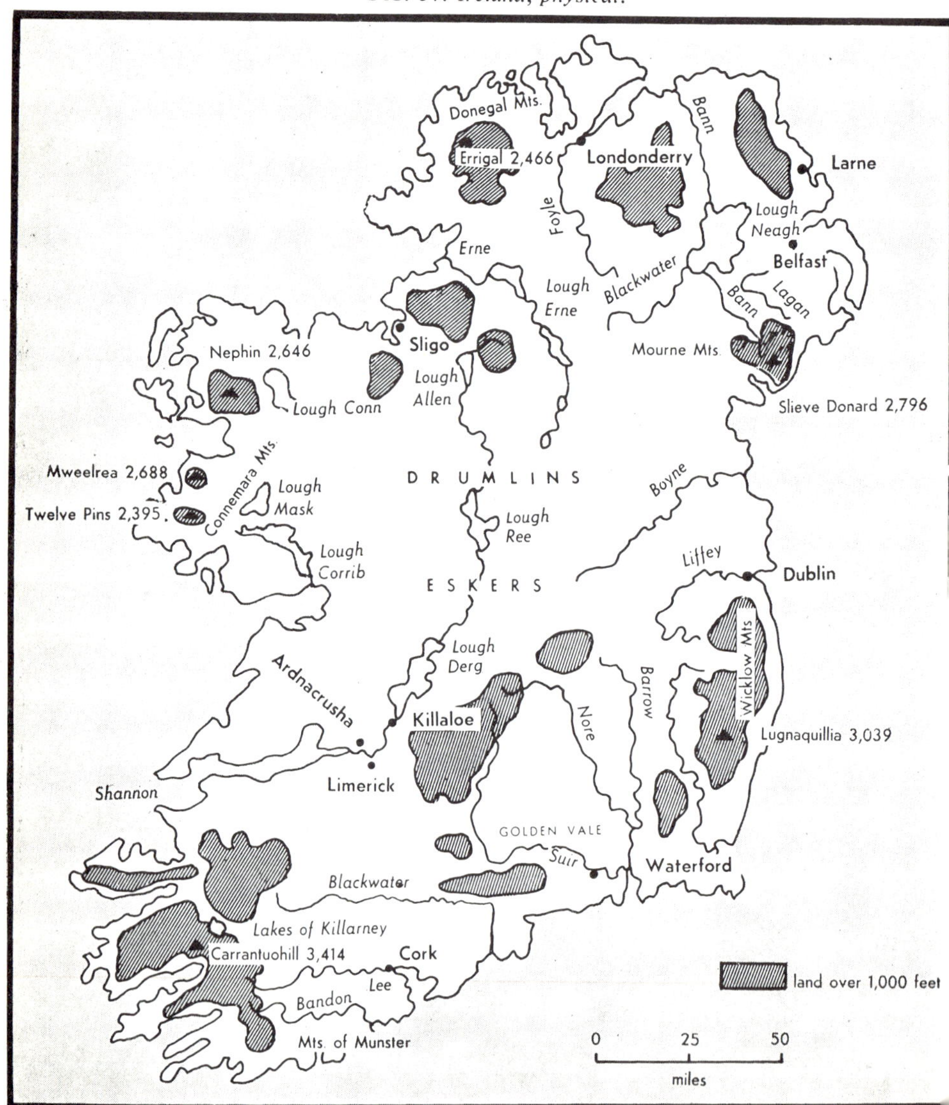

groups fringing the coast between Malin Head and Galway Bay are composed of Pre-Cambrian metamorphic rocks and granites similar to those forming the Scottish Highlands. The lower ground is floored by Carboniferous limestone which is thickly covered with boulder clay and other glacial deposits. In some places the limestone is exposed and forms infertile limestone pavements. Elsewhere the ill-drained glacial drift, the high rainfall and poor run-off give rise to blanket bogs which stretch over both hill and valley, their wetness being maintained by the low evaporation of west Ireland. Only the highest mountains rise above the blanket bog. In such country the drainage is ill-defined and irregular, with a tendency for the rivers to spread out into lakes such as Lough Conn, Lough Mask and Lough Corrib.

In general, the soils are scanty and often infertile. Thus there is little land suitable for cultivation. Life in this westerly part of Ireland is one of extreme difficulty.

A type of farming similar to that of the crofters in the Scottish Highlands is widely practised. The most fertile soil is used to grow crops of potatoes, oats and hay for subsistence. A few pigs, cattle and poultry are also kept. This is subsistence farming, meaning that virtually all the farmers' produce is needed for their own consumption, with little or no saleable surplus.

Sheep are reared, and these have fostered a small woollen industry producing such products as Donegal tweed. This is still to some extent a domestic industry, but there are factories making tweed and carpets at Killybegs in Donegal. There is a small fishing industry off the coast. Some lobsters are exported, and at Sligo there is a relatively new industry of making dehydrated soups from shellfish. Fish-meal is also made at Killybegs.

Tourism is one of the main sources of income. The Government is encouraging the development of the road system and improving hotels in order to attract more visitors to Ireland.

There are no large towns. The most important towns, such as Sligo and Westport, lie on the coast. Low-grade coal is mined at Arigna.

The Central Lowlands

The centre of Eire is not entirely lowland. Towards the south high land, such as Slieve Bloom (1,733 feet) and the Silvermine Mountains (over 2,200 feet), marks the outcrop of Silurian and Old Red Sandstone rocks. The solid rock, forming most of the lowland, is Carboniferous limestone, but it seldom outcrops at the surface. During the Ice Age it was worn down and covered with a thick mantle of glacial drift. Thus, the scenery is very different from that formed by the Carboniferous limestone in the Pennines and Mendips. Much of the glacial drift is boulder clay, and, as a result, these areas are ill-drained. Hence there are good conditions for the formation of extensive peat bogs, and the peat may be over 50 feet in depth. The Bog of

Allen is one of the largest. For centuries the peat has been used as fuel.

In some places the glacial drift takes the form of sinuous sandy ridges called eskers. Where these occur they form important lines of communication across otherwise boggy land.

The Central Lowlands are drained by sluggish streams, the more important of which drain to the River Shannon, the longest river in the British Isles (240 miles). The Shannon meanders across the centre of the plain, broadening into such lakes as Lough Allen, Lough Ree and Lough Derg. The gradient of the Shannon in its first 200 miles is extremely slight, but between Killaloe, on the southern end of Lough Derg, and Limerick, it drops over 110 feet in a series of rapids caused by an outcrop of Old Red Sandstone and Silurian rocks. The water power has been utilized to generate electricity.

Occupations. In spite of large areas of unproductive bogs, many beef cattle are reared on the Central Plain. The export of live cattle amounts to over a third by value of all Eire's exports. Poor communications restrict dairy farming to the vicinity of towns like Dublin, and to the north near the border with Northern Ireland, where communications are better.

There is little agricultural land, except on the plain west of Dublin where potatoes, oats and barley are grown. Elsewhere small farmers raise subsistence crops of potatoes and oats, but yields are not good. Lead, copper and zinc are mined at Tynagh in Galway.

Dublin (pop. 650,000) is the capital of Eire. It is situated on the mouth of the River Liffey and is the Republic's principal port. Railways radiate out across the Central Plain to towns on the east, west and south coasts. Its oldest industries are those of making stout (Guinness), whiskey, biscuits and poplin. More modern industries include the manufacture of electrical equipment, steel, aluminium and copper wire and cables, and other engineering products. Glass containers are also made.

As a port its chief trade is with Britain and Northern Ireland. Live cattle are its principal export, followed by food products (eggs, bacon, etc.), and an increasing quantity of manufactured articles. It imports a considerable amount of raw materials and manufactured goods. Dun Laoghaire (Kingstown) at the entrance to Dublin Bay is the terminus of a ferry service to Holyhead and Liverpool. Dundalk and Drogheda are small ports chiefly engaged in importing fuels—mainly coal and coke—from Britain.

Limerick (pop. 58,000) is situated at the head of the Shannon estuary. Although it has a splendid sheltered harbour it is not a great port. Its hinterland produces agricultural products which are not needed by the American countries which face it across the Atlantic. Since the trade in Irish agricultural products is chiefly with Britain, other ports like Dublin and Cork are better placed. The presence of the nearby Ardnacrusha hydro-electric plant may encourage more manufactures. Its main industries are the processing of milk and the manufacture of butter and bacon. It also has flour-milling, clothing and tanning industries.

Lower down the Shannon estuary, at Foynes, is Shannon Airport, where some trans-Atlantic air flights land. Near the airport is a large industrial estate where a number of firms—including several foreign—have built factories, attracted by the very favourable rents and freight charges which apply in the area as an inducement. These factories manufacture a variety of goods, including knitwear, transistors, pianos and machinery; and industrial diamonds are processed.

Galway (pop. 25,000), a small port on Galway Bay, is too shallow for modern shipping. Industries include flour-milling, clothing and carpets. There is also a small timber and furniture industry.

South Ireland

The south part of Ireland is structurally connected with Wales. In the south-east the Wicklow Mountains, like Central Wales, are mainly composed of Lower Palaeozoic rocks. However, a difference exists in that the northern and western part of the Wicklow Mountains are composed of the largest granite intrusion in the British Isles; although the mountains rise to a little over 3,000 feet, a great deal of the surface is moorland at a much lower elevation. Deeply cut wooded valleys dissect the highland and their beauty attracts many visitors. As in Central Wales, the chief industry of the moorlands is sheep-rearing.

There is an unworked, though potentially fairly rich, copper mine near Avoca; small quantities of lead and zinc are produced in Tipperary. Westward, the surface is made up of parallel ridges which follow an east-west Armorican trend, similar to the structure of the rocks in South Wales. The ridges are composed of Old Red Sandstone, separated by drift-filled valleys floored by Carboniferous limestone. Where these ridges and valleys reach the sea in the south-west they form an excellent example of a ria type of coastline in Dingle Bay, Kenmare River and Bantry Bay. Behind these inlets lie some of the most admired mountain country in Ireland. The Killarney lakes, lying on the eastern side of the Macgillacuddy's Reeks and Purple Mountains, are a well-known tourist centre. Unfortunately, in the past this has not prevented large-scale unemployment and population loss. However, new industries have recently been attracted to the area.

Except for the highest parts, most of the surface of south Ireland is covered by a thick mantle of glacial deposits, which are better-drained and therefore more fertile than those of the Central Plain. The drainage of the area is interesting since it forms a typical trellised or grid pattern, the transverse-consequent streams making useful gaps through the ridges, and the intervening lowlands being drained by subsequent streams.

Agriculture. On the eastern lowlands, where the rainfall is relatively low, there is arable farming. Crops of barley, oats and roots are grown. Some wheat is grown in the Nore and Suir valleys, since this area has Ireland's highest average summer temperatures. Most farms keep cattle, pigs and poultry.

As one goes westward the rainfall increases, and dairy farming becomes the characteristic occupation. The lowlands have some of the best pastures in the British Isles: the Golden Vale near Tipperary being particularly noted for its dairy produce. Much of the milk is made into butter and cheese in factories, and the skimmed milk fed to pigs. Bacon is produced in Limerick and Cork and other market towns. Most farms grow a few crops such as oats and potatoes.

Around the small town of Castlecomer, between the valleys of the Rivers Nore and Barrow, a small downfold contains deposits of coal. This is the main part of the so-called Leinster coalfield. The coal, nearly all anthracite, is unsuited to domestic use. Production can be over 100,000 tons in a year.

Cork (pop. 125,000) is a port situated on the point where the River Lee enters Cork Harbour, a large sheltered inlet. Its outpost of Cobh (Queenstown) is a port of call for trans-Atlantic liners, and next in importance as a port after Dublin. It exports live cattle, bacon and dairy produce to Britain. Imports include crude oil, coal, cattle feed, foodstuffs, etc. A large oil refinery, opened in 1959, is situated at Whitegate on the eastern side of Cork Harbour. Cork also has engineering and steel works, and builds and repairs ships. Imported coal supplies an electricity power station.

Waterford (pop. 30,000) is a small port on the River Suir where it widens to form a large sheltered harbour. Besides exporting dairy produce the town is well known for the manufacture of high-quality glassware.

FIG. 38. *A typical small Irish farm, Kerry.*

Wexford (pop. 12,000) exports live cattle to Britain and nearby is Rosslare, a packet station for travel to Fishguard in Pembrokeshire. Other towns situated inland, such as Kilkenny, Tipperary, Mallow, Carlow, Clonmel and Fermoy, are market towns commanding gaps through the east-west ridges. They are all situated on railways which connect the south coast with the Central Plain.

Fuel and Power. Peat has always been the principal fuel used in Irish farms and cottages. In the 1930s the Government formed a board known as the Bord-na-Móna, to investigate ways of making better use of Ireland's extensive deposits of peat. Modern cutting machines are now used in many parts of the Bog of Allen. Peat is cut into sods and dried more quickly and efficiently than by the hand-digging process.

A plant at Lullymore makes peat briquettes, and other such plants are planned. The dried sods and briquettes are used as fuel in electric power plants at Allenwood, Portarlington, Lanesborough, Lullymore and Ferbane in the Bog of Allen. Hand-cut peat is used at the Gweedore (Donegal), Screeb (Galway), Milltown Malbay (Clare) and Cahirciveen (Kerry) power stations.

Hydro-electricity. The first hydro-electric power station was completed in 1929 at Ardnacrusha on the Shannon, utilizing the rapids between Killaloe and Limerick. Since then the following four other rivers have been harnessed and the possibilities of others are under investigation:

Liffey: near the Pollaphuca Falls and lower down at Leixlip, where the river has a fall of about 60 feet.

Erne: between Belleek and Ballyshannon; and between Cliff and Cathaleen's Falls.

Clady (*North Donegal*)*:* water from the Clady is diverted and piped to a power station near the mouth of the Gweedore River.

Lee: at Carrigodrohid and Inniscarra.

Electricity is produced by the use of coal and oil near Cork and Dublin.

By 1965, of all the electricity generated in Eire, 24 per cent was hydro-electric, and 32 per cent from peat-burning stations. A network of grid lines distributes the power over the country. Since 1946, some 300,000 rural homes have been connected with the supply, but the use of electricity on farms has not progressed as fast.

New Industries. The constant drain on the population of Ireland caused by the high level of emigration greatly concerns the Government. A relatively large proportion are farmworkers. Many of the farms have to contend with difficulties caused by wet climate, poor soils, inadequate transport and lack of capital. Thus, they maintain only a low standard of life. This encourages depopulation; the Irish who leave usually go to London, Liverpool or other industrial areas in Britain.

In 1958, in order to try to stem emigration, raise the standard of living and attract highly trained Irish workers back from abroad, the Irish Government initiated a scheme to encourage industrial development in Eire. It announced large tax concessions and substantial money grants to attract new foreign industries to the country. As a result, many foreign firms have set up factories in different parts of Eire.

A great variety of products are made. The volume of industrial output increased by 58 per cent between 1958 and 1964. Efforts are also being made to improve the tourist industry by improving communications and hotels. Receipts from tourists work out at over four times as much per head of population, as compared with the United Kingdom.

Trade. About two-thirds of Eire's exports go to Britain. Of this total about three-quarters are agricultural products, including live animals and processed products. Eire's chief imports are manufactured goods and raw materials, including tropical products and coal and oil. Britain supplies about half of the imports.

QUESTIONS

1. Give an account of the rise and development of the textile industry of Northern Ireland; describe the chief products and name the towns engaged in the industry.
2. Belfast and Dublin are ports on the east coast of Ireland. Compare and contrast their sites and port activities.

REVISION SUMMARY OF THE BRITISH ISLES

GENERAL

THE LAND

CLIMATE, SOILS AND VEGETATION

Page

(4) Variations in the length of the **growing season** and effect on agriculture. 30

(5) **Frost** intensity and frequency—effect on agriculture and transport. 31

(6) **Urban climates**—"man-made" modifications to the regional climate; pollution of the atmosphere. 31

B. SOILS

(1) Basic division between podsols and brown earths but many local complexities through rock type, glacial deposits, farming activities. 32

C. VEGETATION

(1) Almost complete **lack of "natural" vegetation** through human interference. 34

(2) **Changes in vegetation** since the last glaciation resulting from climatic change. 34

(3) Factors influencing distribution of types of vegetation at the the present day. 35

POWER, MINERAL AND WATER RESOURCES

A. FUEL AND ENERGY SUPPLIES

(1) Distribution of **coalfields** on edges of upland areas. 37

(2) **Uses of coal** for industry, power stations, domestic consumption and gas industry. 38

(3) **Modernisation of collieries** and concentration of mining on the most productive and cheaply worked coalfields. 38

(4) **Generation of electric power**—increasing use of hydro-electricity, oil and nuclear power stations. 39

(5) Development of the resources of **gas** in the North Sea. 40

(6) Problems of **water resources**—unequal distribution; pollution; need for reservoirs and location in upland areas of high rainfall. 41

B. MINERAL RESOURCES

(1) Long history of mineral exploitation but overwhelming importance of **iron ore** from sedimentary rocks of lowland Britain. 42

(2) Value of resources of **salt,** anhydrite, fluorspar barytes. 43

(3) Fortunate occurrence of **limestone** in relation to coal and iron for industrial development. 44

(4) Increasing importance of reserves of **sand and gravel** and decline in use of natural building stone. 44

(5) Occurrence of mineral resources in lowland Britain reinforcing the importance of the **lowlands** in the economy of the country. 45

Page

LAND-USE

THE CORE AREA: (i) THE MIDLANDS

Page

THE CORE AREA: (ii) SOUTHERN ENGLAND

(1) Dominated by the **London** conurbation; importance of the **Thames estuary;** diversification of industry. 64

(2) Problems from concentration of population in London area. 66

(3) Vale and scarpland relief; wide variety of rocks and soils but limestone, chalk and clay of major importance in land-use. 67

(4) **The Weald**—dissected anticline with inward-facing escarpments. 67

(5) **Agriculture** in Southern England related not only to natural resources of favourable climate and soil but also to demands of London conurbation. 68

(6) Distribution of industry and variety of employment available —metal-using industries—**ports of Southampton and London.** 70

THE CORE AREA: (iii) LANCASHIRE AND YORKSHIRE CONURBATIONS

A. LANCASHIRE

(1) Links with the Midland and Southern England. 73

(2) **Lancastria,** a lowland of sedimentary rocks containing coal and salt. 74

(3) Agricultural origins but development of textiles (cotton) especially in Industrial Revolution; importance of links with **Liverpool** and availability of water from Pennine upland. 75

(4) Problems created by concentration on **heavy industries**, coal mining and textiles. 75

(5) The domination of this area by the **Manchester and Merseyside conurbations.** Importance of new industries, including chemicals, and improvements in communications. 76

B. YORKSHIRE

(1) The location of the Yorkshire **woollen textile** industry in the valleys of the Pennines. 82

(2) Characteristics of the woollen industry. 82

(3) Dominance of the area by the **West Riding conurbation**—specialisations in woollen textiles between different centres. 82

(4) The heavy **iron and steel industry** of the Yorkshire coalfield. 83

(5) Value of the **Humber estuary** to the industrial development of the Yorkshire conurbation. 84

PERIPHERAL INDUSTRIAL AREAS

A. SOUTH WALES

(1) Heavy industry based on the coalfield using, originally, local supplies of iron. 85

PERIPHERAL RURAL AREAS: (i) SCOTLAND

A. THE HIGHLANDS

B. SOUTHERN SCOTLAND

Page

PERIPHERAL RURAL AREAS: (ii) ENGLAND AND WALES

A. THE LAKE DISTRICT

(1) Uplifted glaciated massif with radiating valleys and important lakes. 107

(2) Little agricultural potential other than sheep rearing but very important for water supply and recreation. 107

B. THE WELSH MASSIF

(1) **Upland** area with highest mountains in the north-west. Glaciated. 107

(2) **High rainfall,** leached acid soils restrict agriculture to lowland fringes on west and north and to valleys. 108

(3) Former importance of minerals including slate. 109

(4) Modern land-use for **recreation, water supply** and **electricity** generation. 109

C. SOUTH-WEST ENGLAND

(1) Upland area—*not* glaciated. Unique characteristics of favourable climate on lowland fringe. 110

(2) High **moorland** of resistant rocks. 111

(3) **China clay** industry; tin-mining. 111

(4) **Mixed farming,** dairying, fruit-growing, flowers on coastal fringe. 112

(5) Importance of **tourism** to the economy of the south-west of England. 112

(6) Rias, harbours, fishing; command of the "Western Approaches" by **Plymouth.** 112

D. EASTERN ENGLAND

(1) Zone of vales and scarplands with, to the east, the great lowland of **East Anglia.** 113

(2) **Arable farming** highly scientific, mechanised, and intensive. Introduction of new **light industries.** 113

(3) **East Anglia**—former isolation by forest and fen; importance of the **Norfolk Broads.** 113

(4) **Cereals**; importance of barley—increasing cattle population, and fewer sheep; sugar beet. 115

(5) **Factory farming** of poultry and pigs. 115

(6) Industry still mainly related to the area's agricultural background. 116

(7) **The Fens** as a unique area in Britain; reclaimed land; **drainage problems;** silt fen and peat fen. 116

(8) Rich and intensive **arable agriculture** of the **Fenland** with few industries, mostly related to processing of agricultural produce. 118

Page

IRELAND

NORTHERN IRELAND

A. GENERAL

Geologically connected with S.W. Scotland. Large areas of Antrim covered with glacial deposits. The few minerals include bauxite in Antrim and a small coalfield in Tyrone. 119

B. OCCUPATIONS

(1) **Industries:** chiefly located in Lagan and Bann Valleys, Lough Neagh Lowlands, near Londonderry. Textiles (notably linen), shipbuilding, engineering, consumer and manufacturing. 120

(2) **Unemployment** from decline in linen and shipbuilding. 120

(3) **Agriculture:** Mixed farming, sheep on hills, dairy and beef production, oats, barley, potatoes, flax, fruit and market gardening round south end of Lough Neagh. 121

REPUBLIC OF IRELAND

A. NORTH AND NORTH-WEST

(1) **General:** mountains between Galway Bay and Malins Head, **similar rocks to those of Scottish Highlands.** On lower ground limestone thickly covered with glacial drift that gives rise to boggy country. 122

(2) **Occupations:** crofting, fishing, tourism, Donegal textiles. 123

B. CENTRAL LOWLANDS

(1) **General:** lowlands broken by highlands. Glacial drift on low ground means **extensive bogs drained by sluggish streams.** Shannon, low gradient until near mouth—electricity generated. 123

(2) **Occupations:** beef cattle; little good agricultural land. 124

(3) **Towns: Dublin** (capital), Dun Laoghaire, Limerick. 124

(4) **Industry:** brewing, distilling, food, textiles, engineering. 124

C. SOUTHERN IRELAND

(1) **General: structurally connected with Wales.** Sheep on moors. Copper, lead, zinc mined. Indented west coast. 125

(2) **Occupations:** in drier eastern part, arable farming combined with cattle, pigs, poultry. Climate west is wetter and dairy farming general, pigs reared on skim-milk. Some wheat grown. 126

(3) **Towns:** Cork, Queenstown, Rosslare. 126

(4) **Fuel and Power:** peat for power stations. Hydro-electricity in many parts. Coal and oil used near Cork and Dublin. 127

(5) **Industries:** foreign firms' manufacturing and assembly plants. Tourism encouraged. 127

(6) **Trade:** about two thirds of products go to Britain, mainly agricultural. About half of imports come from Britain. 128

EUROPE AND THE U.S.S.R.

CHAPTER 12

THE FRAMEWORK OF EUROPE

EUROPE, unlike all other continents, is a continent without extremes, either of climate, weather or physical conditions. There are no great deserts of sand or ice, no dense jungles, no really high mountains to compare with the vast Himalayas of Asia or the Andes of South America.

Europe is separated from the smooth coastline of Africa by the deep but relatively narrow Mediterranean Sea. Europe contains only about one-seventh of the world's people, but contributes a much larger proportion than this of the world's manufactured goods and agricultural produce.

Europe has a relatively small area and roughly indented coasts. Because the surrounding broad shallow seas are swept by warm Atlantic currents it is unusually warm for its latitude. The tempering influence of the sea, helped by the easterly drift of depressions, also penetrates far inland across the low western coasts. Big areas of water almost surrounded by land, like the Mediterranean, the Baltic and the Black Seas, also help to prevent extremes of climate, for the sea warms up much more slowly than the land, but retains its heat far longer.

Europe is a "low" continent. Six-tenths of its area lies below a height of 600 feet. Despite the diversity of the surface relief, mountains nowhere form a serious barrier to communication as they do in Asia. Europe possesses many areas of fertile soil, but even in early times the vegetation was never so dense as to stop man clearing it.

Europe has gained particularly by the fact that it belongs to the same land-mass as Asia, which was the cradle of civilization. From Asia, civilization passed to the great cultures of Crete, Greece, and in more recent times Rome. But in historic times the energy and initiative of Europeans in exploring and settling other continents has been the major factor in spreading civilization throughout the world.

Relief and Structure

The relief features of Europe as we know them today emerged in their broad outline quite late in geological times. It is probable that they first appeared in roughly their present form about fifty million years ago, in mid Tertiary times, when immense earth-storms took place and the great alpine mountain systems appeared across southern Europe, extending far into Asia. The upheaval of the mountains was so violent that many parts of the surface to the north were shattered and buckled. The worn-away roots of these older mountains were raised up as large plateaux or faulted down and filled with materials eroded from the new highlands adjacent to them. Some of the downcast areas were later covered by new seas in late Tertiary times.

The final detailed moulding of the European landscape was commenced about a million years ago. Great ice sheets formed over the mountains, similar to those of Antarctica or Greenland today. The ice spread south from Scandinavia and Scotland or descended from the higher peaks in the Alps. When it finally melted, deep U-shaped valleys had been gouged out, and great thicknesses of clayey material that had been carried with it were deposited on the plains. In Northern Europe, the ice had scraped bare great areas. With the final melting of the ice, there were changes in the relative levels of land and sea, shaping the outline of the modern Baltic Sea and submerging a great lowland in the area we now know as the North Sea. This last event is known to have been continuing while man first appeared in Britain; and even at the present time we cannot regard sea-level as unaltering.

Ancient Lands of Northern Europe

If we visited Sweden or Finland we would find immensely old crystalline rocks exposed at the surface. This area, called the *Baltic Shield*, was scraped bare by the great ice sheets. Now bare rocks form a low plateau dotted by forest, with peat bogs and lakes which have formed in hollows gouged by the ice. The plateau dips gently south-eastwards to disappear beneath a cover of relatively younger rocks roughly south-east of a line joining the island of Gotland to the Gulf of Finland and lakes Ladoga and Onega. The northern edge of the younger rocks, of Cambro-Silurian age, forms a low limestone scarp. The younger rocks lie in flat sheets over the old crystalline base strata, suggesting that they have been little disturbed since they were laid down.

The second oldest belt forms a broken line of mountains from Spitsbergen through western Norway into north-western Britain. These *Caledonian* mountains first appeared along the eastern edge of a now-lost Atlantic continent, probably about 300 million years ago. Earth forces later changed their soft rocks into hard crystalline materials, and there are remains of ancient volcanoes. The rocks have not only been folded but also large pieces of country have been thrust eastward over the surface of the Shield for tens of miles.

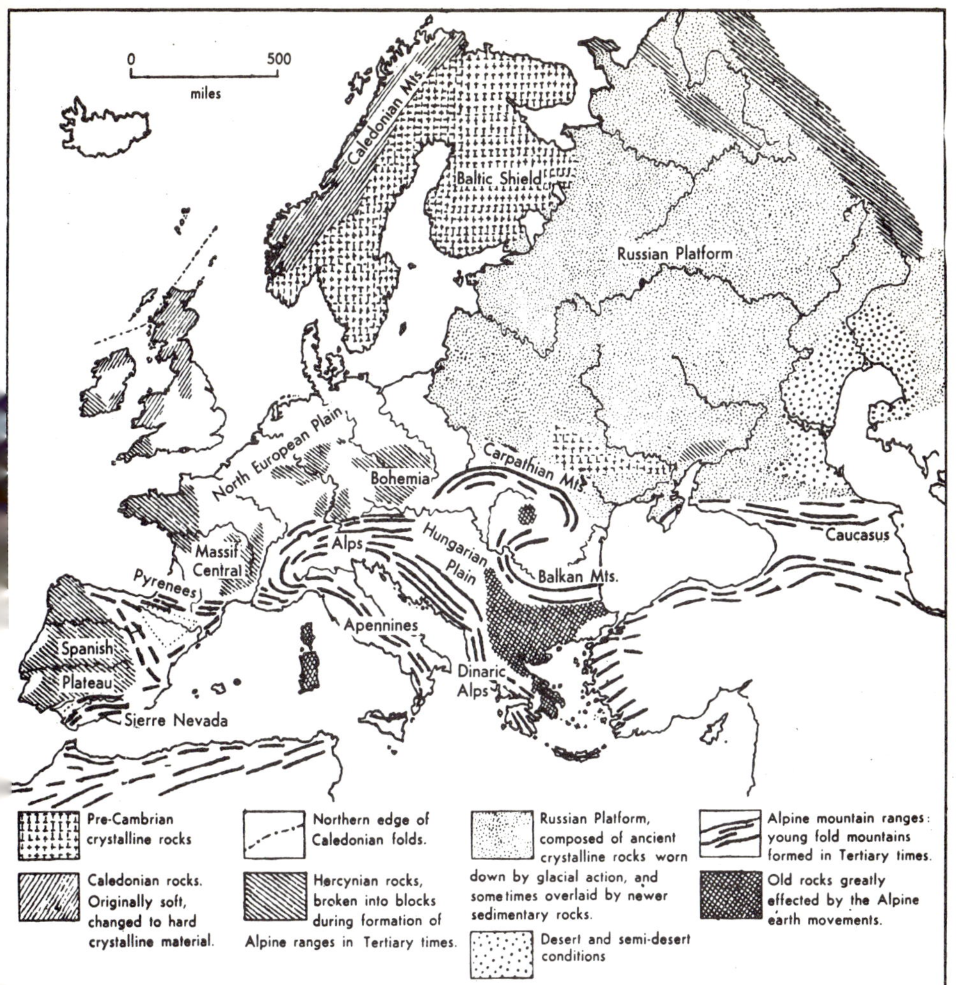

FIG. 39. *The main physical features of Europe as far as the Urals. The Urals are the traditional geographical boundary between Europe and Asia, though this is purely arbitrary since, in human terms, they have never formed a barrier.*

Note the geological relationship between the main land mass and the offshore islands —like Britain, Sardinia, etc.

At several periods, after long ages of weathering into low plateaux, earth movements re-uplifted the blocks, and they were then deeply incised by rivers. The result is that they now have bare rounded summits, all at very much the same height in any particular area, and step-like surfaces on their flanks, into which the U-shaped valleys are cut. Some of the higher parts of the summits have been cut by corries during the Ice Age. A greater amount of uplift in western Norway has left deep cleft-like valleys, often drowned to seaward to form winding fiords and an island-studded coast. Inland, where uplift was less marked, the valleys are shallower and broader. In south-west Norway, the mountains rise above the forests to over 8,000 feet, and there are remains of more extensive former icefields. In the north, the mountains occur as more isolated peaks, much more shattered by frost. To the east, the plateau surfaces descend in broad wooded steps to the Baltic Shield, crossed by swift rivers tumbling over fine waterfalls. Spitsbergen, in the Barents Sea, remains covered by a great sheet of ice.

Iceland is an extensive plateau of young volcanic lava pierced by active volcanoes (e.g. Hekla). The higher regions are ice-covered. Iceland was formed in Tertiary times by successive additions of sheets of lava long after the last of the old Atlantic continent disappeared.

Hercynian Europe

Most of Germany, France and western Czechoslovakia is composed of the remains of mountains formed late in Carboniferous times, about 220 million years ago. Under Holland and Belgium and many parts of France and Germany, these remains are covered by a thick mantle of younger rocks. These *Hercynian* ranges were formed as fold mountains, but were eroded to low plateaux over a great period of time. In places they were submerged

FIG. 40. *A fiord, created by glaciation, and characteristic of the Norwegian coastline. Fiords provide excellent harbours and anchorages, though the morainic or rock bars at their mouths restrict the size of vessels that can enter.*

by warm seas during Secondary times, in which materials eroded from the mountains were deposited as the later Mesozoic (Secondary) sediments.

The force of the gigantic earth movements in Tertiary times that accompanied the building of the Alpine ranges shattered the Hercynian country into great blocks, some being uplifted and others depressed. The effect of these movements on the present landscape is more important than the distribution of the original folded structures. The movements also diverted the rivers flowing across the low plateaux of early Tertiary times. Many abandoned their old valleys or cut deep gorges through the newly uplifted blocks.

Today, the Hercynian country is a landscape of rolling and mostly wooded hills, broad but deep valleys (sometimes with gorge-like sections), large basins and plains of rich farming country.

Bohemia

Bohemia is one of the best examples of Hercynian country. It is a large central basin surrounded by a rim of tilted blocks that form roughly dissected, forested uplands. The tilting of the blocks has left many steep, faulted scarps. The floor of the central basin rises towards the south, where the rivers have cut deep, picturesque valleys. Higher blocks of country (called horsts) stand as forested hills, while in places softer rocks have been eroded into shallow basins. A most impressive scene is where the River Vltava has cut a deep incised meander below the Hradcany castle at Prague. In the lower, northern part of the basin, in the upper reaches of the River Elbe (here called the Labe), there are broad plains covered by sedimentary rocks that make fertile farmland. On the north-east, the forested Sudety mountains are rough country formed of parallel blocks, separated by broad vales where the land has been down-faulted. The high rolling hills of the north-west have a south-facing scarp edge that overlooks the wide depression of the farming country of the Ohre valley. This line of weakness in the crust is also marked by rounded volcanic hills and mineral springs around which many famous spas have grown up. Between these two uplands of the Erzgebirge and the Sudety, the River Elbe escapes to the north through a rugged gorge cut into soft sandstone as the land rose in Tertiary times; but in some places it has cut through volcanic rocks. The south-western edge is an inhospitable hilly frontier country of almost virgin forest. The eastern edge of Bohemia rises into rather higher limestone hills that overlook the wide depression of the Moravian Corridor. This important north-south routeway is fertile grain-growing country, though troubled by droughts.

Hercynian France

Much of the centre of France is formed by a huge plateau, the Massif

FIG. 41. *Rolling forested country in the Beskids, Czechoslovakia. The Beskids are northern outliers of the Carpathians.*

Central, with a general height of 3,000 feet but rising to over 6,000 feet (Mont Doré (6,188 feet) to the highest point in Hercynian Europe. It can be divided into the regions given below.

The central part of the Massif includes two great down-faulted depressions, occupied by the upper reaches of the rivers Allier and Loire, around which lie volcanic materials resting on old crystalline rocks. The volcanic materials weather into rich soils, and there are remains of old volcanoes in sharp rocky pinnacles or broad, rounded domes. The main volcanic region is the Auvergne, where mineral springs have given rise to many spas. In areas of crystalline rocks, like the Forez mountains, the soils are far less rich.

Exposed to moist air from the Atlantic, the western regions of the Limousin and Marche are open plateau formed on crystalline rocks. The soils are poor, and there is much forest and heathland. However, the lower valleys are more fertile: and are known for their sweet chestnuts. Granite here has weathered into chinaclay, used for the pottery industry at Limoges. The southern part of the Massif is a high plateau of crystalline and limestone rocks. Apart from the Montagne Noire and the Cevennes, this part seldom appears as mountainous. The limestone plateaux (called *causses*) are cut by the deep canyon-like gorges of such rivers as the Lot and the Tarn. The surface of these plateaux is scattered with swallow holes which form the entrances to extensive systems of underground caverns. The eastern edge of the Massif lies along a belt of weakness in the earth's surface, extending from the Hessian Corridor depression in Germany through the Rhine rift valley and south into the trough of the Rhône-Saône rivers. This eastern part is an upland region, often composed of granite weathered into rounded surfaces, like the Morvan in the north and the Vivarais in the south. Between the uplands lie depressions which furnish easy routeways into the Massif Central region from the Rhône valley.

The old Hercynian mountains of the Brittany Peninsula have been planed down to an open, infertile surface. Along the coast the succession of headlands and bays marks the differential resistance of the rocks to wave-attack. Lying generally below 600 feet, there are striking similarities between the geology, physical features and climate of Brittany and Cornwall.

The Rhine and its Surrounding Uplands

After the Hercynian blocks were shattered and disturbed in Tertiary times, the cover of sedimentary rocks over much of eastern France and southern Germany was etched into a landscape of scarps and vales. In central Germany there are also many rough forested hills formed of uplifted Hercynian blocks. Between Basel and Mainz, the river flows along the floor of a 150-mile-long rift valley, formed when the central part of a once continuous upland sank just like a keystone falling from an arch. Uplands remain on either bank. The Vosges and Black Forest descend steeply to the floor of the trough along the lines of great parallel faults. On the gravel terraces by the river stand fine towns, like Strasbourg and Freiburg; and the formerly-meandering river is canalized and bordered by meadows, fields, vineyards and orchards.

Between Mainz and Bonn, the Rhine flows through an impressive gorge cut into the bordering highlands. As these hills were lifted in Tertiary times, the river cut down into the underlying soft schists, so that today we may see the remains of its old valley high on the side of the gorge. Its tributaries, like the Mosel and the Lahn, that once meandered to join it, have cut deep incised meanders into the hills. Subsidence between Andernach and Koblenz once formed a temporary lake that was drained as the river cut down into the hills. It was also filled by powdered pumice blown out of the Eifel volcanoes to the west, whose craters remain today as forest-ringed lakes. Remains of volcanic action can also be seen in mineral springs. North of Bonn the river valley widens and the river flows into its terrace-bordered lower flood plain. Here it is flanked by orchards lying south of Cologne, famed for its magnificent cathedral. To the east is the industrial Ruhr area; to the west are large deposits of brown coal (lignite).

The North European Plain

The North European Plain extends east from the North Sea until it eventually merges into the Russian Plains. On the north and west it is submerged beneath the Baltic and North Seas. (A small northern extension of the plain may be found in the Scanian lowlands of south-western Sweden.) The North European Plain has a deep cover of sedimentary rocks that conceal much older rock layers. The surface has been moulded by thick deposits of clays and sands left by the last ice sheets. The solid rocks appear at the surface at only a few places, while large areas are covered by peat or alluvium.

When the Scandinavian ice sheets finally retreated they left great deposits of material scoured from the north. On the southern fringe, from which the ice retreated first, these have been changed by vast volumes of water flowing from the melting glaciers. Moving north across the Plain, the lines of morainic hills are newer. There are great crescents of these hills across Jutland, Mecklenburg and Mazuria, dotted with lakes and with forests on the less fertile stonier parts. These hills were formed about 25,000 years ago, and mark the last period before the final melting and retreat of the ice. North of these hills lies the drowned Baltic coast, smoothed into lagoons behind long spits formed by the easterly drift of the beaches.

The southern part of the Plain has large heathlands formed on sandy materials left by the retreated ice. Farther south, this heathland changes

FIG. 42. *Glacial lake in Finland—the famous Lake Punkaharju. Finland is known as the Land of a Thousand Lakes.*

into a rich and contrasting landscape covered by the fine-grained fertile yellow dust known as *loess*. It is thought this was blown from the sub-arctic deserts that lay around the retreating edge of the ice or, less probably, from unvegetated lowlands now submerged by the sea.

The retreating ice blocked the natural escape of water to the Baltic depression. Meltwater was forced to flow away, either to west or east, in wide shallow valleys. The modern river system still partly follows these routes, but the present rivers have tried to flow northwards to the Baltic. In doing this they have cut through the crescent of morainic hills at several places, marking their courses by sharp right-angled bends.

The Low Countries

A great and almost featureless plain extends from the hills of Artois in France to the mouth of the Elbe River in Germany, along the coast of the North Sea. It was covered by ice as far south as the Rhine at Arnhem, but farther south it consists of sands, clays and riverine mud. In some places there are patches of loess, or the similar but older *limon*, or even sandy heath with occasional bare dunes, like the Kempenland and Veluwe. Elsewhere it is fertile farming country. A large area of land was submerged by the sea after the Ice Age, but now the coast is protected by an almost unbroken lines of dunes.

Possibly as early as A.D. 700 man began to wrest land from the water by draining coastal meres and marshes. Dykes were built across the foreshore to turn it into grazing land. Once land is drained, it dries out and its level falls: to prevent the sea flooding it again constant pumping is needed. Embankments were built to guard against river floods, and settlements hug even the smallest areas of higher ground. About A.D. 1400 several great storms flooded the then freshwater Lake Flevo, turning it into an arm of the sea. The bay so formed, the Zuider Zee, has been reclaimed in recent years by building a 15 mile-long dam across its mouth and pumping dry individually dyked sections covering most of its area.

North of Den Helder the dune belt is broken into the chain of the Frisian Islands. At low tide they are surrounded by wide sand and mud banks. Modern ideas suggest that these islands may have been built naturally on top of offshore sandbanks, rather than being a line of breached dunes.

Lowland France

Northern France occupies a broad lowland roughly coinciding with the basin of the River Seine and its tributaries. It is usually called the Paris Basin, for Paris stands at its centre and is the focus of its routeways. Sediments that had been laid in a gradually sinking basin were tilted around the edges by the earth movements of the Alpine period. Erosion has etched them into a landscape of low scarps and broad vales, while the rivers that drained the basin before the tilting have cut through the scarps in water gaps usually commanded by important towns. Like a series of saucers laid one on top of the other, the age of the rocks gets younger as we travel towards the centre of the basin. This provides a country with many different qualities for farming, and the fertility is made richer by patches of fine yellow limon soil. With such diversity, it is not surprising that this is one of Europe's richest farming areas.

To the east lie the scarps of the Plateau of Lorraine. The area is rich in iron ore. On the west, the Paris Basin passes into a series of marshy valleys separated by rich farmlands. To the north, gentle anticlines have been eroded away to show their hearts, and here there is even greater diversity in the types of rocks found. The Channel coast was formed late in geological times, and its cliffs rise in places above silted-up lagoons. To the west of the mouth of the Seine lies Normandy, with a climate similar to that of Southern England, and farmed for its dairy produce.

South of the Brittany Peninsula, and inland from the coast, are broad lowlands of young sedimentary rocks. These form important farming lands. The coast south of the wide Gironde estuary is flanked by a great expanse of sand (the *Landes*), formerly heath and marsh, but today extensively afforested. At the foot of the Pyrenees the countryside of Lannemezan is a gigantic cone of material carried down by torrents from the mountains in glacial times.

European Alpine Ranges

In the Tertiary era, starting about 50 million years ago and lasting over some tens of millions of years, the fourth major structural element in Europe appeared. This was the series of high ranges of young folded mountains to the south. They extended west to east across Europe in northward curving bows, and comprise giant folds of soft sediments in which are embedded huge blocks of older crystalline country. It is not completely understood how these mountains were formed. It is usually thought that the materials forming them were laid down in an east-west sea occupying a deep trough or furrow in the earth's crust. Earth movements caused these soft sediments to be compacted and uplifted. In this process, the rocks were cast into giant north- or south-facing folds. Immense bundles of these contorted and twisted folds comprise the different mountain systems of the Alpine mountain belt. As soon as the rocks appeared above the sea, the processes of erosion and wearing-down began. It is to these forces and later uplifts that they owe their rugged mountainous character. The greatest influence was the work of ice during the final phase of the Ice Age, when glaciers deepened valleys and cut big corries into the higher slopes. The result was the typical pyramid form of the highest peaks, such as the Matterhorn.

Along the northern edge a broad depression has been filled with waste from the mountains to form the Swiss Plateau, the German Alpine Foreland, and farther east, the Carpathian-Galician-Moldavian Foreland. To the south the Plains of Lombardy and part of the Adriatic Sea may be formed from a similar depression.

Some other areas also appear to have sunk or have been downfolded. For example, the large basins of the Hungarian Plains and Wallachia, or the Black Sea and part of the Caspian Sea. The Tyrhennian Sea is another sunken area that seems to be bounded by faults. Some of these areas are reaiiy pieces of older country that have become incorporated into the mountain belt, thinly masked by recent sediments. The lovely and dramatic islands of the Aegean Sea are the peaks of submerged ranges, while other, partly submerged, ranges form the Adriatic islands of the Dinaric coast of Jugoslavia.

FIG. 43. *The Swiss Alps—the heart of the Tertiary mountain system. The deeply glaciated valleys of the Alps are clearly shown in this example. Note the U-shaped form of the valley (typical of glaciation), and the remains of a pre-glacial valley high up on the slope as a marked terrace. On the left background a stream descends from a hanging valley. The valleys of the Alps form vital route-ways, and are also centres of population.*

(1) The Alps

The Alps proper extend in a great curve from the Mediterranean to Vienna, the mountain belt varying from 80-140 miles wide. It comprises many roughly parallel chains of mountains, with hundreds of peaks rising above 10,000 feet in height. Even though the Alps lie far south (43°-48° N.) they are high enough to carry permanent snow and Europe's largest glaciers. The highest part is in the west, in the Pennine and Savoy Alps, where peaks over 13,000 feet include Mt. Blanc, Mt. Rosa, the Matterhorn and the Weisshorn.

The Alps consist of a southern belt of sedimentary rocks, mostly limestone and sandstone, and a crystalline core that is surrounded by schists and shales. The northern flanks of the crystalline zone are marked by massifs of the older rocks of Mercantour, Pelvoux, Mt. Blanc and the Aar, now exposed as a result of intense denudation. East of Zurich and Lake Como the Alps are less intensely worn away. The deeper structures here are buried by other masses of rock and are seen only in a few small exposures in the Engadine and High Tauern. The great limestone masses of the southern flank of the eastern Alps are very similar in build to the Dinaric mountains of Jugoslavia.

The outer belts of the Alps comprise either low hills of soft sandstone, mudstone and shale, or broad dry limestone plateaux. The wide but relatively shallow valleys opening on to the plains often have their mouths marked by moraine-dammed lakes. Near the Mediterranean the prevalence of limestone and the long dry summers produce harsher climate and scenery: goats and sheep replace cattle; almonds, vines and olives replace the soft fruits of the north. Scantily settled, these Mediterranean foothills are without important routeways. Communications are, naturally, excellent along the Riviera coasts, famous as a holiday area. The Alpine foothills in the Plain of Lombardy are favoured agricultural areas, as the climate combines the best features of Central Europe and the Mediterranean. Near Lake Maggiore lie the southern entrances to the passes which provide three main routes across the Alps: the St. Bernard, Simplon and St. Gotthard.

The Jura is unusual because structure and relief are closely related, which is not usually true in the Alps. The northern edge is a broad northward-dipping plateau, but the south is a series of gently curving ranges, deeply gashed where traversed by rivers.

The most majestic mountains lie farther south between the Splügen Pass and the Dora Riparia River. The peaks are sharp and pyramidal in shape: caused, once again, by intense frost and glacial erosion. The largest snow-fields, feeding the Aletsch and other glaciers, lie in the Bernese Alps, being the remnants of a Quaternary ice sheet that etched the details of the present relief. Valleys are typically U-shaped; sometimes ledges or shoulders of rock occur high up on valley sides, showing the level of the river valleys before glaciation. Where such "hanging valleys" join there is usually a

FIG. 44. *Summer aridity in Greece—the valley of the Erimanthos. In many parts of southern Europe (and the Soviet Union) summer is extremely dry. By contrast dry stream beds, like the one pictured above, become raging torrents with the winter rains. The scrub and trees with patches of cultivation are typical of these areas.*

waterfall. Steep rocky sides are seen wherever ice has planed away the projecting spurs of the former valleys. On the southern slopes, and in the drier east, the work of the ice is less impressive. Very small ice-fields are found in the eastern Alps in the High Tauern and in the Ötztal Alps. On the eastern margins, however, there are several rich and fertile basins—like the country around Klagenfurt. Also to the south are the limestone Dolomites, which have been carved into fantastic pinnacles and are highly coloured by mineral veins.

(2) The Balkan Peninsula

The Balkan Peninsula is composed of older rocks that were so much affected by the Alpine earth movements that they form a part of this Tertiary mountain belt in Europe. The Balkan Mountains and the Stara Planina form a southern continuation of the Transylvanian Alps. The two systems are separated by the Danube that incised the deep gorge of the Iron Gates through them as they were raised. A broad, low, limestone plateau now lies between these mountains and the river. On the east, an ancient block of rocks covered by fertile loess, the Dobrogea, forces the Danube to turn north to its swampy delta. The southern part of Bulgaria is the wide and relatively flat-topped

Rhodopi. It is an ancient uplifted block, with poor mountain vegetation above well-forested slopes. Large basins separate it from the northern mountains. The climate and soil of the higher basins is poor, the lower basins are richer, though sometimes dry, grassland. However, they form an important routeway into Europe from Asia Minor along the Maritsa valley.

(3) The Carpathians

One of the most noticeable things on a relief map of Europe is the giant eastward-curving bow of the Carpathians. These mountains extend from Bratislava, at the foot of the crystalline Little Carpathians, in a great crescent reaching the Danube at the Iron Gates. The outermost northern ranges are the beech- and fir-clad Beskids, gently rounded into hills made of soft sedimentary rocks, separated by broad warm valleys. To the south rise the jagged crystalline crests of the majestic High Tatra, rising to over 8,000 feet. The High Tatra is separated from the Low Tatra by the broad and sunny Liptov Depression. The Low Tatra has a crystalline core enveloped in limestone. It rises to above 6,000 feet. Still farther south are the Slovak Ore Mountains (Slovenske Rudohorie), with crystalline, volcanic and limestone rocks. There are several fascinating systems of caverns and deep canyon-like valleys in the limestone. The Eastern Beskids or Forested Carpathians are a narrow line of rounded hills covered by thick forest and crossed by several passes linking the Russian steppes with the Hungarian plains. Volcanic hills lie between the Carpathians and the plains, like the cupola-shaped Vihorlat or the isolated Tokaj Hills, famed for their wine.

The heavily forested hills of the Eastern Carpathians seldom rise above 5,000 feet. Nevertheless, they stand like a great buttress against the open, featureless steppe that stretches eastwards into Asia. An outer line of soft rocks has been eroded into rolling hills which curve south-eastwards to merge into the higher, jagged, crystalline Transylvanian Alps. West of the Eastern Carpathians lies fertile Transylvania, a large depression filled by Tertiary clays and sands, but cut by rivers into hilly country. On its western side the forests of the Bihor Massif block its natural contact with the Hungarian Plains, while Wallachia on the south can be reached only through deep gorges cut by rivers through the Transylvanian Alps. Most famous of these gorges is the Red Tower Pass.

(4) The Dinaric Mountains and the Island World of Greece

Also a part of the Alpine mountain belt, the Dinaric mountain ranges lie parallel to the Adriatic coast. Some have been partly drowned and are now marked as offshore islands, like the attractive Krk, Cres, Dugi Otok and many others. If a traveller journeyed inland from one of the picturesque old ports, he would climb up green slopes broken by dazzlingly white limestone rocks on to a barren limestone plateau. Its surface is rough and pitted by collapsed caverns and swallow holes and riddled by caves. This area, called the *Karst*, forms an effective barrier between the Adriatic and the Danubian plains. Farther east is a geologically more complex area descending in dry, but usually forested, hills and broad, warm, cultivated valleys to the Drava-Sava lowlands. It is an altogether more thickly settled and more prosperous country. On the south, the high and inaccessible Albanian-Macedonian Highlands rise to the rough Prokletije (North Albanian Alps), and include the remote land of Montenegro. Numerous mountain basins mark richer and more populous parts. Some contain lakes. On the west there is the down-faulted (and once malarial) Albanian coastal plain. On the east, a long depression is occupied by the northward-flowing Morava and southward-flowing Vardar rivers, and makes an easy corridor for contacts between the Aegean Sea and the Danube Plains.

The southern Dinaric-Alpine structures are the mountainous Greek Peninsula, where summer aridity is made more noticeable by great areas of dry limestone. The complexity of the geological history of these mountains is reflected not only in the drowned peaks that form the lovely islands of the Aegean and Ionian Seas, but also in old coasts and erosion surfaces now lifted high above sea level. Earth movements still continue, marked by frequent tremors, hot springs and volcanoes. The active volcanoes of the Kaimeni Islands rise from the flooded crater of Thira.

FIG. 45. *White limestone inland from Kamenjack, Jugoslavia. Unfortunately, this limestone is well jointed and allows the rainfall to run away underground, so the surface looks bleak and arid. Nothing can grow on this ground because of the dryness and intense reflected heat.*

The Plains of Hungary

The Plains of Hungary represent a basin deeply filled by Tertiary and younger sediments. These probably rest on ancient rocks related to the Balkan-Rhodopi blocks. Materials weathered from the surrounding Carpathian and Dinaric mountains lie in immense cones deposited at the mountains' feet. The central plains are flat and monotonous. They were once the floor of a vast inland sea drained late in Tertiary times. The only relief on these plains is either forested hills on older horsts, or volcanic hills. The rivers meander in very shallow valleys, and there are belts of dunes formed from the sand blown from the flood plains of the Danube and Tisza. West of the Danube the country is rougher; and at the eastern foot of the Bakony Forest lies the Lake Balaton, now only thirty feet deep, but a shrunken remnant of a former sea. Tihany, a favourite holiday resort, lies on its shore on a prominent volcanic plug. Another remnant of the sea is seen farther west in the Lake of Neusiedl. East of the Danube, the country is dry and steppe-like, with great grainfields or cattle ranches. New irrigation canals increase the fertility of this dry country.

FIG. 46. *Trees line the roads across the Hungarian Plains. They act as windbreaks and provide much-needed shade.*

Peninsular and Island Italy

The Apennine ranges of the Italian Peninsula are formed of folded, soft sedimentary rocks—particularly limestones and clays—with occasional plateaux of older, harder rocks. On the western side are numerous remains of volcanic activity, and here there are also small coastal lowlands. Few parts have a truly alpine character. The soft rocks weather into rugged outlines and suffer from bad landslides. Erosion turns them into "badlands." In summer the clays crack badly, discouraging the growth of trees. Among the old volcanoes of the west are several picturesque crater lakes, e.g. Lakes Bolseno, Bracciano and Albano. Vesuvius, near Naples, is the only active volcano on the mainland of Europe. Most of Italy's limited mineral wealth lies among the older rocks in Tuscany.

The mountains and the flanking hill country are easily penetrated by many depressions and long valleys: these are richer areas in an otherwise poor pastoral country. A good example is the Arno Trough. This once contained a chain of lakes; now it is rich farming country centred around the beautiful city of Florence. Similar areas are the Tiber valley and the Alban Hills. As we travel south the country becomes poorer and the forests—of sweet chestnut, beech and conifers on the higher slopes—thinner. However, there is an area of rich volcanic soil around Vesuvius, and the agricultural country round Naples is densely peopled.

The southern Apennines are built chiefly of isolated blocks between which lie easy routes across the peninsula: like the Via Benevento. Calabria, the "toe of Italy," is a high crystalline plateau, well watered throughout the year. The woods and moorland are in contrast with both the steep valley sides with their thin vegetation and the luxuriant semi-tropical vegetation of the coasts. The "heel of Italy," Apulia, is a high limestone plateau separated from the other mountains by a broad depression. When irrigated it produces reasonable crops, but there has been severe soil erosion caused by cutting too many trees. The rivers are raging torrents in winter, but dry to mere trickles in the arid summer.

Sicily has a structure like the Apennine ranges. It is known for its volcanoes (Etna, Stromboli, Volcano) that are related to a belt of weakness along the sunken block of the Tyrrhenian Sea.

The Plain of Lombardy

Materials eroded from surrounding mountains have filled the northern end of the Adriatic trough to form the once marshy valley of the meandering River Po. Rivers from the higher and moister Alps to the north have carried down a greater amount of material than rivers from the drier and lower Apennines on the south. This has forced the River Po towards the southern side of the plain. On the east, the plain ends in a swampy lagoon-fringed coast in the shallow headwaters of the Adriatic. Since Roman times rivers have brought down enough alluvial material to push the land about a dozen miles out to sea. Venice is built on low islands of silt so formed. Away from the river the land rises in low gravelly terraces. Where these are cut by the water-table a line of springs occurs, providing an important water supply when the rivers run low in summer. Many prosperous towns stand on these terraces, usually near to where routes from the mountains reach the lowland. The monotony of the farming landscape of the plain is broken by the small hills of Monferrato on the west and the volcanic Monti Berici and Euganei in the east.

Sardinia and Corsica

Sardinia and Corsica form part of a vast sunken block of country that underlies the western Mediterranean. In these mountainous islands there are big areas of crystalline rocks as well as remains of folded mountains. In the past the poor maquis-like vegetation and malarial coastlands kept both islands scantily peopled and poor. They have, however, some mineral wealth, and offer great scope as holiday areas. The small island of Elba is known for its iron-ore.

Iberia

The story of the building of Europe that has been broadly outlined so far is repeated in miniature in the large, compact, south-western peninsula known as Iberia. The core is formed by large blocks of Hercynian country, and even older blocks on the west, as in Galicia. To the north, north-east and south-east these blocks are flanked by young Tertiary fold mountains. When, during the Tertiary earthstorms, the older blocks were uplifted and tilted, the young, almost horizontal sediments covering them were cut into deeply by the rivers, especially the Tagus. Today we can visit these desolate tablelands of the Meseta. The Meseta has a general height over 2,000 feet, the highest parts being twice that figure.

The plateau of the Meseta is interrupted by jagged mountains that rise above the general level. The rough surface and deeply incised narrow gorges of the central Sierras now present many obstacles to communication within the peninsula.

To the east, the mountain-ribbed 2,000-foot-high plateau of Old Castile rises to the high limestone Iberian mountains. These have a strange tabular appearance and fall in broad steps to the Ebro Depression. The southern edge of the Meseta is marked by the up-swelling of the Sierra Morena, rising gradually above the general plateau level of New Castile. These rock-strewn, undulating uplands are rich in copper, lead, mercury and some coal. The southern slopes of the Sierra Morena drop more sharply to the structural trough of Andalusia in the Guadalquivir Depression.

FIG. 47. *Sierra de Guadarrama: the barren granite country north of the Sierra.*

The Pyrenees

The Pyrenees are the best-known Tertiary mountain system of Iberia. They extend from the Atlantic to the Mediterranean and form a physical divide between Iberia and France. Like the Alps, they contain a hard crystalline core flanked by softer rocks, mostly limestone. The highest point is Pic d'Aneto at 11,169 feet. Their northern slope declines steeply to the Basin of Aquitaine, but the southern Spanish slope falls gently to a belt of foothills. Today, only the smallest remnants of glaciers remain in the highest central and western ranges. The wetter northern slopes have large forests, but the southern slopes are drier, with grasslands on the lowest parts. Rainfall declines eastwards, away from Atlantic influences. The lack of well-used routes across the mountains arises less from physical obstacles than from the fact that it is easy to go round the coastal flanks of the mountains. On the west, a continuation of the mountain belt is contained in the roughly dissected Basque Hills and the Cantabrian Mountains. A few high passes across them carry routes from the northern coastlands and coastal hills into the interior. North-west Spain is a dissected granite peneplain called Galicia. On its eastern edge lies the basin of the River Sil, a rich farming land known as El Bierzo.

A broad depression named after the River Ebro separates the Pyrenees from the main body of the Meseta. And the Meseta is separated from the Andalusian Mountains by the depression of the Guadalquivir—which ends in the unhealthy coastal marshes of Las Marismas, on the seaward side of

Seville. These Andalusian Mountains, sometimes called the Betic Mountains, appear to be related to the mountains of North Africa and to the Balearic Islands. Remains of a former thick forest are found on the north and west: an area with fairly high rainfall. Towards the east there is only poor grassland. The central part has many basins and depressions, watered by mountain streams, and intensively cultivated where the floors are flat. The Mediterranean coastland also has small irrigated and intensively cultivated lowlands, as at Malaga and Almeria. There are many majestic old Moorish towns in this region, the chief being Granada. The lofty Sierra Nevada has snow most of the year on peaks above 5,000 feet.

As the Meseta slopes towards the west, the rivers flow to the Atlantic in widening valleys. Along the coast is a lowland—the Atlantic or Portuguese Lowland—wider in the south than to the north. The hillier northern part is moister than the south and so attracts more people. The wetter northern and western coasts of Iberia have a thicker and greener vegetation than the dry Mediterranean coasts, so that the landscape looks less harsh.

QUESTIONS

1. Discuss the various features that distinguish Europe from other continents.
2. Name the main mountain-building periods in Europe, giving some of the areas where examples may be found.
3. Describe the effects of glaciation in Europe, giving good examples of (*a*) a glaciated lowland area and (*b*) a glaciated mountain area.

CHAPTER 13

EUROPEAN CLIMATE

The heating or cooling of the earth's surface by the sun causes large movements of air. The mixing of gigantic air masses: cold in the north and warm over the tropics, moist over oceans and dry over land, contributes to the seasonal climatic changes over Europe. Masses of gently outflowing air develop into great permanent or semi-permanent high-pressure areas, converging in the varying low-pressure areas. When warmer air meets colder air it rises above the colder air. Thus develops a belt of turbulent air movement called a *front*. These fronts are most marked along ocean margins and are most marked and active in winter. From them move a string of eastward-moving depressions. The day-to-day position and strength of these fronts and depressions produce *weather*. The seasonal pattern of the relative strengths of the air masses makes the *climate*.

In winter, the contrast between the temperatures over land and sea in north-west Europe is greatest, because of the warming of the coasts by the North Atlantic Drift. It is unusually warm for this high latitude. Over the Atlantic fringe, cool or even warm moist oceanic air mixes with dry and very cold air from Siberia. This creates rapid circulation and disturbance. A trail of depressions is developed, which draws in "wedges" of the warmer and moist, or cold and dry air to produce variable wintry weather. Over the northern Mediterranean cold, dry Siberian air mixes with tropical air, warm and moist from the Atlantic, or warm and dry from Africa. The resultant fronts cause the winter precipitation. In summer, warm, dry continental air from Asia or Africa extends over the Mediterranean, barring the way to moist oceanic air. In Northern Europe, warm or cool, dry continental air mixes with moister air from the Atlantic or Arctic, forming a weak front. Shallow depressions then move weakly across Britain and north-west Europe.

In Europe the change from one climatic type to another is very gradual, so it is difficult to fix definite boundaries between them (Fig. 49). And within each part there are many local variations. Britain lies in a belt affected by the Atlantic, so that it is generally without great extremes. Frequent depressions on the oceanic margins pass over Britain: producing notably variable weather. The southern part of this belt, in northern Portugal, has rainfall spread throughout most of the year, with mild winters and warm

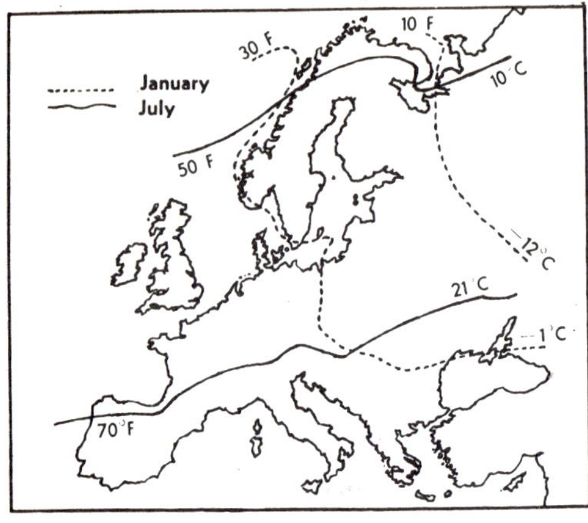

FIG. 48. *Europe: January and July isotherms.*

summers. South of the Duoro River the summer is hot and humid, though without appreciable rain. Thus it is more oppressive than the drier heat of the true Mediterranean climate. If we travel north across Spanish Galicia to the Biscay coast of France, we find winters grow much cooler and summers slightly less warm. In north-west Europe, the coastal lowland is backed by broad plains or low hills, which allow easier penetration by Atlantic air. Sometimes these also permit freezing Siberian air to bring bitter winters, freezing-up rivers and even seashores.

The low dune coast of the Low Countries and Germany is exposed to dangerous storms and floods during abnormally high tides caused by strong on-shore winds.

Despite its northerly position, the Norwegian coast has comparatively mild winters, being warmed by the North Atlantic Drift. High mountains along the coast everywhere receive a heavy rainfall, evenly distributed throughout the year. The shelter they give makes the areas to the east much drier and limits the penetration of winter warmth from the ocean.

The Mediterranean is the most clearly defined climatic region. It has a strong seasonal rhythm. In winter, a trough of low pressure along the northern Mediterranean shore tends to be followed by rain-bringing depressions from the Atlantic. These shed their moisture particularly on the mountains and western sides of peninsulas and islands. The winter on the coast is mild. Blue skies are common in winter, for rain comes mostly in a number of heavy showers. The weather is generally warmer than in Northern Europe, but it may be too cool to sit in the shade on the Riviera even at Easter. Yet in the mountains and plateaux it is cold and there are frequent falls of snow. Local relief may also channel together cold winter air from the high snows: such cold winds are the Mistral of the Rhône valley and the Bora of the north-eastern Adriatic coast. In summer the Mediterranean area is directly under the influence of hot, dry continental air from Asia or Africa. The heat on the coast is tempered by sea breezes by day. There are also persistent north winds, light in the west but strong in the eastern basin, where they are called the Etesian winds. The dryness of the north winds makes them feel cool. Depressions seldom penetrate, and there are blue skies, yet with a parching dryness. The rapid heating and cooling of the land causes a wide daily variation in temperature.

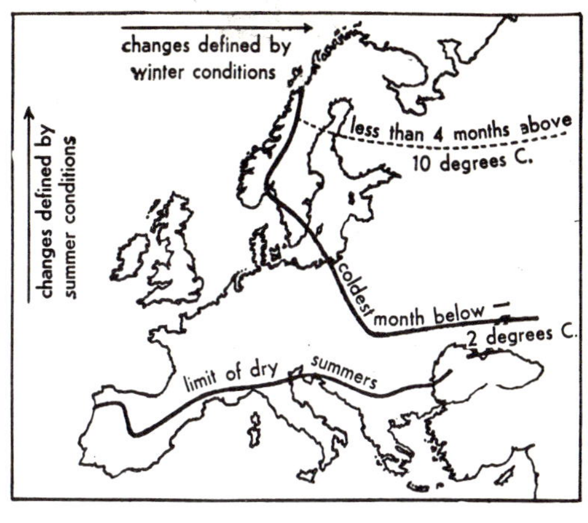

FIG. 49. *Europe: climate distribution.*

The Iberian Peninsula is large enough to have a modified climate of its own. The western and northern borders are moist and have a distinctly Atlantic type of climate. The high plateau has a very low rainfall, with cold sunny winters and hot sunny summers. The climate of the Mediterranean coasts has already been described (see page 158).

A modified Mediterranean climate is also found in the North Italian Plain, with features similar to Central Europe. Its hot summers are broken by thunderstorms, and winter is cold and wet, while autumn is misty and raw. Along the Adriatic and Aegean shores, the Mediterranean climatic régime is restricted to the coast. Because of their height, the mountains produce a more northerly climate in the Balkan Peninsula. There is summer rain, but the winters are often intensely cold and snowy. Summer in the many sheltered basins is very hot, but in winter cold air draining from the surrounding mountains produces very harsh conditions.

The climate of Central Europe is really a transition from the oceanic climate of the Atlantic seaboard to the truly continental conditions over the Russian Plains. There are generally colder winters than in Britain, with frozen rivers and more snow. The summer is markedly warmer, but there are frequent thunderstorms with heavy rain. There are, however, variations from year to year, with quite wet and dull summers or relatively mild winters as fairly common variants. Rainfall and the frequency of the mild spells tend to decrease as one travels eastwards. Hamburg has an almost oceanic climate, but when Berlin is reached there are already colder winters and warmer summers. When one reaches Minsk, continental conditions are well established.

The Central European Hercynian uplands have a considerable variety of climate. The varying relief and the increased amount of sunshine received on south-facing slopes create numerous local climates. The Rhine valley, for example, has warm, early springs, where flowers may bloom long before those on adjacent uplands. The higher areas have long and cold winters, e.g. winter sports continue in the Winterberg Plateau while flowers bloom in the Rhine valley. In other places, relief prevents the entry of rain-bearing air and conditions are rather dry. Parts of central Bohemia and the Moravian Corridor are frequently affected by drought. Surrounded by mountains, the Plains of Hungary have a highly distinctive climate, with very cold winters caused by bitterly cold air flowing from the mountains. On the other hand, the summers are extremely hot, for the mountains stop the entry of both cooler air and moist air. Such hot and dry conditions are akin to the steppes, and the *Puszta* east of the Danube is really a dry, temperate grassland.

A winter ridge of high pressure extending from Central Asia across the Alps, Carpathians and Dinaric mountains is replaced in summer by another tongue of high pressure from the Azores. The mountain belt thus forms a climatic divide between Central Europe and the Mediterranean. Conditions are modified, of course, as one ascends the mountains. In winter, cloudless skies of cold, clear, crisp air are broken only by an occasional passing front. Warm, clear skies—interrupted only by an occasional thunderstorm caused by unequal heating of the mountain air—are typical of summer. In the Alps and Carpathians some peaks are high enough to carry permanent snow, but precipitation tends to decrease east and south. Slopes with a southern aspect are much warmer than north-facing slopes; while eastern mountain basins are dry. Cold air draining to the lowlands in winter leaves the mountains standing in warm, clear air above cold and foggy air.

The mountains and high plateaux of the Balkan Peninsula prevent the entry far inland of Mediterranean climatic influences, but allow deep southward penetration of more northerly features. The summer rainfall nourishes fine forests, but the widespread limestone results in rapid percolation of the water. Thus large areas appear as poor, dry grassland and scrub. In some places melting winter snow and the spring rain raise the water-table creating temporary lakes that dry up again during the summer and autumn.

Everywhere in the mountains, aspect is important. This means that settlements and farmers seek the sunny slopes facing south. North-facing slopes that receive no sun, or slopes that get sun for only a few hours a day, are left as forest and pasture. In many deep valleys there is an unusually early twilight and a rapid fall of temperature as the sun dips below the mountain crest. The angle of slope, determining the amount of sun received, is also an important factor in the development of avalanches. Fortunately, most avalanches follow set paths, and measures can be taken for protection against them. Drainage of cold air in winter and the movement of warm *föhn* winds down mountain slopes also tend to follow set paths and exercise some influence upon the siting of villages and even individual houses.

QUESTIONS

1. What are the main ways in which the climate of Europe is influenced by the North Atlantic Drift?
2. Describe briefly the variation in climatic type experienced across the North European Plain.
3. Discuss the importance of altitude and aspect in the climate of the Alps.

CHAPTER 14

SOILS AND VEGETATION IN EUROPE

VEGETATION is closely related to different types of climate. The climate and the types of rock also affect the development of soils. In Europe, the age-long farming and grazing and the ancient use of forests make it hardly possible to talk of "natural" vegetation. Moreover, man has introduced many plants and trees from other parts of the world. This, coupled with the use of fertilizers and such means of improvement as drainage schemes, has altered the nature of many soils. But despite the strenuous efforts made to overcome the physical conditions of climate and soil, these nevertheless still have a great influence on farming patterns.

Perhaps least modified by man is the tundra of Arctic Europe and its related forms in the high mountains. The plants there are adapted to long, frosty winters and short summers with very long days, and to poor drainage. Small bushes, flowering plants, mosses and occasional stunted trees in sheltered places, are typical. There are large areas of peat bog on the long-frozen subsoil. The tundra is useless for cultivation, but will support reindeer and furry animals. In summer, the tundra is alive with migrant birds from the south, and mosquitoes and midges breed in the stagnant bog-pools. A tundra-like vegetation is found at relatively low elevations on uplands along the Atlantic margins, where there are long, cool and very wet seasons.

To the south, tundra gives way to coniferous forests: the trees becoming larger and more frequent as we travel south, finally developing into the great northern forests. These are the main timber reserves of Europe. Spruce and pine predominate, but to the east there are a larger number of Siberian trees (larch, fir and Siberian spruce). The cool, humid climate and water-logging encourage the development of bogs. Birch and alder tend to grow on burnt or cut-over areas. These sombre, still forests and the dreary bogs are broken in glaciated country by quiet, lonely lakes and small clearings with their log cabins or neat wooden houses. However, the soils are acid and difficult to farm.

Great forests once covered much of Europe. They were composed of many different trees, but deciduous varieties were most common in the south and south-eastern parts. Most of these forests have been cleared: the last great natural forests were cut and colonized in Central Europe in medieval times.

In many parts of the poorer Hercynian uplands and glaciated lowlands forest still occupies over half the area. Large areas have been cut and replanted with quick-growing coniferous trees. In the northern parts of the forest, soils are generally acid with a distinctive ashen-coloured surface mineral layer. Such areas (*podzols*) are not well suited to farming without careful treatment. In the deciduous forest the natural soils are less acid and richer in earthworms, which develop a milder form of humus. They have also been much altered by long cultivation. On limestone rocks are found dark soils, like the productive *rendzina* of Eastern Poland. The most important dark-coloured soils are the black earths (Russian *chernozem*) and related prairie types, that are found extending from the Elbe Basin across Central Europe into the Russian steppe, where they are best developed. They have formed over a fine yellow dust, the loess. This is thought to have been blown out of the great arctic desert that lay round the edge of the retreating ice sheets. The dark soils that overlie this dust are renowned for their fertility, and attract a dense rural population. This is the principal farming belt of modern Europe. A wide variety of arable crops are raised, of which sugar beet, potatoes and grains are most important. The wetter Atlantic margins of the forest belt are extensively used for pasture. This may be carefully managed grassland for dairy or meat cattle, or open pasture on the hills—mostly for sheep. Animals are also grazed on reclaimed coastal marsh. Along the southern fringe of the North European Plain, areas of heathland have been replaced by planted forests. Such forests are also found on coastal dunes and in the once sandy *Landes* of western France.

Where there is a Mediterranean climate, only plants able to avoid or to withstand summer drought can live. Plant devices to stop loss of moisture include small, thick leaves; thick bark; low growth-form; leaves and stems able to hold much moisture; or the ability to exert greater than normal suction. Forests are mostly varieties of pine, cork oak, or the evergreen olive.

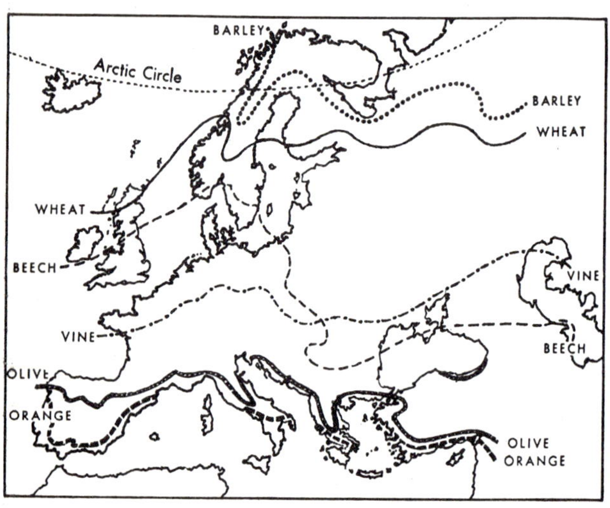

FIG. 50. *Europe: crop and vegetation limits.*

They really appear as parkland and rapidly decay into scrubland. Such scrubs are the thorny thickets of the *Garrigue* and the *Macchia* (*Maquis*). The destruction of the forests during and since classical times by man and his nibbling goats and rooting pigs has caused rapid devegetation and soil-erosion. This has made large tracts useless, and contributed to sedimentation and water-logging in coastal lowlands and depressions, with the consequent development of malarial swamps. Fine forests mentioned by Greek and Roman writers have gone—even some known in medieval Iberia are skeletons of their former glory today, having been cleared by graziers or for metal smelting. Aromatic shrubs and herbs provide food for sheep and goats, but there is too little grass as it withers in summer. Mast for pigs is also less abundant, except in Estramadura in south-west Iberia. Soils, mostly light-coloured red and yellow clay soils, are limy, though poor in humus.

There is a tendency for salt to form easily on the surface (especially in Castile and the Ebro area on younger rocks). Therefore irrigation, which can greatly increase output, has to be practised with great care. During the storms of autumn and winter, the dry, fallow and frequently over-grazed soils are very prone to sheet erosion and gullying.

Much of the hill lands of the Mediterranean region is particularly suited to tree and bush crops. The olive and citrus fruits are admirable crops; the vine is also widely grown.

QUESTIONS

1. Why is it no longer possible to speak of "natural vegetation" in Europe?
2. Write brief notes on the following: tundra, coniferous and deciduous forests, tundra, loess (including chernozem and rendzina soils), podzols.

CHAPTER 15

PEOPLE OF EUROPE

EUROPE contains about 452 million people, a seventh of the population of the world. Compared to other continents it is remarkably densely settled. The population of Europe grew most rapidly in the nineteenth and early twentieth centuries as industrialization spread across the continent from west to east. The rise of industry marked a rise in the proportion of people living in towns and brought a drift from the countryside. The birthrate went up (more people were born) and the deathrate went down (fewer people died —because of the development of medical science).

Population growth has tended to slow down in Europe at the same time as it has speeded up in Africa and Asia. Thus, the proportion of people living in Europe is tending to form a smaller proportion of the world total. Probably about 250 million Europeans and their descendants now live outside Europe. During the nineteenth century, and even into the twentieth century, large numbers of people left Europe to seek greater opportunities elsewhere. The greatest proportion of these emigrants went to North America, but others, particularly those from Mediterranean countries, chose South America. Settlers from the British Isles went notably to the U.S.A., Canada, Australia and New Zealand. Most emigrants went from Ireland, Germany and Scandinavia. There was a large but lesser known movement of people from European Russia into Siberia.

The people of Europe have very mixed racial origins. The north Europeans tend to be tall, long-headed, blond and light-skinned. Around the Mediterranean the colouring is generally much darker. People tend to be shorter, heads either round or long. Central Europe has a large proportion of people of medium build, round to broad heads and brunette colouring.

These are, of course, highly generalized distributions. When we think of Norwegians as tall blond people we forget the pocket of shorter, darker people found around Bergen, possibly descended from Hanseatic merchants. So many different migrations have come to Britain in the course of time that if we look along a crowded street it is hard to find a single feature in common: the British are an extremely mixed population.

About a third of Europe's population speak one of the languages of the Germanic group: for example, English, German, Dutch, Danish, Swedish, etc. Another third speak a Slavonic tongue: Russian, Polish, Czech, Serbo-

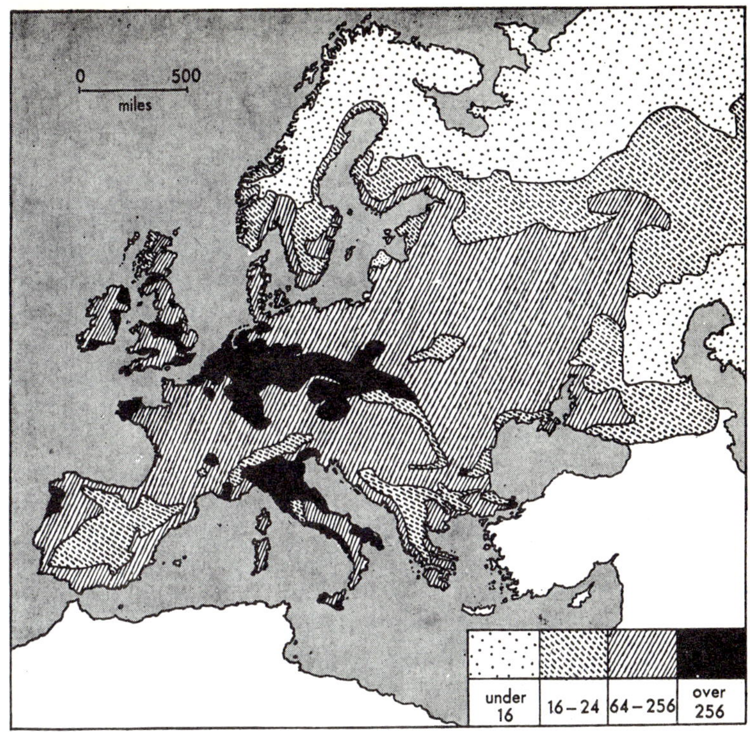

FIG. 51. *Europe: population density per square mile.*

Croat. A quarter of all Europeans speak Romance languages: French, Italian, Spanish, Portuguese, Romanian. The small Ural-Altaic group speaks Magyar (Hungarian), Finnish and Estonian: which are derived from within western Asia, being related to various Siberian languages. The Celtic and Basque tongues are remains of older and once more widely spoken languages now swamped by Germanic and Romance speakers. Modern Greek bears little relation to classical Greek; Latin has been displaced as an international language. The cultural inheritance of Rome left the Latin alphabet dominant. Where the Byzantine (the "Eastern" or "Greek") church was powerful, the Cyrillic alphabet, related to Greek characters, is used, notably in the Russian languages and Serb and Bulgarian. The old Nordic runic alphabet of the Vikings has died out.

Transport and Routeways in Europe

For ages movement in Europe was by water: along the coasts and rivers, crossing from one system to another by portages, or across the narrow seas. Land routes usually focused on waterways. By medieval times the wonderfully engineered and well-maintained Roman roads had been allowed to decay. The Roman roads were military roads, and their decay was inevitable in the absence of any international power dedicated to the maintenance of peace on a basis of strength of arms.

In general, roads were not only bad but became progressively worse until recent times. The real revolution came with the railway. This rescued many areas from stagnation by providing quick, cheap and easy transport. The first short local lines were built in the 1830s; during the next decade lines began to link principal cities. It was in the 1850s that the first main trunklines began to appear in France, Germany and Italy, while Russia (Moscow) was joined to Poland (Warsaw) and Austria (Vienna). Within fifty years, the network as it is today was basically complete. The building of secondary lines continued in Western Europe, and main lines were built in Northern and Eastern Europe up to the First World War. An example is the railway carrying Swedish iron ore to the Norwegian port of Narvik (1903). Between the World Wars few new ines were built, apart from those needed as the result of political changes. An example is the special coal-carrying railway from the Polish coalmines in Silesia to the new port of Gdynia on the Baltic, completed in 1924. In Norway, the Oslo-Bödö line, begun in 1923, was not finished until 1962. Railways built before 1914 naturally fitted into the political geography of the period. After the war, recast national boundaries left many states with railway systems only incompletely expressing their needs. The railways of Austria-Hungary had been focused upon Vienna and Budapest. After 1918, they were shared between several new states. Former trunk routes were reduced to secondary routes or disrupted by new boundaries—like the vital Vienna-Trieste line. Even today, the Polish railway network reflects pre-1914 conditions. In former German territory there are many lines, but in what was old Russian Poland the railway network is comparatively sparse, as in Russia itself. Since 1945 some new railways have been built to complete the systems of Eastern Europe, mostly short linking lines. In Western Europe, despite building to improve junctions and make short cuts, a growing length of secondary line has been closed.

Perhaps the most exciting episode in European railway building came at the end of the last and during the first quarter of the present century, with the completion of the great alpine tunnels.

The railway network is thickest in Western Europe, and thins towards the east, where it forms a wide grid of lines across the country. In Eastern and Northern Europe, and in mountainous areas, a large proportion of the route is single-track. The gauge of railway track over most of Europe is the same as in Britain (4 ft. 8½ in.); but wider in Ireland, Spain and Portugal (5 ft. 3 in.), and in Russia and Finland (5 ft. 0 in.). In Switzerland and Jugoslavia narrow-gauge railway track is quite commonly used.

The network of good motor roads is thickest also in Western Europe, where there is the highest density of motor vehicles. Military considerations have played a large part in this development: particularly in France and, more obviously, in Germany. The poplars lining many French roads were planted

FIG. 52. *Autostrada del Sole—one of Europe's new international network of motor roads.*

to provide shade for marching soldiers. The original German autobahn system was built before the Second World War to be military roads on the old Roman lines (and also to relieve unemployment). Good roads are still rare in Eastern Europe, where private ownership of motor vehicles is small compared with the West, and outside the towns motor traffic is thin. Many country roads are muddy in winter and dusty in summer. New international arterial routes for fast motor traffic are being built all over the Continent, stimulated by the European Economic Community, with its emphasis on unifying Europe. Many of these great new roads—like the Autostrada del Sole—are triumphs of modern engineering.

Today, many airline services not only link together the main European cities, but join Europe to the rest of the world.

There is an ever-increasing demand for petroleum and oil fuels. European production of natural or artificial petroleum are completely insufficient, and most of Europe's oil is shipped from abroad. To ease the problem of distributing the oil from the refineries to the various consumers, pipeline building has begun. Most lines are in Western Europe, but a line has been built from the Russian Ural-Volga oilfield to Eastern Europe.

Much of the traffic in heavy, bulky goods is handled by the railways, but river and canal boats, much larger than those seen in Britain, are also important. Duisburg on the Rhine is one of the world's largest river ports on one of the world's busiest rivers. It is linked by large canals to waterways in the Low Countries and France, and to the Elbe and Eastern Europe by the Mittelland Canal and similar canals. Traffic on the River Elbe to Hamburg has declined because of the political divison of Eastern and Western Europe—traffic being diverted to the Oder and Vistula.

Heavy traffic is carried by the rivers of the Seine basin and by the Scheldt and Meuse (Maas), which are also joined together by canals. Important schemes to make the Rhône more easily navigable have been begun, with the aim of enabling barges to reach Lake Geneva. Closer economic links between France and Germany have called for the canalization of the winding River Mosel (Moselle). Traffic on the Danube, which declined after 1918, and again after 1945, is reviving. It is once again possible to go by passenger ship from Vienna to Izmail.

Many mountain rivers, particularly in Scandinavia and the Carpathians, are used for floating timber. Large log rafts may be seen on the Vistula.

Besides ships that handle the coastal and entrepôt trade of Europe, the ports of Europe handle vast quantities of goods being imported from or exported to other parts of the world. Vast new port installations—Europoort—are being built at Rotterdam. Together with the existing facilities at Rotterdam they will make up the largest port area in the world. Europoort will be the European Economic Community's largest and most important

FIG. 53. *One of the world's busiest waterways: the Rhine.*

export-import centre. Among the ports being expanded in Eastern Europe are Gdynia-Gdansk in Poland and Rostock in Eastern Germany.

Europe's Mineral Wealth

In the early days of the Industrial Revolution, industry used local mineral resources. Nowadays an increasing percentage of the raw materials has to be imported from outside Europe. This is because European deposits have become inadequate or entirely exhausted. Western Germany, France and Belgium are the most important refiners of imported ores, and Italy is developing rapidly along these lines. Coal is plentiful, the bulk coming from the big fields along the northern edge of the Hercynian massifs. The largest field is the Ruhr, producing valuable coking coals for iron smelting. Another important German field is the Saar. The Belgian coalfield in the Meuse valley is hard to work, and already a part has been closed, and Belgium depends increasingly on the northern Kempenland field. The northern fields of France are also difficult to work and its other coalfields are small. As a major industrial country, France is able to supply less than a half of its coal requirements. The main producer in Eastern Europe is Poland. The fields are around Walbrzych and in Upper Silesia, the southern margin of which is worked in Czechoslovakia. Poland and Germany are the principal exporters of coal in Europe. Britain was once the main supplier, but now exports very little. The United States exports a considerable quantity to Western Europe. The inferior brown coal (lignite) is found in large quantities in Eastern Germany, where it is used for power stations and as the basis of the chemical industry. It is also mined near Cologne in Western Germany and in Bohemia (Czechoslovakia), Poland and Austria. A little comes from Southern France.

Coal now supplies less than half the energy needed in Western Europe, though its importance as a source of chemical byproducts (including rayon and nylon) is greater than ever. Its place as a fuel is rapidly being taken by petroleum. Western Europe largely lacks its own petroleum deposits (though there are small oilfields in Northern Germany, Southern France and Eastern Poland), and the bulk is imported from the Middle East. More petrol will eventually come from North Africa: Algeria and Libya. Reserves of oil may also be discovered under the North Sea, but so far exploration is at an early stage. Italy imports petrol from Russia, the main supplier to Eastern Europe. The Romanian oilfields around Ploeisti were once big producers, but their output has fallen, and the fields are in the process of redevelopment. Natural gas is of increasing importance. Big natural reservoirs have recently been discovered under Holland and the North Sea. Italy has its own reserves. France produces gas on the Lacq oilfields, where the high sulphur content of the petroleum makes France a major world sulphur producer. Petroleum, like coal, is also an importance source of chemical byproducts.

Electricity production constantly increases. Thermal power generation is important, and water power in mountainous countries: Norway, Sweden, Switzerland and Italy. France is increasing its hydro-electricity output. Nuclear power is less developed in most European countries than it is in Britain, though stations have been built, notably in France and Italy.

FIG. 54. *Desulphurication plant on the Lacq oilfields, France.*

FIG. 55. *Iron ore mines at Kiruna in the Swedish Arctic produce 20 million tons of ore annually.*

There are many small deposits of a wide range of metals, but today only bauxite (in southern France and Hungary) and iron ore (especially in France, Sweden and Spain) are worked on a large scale. Since 1945 uranium mining has been important in the Erzgebirge and other places in Eastern Europe. Italy and Spain are important mercury producers, while the mineral salts in north Germany have contributed to the country's large chemicals industry.

Land Use in Europe

If we travel widely in Europe, we see that the great diversity of climate and relief, soil and vegetation, is reflected in the many different types of farming. Some of the most important crops are not native to Europe but have been introduced from other continents. The potato, mainstay of diet over large areas, was brought from the Americas. Many fruits grown on the Mediterranean coasts are native to Africa or Asia. The limits of farming have spread into formerly unusable areas by improvements in technique and strains of crops. Many serious deficiencies and weaknesses nevertheless remain. Despite a large trade in food between farming and industrial countries much food must be imported from others outside Europe. The spread of industry and rising living standards have not been equalled by a similar rise in farm output sufficient to make Europe independent of such imports.

As we travel east and south so the proportion of people engaged in farming rises, and the output per farm worker and the technical level fall. In north-western Europe, notably in Denmark and Holland, the worker's output is highest and crop yields per acre greatest. There is also the best output of animal products. In this area there is the most careful control of plant strains and animal breeds, and most intense use of fertilizers and machinery. Even in such otherwise advanced countries as France and Germany villages still often divide their fields into strips shared among peasant farmers. To remove this hindrance to modern farming, compact holdings workable by modern methods are devised by re-allotment of land. And individual owners co-operate together to increase efficiency, buy or hire machinery and generally improve the yield and quality of the crops and, consequently, the prices they receive for them.

The differences between the Communist countries of Eastern Europe and the capitalist democracies of the West are very marked in their agricultural systems. In Communist countries, private ownership of farms is replaced by state farms, collectives or co-operatives. State farms are similar to the private farms of the West, except that the labourers (who have no land of their own) work for the State, whose control is exercised by an appointed manager. Both managers and labourers are paid fixed wages, and all the farm's profits pass directly to the State. On co-operative farms and collectives control is in the hands of committees of workers. On co-operatives the labourers work individual areas of land themselves, though machinery is owned, and such major decisions as choice of crop varieties, are made in common. The produce is sold to the State. On collectives the individual farm workers have only very small plots of land for themselves—used for vegetables. The bulk of the land is held in common and worked in common. Once again the produce is sold to the State, but, in the case of collectives, the money received is shared among all the farm workers, according to their contribution in labour.

The collective farms have proved both very unpopular among the peasants and inefficient, and in Poland and Jugoslavia the co-operative system is preferred. Even in the Soviet Union (where collective farming originated) there has been a tendency to switch to state farms.

Southern Germany, central France and most of the Italian Peninsula could all profit from larger holdings, more machines and more fertilizers. Too many people try to live from the soil, and other forms of employment are needed in rural areas. In Italy and Spain, a large increase in the farming area could be achieved by combating soil erosion or by irrigation. Similar problems are also found in eastern and south-eastern Europe.

The Atlantic coastlands are mainly used for raising animals for meat and milk, but the sunny and relatively moist Atlantic slopes make Portugal a famous wine producer, as is the Portuguese island of Madeira. The rich alluvial soils of Belgium and Holland grow vegetables and flowers. As we travel inland, the importance of arable farming grows. In the north, cereals like rye and barley are grown; but farther south, wheat and sugar beet become important. These are grown notably on fertile loessic soils along the northern edge of the Hercynian uplands. Everywhere the potato is important for food, fodder, alcohol or starch. The eastern loess belt in Polish Galicia and the Russian black earth lands is a great grain country, with important sugar beet and potato crops as well. The wet meadows of the north-eastern forest lands grow rye, flax and hay and also produce honey. The rich warm plains of Hungary produce wheat and coarse grains like maize, but east of the River Tisza they are too dry, unless irrigated. Consequently the Hungarians have built a large canal to carry water to the Hortobagy Steppe, once the home of shepherds and horse breeders. Throughout the Balkans and Wallachia maize and wheat are grown, along with tobacco, limes, and sunflowers—principally for oil. Mediterranean Europe and the Dinaric and Balkan mountains have

FIG. 56. *Portugal: the Duoro vineyards.*

large pastoral areas. Sheep and goats are found everywhere in the eroded mountains of Italy and Spain. The Mediterranean coastlands or watered mountain basins can produce rich crops of valuable citrus fruits, or rice in delta areas. The olive is perhaps the most typical crop of the Mediterranean lands. Early vegetables grown in the warm, moist spring are exported, to feed the industrial populations of Western Europe.

The vine is grown throughout Europe south of about 50° N. Almost half the world's wine comes from France and Italy. Spain, Portugal (including Madeira) and Jugoslavia are also important producers. Germany is known more for the quality than the quantity of its wine. Despite her big production, France imports much wine from Algeria for everyday purposes and blending. The most famous wines of Eastern Europe come from the loess-covered slopes of the Tokaj Hills in Hungary. Oddly enough, the best quality wine comes from very poor soil: chalky, or stony, or sandy. However, good drainage is vital. The vine needs much sunshine so south-facing slopes are best, but rainfall is important in the earlier growing season. Areas prone to late frosts are unsuitable. Apart from domestic consumption, wine and brandy is a very important export for wine-producing countries.

Fishing

There are good fishing grounds in the North Sea, the Atlantic, the Baltic, the Mediterranean, the northern part of the Black Sea and in Arctic waters. Europe has the most indented coast of any continent, and there are many hundreds of fishing ports, big and small. In northern waters herring, cod and flat fish are the main catch. In southern waters pilchards, sardines and tunny are important, and the bulk are canned. (The rivers also provide much fish—particularly salmon and eel from north-west Europe.)

The inshore waters of all the European coasts provide plentiful shellfish. These are mostly eaten quickly, but an increasing volume of fish caught is preserved in some way: by canning, salting (notably cod and herring), smoking (herring and haddock) or freezing.

QUESTIONS

1. Describe the chief linguistic and racial groups of Europe, and discuss the main European population trends since the nineteenth century.
2. What were the main stages in the building of the European railway network? In what ways does this network reflect the political development of Europe?
3. Describe Europe's main mineral resources and the extent to which they are exploited.

CHAPTER 16

COUNTRIES OF EUROPE

France

FRANCE is our nearest European neighbour. In area it is twice as large as the British Isles, but the population is only 50 million. This is very evenly distributed over the country though population is sparse in the Alps, the Jura, the Pyrenees and the less hospitable parts of the Massif Central. Paris, the capital, on the River Seine, is the only French city with over a million inhabitants. The town itself has 2·6 million people, compared with the 889,000 of Marseilles, the second city of France. Together the northern industrial towns have a population of under 1 million. The Paris conurbation contains over 8·2 million. Paris is, of course, the artistic and cultural centre of France and Europe, and possibly of the world. It attracts hundreds of thousands of students, tourists and other visitors.

One in five of the French labour force works on the land. Almost a third of France is cultivated, and a quarter is pasture. Standards of farming, particularly in the south and centre, are low. The broad and generally fertile basins of the major rivers are extensively cultivated. The diverse character of the Paris Basin is reflected in farming: Beauce grows wheat, while northern parts are known for sugar beet. Sheep are kept in the poorer parts and there are large forests. North-eastern France grows potatoes and barley as well as sugar beet, while Normandy's rich meadows are noted for butter, milk and cheese—such as Camembert. Brittany grows early vegetables and flowers. Major wine areas are the Garonne basin around Bordeaux, the Burgundy region round Mâcon and Beaune, the Champagne country north-east of Paris, Alsace and the Loire valley (also noted for its fruit and fresh-water fish). Provence and Mediterranean France generally are known for early vegetables and flowers, fruit, nuts and olives. Much mediocre wine is also produced.

The mountains are livestock-rearing areas: in the Pyrenees and Alpes Maritimes sheep are important, as they are in the Massif Central (particularly in the limestone Causses, their milk being used to make Roquefort cheese). The Savoy Alps and the Jura are noted for dairy cattle. In the Massif Central, Charolais is noted for its white beef cattle. The patches of volcanic soil of the Massif Central are noted for good grazing, but much is poor grass in the Morvan and Vivarais. In the north-west, around Limoges, sweet chestnuts are grown. East of the forested Vosges on the Rhine Rift Valley lies the garden-like plain of Alsace, known for its orchards, beer and wine.

France has insufficient coal, particularly for metallurgical use, and has to import it. The main coalfield is around Lille, Lens and Valenciennes in the north, but small deposits are found in the south and in the Massif Central. Lorraine in the east is rich in iron ore—notably around Longwy, Briey and Nancy. Iron and steel works are concentrated in Lorraine, e.g. the Orne and Fentsch valleys, Thionville, Maizières; and on the northern coalfield at Denain, Douai, Valenciennes. Steel is also made at St. Etienne, St. Nazaire, Le Creusot. Petrol is refined at St. Nazaire and near Marseille, being distributed by a growing network of pipelines. There is a very large chemicals industry: both oil refineries and steelworks produce chemicals as a byproduct, and the oil found at Lacq is rich in sulphur. Bauxite (aluminium ore) is mined in southern and south-eastern France, and in fact is named after the town of Les Baux in Provence, where it was first discovered.

Lack of coal has stimulated development of hydro-electric power, as has France's supplies of aluminium ore (which needs abundant cheap electrical power to refine: hydro-electricity is the cheapest to produce of all). This power is generated principally in the Alps and Massif Central. This new source of power, together with petrol refining and its chemical by-products, have attracted industry to the region around Lyon and Marseille.

FIG. 57. *Power supplies have always been a major French problem. To augment coal-generated electricity, hydro-electric stations like this one on the Durance have been built. They are very important in supplying electricity to the growing industrial capacity of southern France.*

The Lille-Roubaix-Tourcoing area and Rouen remain the centre of the French textile industry, as they have been for hundreds of years. Today, an increasing percentage of textile production is in man-made fibres. This applies both to the textile centres of the north and the traditional silk industry of Lyons, in the south.

Grenoble is a centre of the electro-metallurgical industry; Paris and Rennes of the car industry; Clermont-Ferrand of tyre manufacture; Douai, machine tools; Strasbourg, electrical goods; and St. Nazaire and Brest, shipbuilding. In recent years the aircraft industry has grown into one of France's most successful exporters. French civil and military aircraft have been sold all over the world. The industry is widely distributed, but the two towns most concerned are St. Cloud, near Paris (military aircraft), and Toulouse, in the south (civil aircraft).

Netherlands

The Netherlands lie on the low North Sea coast, two-fifths being below sea level and half having to be protected against flooding. The history of the country is closely bound with land reclamation and protecting the coast and reclaimed areas from sea and river flooding. The most famous modern example of this struggle is the reclamation of the former Zuider Zee, beginning in 1920. Part of the Zuider Zee remains as a freshwater lake, its name having been changed to Ijsselmeer, the rest is drained land. Since the terrible floods of 1953 work has also been on foot on a far greater project than even the Zuider Zee reclamation. This entails a barrage of dams and sluices round the whole Scheldt estuary area of Zeeland. This will, simultaneously, end the flood danger in this most exposed area of the Dutch coastline, improve

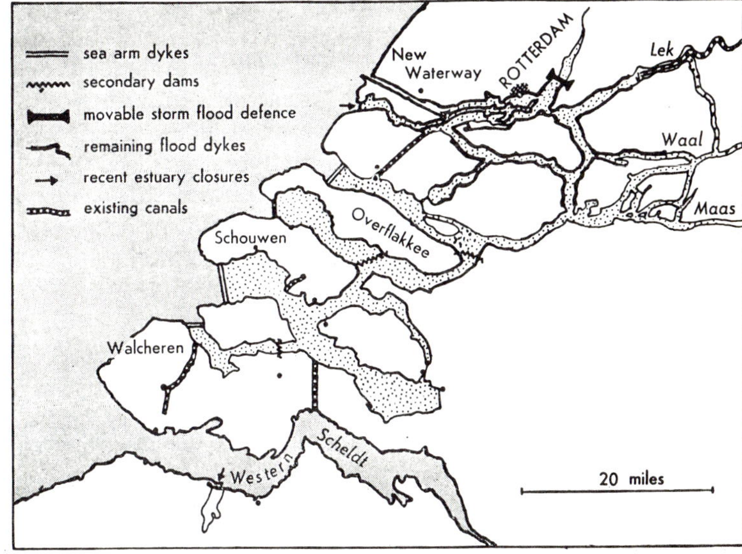

FIG. 58. *The Delta Project, Netherlands.*

FIG. 59. *Stage in the construction of part of the Delta Project.*

communications which have always been difficult in the region—thereby hampering settlement and development and enable the area to be opened up for development, and relieve the present chronic shortage of fresh water in the area. (Much ground water is now salty. As fresh water has been pumped from reclaimed areas, it has been replaced underground by sea-water seeping landwards, which can prove detrimental to farming.)

The population of the Netherlands is 12·7 million, the area being one of the most densely populated in Europe. Fifteen cities have more than 100,000 people, and three more than 500,000. Seven-tenths of the country is farming land, two-fifths arable and three-fifths pasture. The rich, moist meadows support a large dairy industry—famous for milk, butter and cheese.

Vegetables, flowers and bulbs are grown on a large scale (notably in Holland province and the Aalsmeer and Haarlem districts). Most of the arable land is in the north-east, south and south-west. Pig breeding is important around Rotterdam and Utrecht.

Coal is mined in Limburg, near Maastricht; salt at Hengelo and Delfzijl. Oil is worked near the border with Germany, and large deposits of natural gas have been located under Groningen province. Imported crude oil is refined near Rotterdam, and a chemical industry, based largely on oil and other imported materials, is also established there. The shipbuilding and engineering industries are established between Schiedam and Dordrecht and at Amsterdam. Utrecht is another town with engineering industries, and the Philips works at Eindhoven is one of the world's largest centres of electrical engineering. Arnhem is the centre of (imported) metal ore refining.

The chief port of the Netherlands is Rotterdam, which is being enlarged by the construction of the adjacent Europoort. Europoort is specifically designed to serve the import-export trade of the European Economic Community. The Rotterdam-Europoort area combined make up the largest dock area in the world.

Imported raw materials, many of them coming from former colonial territories, support the confectionery, tobacco and diamond industries. These industries, which are economically very important, are centred chiefly on Rotterdam, but they are also established at Amsterdam, which is connected to the open sea by the North Sea Canal.

Luxembourg

The Grand Duchy of Luxembourg is a small state in the Ardennes. It is closely associated with Belgium and Holland in the Benelux economic union. The Benelux association is important, in that it was the "pilot" for the larger and later European Economic Community. It has 335,000 people, of whom 77,000 live in Luxembourg City. The most fertile area is the southern Gutland or Bon Pays of the Lorraine scarplands. Forestry and patches of cultivation are found in the Ardennes to the north. Surplus farm produce goes to Belgium and Germany. Iron ore is mined in the south and coal is imported from Germany, big steelworks are found in this area.

Belgium

Belgium was created after the Napoleonic wars. It is densely settled and has a population of 9·6 million. Brussels, the capital, with its suburbs, has over a million people. The Flemish people in the north, the slight majority, speak a language related to Dutch, but the Walloons south of a line from Kortrijk through Waterloo to Liège, speak a French dialect. There has been a certain amount of friction between the two communities in recent years.

The coast and its marshes are less threatened by the sea than in the Netherlands. Central Belgium is a rolling loessic plain rising towards the Meuse. In the south-east are the forested uplands of the Ardennes. There are coal measures along the Meuse valley and in the Kempenland heaths of the north, but Belgium has to import the bulk of her raw materials. Heavy industry is mostly concentrated in the Meuse valley (e.g. Seraing, Liège, Charleroi). The Flanders plain is a great textile region (notably for carpets). Non-ferrous imported ores are refined (Liège and the Kempenland). Engineering is found at Seraing (locomotives), Antwerp, Ostende and Hoboken (shipbuilding), Brussels, Liège and Charleroi. Ostende is a major cross-channel port.

Belgium is intensely cultivated and, except for wheat and cheese, is self-supporting as far as food supplies are concerned. Eggs, vegetables and flowers are exported. The chief wheat-growing areas are Hainaut and Brabant; rye is the principal crop in the poorer Kempenland; while in East and West Flanders and near Antwerp potatoes and sugar beet are grown. Cattle are kept everywhere, but sheep are mainly confined to the northern heaths. Pigs are the typical livestock on most Belgian smallholdings, especially in Brabant. Ghent, Brussels and Antwerp are the main market-gardening areas.

Norway

Norway, on the western side of the Scandinavian peninsula, has the benefit of the warmer waters of the North Atlantic Drift. Consequently its winters are milder than those of Sweden, and its coasts are never frozen. Most of Norway is composed of mountains, and its population is only 3·8 million, half a million of whom live in Oslo, the capital.

With the population concentrated along the coast, fishing is a major industry (employing about one quarter of the working population) and whaling is very important, notably in Antarctic waters. A great deal of the fish caught is processed or frozen for export.

Farming is notably pastoral, with animals stall-fed in winter and grazed in summer on mountain pastures which may be many miles from the valley farms. Reindeer are the mainstay of the economy in northern Norway.

Arable farming and hayland is found only in the deeply incised valleys or the narrow coastal strips by the fiords. Mountain meadows, called *saeters*, were formerly much used to provide summer grazing for herds, but their use is declining.

Norway has the third largest merchant fleet in the world, after Britain and the United States, with many ships for specialized freights: for example, oil and timber. Like Sweden, most of Norway's power comes from hydro-electric stations. Two-fifths of all the energy produced goes to supply Norway's principal industries: chemicals, metallurgy, aluminium (made from imported ores) and fertilizers. Coal is mined in Spitsbergen.

The port of Narvik is the terminal of the important iron ore-carrying railway from the Arctic mines of Sweden. Unlike Swedish ports, Narvik is ice-free all the year.

Sweden

Sweden is the largest and industrially most important Scandinavian country. It has 7·9 million people, over a million of whom live in the capital, Stockholm, and its suburbs. Western Sweden is mountainous, and the rivers of this region provide 92 per cent of her energy; most of the sites for hydro-electric power stations have been utilized; future stations will be thermal.

Over half of Sweden is covered by forests and paper, woodpulp, timber, cellulose, etc., are important, notably around Sundsvall and in other coastal areas.

Within the Arctic Circle, at Kiruna, vast quantities of rich iron ore are mined. In summer most of this ore is shipped out via the Gulf of Bothnia, but when this is frozen in winter Narvik, in Norway, is used. Swedish industry is extremely highly developed, with very high export levels. Emphasis is put on very high quality of design and manufacture, for which Swedish goods have a world-wide reputation. Most of Sweden's iron and steel comes from

Domnarvet. The biggest centres of Sweden's manufacturing industries are: Göteborg (ball-bearings, shipbuilding, cars, precision instruments) and Malmö (shipbuilding and textiles). Other centres of Swedish industry are Norrköping (cars), Örebro (railway equipment), Bofors (armaments) and Falun (chemicals and metals). The products of the Swedish car industry have been very well received abroad, and Sweden now exports over 140,000 cars a year.

Less than one-eighth of Sweden can be farmed. Wheat, sugar beet and other cereal and root crops are grown in the south (western Skane) and also, to a lesser extent, in central Sweden. Northern Sweden is mainly a livestock-rearing area, but rye and oats are grown in the north, where special crop varieties suitable for the very short growing season have been evolved. Skane is a centre of pig farming, as sugar beet waste is available for fodder. In the far north and in the mountains semi-nomadic Lapps breed reindeer. Most Swedish farmers supplement their incomes by exploiting the forests.

Denmark

Denmark is a generally flat, sometimes rolling country, underlain by boulder clay, and spanning the Jutland peninsula and the adjacent islands. (The enormous Arctic island of Greenland is also a province of Denmark. It is largely uninhabited.)

Denmark's population of 4·8 million is the most densely concentrated in Scandinavia. The historical development of European trade and its position on the Baltic have made Copenhagen one of the most important commercial centres in Northern Europe. With a population of 1,378,000 including suburbs it is the largest city (in extent) in Scandinavia.

Except in the coastal areas, and on soils formed from the youngest boulder clays, Denmark is not a particularly fertile country. However, the co-operative farming system produces very high yields of crops and livestock.

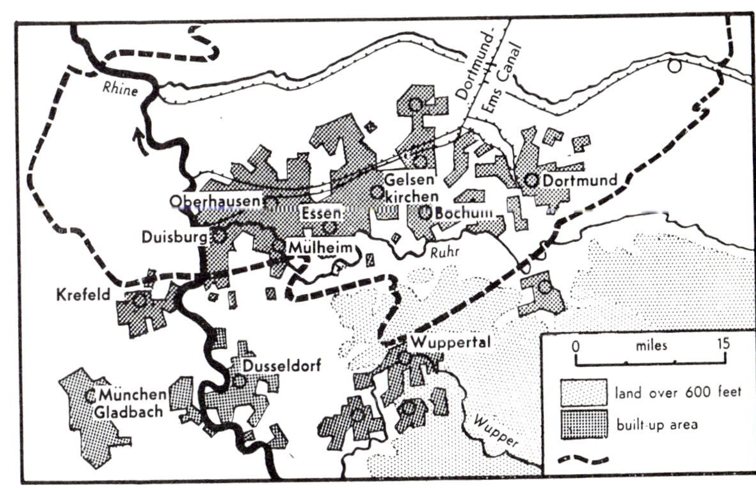

FIG. 60. *Simplified map of the Ruhr, Western Germany.*

Pigs, cattle (particularly dairy cattle) and poultry are the basis of the Danish agricultural industry; and pork and bacon, butter, milk, cheese and eggs constitute the country's chief exports.

Much fertilizer and animal feedingstuff is produced, but industry is dominated by food processing: the making of butter and cheese, canning, bacon-curing and sugar manufacture (from beet).

There are also important shipyards at Copenhagen and Odense. In recent years Danish domestic goods: furniture, pottery, glassware, textiles etc., in common with similar products from the other Scandinavian countries and Finland, have attracted world-wide demand, and have become important exports.

Unlike other Scandinavian countries Denmark is too flat to be able to produce hydro-electric power, and she "imports" electricity from Sweden.

Finland

Finland is usually associated with Scandinavia though, in fact, the Finns are of very different racial stock, and the Finnish language is completely different from any other in Scandinavia. However, some Swedes live in the south-west coastal areas of Finland, and both Swedish and Finnish are official languages.

The population of Finland is 4·7 million, 696,000 of whom live in Helsinki, the capital.

Over 10 per cent of Finland is lake and river, and only 9 per cent is cultivated: the rest is forest, swamp and lake. The Finnish climate is markedly continental, and the coasts are frozen in winter.

From the twelfth century until 1809 Finland was under Swedish control. From then until 1918 it belonged to Russia.

Farms are situated in forest clearings. In the east they are primitive by modern standards, but they are much more up-to-date in the south-west. Late or early frosts are a constant danger and can cause serious crop failure. Oats, wheat and barley are the main grains, and potatoes the principal root crop. Much hay is produced for animal fodder during the long winter months. Wolves and lynxes are a menace to animal farming in the north and east, though their furs are valuable.

Ninety per cent of Finland's exports are forest products: wood and wood products, cardboard, wood pulp, paper, cellulose. But Finland's best-known exports are her domestic goods: glassware, pottery, furniture and textiles, which have an international reputation for design and quality.

At Outokumpu there is the largest copper field in Europe. The ore is very complex, and other minerals, such as nickel and cobalt, are extracted from it. There is also some iron and molybdenum.

The principal source of power is hydro-electricity.

Iceland

Iceland is a large island in the North Atlantic structurally, culturally and economically associated with Europe. It is really a gigantic desolate plateau of young volcanic rocks, with a rough coastline. There are many volcanoes and hot springs (geysers) as well as icefields and glaciers. It was first settled by Vikings in the ninth century, and from 1381 to 1944, though completely independent, had the same sovereign as Denmark. In 1944 it became an independent republic. The 200,000 people live mainly on the coast, the interior and southern coast being largely uninhabited. Reykjavik, the capital, has a population of 80,000. Most people depend on fishing or sheep farming on poor upland pasture. Sixth-sevenths of the land is, however, unproductive. Industry is mostly concerned with fish processing. Much use is made of the country's hot springs for domestic heating.

Germany

Modern Germany comprises two states: the Federal Republic (West Germany, capital Bonn) and the German Democratic Republic (East Germany, capital Berlin-Pankow). The Western republic has a population of 59·9 million people, but only 17 million live in the eastern state. Berlin

FIG. 61. *Locomotive shed of the Krupp factories at Essen.*

(3·2 million people) is a divided city: though it lies in East Germany, the western part of Berlin is garrisoned by American, British and French troops. (This curious situation was caused by the relative positions of the Allied armies—Soviet and Western—in 1945.) Hamburg and Munich each have over a million inhabitants.

West Germany. West Germany is one of the greatest industrial powers in the world. It has large coal resources in the Ruhr, Aachen and Saar fields. There is also iron ore near Hannover and in the Siegerland. Lignite is found near Cologne, and oil near Hamburg, Hannover and Münster. The major concentration of heavy industry is the Ruhr (Duisburg, Oberhausen, Essen, Bochum and Dortmund), supplied by Swedish and African ore, brought mainly by waterway. There is also an enormous concentration of engineering (e.g. Krupp at Essen). Mönchen Gladbach, Rheydt and Viersen are big textile towns. Chemicals are manufactured in the Ruhr and also at Leverkusen, Hoechst and Ludwigshafen-Mannheim. Cologne makes cars and cosmetics; films, clothing and precision goods are made around Stuttgart; the enormous Volkswagen car works are at Wolfsburg. Hannover has large rubber works. Hamburg, Bremen and Kiel build ships. Various industries are found in Nürnberg, Augsburg and Munich, the last also known for its beer.

East Germany. Before 1945 eastern Germany had engineering, chemical and textile industries in Saxony and Thuringia, though the north was chiefly agricultural. There is little good coal or iron ore, but there are large deposits of lignite (Halle, Senftenberg) and mineral salts (Saxony), and some non-ferrous metals and uranium (near Gera and Aue). Since 1945 these have been used as the basis of further industrial development. Electric power generated from lignite, large quantities of water from the Saale and Elbe rivers plus local mineral salts have encouraged growth of a giant chemicals industry: principal plants are Leuna, Bitterfeld and Piesteritz. Using Russian or Polish coal and Russian or Swedish ore, iron and steel plants have been built at Eisenhüttenstadt on the Oder and at Calbe; others are at Berlin, Unterwellenborn and Riesa. Engineering is very important: Halle makes railway engines and wagons, Jena produces precision instruments, Karl-Marx-Stadt milling machines, while Leipzig makes machine tools. Textiles are made in many towns (e.g. Karl-Marx-Stadt, Plauen and Leipzig). Porcelain comes from Meissen. Rostock, the chief port, has considerable shipyards.

Agriculture. About a quarter of the area of Germany is forest. Arable farming is most important; in the north rye and potatoes are grown, but in central Germany sugar beet is cultivated, while wheat and other grains grow in the south. Oldenburg and Friesland are noted for dairy cattle, also found in Schleswig-Holstein. The vine grows in the Rhine and Mosel valleys. Only on the higher uplands are sheep found, but pigs are kept everywhere and fed from waste farm produce.

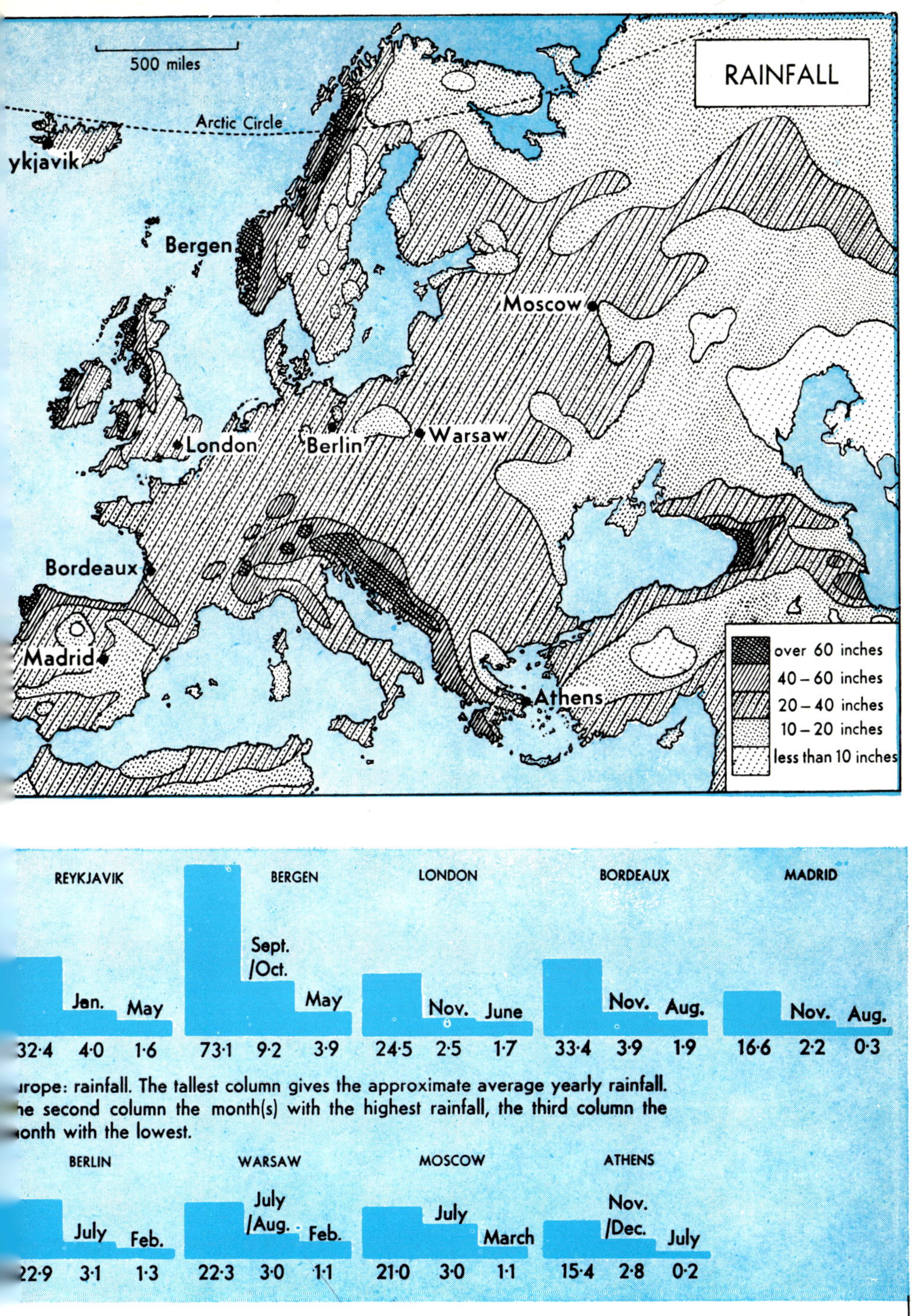

RAINFALL: ZONES AND SELECTED STATIONS IN EUROPE

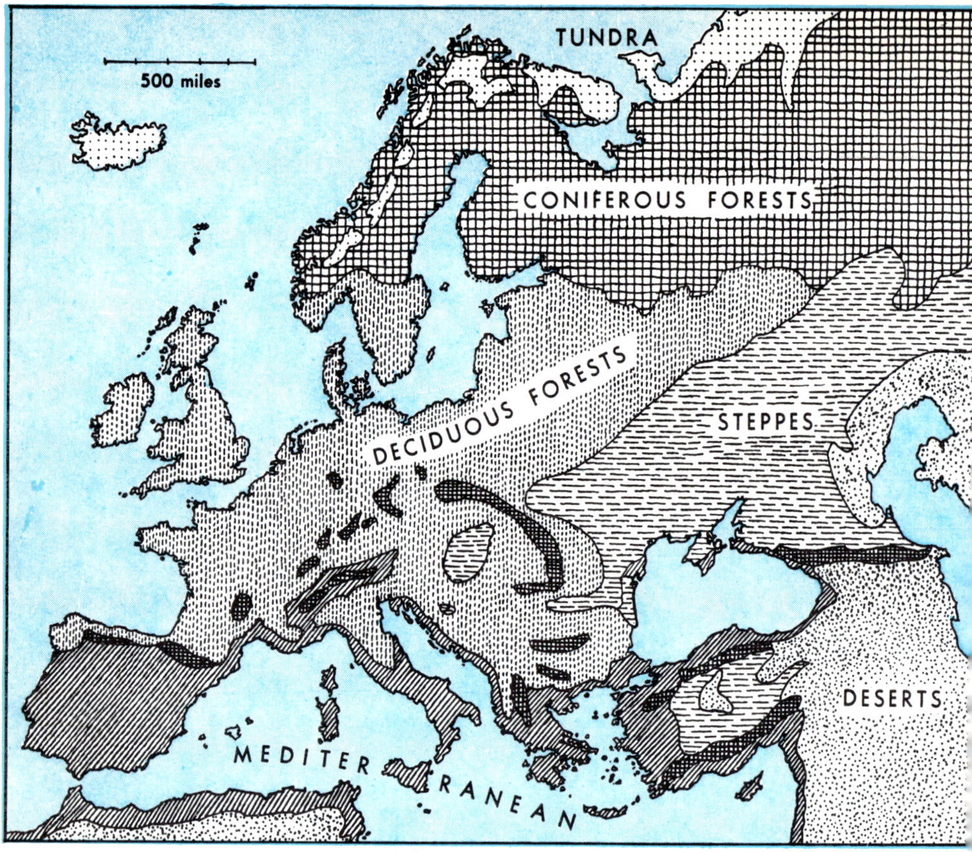

500 miles

TUNDRA

CONIFEROUS FORESTS

DECIDUOUS FORESTS

STEPPES

DESERTS

MEDITERRANEAN

coniferous forest vegetation | deciduous forest vegetation | Mediterranean type vegetatio

tundra

natural steppe

cultivated steppe

VEGETATION: NATURAL ZONES IN EUROPE

Austria

The small modern state of Austria is all that remains of a once great Central European empire. Nearly a quarter of its 7 million people live in Vienna, the capital. The language spoken is German. It is a mountainous land with small lowlands and hill country along the Danube and around Vienna. The isolated mountain valleys each have separate customs and costumes. Farming is limited to valley floors and the eastern lowlands, while cattle are grazed on mountain pastures in summer. The Danube valley has fruit growing and vineyards. Large forests support various wood-using industries. There is some coal and lignite, and iron ore comes from Eisenerz. Some oil is found near Vienna. Austria is also the major world producer of graphite. Mountain rivers provide hydro-electricity, and the main railways are mostly electrified. Vienna has big engineering works, while there is a steelworks at Linz. Graz has metal-working, paper and textiles. Vienna is noted for luxury goods. Tourism is a major industry, in both summer and winter. Innsbruck, Kitzbühel, Portschach are important resorts, while Bad Gastein and Bad Ischl are noted spas. Salzburg has a famous annual music festival.

Switzerland

Switzerland, a small, mountainous central European country, has a population of 5·4 million, and is a remarkable example of how different national groups can co-operate successfully and peacefully. Four different languages are spoken: about a fifth of the population, mostly in the west, are French-speaking. Well over two-thirds, in the north and centre, speak German. A tenth, chiefly in the southern canton of Ticino, speak Italian. And a very few in the south-east speak Romansch.

The largest city is Zürich, with 664,000 people, but the capital, Bern, has only 255,000. Switzerland is a confederation of cantons, and is a strictly neutral country, being unassociated with any other Power or group of Powers.

Economically, the most important part of Switzerland is the glaciated northern plateau which contains all the major towns. To the north-west is the rolling mountainous Jura, to the south majestic alpine ranges with some of Europe's highest peaks (e.g. the Matterhorn, 14,705 feet).

Dairying is the most important branch of farming, either from stall-fed cattle in the northern plateau area or from grazing on the alpine meadows in the south. Pigs and poultry are widely kept. Most of the country is unproductive, being mountain.

·Milk products and foodstuffs—chocolate being the best known—are important export items. The forested lower mountain slopes provide wood for sawmills. Salt is mined in the Vaud and near Lake Constance. Energy comes from a highly developed system of hydro-electric stations in the Alps and on the Rhine.

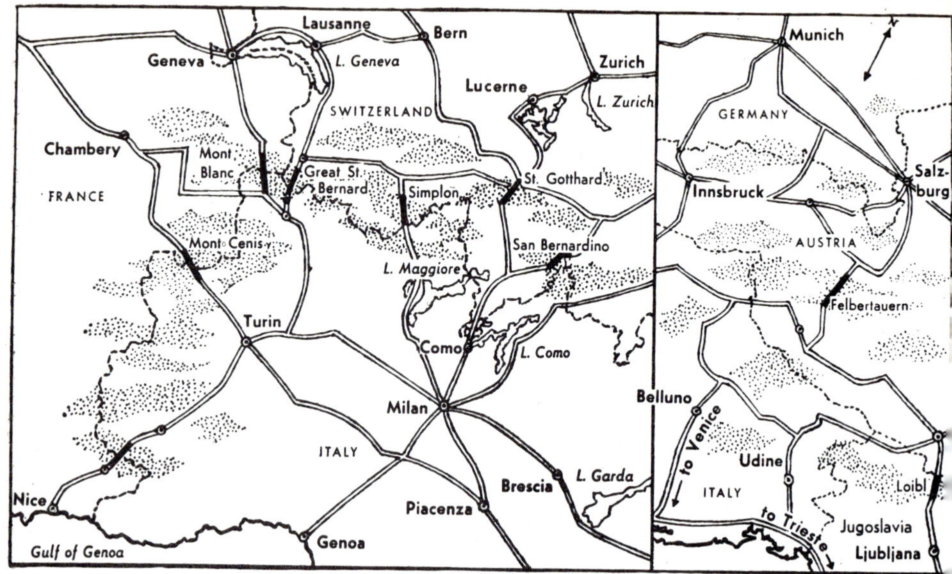

FIG. 62. *Alpine tunnels and main routes.*

Switzerland is a key area in Europe's communications system. International railway routes pass through the Alpine tunnels (Lötschberg-Simplon, St. Gotthard) and, in addition, between 1964-65 two new tunnels have been opened to carry *road* traffic through the Alps. The Great St. Bernard is 3½ miles long, and links Switzerland and Italy. The San Bernadino is 4 miles long, and links the Swiss cantons of Graubünden and Ticino, thereby creating an almost straight route from Germany to Italy. (There is also a third Alpine tunnel, the Mont Blanc, 7 miles long, linking France and Italy.) Because of Switzerland's central position and importance in the European railway system, she is sometimes referred to as "the turntable of Europe."

An oil pipeline runs through the Great St. Bernard tunnel from Genoa to a Swiss refinery near Aigle in the Rhône valley. Another runs through the San Bernadino from Genoa to Ingoldstadt in West Germany. The principal value of the tunnels, however, is to provide trans-alpine road links when the high alpine passes are blocked with snow in winter.

Swiss industry, concentrated in the northern towns, specializes in high-value precision products using imported raw materials. Watch and clock making is found in the Jura (La Chaux de Fonds); engineering at Oerlikon, Zürich and Winterthur; textiles at St. Gallen and Zürich; chemicals at Basel; and foodstuffs at Lausanne. Switzerland—particularly Zürich—is also one of the most important banking centres in the world.

Switzerland's glaciers (the largest in Europe), mountains, ski slopes and lakes make it one of the most popular summer and winter holiday areas in Europe. Among the most famous resorts are St. Moritz, Zermatt, Montreux, Lugano and Lucerne.

Poland

Poland is a remarkably flat country straddling the almost featureless North European Plain. However, its southern borders lie in the magnificent Carpathian mountains, and in the north are low morainic hills with lovely lakes along the Baltic shore. Poland is slightly greater in area than the United Kingdom, and has a population of 39·1 million people. The capital, Warsaw, has 1·3 million people, and Łodz 749,000. A cluster of large towns lies on the Upper Silesian coalfield.

The poor glacial soils of the North European Plain are not very productive; the best farming country lies on loessic soils in the Silesias and in Galicia (Krakow-Rzeszow), where sugar beet and wheat are grown. Rye, oats, barley and potatoes are more important in the north; but everywhere large numbers of pigs, geese and poultry are kept. Most farms are small and peasant-owned.

In Upper Silesia and near Walbrzych there is coal; lignite is found in Poznan province and in Lower Silesia. Iron ore is mined near Czestochowa and Kielce. Salt is worked on the Vistula near Bydgoszcz. Upper Silesia is the main heavy industrial area, with iron and steelworks at Chorzow, Sosnowiec and Zabrze. A large works has been built at Nowa Huta near Krakow. Using Hungarian bauxite, Skawina makes aluminium. Engineering is concentrated in Warsaw, Wroclaw, Chrzanow, Poznan and the Baltic ports (shipyards). Lodz and Bialystok are textile towns, as are some Lower Silesian centres.

Czechoslovakia

Czechoslovakia, which was one of the national states created after the First World War, has 14·3 million people. Its capital, Prague, has a population of 1,030,000. The Czech lands (Bohemia) are upland country, while Slovakia is mountainous.

It is an industrial country. In Silesia and near Prague and Kladno there is coal; lignite is found in the Ohre valley. Near Prague iron ore is also mined, and in the Bohemian Ore Mountains non-ferrous metals and uranium are found. Big steelworks are sited at Moravska Ostrava, Trinec and Karvina. Plzen and Prague have big engineering works and there is a motor car (Skoda) factory at Kolin. Electrical goods come from Brno. The main chemicals plants are Zaluzi, Bratislava, Usti, Pardubice and Ostrava. Textiles are important in northern Bohemia (Liberec), Brno and at Ruzomberok in Slovakia. Gottwaldov has giant shoe factories. Northern Bohemia is renowned for its traditional industry of glassmaking. Industry in Slovakia is relatively underdeveloped, but there is a big potential, and recently a large steel plant has been built at Kosice.

Two-thirds of the agricultural land is arable. Sugar beet is important in the Elbe valley and the lower Vltava. Olomouc has many refineries. Hops for beer come from Zatec and Plzen (the original home of "Pilsner" lager). Potatoes are important in southern Bohemia. Southern Slovakia is known for maize and pig rearing, though the Carpathians are mainly a sheep and cattle-rearing area, with forestry an important industry. Unfortunately, much of the best farmland suffers from frequent acute drought.

The beautiful Carpathians attract many tourists, particularly from Eastern Europe and the Soviet Union, who also visit Bohemia's world-famous spas, Karlovy Vary and Marianské Lazne (formerly Carlsbad and Marienbad).

Hungary

Hungary lies in the monotonous plains of the middle Danube, broken only by low hills like the Bakony Forest and the Mecsek. Hungarians—Magyars—are descended from Asiatic nomads who settled in the open Puszta in the ninth century A.D. 18 per cent of the 10·2 million people in the country live in Budapest, the capital.

Most people are engaged in farming, but industry is growing rapidly. There is little forest, except in the Bakony Forest and in the northern Carpathian outliers. West of the Danube grains, sugar beet and hard fruit are important crops; east of the Danube grain and livestock predominate. Principal crops are: sunflowers, maize, tobacco, paprika, and rice along the Tisza river. Large numbers of pigs and poultry are raised; meat products, notably sausages, form a tenth of all exports. Two-thirds of its area is arable land, a half devoted to grain crops. Wine is a noted product of the

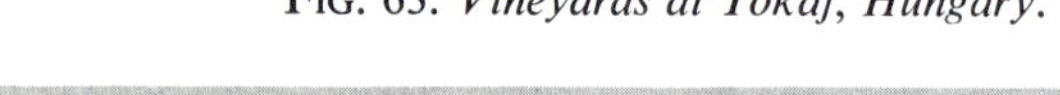

FIG. 63. *Vineyards at Tokaj, Hungary.*

northern shore of Lake Balaton and from the Tokaj Hills, and is a valuable export.

Hungary has little coal but fair amounts of lignite and large quantities of bauxite. Some uranium comes from near Pecs. Aluminium making is mostly sited in the northern Bakony Forest, where lignite and bauxite are found. Iron and steel is made in a big new works at Dunaujvaros on the Danube, using mostly raw materials from Russia; steel is made in Budapest and Miskolc and there are foundries at Gyor. The Bakony Forest region, Budapest, Dunaujvaros and Pecs are chemicals centres. Engineering is chiefly centred in Budapest and Gyor, the chief products being river boats, motor vehicles and electrical goods.

A striking feature of the Hungarian landscape east of the Danube is the immense townlike villages set in the open Puszta. These date from medieval times and were intended as a defence against Turkish raids.

Bulgaria

Until recently Bulgaria was one of the most backward countries of the Balkans. It has 8·3 million people, but only a third live in towns. Most towns are small, but the capital, Sofia, has 859,000 people. Bulgarians are mostly Orthodox Christians and their language belongs to the Slavonic family, written in the Russian alphabet.

Two-thirds of the country is upland and mountain. The climate consists of long, hot summers with thunderstorms, and cold winters, with snow in the mountains. The low rainfall and high evaporation in summer make irrigation desirable, and numerous dams are under construction. The soils are moderately fertile, and there are patches of rich black earth, mostly in the north.

Most people are farmers, and almost half the country's area is arable land, but yield is low. There is a real need for more fertilizers, and a big new works has been built to make these at Dimitrovograd. The most important crops are wheat, rye, maize and rice, followed by fruit (e.g. melons, grapes, cherries) and vegetables (e.g. tomatoes). Fruit in the form of jam and preserves is a well-known Bulgarian export. Sunflowers and tobacco are also important. Roses are grown for Attar of Roses for perfumes, and cotton is also produced. Sheep are the main livestock, but pigs are also common.

Natural resources are coal, salt, iron ore and manganese. Industry is centred in Sofia, Plovdiv and Varna (the chief port). Machine tools and farm machinery are built. Dimitrovo (Pernik) has an iron and steel works.

Romania (Rumania, Roumania)

Romania, a south-east European country, is slightly larger in area than the United Kingdom but has only 19 million people. Its capital, Bucharest, has $1\frac{1}{2}$ million people. The Romanian language is derived largely from the Latin of Roman settlers in Dacia influenced by later Slav migrants.

The south (Wallachia) and east (Moldova) are the broad plains of the Danube, Siret and Prut valleys. Similarly, the western part of the country lies in the plains of the Tisza; but the centre and north are the uplands and mountains of Carpathia, amid which lies the lower hill country of Transylvania. The southern plains have hot, dry summers and cold winters, with features of the dry steppe on the east. The winters are milder in Transylvania, though the summers are cooler. The best soils are in the Danube and Prut plains and in Transylvania. Poor soils, mostly forested, cover the higher mountain areas.

Two-thirds of the labour force are engaged in farming. Over 40 per cent of the agricultural area is arable. Three-quarters of the arable is used for grain (maize, wheat, rye), though yields are poor. In the Danube and Tisza plains some rice is grown. Flax grows in the mountains and cotton in Muntenia. Moldova and Dobrogea are noted for sunflowers. Transylvania raises sugar beet and potatoes. In Wallachia, especially near Bucharest, vegetables are an important crop. On southern slopes, vines are grown, but their wines are little known in the West. Pigs are kept in the grain-growing areas, and cattle and sheep in the mountains and in Dobrogea.

Romania was once an important oil producer, but output has not quite kept pace with world expansion. The deposits are in the Carpathian foothills: Ploeisti, Bacau and Oltenia. There are large reserves of natural gas. Coal comes from Petrosani and lignite from the Banat. Iron ore and manganese are also found in Carpathia: the iron and steel plants are at Resita and Hunedoara. Copper, lead and zinc are mined in the eastern Carpathians, and gold and silver in the Bihor Massif. Main engineering centres are Bucharest, Brasov, Resita; and chemicals, notably fertilizer, also come from Bucharest, Brasov, Ploeisti and Bacau. Wood-working industries are important, especially in the Transylvanian Alps.

Italy

Italy is a long peninsula jutting south-eastwards into the Mediterranean Sea. Its northern boundaries lie in the Alps, and its territory includes the islands of Sardinia and Sicily. It still has active volcanoes—Etna, Vesuvius. Somewhat larger than the United Kingdom in area, the population is 51·5 million people. The capital, Rome, has 2·6 million people, and Milan 1·7, Naples 1·3 and Turin 1·1 million.

Twenty-six per cent of the working people are farmers, and roughly a third of these are landless labourers. The hot, dry summers of the centre and south do not favour farming without irrigation, but the mild wet winters help many crops, and the modest home demand for meat is met by the northern areas.

FIG. 64. *Turin—the great* FIAT *motorworks.*

Most farms are small. Wheat is grown everywhere except on the highest land, and Italy is 90 per cent self-supporting in wheat. Soft durum types are grown, suitable for the various pastas (spaghetti, etc.) which form the basis of Italian diet. Northern Italy is a maize producer, while the olive is essentially a southern crop, though it also grows in Liguria and Tuscany. The vine grows everywhere, but the most important areas are Piedmont, Tuscany, Campania and the Trentino. Italy's production of wine is only slightly less than that of France. Venezia and Emilia produce sugar beet. Citrus fruits grow on the coasts and also in some sheltered valleys in the Alps and Apennines. The Riviera di Fiori is noted for flowers.

Soil erosion is a real problem in Peninsular Italy, where heavy downpours cut into the soft, parched ground, made up of clays and sands. The richest areas are fertile silt-covered mountain basins or volcanic soils (e.g. around Vesuvius), but yields are greatest in the Plains of Lombardy. Pastoralism is the main occupation of the mountains and has often caused deforestation.

For energy, Italy depends on hydro-electric power and natural gas. The bulk of Italy's oil is imported and refined, chiefly at Savona and Cortemaggiore. Sulphur and mercury are found in Tuscany and both are exported. There is an iron and steel industry, using imported raw materials, established at Genoa, Piombino, Venice, near Naples and Taranto. Scrap is used in the northern industrial towns to make steel and steel products. Milan, Turin, Brescia and Modena build motor vehicles, and Turin is one of the greatest centres of motor car manufacture in the world, and the car industry is one of Italy's biggest exporters. Savona, Genoa, La Spezia and Livorno have shipyards; Milan, Turin and Ivrea construct electrical goods and calculating machines. There is a large textile and clothing industry in northern Italy (Milan, Como and Legnano). Silk culture is carried on all over Italy, the manufacture of silk textiles being concentrated in the north. Electro-chemicals and electro-metallurgy are found at Bolzano and Trento. There is considerable production of volcanic sulphur and a little oil is found in Sicily.

Tourism is very important, partly because Italy has much to offer by way of weather, landscape, sea and beaches, historical remains and art treasures, and partly because Italy imports far more than she exports, and tourism is therefore encouraged as a source of income to make up the trading deficit.

Jugoslavia

Jugoslavia is probably better known to Western Europeans than any other Communist country. It has always been independent in outlook and welcomes foreign visitors. The population is 18·5 million, but less than a fifth live in towns. The capital, Belgrade, has 843,000 people. The principal language is Serbo-Croat: in the north it is written in Latin characters, but in the south the Russian (Cyrillic) alphabet is used. Other related languages used are Slovenian (Latin alphabet) and Macedonian (Cyrillic alphabet); Albanian is spoken in the Kosovo mountain district. Most people in the north are Roman Catholic, but southern Christians are mostly Orthodox. There are also some Moslems.

The northern and eastern borders lie in the hills and plains of the

FIG. 65. *Belgrade, capital of Jugoslavia, had to be rebuilt after the destruction in the Second World War.*

Pannonian Plain. Between this plain and the Adriatic coast and islands are rugged ranges and high, barren, inhospitable plateaux—mostly of limestone (the Dinaric mountains and the Karst). Long, hot summers with severe winters are experienced in the Pannonian Plain. Cooler summers with severe winters with much snow are common in the mountains. The Adriatic coast has hot, dry summers and relatively mild winters.

Jugoslavia is still an essentially agricultural country. A third of the area is forest and a tenth waste. A quarter of the area is pasture. Too many people, particularly in the south, try to live from the land, and 80 per cent of the agricultural community are small farmers. The high limestone plateaux are rough pasture or forest, with patches of maize, tobacco and tree crops in basins. Sheep are the main livestock, though pigs are widely kept in Serbia. Amid the large forests on the northern side of the mountains are found orchards, maize and wheat. On the south, citrus fruits and olives will grow, and there are many vineyards. Slovenia is well forested, with crops of oats, rye, maize and wheat, and dairying is also important. Cereals, vineyards and plum orchards (for *slivovica*—plum brandy) are typical of the Pannonian Plain, and large numbers of pigs are kept. Sugar beet and sunflowers are also grown here.

Lignite is found in Slovenia, Serbia and Bosnia: used mostly for power stations, and there is a growing use of Jugoslavia's large hydro-electric potential. Iron ore comes from Vares and Ljubija in Bosnia. Coking coal is brought from Poland to the steelworks of Zenica. Ships are built in Rijeka and Pula; engineering is also found in Maribor, Belgrade, Bosanski Brod, Ljubljana and Osijek. Textiles are made in Tetovo, and the food processing industry is widely distributed throughout the country. Aircraft are made at Zemun. Mining is important: copper from Serbian Bor; bauxite from Dalmatia; lead and zinc from Trepca; mercury is found in Idrija; and there are salt pans on the Adriatic coast. Sisak works petroleum. Raw materials for chemicals provide the fertilizers made at Maribor and Sibenik, and salts mined at Tuzla are also used. Paper is made in Ljubljana.

The lovely old towns and fine resorts of the Adriatic coast attract many tourists from all over Europe. Tourism is, in fact, a major industry.

Greece

Modern Greece is a poor country. Yet, in ancient times, it was the cradle of European civilization. The population is 8·6 million, nearly 2 million of whom live in Athens, the capital. 56 per cent of the population are town-dwellers, and 44 per cent live in villages and rural communities.

Only a quarter of the area is cultivable. Farming concentrates on tobacco, olives and viticulture. Rice is widely grown in central and southern Greece, olives in the south. Sheep and goats are the main livestock, notably in the mountains. In a strongly Mediterranean type of climate, with large areas of limestone, irrigation is important in cultivating fruits and vines. Much has been done to drain formerly malarial coastal lowlands or lake basins, where some cotton is grown. The islands produce vines, currants and citrus fruits.

Crete, once the home of an even older civilization than that of the Greek mainland itself, is essentially pastoral, but can produce rich crops when irrigated. Northern Greece grows wheat, maize, sunflowers, tobacco, plums and nuts.

Athens and Piraeus are the main industrial centres, mostly making handicraft goods. Food-processing, textiles and engineering (notably shipbuilding) are expanding industries. Ptolemais has a large fertilizer plant. Tourism is very important, thousands of visitors being drawn to the historic sites of Athens, Delphi, Mycenae (Crete), etc. Greece has a strong maritime tradition, and the Greek merchant fleet is the sixth largest in the world.

Cyprus

Cyprus is a large, mountainous island in the eastern Mediterranean. Of the population of 614,000 about 77 per cent are Greek and the rest Turkish. The economy is largely based on Mediterranean-type farming and about half the area is used for this purpose. Products are wines, raisins, vegetables (notably potatoes) and fruits. Mining of iron and copper pyrites and related minerals is important (Cyprus is second world producer of pyrites). The political future of Cyprus has long been subject to dispute—formerly between Britain and the Greek community, more recently between Greece, Turkey and the Cyprus government itself.

Albania

One of the smallest, most isolated and least known countries, Albania is a rough mountain land with a narrow malarial coastal lowland. There are 2 million people, mostly in the foothills. The capital, Tirana, has 169,000 people. Albanian is an ancient Aryan language quite dissimilar to Greek or Slav. Most people are Moslems.

Agriculture and pastoralism dominate the economy. The only really fertile areas are the Adriatic foothills and the basin around Korce: maize, wheat, tobacco, olives and citrus fruits are the main crops. Some primitive irrigation systems are found, but elsewhere an almost nomadic pastoral economy with sheep is found. Forests of deciduous and coniferous trees are a largely untapped source of wealth. Industry is chiefly processing foodstuffs, some textiles and shoe-making. Minerals are chrome, iron-nickel ore (exported to Czechoslovakia). A little oil is worked and refined.

Despite the efforts of the Communist régime since 1947, Albania is still a backward country.

Spain

Spain is approximately twice as large as the United Kingdom in area, but it has a population of only 32·4 million. Madrid, the capital, has a population of nearly 3 million, while Barcelona has 1·7 million people. The smaller towns tend to be agricultural centres rather than truly urban ones.

By Western European standards Spain is a poor country: one third of the labour force is engaged in agriculture, and agricultural produce—particularly vegetables (for example the famous Spanish onions), wine, olives and olive oil and oranges—makes up over half Spain's exports. Farming still largely depends on manual labour.

Steel output is equal only to one of Britain's major works. There are only half a million cars and the railway network is thin and antiquated.

Comprising mostly a high plateau ribbed by higher mountain ranges, most of Spain is dry and depends on pastoralism and tree crops, though it is self-sufficient at the present low standards in most essential foods. Wheat is the main crop of the Meseta plateau, where sheep and goats are kept. The more moist northern coast grows crops like potatoes and onions. The irrigated lands of the Mediterranean coast are the main producers of citrus fruits, melons and vines. The Guadalquivir valley (Andalusia) is known for its wines and its oranges. The most famous wine of Spain is produced here: sherry, which is made around Jerez de la Frontera. Olives are an important crop south of the Tagus, on the eastern coast, and in the Ebro basin. Rice is grown on irrigated land in the Ebro delta of the coastal parts of Valencia, and tomatoes are increasing in importance. The main hope for agricultural expansion depends on irrigation: big schemes have been developed in Badajoz province, in Aragon and in Jaen province, while sugar beet growing has been encouraged by irrigation in the Ebro basin.

Once an important iron ore producer, the mines of the Bilbao area have declined. Spain is a major European producer of mercury and zinc and a wide range of other minerals, though not in large quantities. Salt and pyrites are raw materials for the chemicals industry. There are supplies of poor coals and lignite, but hydro-electric development is increasing, along with more irrigation projects. Steel is made at Avilès near the Oviedo coalfield. Most of the limited industry is in the Basque provinces, Barcelona (engineering, textiles), Madrid (engineering) and Valladolid. The works are usually small, and all industry is hampered by the lack of good markets.

Portugal

Portugal, the second, smaller and even poorer Iberian country, has a population of 8·7 million. The capital, Lisbon, has 802,000 people.

Most people live in the western coastal lowlands and lower valleys, notably around Lisbon and, farther north, around Oporto and the Douro valley,

FIG. 66. *Cork is an important natural product in both Spain and Portugal.*

which is the principal wine-growing area. The mountains along the frontier with Spain, with their deep gorges, are sparsely settled.

In northern Portugal too many people try to eke a living from the soil, this being one of the most intensely settled and farmed areas in Europe. One of the prime reasons for the situation is the backward state of Portuguese industrial development, so that there are few alternatives to farmwork.

Crops grow nearly all the year, though irrigation is needed in summer. Maize, vegetables and potatoes are common crops, while hay and green fodder crops are grown for cattle-feed. In the mountains sheep are kept on pastures amid patches of scrub oak and chestnut. The Douro valley is famous as the home of Port wine, which is exported via Oporto to Britain and elsewhere. Another famous Portuguese wine known all over the world comes from the island of Madeira.

In the south, saltpans and coastal forests occur, while cattle are grazed on marshier ground. Crops such as maize, rice, fodder, garden crops and grapes are common. The farther south one travels, the more important irrigation becomes. Cultivated woodland in the southern province of Alentejo is used as pannage for pigs and for cork oak; it is also occasionally cropped with cereals. Extensive wheat cultivation follows long fallowing and grazing by sheep. Algarve has typical Mediterranean tree crops—figs, almonds, olives, as well as pastoral areas. Fishing, notably for sardines, is important along the coast, though much of the olive oil used in canning is imported. Deep-water fishing is also important, the principal catch being cod.

Lisbon and Oporto have some small industries (chemicals and engineering) and a steel mill has been built. Hydro-electricity has been developed on tributaries of the Douro and Tagus. Pyrites are mined in Alentejo province, and wolfram comes from near Oporto, where there is also a little coal.

QUESTIONS

1. Write brief notes describing and comparing: (*a*) Spain and Portugal; (*b*) Benelux countries; (*c*) the Scandinavian countries.
2. What are the main contrasts between the industrial economies of the two German states?
3. Discuss the positional and economic importance of Switzerland.
4. Briefly compare the industrial and agricultural economies of France, Italy and any third Western European country with those of Czechoslovakia, Jugoslavia and any third east European country.

FIG. 67. *Europoort (Rotterdam) is the greatest port in Europe, and the largest in the world.*

CHAPTER 17

UNION OF SOVIET SOCIALIST REPUBLICS

THE U.S.S.R. covers one-sixth of the earth's land surface and extends across Eastern Europe and Northern Asia. Its area is ninety times that of Britain. There are immense contrasts of climate and conditions: in the north are Arctic deserts, in the south deserts as scorching in summer as the Sahara. It is a country of great plains open in the north to the Arctic Ocean; but ringed in the south by a high mountain wall, which prevents the entry of warm influences from the south. So great is the distance between the east and west of the U.S.S.R. that the people of Moscow are eating their lunch when their comrades in Vladivostok have finished the day's work.

Because it extends across the heart of a vast continental block, the Soviet Union has a climate marked by seasonal extremes. The greater part of the year is very cold, but in summer the land warms up very quickly and the weather is hot. Another great problem is the vast distances that separate the major regions. Immense areas of the north and east are so remote that they are still relatively unknown. It is only in the last thirty years that proper exploration has been completed. The 236 million citizens of the Soviet Union are made up of over 160 different national groups, though the Russians and their cousins the Ukrainians and Byelorussians form three-quarters of the total population. Three out of every four people live in the western European part. Few people, therefore, live in the much larger area of Asiatic Russia, which has vast areas of almost uninhabited country in the forests and tundra of Siberia and the deserts of Central Asia. Siberia is a pioneer country, in many ways reminiscent of the American West a century ago.

European Russia

Most of the Soviet Union west of the Ural mountains is a vast plain. Ancient crystalline rocks appear at the surface in the northern Baltic Shield and the southern Podolian-Azov shield. Elsewhere they are covered by little-disturbed sedimentary rocks. Along the southern edge of the Baltic shield the sedimentary rocks end in a low north-facing scarp. The northern part of the plain is a dreary land of forest and bog, with many patches of bare rock. The north-western part of the plain has freshly marked glacial features. Great morainic hills extend along the borders of the Baltic republics from Minsk to Moscow, and form the main watershed of the plain. South-flowing

FIG. 68. *View from the Kremlin over Moscow from Lenin's Tomb and Red Square to St. Basil's cathedral. The Kremlin is the ancient fortress heart of Moscow. The city itself grew up at the centre of river routeways across the European Russian Plain.*

rivers are sluggish and meander southwards across the immense, featureless plains. Their right banks are, nevertheless, mostly marked by steep bluffs. They cross the hard rocks of the Podolian-Azov shield in great eastward-curving bends. Where the Dnieper crosses these old rocks it flows very swiftly over rapids. The southern shield has also ponded back the middle reaches of the rivers, so that poorly-drained, swampy country is found in the Pripyat and in the Oka-Don Lowland. North-flowing rivers are shorter and relatively swift.

The great Quaternary ice sheets sent huge lobes of ice down the main valleys. Beyond the edge of the ice the fine, dust-like loess (see page 162) was deposited, on which developed a rich, dark soil that makes the Ukraine the best wheatland in the Soviet Union. East of the Volga, low plateaux lie along the foot of the Urals, and the river flows close to its high right bank, to empty into the Caspian Sea (which is shrinking—though why is not really understood). The great bend in the Volga at Kuybyshev is thought to be caused by the river having found a new channel after ice had blocked the original one. In glacial times, the Caspian Sea was much larger and may have flowed into the Sea of Azov through the Manych valley. It is separated from the Sea of Azov by wide grain-growing plateaux in Krasnodar and Stavropol.

FIG. 69. *The Ukrainian Black Earth lands are a major Soviet granary. Here on their northern edge, modern combine harvesters gather the crop. In the background are seen shelter belts to break the harsh winds which carry away topsoil.*

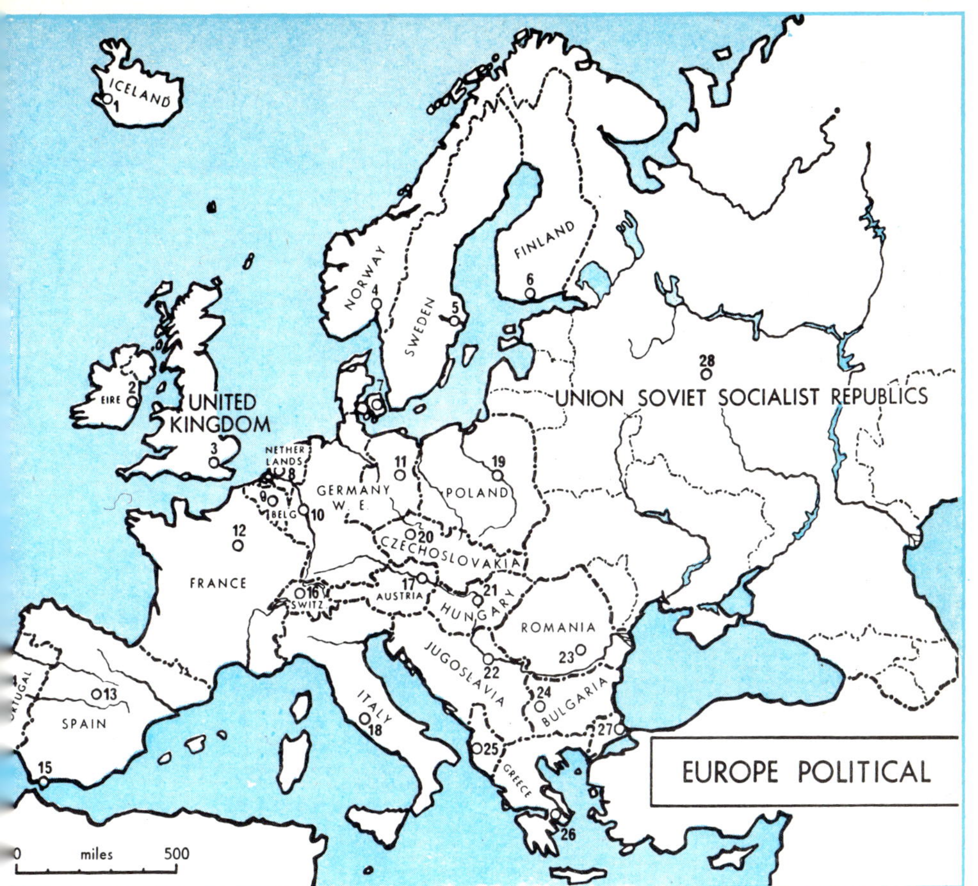

(1) Reykjavik
(2) Dublin
(3) London
(4) Oslo
(5) Stockholm
(6) Helsinki
(7) Copenhagen (Denmark)
(8) Amsterdam
(9) Brussels
(10) Bonn
(11) Berlin
(12) Paris
(13) Madrid
(14) Lisbon
(15) Gibraltar
(16) Bern
(17) Vienna
(18) Rome
(19) Warsaw
(20) Prague
(21) Budapest
(22) Belgrade
(23) Bucharest
(24) Sofia
(25) Tirana (Albania)
(26) Athens
(27) Istanbul (Turkey)
(28) Moscow

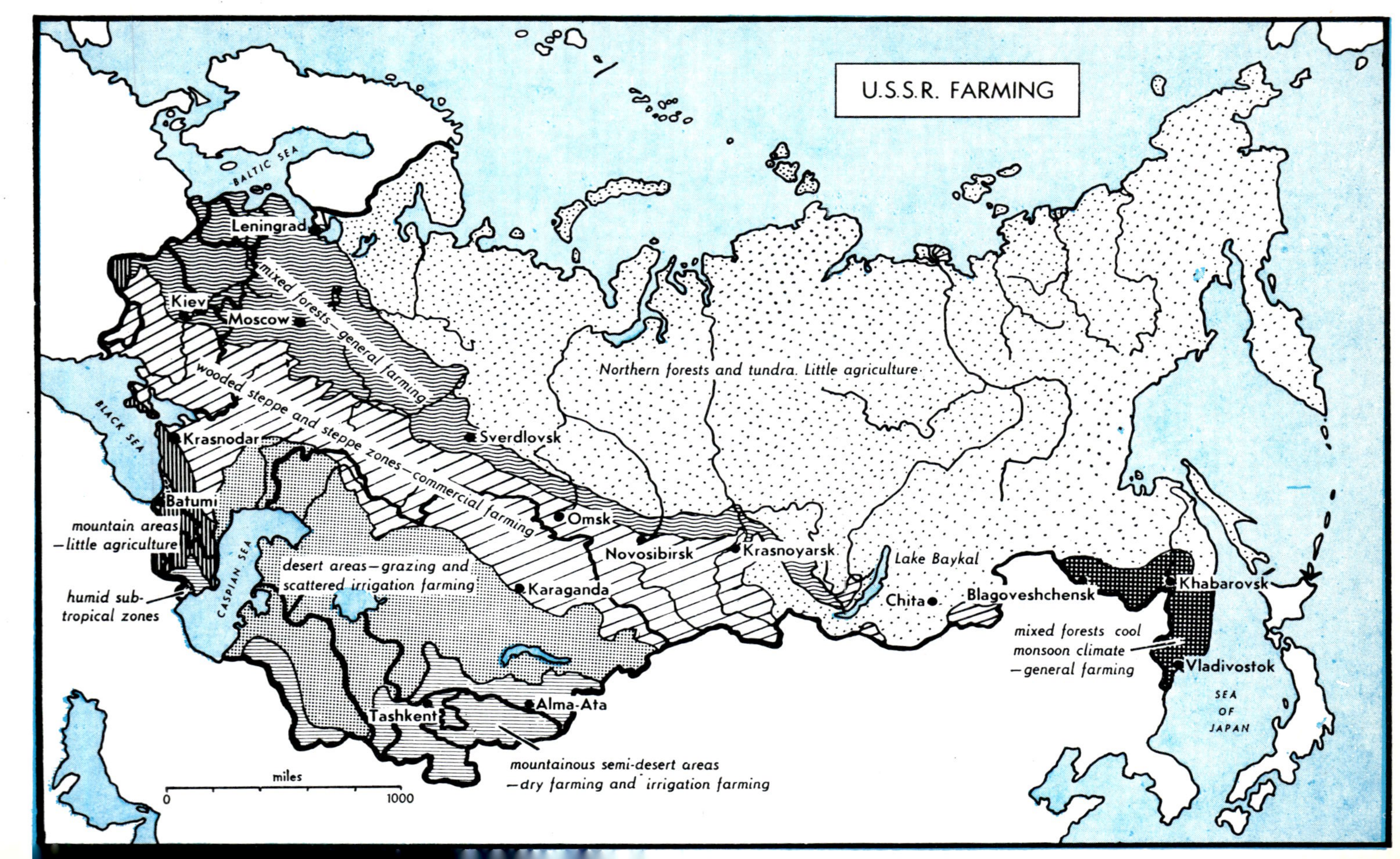
U.S.S.R. FARMING
BALTIC SEA
BLACK SEA
CASPIAN SEA
SEA OF JAPAN
Leningrad
Kiev
Moscow
mixed forests—general farming
wooded steppe and steppe zones—commercial farming
Northern forests and tundra. Little agriculture
Sverdlovsk
Krasnodar
Batumi
mountain areas —little agriculture
humid sub-tropical zones
desert areas—grazing and scattered irrigation farming
Omsk
Novosibirsk
Krasnoyarsk
Karaganda
Lake Baykal
Chita
Blagoveshchensk
Khabarovsk
Vladivostok
mixed forests cool monsoon climate —general farming
Tashkent
Alma-Ata
mountainous semi-desert areas —dry farming and irrigation farming
miles
0
1000

Asiatic Russia

A similar but even more dreary plain, called the West Siberian Lowland, lies east of the Ural mountains. It is a vast and seldom-visited wilderness of swamp and forest, crossed by huge sluggish rivers like the Ob, Irtysh and Tavda. On its southern edge, the forests change into parklike country.

The Ural mountains form parallel ranges of low rounded outline extending some 1,500 miles from north to south. They separate the two great areas of plains, but in the south the Urals die away into the plains in the low Mugodzhar Hills. These arid hills overlook the land gate from Asia into Europe on the northern shore of the Caspian. In the far north, the Urals are also much lower and partly drowned by the sea, but can be traced in Vaygach Island and in the Arctic, ice-girt islands of Novaya Zemlya. The steeper eastern slope of the Urals is rich in minerals. The low central part is easily crossed and a focus for main railway routes from Europe into Siberia via Sverdlovsk and Chelyabinsk.

The Turanian Plain lies to the east of the shallow Aral Sea. It is bordered on the west by the scarp edge of the Ust-Urt plateau and it rises eastwards into the Kazakh Upland, with its isolated hills. In the south-east it is marked by a deep, young trough, in which lies the shallow Lake Balkhash. In these arid regions wind has moulded the landscape; but there is evidence of once greater moisture in the presence of salt lakes, dried-out river courses and the shrunken outlines of the Aral Sea and other lakes. During this wetter period, water from the West Siberian Lowland flowed through the Turgay Depression into the Aral Sea and along the now dried up Uzboy *wadi* into the then much larger Caspian Sea. Today, there are large tracts of sandy, clayey or salty desert or poor steppe: like the Bet-Pak-Dala, the well-named Plain of Misfortune, in the east, near Lake Balkhash.

Siberia

Central Siberia is formed of great blocks of ancient rocks (some of the oldest rocks in the world are found in the Anabar shield) which have been cut into a dissected plateau by rivers. But, to a traveller along the valleys, the country looks mountainous. It is rich in minerals and coal. A broad depression separates the central Siberian Plateau from the mountains of the Arctic wilderness of the Taymyr Peninsula, and the islands of North Land (Severnaya Zemlya, only discovered in 1913) that are slowly rising from the sea. Southern Siberia is built of vast tilted blocks of country that give it a very rough topography. There are traces of quite recent volcanic activity (e.g. in the Aldan Plateau), and many recently formed troughs: in one of which lies Lake Baykal, the world's deepest lake. On thc east, another trough filled with young sediments (followed by the river Lena) separates the ancient land from the younger mountains of north-eastern Siberia. These young folded mountains contain huge uplifted but unfolded pieces of country, like the Kolyma Basin and the Yukagir Plateau. The true nature of these mountains is still being discovered, for they were not accurately explored until 1926. Still younger mountain ranges, with several volcanoes, form the large Kamchatka Peninsula and the Kuril Islands, as well as the lozenge-shaped island of Sakhalin. The partly drowned mountain system that is marked by the Kuril Islands really forms an "outer" coast to Pacific Siberia, separating the basins of the Pacific and Okhotsk Seas.

The Mountain Crescent

A belt of young Tertiary mountains lies around the southern and eastern borders of the Soviet Union. On the south-west are the low, forested Carpathians. Mountains also form the southern Crimea, where they rise in white cliffs above the waters of the Black Sea. Mountains occur again in the majestic snow-clad Great Caucasus of truly alpine ruggedness. Here Mount Elbrus rises to 18,480 feet. The southern part of the Caucasian isthmus is formed by the lower ranges of the Little Caucasus and the volcanic plateau of Armenia, and is deeply scored by such rivers as the Araks. Between the Little and Great Caucasus lies a broad depression blocked in the centre by the rough Suram mountains. The western depression, Kolkhida, is moist and luxuriant, but the eastern Kura depression is poor steppe, and sometimes semi-desert.

FIG. 70. *Volcano in Kamchatka.*

FIG. 71. *Mount Elbrus in the Caucasus.*

The Soviet border in Central Asia and in southern Siberia is marked by ranges of young folded mountains in which lie blocks of older country. In Central Asia the foothills and surrounding plains are watered by rivers from the snows of the high mountains. These sustain rich oases on the fertile loessic soils, and fertile basins are also found within the mountains, like that around Fergana. The highest parts are found in the high plateau-like Pamir and in the sparkling peaks of the Tyan Shan. On the north of the Tyan Shan two depressions, the Dzungarian Gates, form a key routeway from Chinese Central Asia into the plains of Russia. Southern Siberia consists of massive uplifted blocks of country cut into jagged mountains or bare rounded summits, so often hidden in cloud. Such country is to be found in the Altay and Sayan mountains. The lower slopes of the mountains have magnificent forests, and around their feet lie broad grasslands.

Several large lakes lie in structural basins within or around the mountain belt. Examples are Lake Balkhash, Lake Issyk Kul and Lake Baykal. Lake Teletsk is in a narrow, glacially-deepened trough.

QUESTIONS

1. Using maps wherever possible, give a brief account of the main physical divisions of the U.S.S.R.

CHAPTER 18

CLIMATE, SOIL AND VEGETATION AND THEIR EFFECT ON RUSSIAN LIFE

In a country of vast plains where there is little contrast in relief, climate naturally has a major effect on soil and vegetation, and plays an important role in everyday life. Russian geographers divide their country into broad regional belts from north to south, based on climate, soil and vegetation. In the heart of such a large continental block, the Soviet Union has a remarkable rhythm in the seasonal round. It can be said with a certain truth that it has climate but little weather!

The central lowlands, open to the lowland north, allow a deep penetration of cold, so that winter is everywhere long and bitter. It lasts from four months in the extreme south to over nine months in the far north. It is made bearable by long periods of clear, sunny weather, unfortunately punctuated by savage blizzards. A wet and short spring marks the great thaw, with gigantic floods that reduce the countryside to a quagmire, making it hard to move about. There follows a short summer, warm in the north but really hot in the south. The continentality of the country causes a wide daily variation in temperature. In the Ukraine, after a sweltering day in the steppe, the traveller may sit and feel chilled in the cool of evening. Violent thunderstorms are common. Autumn is so short as to be almost unnoticed, then winter grips the country once again. In the southern deserts the quickly generated summer warmth that makes them almost "Saharan" in summer, rapidly dissipates in autumn, and winters are abnormally cold for the latitude, because the high southern rim of mountains excludes the warm influence of the Indian Ocean.

Moving from farthest north, Arctic stony desert passes into tundra. Here winter lasts until June and returns by September. Being so far north, winter is almost continual darkness; summer is nearly continuous daylight, which compensates plants for the shortness of the season. The tundra is a desolate land of bog and stunted plants that bloom briefly but brilliantly in the short summer. The summer air is alive with insects and birds, and occasional polar bears or an Arctic fox may be seen. Reindeer wander into the summer tundra, followed by wolves which prey on the herds.

Southwards, the traveller would begin to see more and more trees. Slowly he passes into the coniferous forest belt, the *tayga.* It is gloomy country, with rotting trees and patches of bog. Summer lasts longer and is warmer than in the Arctic, but it is as cold in winter. The summer air is often filled by acrid smoke from great forest fires, particularly in drier eastern Siberia where there is also less bog.

Permafrost

The greater part of Siberia has permanently frozen ground: *permafrost.* Summer is too short to thaw more than the surface layers, for frost penetrates deeply into the ground because of the very thin snow cover. In the tundra nearly all ground suffers from a very deep permafrost. In the northern tayga there are islands of unfrozen ground (notably on well-drained sands), and the farther south we travel the larger these unfrozen patches become. On the southern fringe of Siberia there are only a few patches of frozen ground. Water from deep beneath the frozen layer sometimes breaks through and wells out at the surface. Trees on the frozen ground are tilted and twisted into "drunken forests." The effects of permafrost on roads, bridges and buildings can be severe and dislodge foundations, unless special precautions —elaborate and very expensive—are taken. Relatively few people live in the northern forests, but Russian colonists have settled along the rivers. Native hunters and herders live in the forests but, like the fur-bearing animals, Russian colonization has driven them deeper into the remoter forests. So far the almost limitless reserve of timber has only been scratched, though much of this wild forest consists of over-ripe or deformed trees that may never be touched.

FIG. 72. *Pack-ice on Lake Baykal, Siberia.*

Farming Lands of the Mixed Forests and Wooded Steppe

The soils of the coniferous forests are cold and acid, and the climate is poor for farming. In European Russia however, there is a broad wedge of mixed and deciduous forest that tapers eastwards. The climate is less severe and the soils easier for the farmer. Much ground has been cleared, and even some bog has been drained. It remains a poor country, and the Russian farmer, despite modern machinery and improved techniques, reaps a meagre and uncertain reward. He still works a wet, cold, poor soil that lacks adequate nutrients. Large forests remain, though towards the south the forest thins and trees are scattered over the grasslands. The countryside consists of large fields and villages broken by clumps of trees. It is the natural parkland of the wooded steppe, now one of Russia's main farming belts. The soils are more rewarding and include some fine, dark earths, but in the valleys there are wet meadows and bog. Even in Siberia, this wooded steppe is being increasingly colonized for farming, and colonists are moving north from it along the rank meadows of the great north-flowing Siberian rivers.

North of the mixed forests, crops are chiefly flax, rye and other northern cereals, and dairying. To the south sugar beet, potatoes, wheat and stall-fed cattle or pigs are predominant, while in the far south wheat and maize for grain are common. Maize is now being grown everywhere for fodder. The forests are famed for their mushrooms, wild strawberries, and fine honey.

The Steppe

As we travel south through the wooded steppe, trees become fewer until we are in a vast grassland, the true steppe. In the north grass forms a continuous cover, but towards the dry extreme southern fringe there are clumps of grass between bare patches. The best steppe has black soils developed over loess that retain their fertility for long without much fertilizer. In European Russia and in Siberia nearly all the best land has been taken for cultivation, for it is ideal grain country. (Within the last ten years vast areas of virgin steppe have been ploughed for cultivation—not always very successfully.) It is now rare to find natural steppe here. The Ukrainian steppe, a vast open land, is an undulating ocean of giant fields. From the fields it is hard to see any villages, because they nestle in the valleys and gullies where there is water. Many parts of the steppe are badly gullied, because bad farming has exposed the soils to attack by the rain and wind. The soil is blown away from fallow lands by winter winds, or summer thunderstorms, and spring showers cut gullies between the crop rows on the parched ground. Winter is bitterly cold and strong winds from Siberia sweep across the open land carrying away the snow and exposing the soil to frost. Long periods of clear blue skies in still air make it more bearable than the overcast skies and greater humidity of the forest belt. Nor does it suffer from the raw and unhealthy incursions of Atlantic air in early winter that are experienced around Leningrad and Moscow. Summer is longer, drier and hotter than in the forest belt. The skies are often heavy with dust until cleared by a late afternoon thunderstorm, though the major rainfall is in spring.

The steppe lands are a great granary—despite the uncertainty of their rainfall. However, there is a need for careful farming if soil erosion, or surface salt formation in the south, are to be prevented. To help the farmer, tree shelter belts have been planted. Ponds have been built in gullies to collect any available source of moisture: winter snow, spring rain and summer thunder rain. The face of the steppe is changing, yet its vastness—one undulating wave passing into the next until the distant horizon—makes a deep impression on the traveller's mind.

Deserts of the Soviet Union

The grassy tussocks grow fewer as steppe merges into semi-desert. Around the southern shore of the Aral Sea, and in places around the Caspian, true desert is found. If it can be watered this becomes a rich country. Oases along the rivers have long been cultivated, and the areas where mountain streams had their deltas on loess-covered footslope plains were early centres of civilization. In these riverine oases conditions are similar to those along the Nile. Contrasting with the desert of much of the region, jungle can still be found in the untamed delta of the Syr Darya.

Beyond the richness of the oases, in the true desert-steppe, there are only the tamarisk and groves of weird, dead-looking saxsaul. The brief rains in spring bring a carpet of bright flowers. Many ancient oases were devastated by medieval Mongol raids. Those surviving have been improved since the Russian conquest by replacing ruined or outworn irrigation systems or by

FIG. 73. *Levelling land for a new State Farm in the Syr Darya region.*

FIG. 74. *Mechanized harvesting on a tea plantation in Georgia.*

building new ones. A notable example has been the elaborate system of canals built in the Fergana basin in order to expand cotton growing.

Small Areas of Special Note

The south-west corner of the Caucasian isthmus, the Kolkhid Lowland of Georgia, has an intensely warm and humid climate where tea and fruit crops may be raised on soils similar to sub-tropical red soils. In the Far East, in the Amur Basin, temperate monsoonal influences penetrate, and vegetation is composed of species found in North China and Manchuria. It is from here that the Soviet Union can obtain rice and soya beans. The forests of the Sikhote Alin and Khingan mountains contain trees not found elsewhere in the country.

QUESTIONS

1. Give a brief description of the characteristic annual climate over much of Russia.
2. What is permafrost? Describe its effects on economic development.
3. Describe the steppe lands, and give details of their utilization today, and how they may be developed in the future.

CHAPTER 19

THE SOVIET PEOPLE

Historical Development

THE Eastern Slav tribes, from whom the main divisions of the Russian people grew, spread outwards from the Pripyat marshes between the second and fifth centuries A.D. They rapidly occupied almost empty forest. By the tenth century, with Viking help from Sweden, they had established their own states and accepted Byzantine (Greek) Christianity. However, they were pushed from the steppe fringe by rising nomad power: culminating in their subjugation within the Mongol empire in the thirteenth century. As a result their colonization was turned north and north-eastwards against a thin population of Finnish peoples in the forests. From the fifteenth century on, under the leadership of the Moscow princes, the Slavs again began to extend their power. Late in the sixteenth and during the seventeenth century, they spread rapidly across Siberia. Expansion westwards was slower, for Poland remained strong until partitioned in the eighteenth century. But at this time too, Russia reached the Black Sea and the foot of the Great Caucasus against the failing Turkish power. Early in the eighteenth century the Russians defeated the Swedes and reached the eastern Baltic, where they founded St. Petersburg as a "window on the world." Finland passed to the Tsar in 1809. The nineteenth century was marked by a quick spread across the Central Asian desert to the mountain foot, completed in the last fifteen years of the century, much to the distaste of Britain. Early in the nineteenth century there had been a relatively peaceful absorption of Transcaucasia. From 1858, as the Chinese gradually lost their hold of their vast but vaguely defined empire, the Russians spread into the Amur basin. They also established spheres of influence in Tuva, Mongolia, Manchuria and even in western China. Russian expansion in Manchuria was halted by the war with Japan in 1905. After the First World War and Revolution, the defeated Russian State lost possession of Finland and the Baltic states, as well as a large western area to Poland and Romania. But, since 1945, Russia has retaken most of this territory in Europe, and strengthened her position in Eastern Asia.

The Soviet Union embraces over 160 national groups—though the Slavonic Russians (Great Russians, Ukrainians, Byelorussians) form over three-quarters of the total population. While local language and culture is preserved and fostered, Great Russian is the state language, written in the Cyrillic alphabet. Only a minority of Soviet citizens are observing Christians—mostly following the Russian Orthodox Church. The peoples of Central Asia and Caucasia are largely Moslem. The primitive tribes of Siberia are pagans.

Fifteen of the principal national groups have their own republics. These form the Union of Soviet Socialist Republics. These republics have a large measure of autonomy and a "common frontier" with countries outside the Soviet Union. This is in case they wish to exercise their theoretical constitutional right of leaving the U.S.S.R. Other smaller and less developed groups have fewer measures of autonomy, according to their size and achievements. Even within these republics there are large minorities of Slavonic settlers where the titular group is not Slavonic: e.g. Slavonic settlers actually outnumber Kazakhs in the Kazakh Republic.

Though the *Politburo* in the Kremlin fixes most policy, there are two bodies corresponding to those in our parliamentary system: the Soviet of the Union is elected by all citizens with a vote, who send one Deputy for every 300,000 voters. The Soviet of Nationalities has twenty-five Deputies from each of the fifteen union republics, eleven from each autonomous republic and so on down the scale. Legislation must be passed by both Soviets to become law.

Population Distribution

We can see from Fig. 76 that Soviet population is concentrated in European Russia, which contains three-quarters of the total. In fact, south of Leningrad and west of the Volga, two-thirds of the Soviet population live in one-sixth of the country's area. Nevertheless, even within this area, densities are only moderate, with somewhat more thickly-settled areas around the larger towns. Such clusters stand out around Moscow, Leningrad, Kiev and Kharkov, as well as around Gorki on the Volga, and at favoured places on the Baltic coast. Denser population is also seen on the Donbass coalfield and around the towns of the Dnieper bend. Densities in the countryside are greatest on the better farming lands of the south, notably in Volhyniya and Moldavia. The swamps of Pripyat and of the Oka-Don lowlands stand out as sparsely-settled territory. The belt of relatively well-settled country extends east of the Volga into the new oilfields. In the Urals people live notably in the central, lower part, and in the industrial towns of the eastern slope. An area of thicker settlement also exists in western Siberia, in the industrial towns of the Kuzbass coalfield and along the Trans-Siberian railway. This distribution suggests a settled triangle that contrasts with the sparseness of the rest of the U.S.S.R. This triangle lies in the best farming country, like a great wedge between the northern coniferous forest and the dry southern steppe. Outside this wedge, there are small patches of densely-settled country in northern Caucasia on the rich wheatlands around Stavropol and Krasnodar, but towards the Caspian there is only empty steppe. The good farming lands of Transcaucasia, with its fruit orchards and tea gardens, are well settled, as are the fertile oases of Central Asia. Small centres of population occur along the Trans-Siberian railway in southern Siberia and the Far East, notably around Vladivostok.

Elsewhere densities are low, seldom exceeding three people per square mile. This is true of northern and north-eastern Siberia, the deserts and poor steppe of Central Asia, and in even the poorer parts of north-eastern European Russia. Yet everywhere small groups of settlers are found along rivers and in mining and trading settlements.

Townspeople of Russia

The total population of Russia rose from 147 million in 1926 to 236 million in 1968, despite the loss of over 20 million dead in the Second World War. At the same time, townspeople increased from 26 million to 131 million, while, in contrast, country dwellers decreased by over 13 million. Thus, by 1968, more than half the Soviet population lived in towns. The large towns have continued to grow and nine cities have more than a million people, including Novosibirisk, founded in 1894. Great new cities have risen since the 1930s, for example: Karaganda (505,000), Magnitogorsk (360,000), Komsomolsk (210,000), Norilsk (127,000), Vorkuta (55,000), Angarsk (a phenomenal increase to 187,000 since 1948) and many others. The centres of big towns contain many fine buildings, but there is still a severe housing shortage and wooden houses are common in suburbs, especially in the

FIG. 75. *Outskirts of the new town of Divnogorsk, on the Siberian "frontier."*

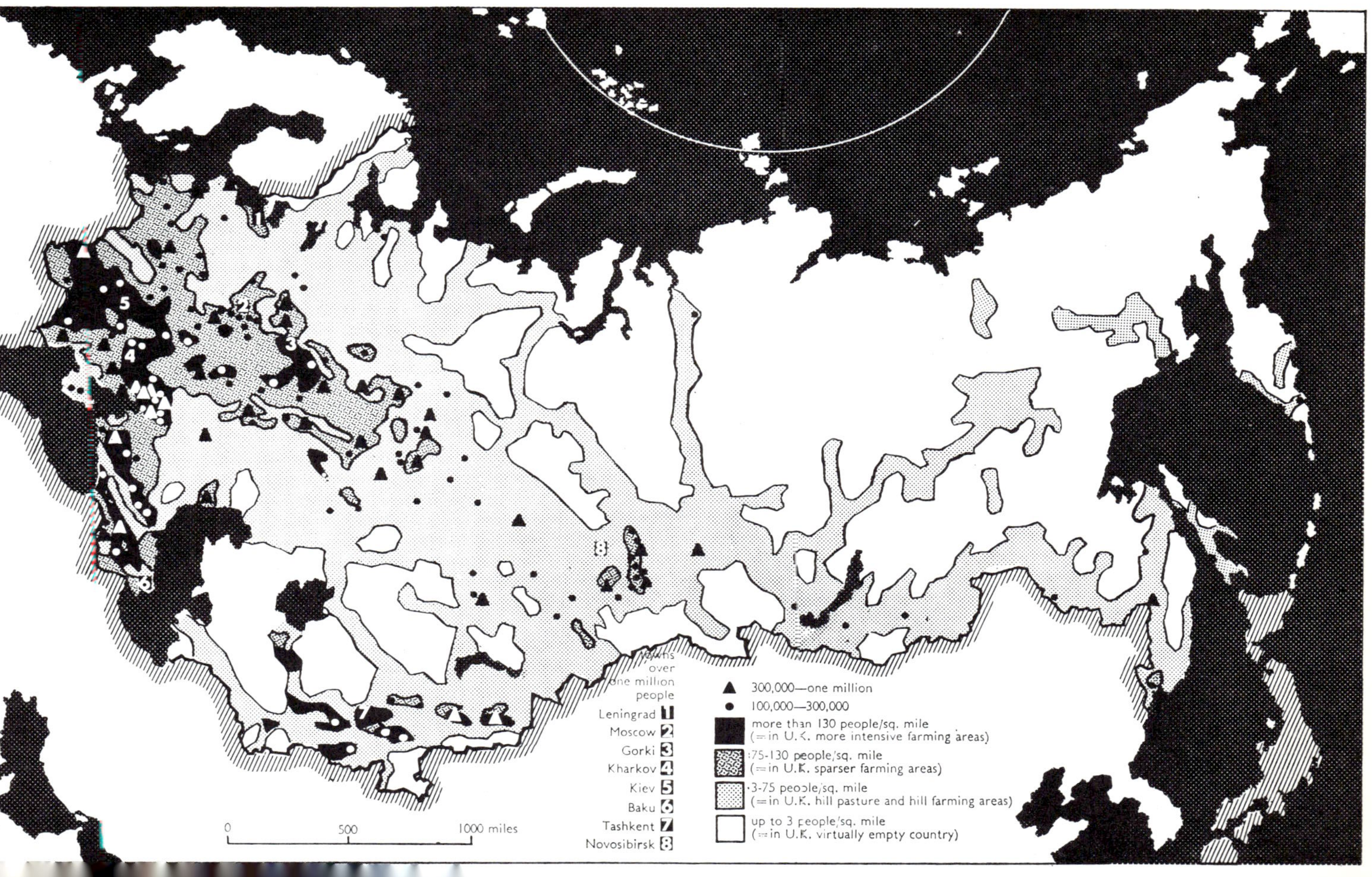
Towns over one million people
Leningrad 1
Moscow 2
Gorki 3
Kharkov 4
Kiev 5
Baku 6
Tashkent 7
Novosibirsk 8
300,000—one million
100,000—300,000
more than 130 people/sq. mile
(= in U.K. more intensive farming areas)
75-130 people/sq. mile
(= in U.K. sparser farming areas)
3-75 people/sq. mile
(= in U.K. hill pasture and hill farming areas)
up to 3 people/sq. mile
(= in U.K. virtually empty country)
0
500
1000 miles

FIG. 77, TOP, *Contrasts in style: the elegance of eighteenth-century Leningrad, and* FIG. 78, BELOW, *a new housing estate in the Moscow suburbs.*

smaller towns. Replanning of towns destroyed during the war has been generously carried out. The new towns are usually planned so that their streets focus on the entrance to their main works or mine. In every town there is abundant greenery and large parks. Unfortunately, workmanship is often shoddy, and architecture frequently oppressive. Nevertheless, everywhere the Russians are feverishly building new homes and public buildings.

QUESTIONS

1. Give an account of the national and cultural diversity of the Soviet people, and describe the ways in which they are combined together within the U.S.S.R.
2. What are the most densely settled areas of the U.S.S.R., and how is the distribution of population changing?

CHAPTER 20

RESOURCES, INDUSTRY AND OCCUPATIONS IN THE SOVIET UNION

THE U.S.S.R. is exceptionally rich in minerals. It has large resources of coal and growing reserves of petroleum. There are rich deposits of iron ore and its alloy metals, besides non-ferrous metals like lead and copper. Plenty of minerals for the chemicals industry are found. There is abundant water power to generate electricity. However, some minerals—like tin and tungsten—are scarce, but these can usually be obtained from other Communist countries. Unfortunately, some mineral deposits are hard to work or are found in inhospitable, remote regions.

Coal still supplies about half the energy. The main fields are the Ukrainian Donbass, the Siberian Kuzbass and the Karaganda in Kazakhstan. Along with the poorer Ural fields, these contribute three-quarters of the output. Another tenth comes from the Moscow lignite field. Of growing importance is the Vorkuta field in north-eastern European Russia, while immense reserves are said to exist in central Siberia. The Donbass, Kuzbass and Karaganda fields supply the Ural area, which has not enough coal of its own. Coal from the Donbass and Vorkuta is also sent to Moscow and Leningrad.

Oil and Natural Gas

The last ten years have completely changed the pattern of Soviet petroleum supplies—now thought to be abundant. The major reserves lie in the newly developed fields between the Volga and the Urals, which contain three-quarters of the total reserves. A half of Soviet production may now come from the northern wells of these fields. However, the older oil centres of Baku and northern Caucasia remain important, and oil is now also extracted from wells drilled in the floor of the Caspian Sea. Other fields supply only small quantities. Pipelines are being built to speed the distribution of petroleum. The refineries are mainly along the Volga (served by tankers), in northern Caucasia and near Baku. New refineries are being built in Siberia (Omsk, Angarsk), where there are synthetic petrol plants removed from Germany after 1945. Natural gas comes not only from the oilfields already mentioned, but also from new sources. Underground gas reservoirs are known on the lower Ob river, near Bukhara, and around Lake Issyk-Kul.

Electricity

From the early days of the Revolution electricity generation for industry and farming has been considered of primary importance; yet only two-thirds of the Soviet farms have current supplied. Most current is generated by thermal plants, found in all the bigger towns. Atomic generators have been built at Obninsk (Moscow), Voronezh, Ulyanovsk and at Byeloyarsk in the Urals. Of the larger hydro-electric plants, the first was commissioned in 1926 in Volkhov, near Leningrad. Most impressive are the Dneproges dam on the Dnieper and the huge barrages on the Volga at Volgograd, Kuybyshev and Gorki, all built since 1945. Larger plants are to be built in Siberia, eventually to supply current to European Russia.

The mineral deposits of the Soviet Union are widely scattered. The picture is nevertheless dominated by the Urals, and the many rich deposits of the eastern Ukraine, western Siberia and the Altay country. Other important workings occur in the Central Asian and Caucasian mountains and the deserts of Kazakhstan. With ruthless determination, the Russians have started mining in the Arctic as well as in the arid desert.

FIG. 79. *Part of the Tsimlyanskaya hydro-electric station on the Volga-Don canal.*

Leningrad
Moscow
Donbass
Ural-Volga
oilfield
Karaganda
Kuzbass
Central Asia
0
500
1000 miles
engineering
chemicals
iron and steel
textiles
coal mining
petroleum
iron ore mining
other minerals mining

Industry

Modern industry did not really begin in Russia until after the serfs had been freed in 1861; it was still in a comparatively undeveloped state by the outbreak of the First World War. Then the disorganization and destruction during the Revolution and Civil War (1917-20) meant a new start for industry. Regions which already had industry were quickly repaired and restarted. Emphasis was put on building industry in eastern parts of the country where it did not already exist. The first stage was to build a big industrial base in the Urals and western Siberia. Though 1,500 miles apart, they were to supply each other with coal and iron ore. Coking coal, in short supply in the Urals, was sent to them from western Siberia, where it was thought—wrongly as it turned out—that there was little iron ore. Big new steelworks were built at Magnitogorsk (in virgin steppe) and Nizhniy Tagil in the Urals, and in the Kuzbass at Novokuznetsk. The plan did not work well, as the railway across the steppe could not cope with the sudden growth of heavy traffic. Coal was also supplied from Karaganda, while in the Kuzbass iron ore was found in the mountains to the south.

Lack of industry in eastern Siberia and poor communications from the west had been important factors in the defeat of the Russians by the Japanese in 1905. Under the Soviet régime new industries were built in the Far East, notably a steelworks at the new town of Komsomolsk, while a big railway works was opened at Ulan Ude. In 1931, a railway was completed from western Siberia to the Central Asian oases, so that wheat, coal and steel could be exchanged for cotton and raw minerals. Textile mills were built at Barnaul in Siberia, and a steelworks at Begovat—using Siberian pig iron and local scrap—opened in Central Asia. The supply of Siberian wheat meant that more land in Central Asia could be used for cotton-growing.

Discovery of oil between the Volga and the Urals brought new industrial growth along the river. In 1941, the temporary loss to the Germans of important industrial resources in the Ukraine and northern Caucasia, forced the U.S.S.R. to develop new areas. Coal from Vorkuta in the Pechora basin began to replace the lost coal of the Donbass, while oil from Ukhta replaced that formerly obtained from the North Caucasian wells. Karelia sent iron ore to offset the loss of Krivoy Rog. Plant and workers, moved from the path of the advancing Germans to the Urals and western Siberia, also helped the growth of industry in these interior regions. Though some returned to their old areas after the war, many remained in the more easterly regions. In the late 1950s, the growth of southern Siberia as a heavy industrial centre was foreseen. Iron ore would come from the Angara-Ilim basins and coal from southern Yakutia, while vast hydro-electric stations would supply current. Nevertheless, three-quarters of Soviet industrial capacity and output is still west of the Urals, chiefly in the Ukraine and around Moscow.

Three-quarters of iron and steel output of the U.S.S.R. comes from the Ukraine and the Urals. In the Ukraine the main producers are towns of the Donbass coalfield and the Dnieper bend, while Zhdanov on the shore of the Sea of Azov is also important. They use Donbass coke and mostly Krivoy Rog ore, but some ore comes from Kerch in the eastern Crimea. In the Urals the largest works are Nizhniy Tagil, Magnitogorsk and Chelyabinsk. Ore comes from local mines or from new mines in Kazakhstan. Coal is brought from the Kuzbass and Karaganda, and is sometimes mixed with inferior Ural coal for coking. A tenth of the production of iron and steel is from western Siberia, particularly the big works of Novokuznetsk on the Kuzbass coalfield. Slightly less comes from Tula and Lipetsk, and also from Moscow, which makes special steels. Besides the new steelworks of Komsomolsk in the Far East and Begovat in Central Asia, a new works has been built at Cherepovets on the Rybinsk reservoir 250 miles east of Leningrad. It uses Vorkuta coal and some Karelian ore. Temir Tau, opened in 1961 near to Karaganda, uses local coal and ore.

Moscow and the towns around it dominate the manufacturing industry. High-quality engineering, electrical goods and motor vehicles are especially

FIG. 81. *Opencast coalmining in the Karaganda region.*

FIG. 82. *Soviet oilfields and pipelines.*

important. The greatest concentration of textile mills still lies east of Moscow. Smaller concentrations of manufacturing industries are also found in all the larger towns in European Russia, notably Leningrad, Kiev, Minsk, Riga, Kharkov and Volgograd (formerly Stalingrad). On coalfields and orefields engineering industry tends to specialize in the heavier articles. The large industrial towns of European Russia also manufacture the lighter chemical products, but on the orefields and coalfields are made the basic, heavy chemicals. Moscow is very important for light chemicals, drugs and cosmetics. Leningrad makes natural and synthetic rubber articles. Apart from Berezniki and Solikamsk on the Kama salt fields, the Volga basin is emerging as a major producer of chemicals, usually associated with the increasing output of petrochemicals from the new oil refineries. New industrial centres include the big farm machinery works at Tselinograd (for the Kazakh wheatlands), and the tractor works of Rubtsovsk in the Altay. Machinery for diamond and gold-mining in eastern Siberia is made locally. There is also a new chemicals industry on the Baykal coalfield at Cheremkhovo, and nearby at the new town of Angarsk.

Farming

Even after the serfs had been freed in 1861 the Russian peasant remained a backward farmer. He had few means with which to improve his methods or his equipment. Traditionally, the forest belt always was short of food, because of its poor land and uncertain harvests. It imported food from the southern grain-growing steppe, where production rose as Russian colonization progressed in the eighteenth and nineteenth centuries. Many peasants in the forest belt made a living by hand-producing certain articles as well as by farming.

After the Revolution, the growth of industry and of the number of people in towns drew many peasants from the countryside. A new solution to the problem of how to feed them had to be found. Rather against their will, peasants were gathered into collective farms. In some regions where peasants opposed the new farms, or where farming was newly commenced, state farms were started. Such state farms still contribute a major part of the food on the Russian market. Peasants still own about half the animals and grow a large share of most food crops. Their own produce can be sold as they wish. The extent of this private farming and Government toleration of it varies greatly from year to year. Farming has been hampered by the unpopularity of the collective farm and the conservative attitude of the peasant. Even a short journey in the Russian countryside shows that standards vary much from farm to farm. It is still possible to see old-fashioned and the most modern methods and equipment side by side. Many of the ablest people have also been drained away to the towns. Further, Government planning has often neglected farming in favour of industry or transport. Nevertheless, new methods have been introduced, as well as new crops and better strains of plants and animals. There has been provision of machinery, shelter belts and irrigation. But nothing can be done about the harshness of the physical conditions in which the Russian must farm.

Mixed farming predominates in the triangle of well-settled country described on page 214. In the northern forest lands it is concerned with dairying, rye and flax cultivation. Increasing areas are being sown with rapidly ripening types of wheat and, in Byelorussia, maize is grown for fodder. Standards are being improved by using more fertilizer and by draining land adequately. In the southern wooded steppe there is much arable farming: with sugar beet and potatoes as important crops; but as one travels south, more fields are sown with wheat and other grains. It is common to see huge combine harvesters at work in the vast fields at harvest time. The chief crop of the true steppe is spring wheat, but crops such as maize and oilseeds are often seen. Throughout the southern farming belt there are extensive fields of bright yellow sunflowers, grown for their oily seeds. In the western Ukraine sugar beet is an important crop, and the skyline is broken by the tall, thin chimneys of refineries. Many animals in the wooded steppe and steppe are stall fed on sugar beet waste or coarse grains, whereas the haylands and rank meadows of northern Russia are used for grazing dairy animals. New crops,

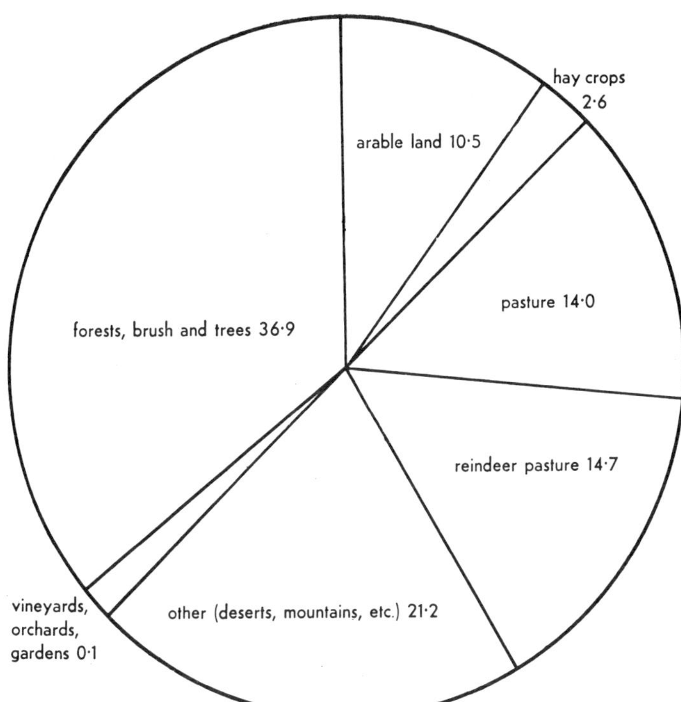

FIG. 83. *Land use in the U.S.S.R.*

like the rubber-bearing dandelion, *kok sagyz*, discovered in the Central Asian mountains in 1934, have been introduced. Tobacco is an important crop, while potatoes are often grown for starch or alcohol. In some favoured areas, like Moldavia and the Crimea, the vine is cultivated, the Crimea and Caucasia supplying the best wines. Naturally, farms near to big towns produce fresh milk, meat and vegetables for local consumption, and those farther away tend to send their produce for preservation or for canning. This is true of areas around Krasnodar and Stavropol in northern Caucasia, for example, where Krimskaya has one of the largest meat-packing plants in Russia. Russians use a lot of sour cream and milk, and they cook in sunflower oil, whose seeds they also chew—rather like chewing gum. Honey is also very popular, with some of the best coming from the upper Volga.

Farm settlement in the Siberian wooded steppe in the nineteenth century also spread along the rank meadows by the great rivers. Even before 1914 there was an important dairying industry here. Russian and Ukrainian farmers who settled in eastern Siberia and the Far East have had to adopt Chinese methods of cultivation, and even Chinese crops: like soya beans and perilla. Chinese and Koreans grow rice around Lake Khanka, while Russian settlers dairy farm in the Zeya-Bureya valleys.

Population growth, particularly in towns, and the increased consumption of food related to the undoubted rise in living standards, have placed an ever greater burden on the Russian farmer. For political reasons the Soviet Union has been reluctant to buy food supplies from non-Communist countries, although wheat was bought from the U.S.A. in 1963 and from Canada in 1964. Most important has been the expansion of bread grain and animal fodder cultivation. In the immense area of unused steppe extending from northern Caucasia across the southern Urals into western Siberia and northern Kazakhstan to the foot of the Altay, the Soviet authorities have conducted, since 1950, a gigantic ploughing campaign. This "Virgin Lands" scheme covers an area as large as peninsular Europe. It has been a bold move and not altogether a successful one, because rainfall is low and unreliable. Strong winds in winter and spring sweep away loose, friable soil, or crops may be washed out by thundery downpours in summer. Both spring and autumn are really too short to allow the farmer enough time to achieve all his tasks. Despite such serious natural hazards, colossal state farms have been formed, and over three-quarters of a million people brought from other parts of the country to work them. There is a plan to build a canal from the Ob to Karaganda to water the drier south-eastern corner. It is a gamble to reap a relatively low-yielding harvest against all these hazards and problems, and so far the results have fallen short of expectations, but the scheme, even if only in modified form, is essential if the growing industrial population of the Urals and Siberia is to be fed.

To reduce the burden of carrying food to the mining settlements and to avoid using the Northern Sea Route of the Arctic, special "Arctic" farms have been started. These also provide the fresh foods so vital in maintaining health

FIG. 84. *Contour-ploughing for fruit growing in Uzbekistan. The trees will also reduce erosion and retain moisture.*

FIG. 85. *A maize-grading factory in the Alma-Ata region, Kazahkstan.*

by the production of meat, milk and vegetables. They have had to be very carefully sited to take every advantage of slightly easier local conditions. At Igarka, for example, a sandy island free of permafrost in the middle of the wide Yenisey is used, for the warmth of the river water gives a local temperature several degrees above the prevailing average. Even with the very high costs involved in "Arctic" farming, it is easier (and often cheaper) to produce food this way than carry it along slender and difficult routes often closed for long periods in winter.

Small areas of unusually favourable soil and climatic conditions help to produce crops that make the U.S.S.R. largely independent of outside supplies. One of the most important of these areas lies at the foot of the Central Asian mountains and in sheltered basins within the mountains. Well-watered areas on fertile loess soils can produce valuable crops like cotton, mulberry and rice and sub-tropical fruits. Along the rivers that drain into the desert sands, oasis conditions and irrigation produce about a tenth of world cotton output. On moist but non-irrigated lands around them, crops of grains and grasses and, in some places, sugar beet, are raised.

A second important area lies on the sticky red soils of humid western Transcaucasia. Here tea and citrus fruits are grown, as well as valuable tung oil for paints and varnishes. (Although the fruits of this region are usually tinned, dried or otherwise preserved for shipment elsewhere, so vast are the distances, and so high the transport costs, that in the heavily populated area of European Russia they are luxuries, as are lemons, grapefruit and peaches. Cocoa, which has to be imported, is a great luxury. Tea is commonly drunk, imports from China and India augmenting home production. Imported coffee is a popular beverage.)

Pastoralism in the Soviet Union

In Central Asia and southern Siberia, as well as in the Caucasian isthmus, pastoralism remains important. In mountains and plains wherever the steppe has not been occupied by farmers, nomad tribes raise animals for meat, hide, wool and tallow—as they have done for ages. Cattle are kept mostly near the fringe of the settled farming country. Their meat and hide is processed locally and sent away to other parts of the country. Sheep are more widespread. A breed important for its fine wool is the curly Astrakhan sheep, while other breeds have been improved by crossing with European types. Horse raising was once more important, for until recently the Soviet army had many fine cavalry regiments. Russian strains of Arab horses and the small Mongolian horse are bred. In western Turkmenistan Bactrian camels are raised, and in the Altay and Sayan mountains natives breed maral deer and yaks. Nomads of the northern forests raise reindeer and dogs for sledging and hunting.

In the last forty years the life of the nomad herder has been made easier. Wells have been dug in arid lands; fodder has been provided in both the northern tundra fringe and in the steppe; weather forecasting has been introduced, and aircraft used to maintain contact with the nomads. These factors have helped to eliminate hunger among the flocks and do away with the long drives between pastures, so reducing the annual loss of stock. The nomad is no longer at the mercy of unexpected blizzards or droughts. Unfortunately, in the early days many nomads rebelled against the new methods, and thousands of valuable animals were lost.

One of the earliest attractions of Siberia was its wealth of fur-bearing

FIG. 86. *Arkhar-Merino sheep in Kazahkstan.*

FIG. 87. *A maral farm.*

animals, which slowly retreated into the most remote corners of the forests. To counteract their reduction in numbers and the difficulties of hunting in the remotest forests, farms have been established to breed animals for furs; and it is now forbidden to hunt certain of the rarer types. The Soviet authorities have also established special reserves where the fast-disappearing animals of the steppe may be preserved. Similar reserves exist in other areas: for example, in the Belovezh Forest of western Byelorussia, home of the last European bison.

QUESTIONS

1. Describe the growth of the Soviet petroleum industry, name chief areas, and show the importance of pipelines.
2. What are the chief iron-and-steel-producing areas in the U.S.S.R.?
3. Describe the Soviet system of agriculture. What are the chief areas, and the types of farming involved?

CHAPTER 21

TRANSPORT AND COMMUNICATIONS

To keep contact over great distances and under harsh climatic conditions has always been one of Russia's major problems. Rivers were the traditional routeways along which people and goods were moved. The portages where goods crossed from one river system to another were usually commanded by fortified towns of great historical importance. Both land and water transport are more hampered by climatic influences than by difficulties of relief. The railway has been found most suitable as a means of transport, because it can overcome not only the hindrances of climate but also the great distances that have to be covered. Today, well over three-quarters of the goods and passengers move by train. Aircraft are used for special goods and for a growing number of passengers. Pipelines are also being built. Although the U.S.S.R. has a very long coastline, most movement is within the continental heart of the country, so coastal sea routes are relatively unimportant.

The first short railway was built in 1837. Yet by 1860 only two main lines existed: from St. Petersburg (now Leningrad) to Warsaw and to Moscow. Thereafter building was more rapid, and by 1900 a widely spaced grid of railways covered the country west of the Volga. By 1910 the first links into Siberia as far as Lake Baykal, and into Central Asia as far as Tashkent, were complete. After the Revolution, several important trunk lines were built. Most important was the completion in 1931 of the Turksib railway, from Semipalatinsk to a point near Tashkent. Another vital link from the Urals to western Siberia, the South Siberian Railway, was begun before the war, but not completed until 1955. It runs from Magnitogorsk through Tselinograd and Barnaul to the Kuzbass. During the Second World War a railway was built into north-eastern European Russia to the coalfield near Vorkuta. Much effort was also spent in improving technical standards on the main railways. Soviet railways are designed primarily to move heavy, bulk freights: these are coal, ores, cement and timber. Passengers are of secondary importance, and mainline expresses move little faster than the goods trains.

The great open plains or rolling hill country mean gentle gradients, but there are more serious obstacles to railway building than relief. Marshy

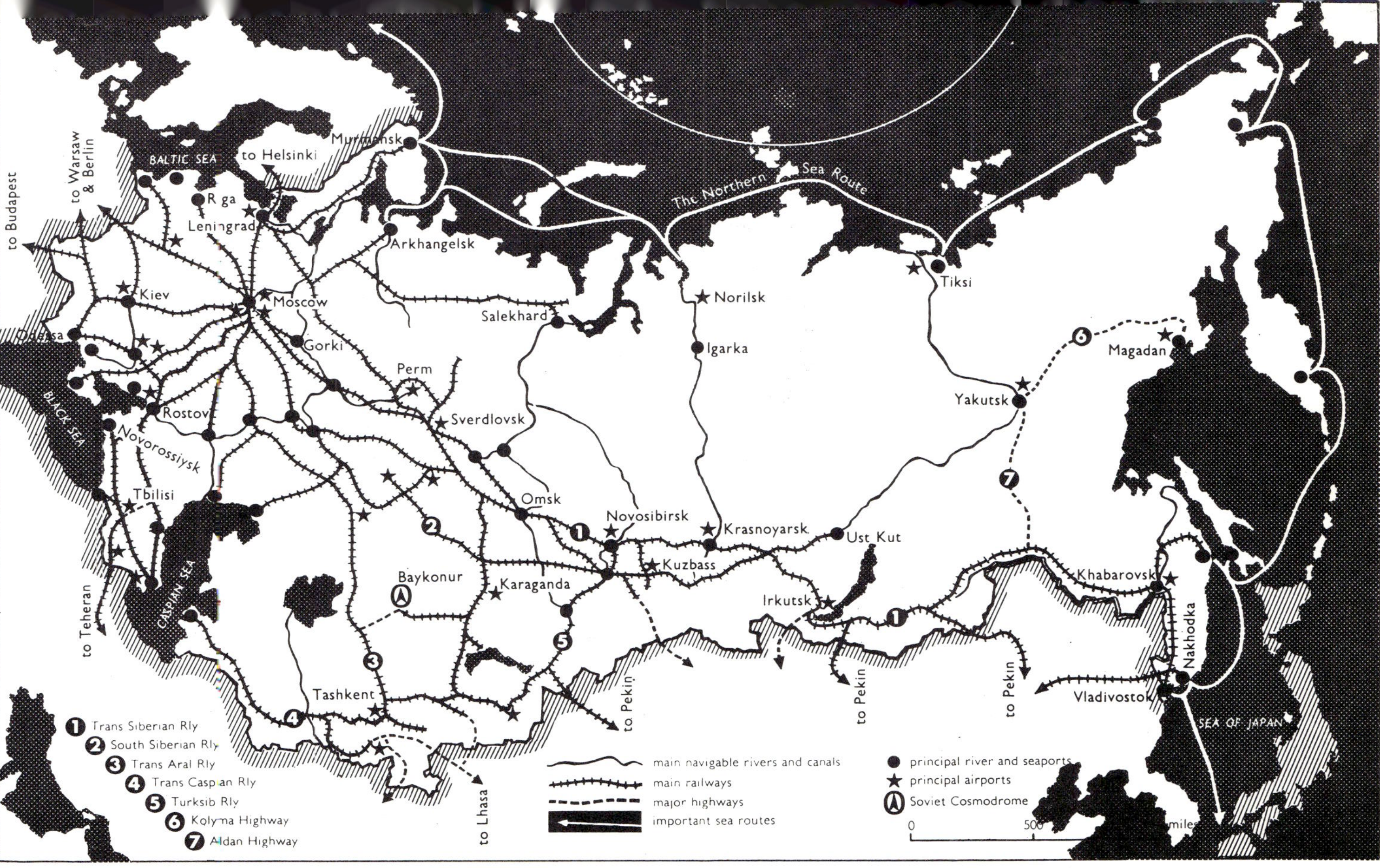

FIG. 88. *Main land, water and air transport routes of the U.S.S.R.*

FIG. 89. *The Trans-Siberian railway, near Lake Baykal.*

ground makes it hard to lay a stable roadbed; wide rivers with broad flood plains demand excessively long bridges; erosion gullies in the steppe mean many short bridges and embankments, and there is the danger of washouts in thunder showers. Once the railway gangs enter the country east of the Yenisey river they face the hazard of permafrost, which can deform the track and tilt bridges. In the arid south, water shortage is a real problem. In the Russian countryside shelter belts along the railways are a common sight. They prevent drifting snow or sand from blocking the track, and break the force of winter gales. In such a flat country tunnels are rare. Steep gradients are found on a few short branches into the Caucasian mountains and on the main Baku-Batumi line through the Suram Pass. Some heavily graded lines are also found in Central Asia and in Transbaykalia near Ulan Ude, but there are a few steep climbs in the Urals and the Donbass coalfield.

Two-thirds of the traffic is concentrated on the quarter of the railway mileage which has multiple track. The single track lines are often very lightly built and carry extremely light traffic. On most branch and secondary lines there is only one passenger train a day in each direction. The busiest routes are from Leningrad through the industrial districts around Moscow, and those to the Donbass and Ukrainian industrial cities in the south. Other very busy lines are from the Moscow region across the Volga into the Ural region, and then eastwards into western Siberia. More traffic is also flowing into the Eastern European countries. Brest is the frontier station for Poland and Germany, while Chop serves Czechoslovakia and Hungary. Ungeni is the station for Romania and Bulgaria. Special facilities at these stations help the rapid transfer of merchandise from Russian broad gauge (5 ft. 0 in.) wagons into European standard gauge (4 ft. 8½ in.) wagons. The bogies of through passenger carriages are also changed here by special machinery. Similar equipment exists on the new railway from Ulan Ude in Siberia via Ulan Bator (Mongolia) to Peking, as well as on the Manchurian border for trains to Harbin and Pyongyang. Many main express trains have recently been speeded up, but it still takes nine days to travel from Moscow to Vladivostok. Of course, it is quite a long way—5,800 miles!

Great Rivers of the U.S.S.R.

During the long winter the rivers are frozen. In spring the thaw causes great floods. On the Siberian rivers that flow northwards, the upper reaches thaw earlier than the lower, northern, reaches. Water flows over the still-frozen parts and floods immense areas of country. In July and August the dryness of summer causes the water level to fall and hinders shipping again. Rivers flow generally northwards or southwards serving the limited areas of their own basins. Conversely, movement in Russia is mainly eastwards or westwards, and the use of rivers is therefore limited. It is too costly to trans-ship goods across portages from one system to another.

The most used river system is the Volga and its tributary, the Kama. It carries two-thirds of all Russian river traffic. Though the Volga flows into the land-locked Caspian Sea, since 1955 it has been linked by a canal to the River Don and so into the Sea of Azov. The canal, across the narrow neck of land that separates the two rivers near Volgograd, corrects one of nature's errors. An old canal system in the north has also been enlarged, so that Volga boats can now reach the Baltic and White Seas. A canal also makes a short cut from the Volga to Moscow. Another important river is the Dnieper. Before 1939 a large dam near Zaporozhye made it easy for boats to sail from

FIG. 90. *A typical Siberian river—the Yenisey.*

the Black Sea up the river to Kiev. Now it is proposed to rebuild an old canal so that boats will be able to sail from the Dnieper to the Bug and the Vistula and so to the Baltic Sea. The Northern European rivers and the great rivers of Siberia are only of local importance. The short, swift rivers of Caucasia are too difficult to sail on, as are the silt-laden rivers of Central Asia.

Cement, oil, building materials and timber are the main goods carried on the rivers. Oil is particularly important on the Volga. Great rafts of logs may be seen on most rivers, but they are most common on the Northern European rivers, and on the rivers of Siberia. Tugs pulling strings of barges, as well as proper river cargo vessels can be seen. There are still many passenger ships. Large, well-furnished passenger ships and fast hydroplanes ply on the Volga and other rivers.

Russian Roads

In former times the only roads in Russia were the portages between one river and the next. The first proper paved roads did not appear until the eighteenth century. Even nowadays most Russian roads are still gravel or earth-surfaced. A few well-surfaced major highways have been built since 1945 to join together the great cities of European Russia. Roughly surfaced but well-engineered highways are also to be found in Siberia and Central Asia. They lead from railway terminals to the remoter parts of the country.

FIG. 91. *The Volga, "the mother of Russian rivers," is an important economic highway.*

In winter Siberian rivers are frozen hard and are sometimes used like roads. After their conquest, in order to control the unruly mountain peoples three finely engineered roads were built across the Great Caucasus. Some rough roads leading from Manchuria and Mongolia into southern Siberia are great trading routes. The Central Asian oases are linked to each other and to China by the ancient Silk Road, though traffic today is mostly by motor lorries. Nevertheless, it is still impossible to travel right across Russia by road. Road traffic in the U.S.S.R. is usually light and consists mostly of buses and lorries. One of the most obvious differences between everyday life in Western Europe and the Soviet Union is the very few private cars in the U.S.S.R. Transport in towns is usually by tram, bus or possibly taxi.

Seas Around the Soviet Union

The seas around the U.S.S.R. all suffer from the disadvantage that they are frozen for part of the year. Where the ice is thin and does not last long, ports may be kept open with icebreakers. At many ports, however, the ice lasts a long time and becomes very thick, so that it cannot be easily broken. Other hazards in Arctic seas and in the North Pacific is that the mixing of cold and warm waters in summer produces frequent fog, and there is the danger of icebergs.

The busiest Russian sea traffic is across the Black Sea from the Caucasian to the Ukrainian ports. There is also shipping entering the Sea of Azov. Unfortunately, the ports along the Ukrainian shore, where the water is shallow and fresh, freeze up in winter.

Another busy area is the inland Caspian Sea, where the direction of trade is generally from southern ports to the mouth of the Volga and Astrakhan. The depth of water in the sea has been falling for many years, and the northern part is now shallow; since sea-going ships cannot sail into the mouth of the Volga, they transfer their cargoes to river boats at an artificial island port well out to sea. The shallow northern Caspian is also frozen in winter. The Russian ports on the Baltic coast are similarly gripped by ice in winter, though it is only very thick and long-lasting in the narrow Gulf of Finland, where Leningrad, one of the country's chief ports, is sited. The former German ports of Kaliningrad (Königsberg) and Baltiysk (Pillau) can usually be kept open easily, and have become major Russian naval ports.

At the end of last century, Nordenskjöld sailed in his little *Vega* along the northern coast of Russia from west to east and completed the centuries-old search for a North-East Passage to the East. Under immense physical hardship and great cost, the Soviet authorities have developed this route from their western to eastern ports. For a few months in summer, ships can pass through. Navigation is not easy, for there is much fog and many shallows. Passage through the narrow Vilkitskiy Strait is often hard because of grounded icebergs. In winter, Murmansk in the Barents Sea is kept open by the warmer water of the North Atlantic Drift, but ports on the White Sea and even in the Barents Sea are closed by ice. Farther east, the ice lasts longer: it is often June before ships can enter the Kara Sea. Similarly, the ports of the Okhotsk Sea are closed for much longer than ports on the Pacific coast. The principal Pacific ports of Russia are Vladivostok and the new port of Nakhodka. Other ports are relatively unimportant, except as fishing harbours.

Because of the great distances and harsh terrain of Soviet territory the aeroplane is very useful. In winter, when land and sea are ice-bound, there is good flying weather in the clear, anticyclonic conditions. All major towns are now linked together by airlines. They also serve remote outlying settlements in the desert and tundra, or observation stations in the Arctic. In 1967 the Russian airline systems carried 55 million passengers.

QUESTION

1. What chief factors have influenced the Soviet transport system? How have they affected the development of (*a*) roads; (*b*) railways; (*c*) airways?

FIG. 92. *The nuclear-powered* Lenin, *the world's greatest icebreaker.*

AFRICA

CHAPTER 22

GENERAL

AFRICA is the second largest of the continents. It is nearly three times as large as all Europe, having a total area of about 11½ million square miles. It extends from approximately 37° N. latitude to nearly 35° S. latitude, so that the Equator almost bisects it, and most of its area lies within the Tropics.

Africa is a very compact continent with a notably even coastline and few large inlets or bays. As a result its coastline is actually shorter than that of Europe (15,000 miles to Europe's 19,000 miles).

Structurally, most of Africa consists of a huge plateau of ancient rocks. The plateau is higher in the south and east than in the north and west, and the coastal plains are mostly narrow. There are only two extensive ranges of fold mountains, one in the extreme north and one in the extreme south. In East Africa faulting of the plateau surface has formed the great system of rift valleys extending from Nyasaland to the Jordan valley. In South Africa, the uplifted edge of the plateau has the appearance of an eroded mountain range rising in the Drakensberg to over 11,000 feet. Volcanic activity has provided the greatest heights on the continent, Mt. Kilimanjaro reaching over 19,000 feet. This peak, although almost on the Equator, is permanently capped with snow and ice. In Ethiopia a great mass of land has been uplifted to an average of over 6,000 feet in height over an area of 300,000 square miles.

Drainage

The longest African river, the Nile, flows northwards for some 4,200 miles carrying the waters of a great system of freshwater lakes across the world's largest desert to the Mediterranean Sea. In the centre, the mighty Congo drains an area of 1½ million square miles of forest and grassland. These, and most other African rivers, are obstructed by waterfalls and rapids owing to the descent from the plateau to the coastal plain. In addition, water levels in most rivers vary greatly from one season to another. Because of these disadvantages the value of the rivers as highways of transport is much reduced. On the other hand, the waterfalls have given Africa greater resources of potential water power than any other continent, except perhaps Asia. The falls also hindered exploration and penetration of the interior.

Climate

The maps on pages 269-70 show the main facts of African climate. The temperature maps show that there are no "cold lands" in Africa, except for very small areas on the highest mountains. The area of greatest heat moves with the overhead sun, so that in January the hottest areas are south of the Equator, and in July the hottest areas are north of the Equator. The greater mass of land in the north becomes hotter than the narrower southern part.

This movement of the "Heat Equator" causes variation in pressure. In general the hotter areas have the lowest pressure, and winds tend to blow inwards towards the low pressure. These in-blowing Trade Winds import vast quantities of moisture from the warm oceans over which they have passed. In the centre of the hotter areas lies a zone of calms or light breezes where, due largely to convection, the converging masses of warm, moist air rise upwards. The subsequent expansion and cooling gives rise to heavy rainfall in these regions. The rainfall maps show the movement of the area of maximum rainfall north and south. The area with the heaviest rainfall corresponds with the overlap of the seasonal rainfall areas. To the north and south of this belt of maximum rainfall the precipitation becomes more and more seasonal and the total less.

The large northern desert area is caused by the fact that the Sahara is a region of high pressure in winter and therefore of out-blowing winds. The high pressure is caused by the masses of descending air which are being warmed by compression. They have no tendency to give up moisture; in fact, they tend rather to absorb moisture. This is, therefore, an area of clear, cloudless skies, with hot days and cold nights. In summer, a low-pressure area develops, causing in-blowing winds (e.g. the Guinea monsoon). However, the Sahara is so far from the sea that most of the rain is dropped before it reaches the Sahara proper.

The Kalahari desert of the south-west is caused by the fact that the winds from the east have lost most of their moisture on the high lands of south-east Africa. The aridity of the coastal region is accentuated by the fact that the Benguela current cools any air masses that happen to move inland from the Atlantic, and this tends to inhibit rainfall on the adjacent coast. Note the relatively small areas of winter rainfall in the extreme north-west and south-west. The rain of these areas is brought by the depressions of the westerly wind belts. Like the areas of high and low pressure, and the "Heat Equator," the wind belts move north and south with the overhead sun.

As a result of the distribution of temperature and rainfall over the continent, different parts have different climates and vegetation (Fig. 93).

Climate and Vegetation Zones

The Equatorial Zone. Temperatures are high throughout the year—the daily average not varying much from 80° F. The rainfall is heavy, usually about 80 in., spread throughout the year but frequently with a season or seasons of heavier rainfall. The resultant vegetation is dense, steamy forest with a great variety of tall hardwood trees, laced together with lianas. The environment is neither favourable to man nor agriculture: the forest is difficult to clear, soils become sterile after a few years of cultivation and new clearings have to be made. Diseases of all kinds flourish. Wild animals are scarce but insect life is abundant.

Tropical Summer Rain Zone. Here temperatures are still high, but there is a distinctly hotter season. The main difference is between the wetness of the hot summers and the dryness of the warm winters. The resultant vegetation is savanna grassland, for forest cannot flourish where there is such a pronounced dry season. The landscape has the appearance of a vast parkland with trees dotted about among the dense tall grass. The grass dies down during the dry season and springs up again, fresh and green, with the coming of the summer rains. This is the African "big game" country. The inhabitants of this vast grassland live by agriculture and by herding cattle. Some tribes, such as the Masai of East Africa and the Fulani of northern Nigeria, are purely herdsmen and depend entirely on their cattle. Parts of the savanna are favourable to more intense agriculture and have been settled by European farmers, e.g. in Rhodesia and South Africa.

Desert and Semi-Desert Zones. The areas shown as desert or semi-desert are those with less than about 16 in. (400 mm.) of rain in a year—in a continent like Africa where evaporation is high, this is the minimum necessary to support a continuous cover of vegetation. It should be borne in mind, however, that there is no sharp line between one type of climate and vegetation and another—as the rainfall decreases, so the vegetation becomes gradually more scanty. As well as having a low rainfall these areas are also marked by a wide range of daily temperature, the intense heat of the day usually being followed by a sharp fall in temperature at night. This is caused by the lack of cloud cover and rapid radiation.

Population is largely confined to oases, but a few nomadic herders wander along the wetter margins, their flocks and herds depending on the permanent scrub and the grass which springs up after the occasional rains. In the Saharan oases, date palms and a wide variety of other crops can be grown with the aid of irrigation. In the smaller Kalahari desert area of the south-west there is more vegetation and there is sufficient wild game to support the primitive Bushmen.

Mediterranean Zones. These areas have typical Mediterranean climates, with hot dry summers and mild winters. Most rain falls in winter, but the

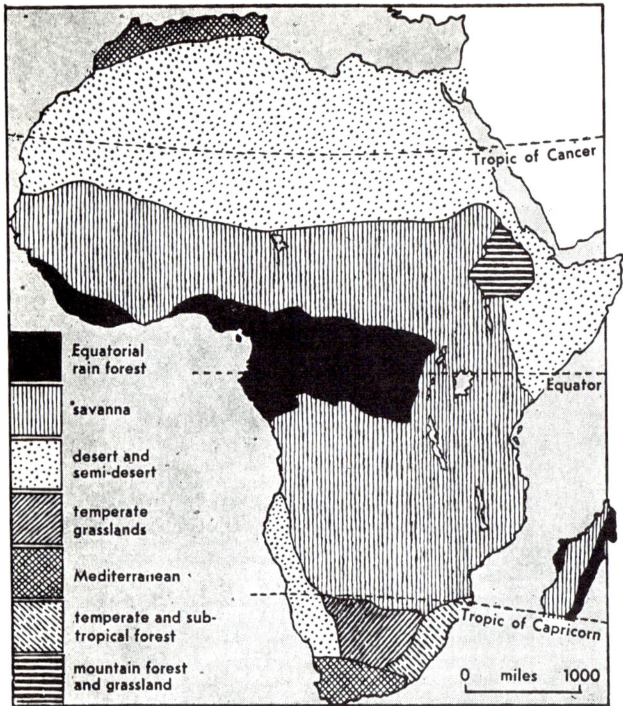

FIG. 93. *Africa—zones of natural vegetation.*

total rarely exceeds 30 in., except on mountains. Like other Mediterranean areas, these regions support a dense population who produce large crops of wheat, fruit and vines.

China-type Zone. This region, along the east coast of South Africa, has hot summers with abundant rainfall. Winters are warm with a much lower rainfall. The heavy summer rains are due to the strong on-shore winds drawn into the hot interior at this season. Rainfall decreases inland, but is over 30 in. nearly everywhere. The natural vegetation along the coast is forest, but much has been cleared and replaced by scrub. Inland, savanna vegetation is more typical. This is an area of white colonial settlement, sugar cane and sub-tropical and tropical fruits being the most important crops. The Africans in this region are chiefly pastoralists.

QUESTIONS

1. What features of the African coastline and of African rivers were a hindrance to early European explorers?
2. Write a reasoned account of the distribution of rainfall in Africa.
3. Find the full meaning of the following terms: rift valley, potential water power, Trade Winds, convection rain, monsoon, pastoralism.

CHAPTER 23

SOUTH AFRICA

THE Republic of South Africa consists in the main of a tilted plateau, higher in the east than in the west. It derives the greater part of its rainfall from the south-east Trades, although the southern tip receives rain in winter as the westerly wind belt moves north. In consequence the rivers originate in the higher, well-watered eastern districts, and have long courses flowing to the west.

Those flowing east have shorter, steeper courses. On this plateau and the narrow coastal plains has developed the most economically important country in Africa. It is the most advanced in agriculture and in industry, and it has the best transport system. Its people, both negro and white, have the highest standard of living on the continent, 50 per cent higher than that of their nearest rivals. Since 1945 the whole economy, and industry in particular, has made great strides, the total value of industrial production now exceeding that of agriculture and mining combined.

South Africa is a large country—nearly 500,000 square miles—five times as large as the United Kingdom (South-West Africa, which is administered by the Republic, covers another 318,000 square miles) but has a relatively small population—about 18½ millions in all. The racial composition of the population in 1966 was: Europeans, 3½ millions, 19 per cent; Bantu (Africans), 12½ millions, 68 per cent; Coloured (mixed race), 1¾ million, 10 per cent; Asians (chiefly Indians), ½ million, 2¾ per cent.

Relief and Climate Regions

The Cape Ranges. The ranges and plateaux of the south-west have a Mediterranean type of climate. This is because in winter the westerly wind belt with its associated depressions move north bringing rain. In summer the south-east Trades bring rain only to the eastern part of the south coast. Winters are mild and summers are hot and sunny. Here are figures for a typical station:

Cape Town. January average 21° C. (69° F.), July 13° C. (55° F.); rainfall 25 in. (635 mm.).

After clearing the dry scrub vegetation the European settlers found the area very suitable for growing both deciduous and citrus fruit and vines, and for the cultivation of wheat. Sheep are reared for wool, especially in the drier areas.

The East Coast Hill Region. This is a land of hummocky hills, chiefly grass-covered with thickets of bush in the hollows. A series of escarpments culminate westwards in the great wall of the Drakensberg.

Summers are hot and wet—rainfall usually totals more than 30 in. (762 mm.), brought by the south-east Trades which have blown over the warm Mozambique current. Winters are comparatively dry and mild. Most South African rivers flow only during the wet season—here the flow is maintained throughout the year and some, like the Great Fish River and the Sundays River, have been dammed for irrigation. This region contains a greater density of Bantu population than any other part of South Africa, and their most important resource is their cattle. Large quantities of mealies (maize) are grown as a subsistence crop. Wattle bark plantations have been established on the Natal Uplands. Nearer the coast, where frost is unknown, sugar cane, pineapples, bananas and oranges are grown.

Durban. January average 25° C. (77° F.), July 18° C. (65° F.); rainfall 40 in. (1,016 mm.).

South-West Africa. This area, chiefly the interior highlands of the south-west together with a coastal zone of pure desert, is very sparsely populated. On the interior highlands, rising to over 7,000 feet, the average rainfall of 10-15 in. (254-380 mm.) supports light vegetation. This allows cattle ranching of the extensive type, especially as underground water is often available.

Port Nolloth. January average 15° C. (59° F.), July 13° C. (55° F.); rainfall average 2 in. (50 mm.).

The Kalahari and Orange River Basins. With an average elevation of 2,000-3,000 feet this region is surrounded on almost all sides by land at least 2,000 feet higher. Owing to its interior location and the surrounding rim of high land, rainfall is deficient, the average being 5-10 in. (50-250 mm.). There is a partial covering of grass and scrub, and this, together with underground water in some areas, allows the rearing of cattle and sheep. But the number of animals per square mile is low. The primitive Bushmen (the original inhabitants of the area—having been here long before the arrival of the negroes), who live by hunting and gathering, appear to be dying out.

The Veld (High Veld, Upper Karroo and Transvaal Basin). The interior of South Africa is occupied by a high plateau, the altitude ranging from 3,000 to 6,000 feet. Because of the altitude, temperatures are lower than would be expected for the latitude. Rainfall decreases westwards from about 40 in. (1,016 mm.) on the escarpment to less than 15 in. (380 mm.) in the south-west. However, rainfall is notoriously variable, and may fluctuate by 20 in. (500 mm.) from one year to another. This variability can cause disastrous droughts, exaggerated by the high evaporation rates of the hot summers.

Johannesburg. January 18° C. (65° F.), July 10° C. (50° F.); rainfall 31 in. (787 mm.).

The surface of the Veld is usually level or gently undulating, with an occasional low ridge or isolated hill (*kopje*). The gold-bearing ridge of Witwatersrand is a typical example. The main crops of the Veld are maize and wheat, but sorghum (a·grain), cotton and tobacco are also grown. Near the towns there is market gardening and vegetables are grown in large quantities. Many dams have been constructed to conserve water for the dry season. Large numbers of cattle and sheep are reared, more cattle being reared in the north and more sheep in the south. Merino wool is South Africa's most valuable agricultural export.

Minerals and Industry

The table shows the value of the most important minerals in 1967.

Gold	£443,794,000	Diamonds	£34,302,000
Copper	£56,816,000	Iron Ore	£15,650,000
Coal	£45,116,000	Asbestos	£15,410,000

Figures are no longer available for uranium production, probably second only to gold in the value of output.

Other minerals produced in important quantities are platinum, manganese, antimony, corundum, silver, lead, titanium, kaolin, nickel and molybdenum.

Gold is South Africa's largest single earner of foreign exchange. The chief mining area is called the Rand (Witwatersrand). The gold occurs as fine particles in a hard conglomerate rock called "banket." This rock is obtained from deep mines and is crushed by heavy machinery before being treated with cyanide to dissolve the gold. Today, more than fifty large gold mines, producing 75 per cent of the Western world's gold supply, are spread along a 300-mile arc in the Transvaal and Orange Free State.

Uranium is derived as a byproduct from gold residues. Since the raw material is provided by the gold mines, this is a very profitable industry. Practically the total production is exported.

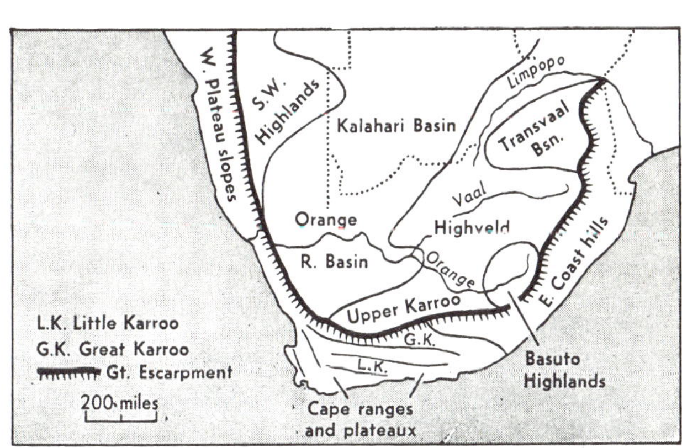

FIG. 94. *Regions and sub-regions of South Africa.*

Coal. South Africa probably has 80 per cent of all the coal in Africa—enough to last for hundreds of years even if production is increased. The largest fields are in the Transvaal, but there are also important fields in the Orange Free State and Natal. The coal is normally in thick seams at shallow depths—this, together with cheap labour and a high degree of mechanization, gives the country the cheapest coal and coal-generated electricity in the world. Production is now about 49 million tons a year.

Diamonds produced a "Rush" in 1869 after the discovery of the "Star of Africa" diamond. Soon afterwards the famous "pipes"—ancient volcanic necks—were discovered, and some of these were found to contain large quantities of diamonds. The main areas of production today are Kimberley, Luderitz and Pretoria.

Industries

Before the First World War South Africa was chiefly an agricultural country. Both wars gave a great impetus to manufacturing. By 1934 steel was being made and this led to the growth of engineering and metal industries. During the Second World War many imports were cut off, and this led to the local manufacture of supplies for agriculture and the mines. After 1945 many overseas companies established branches in South Africa, so that today a large proportion of its consumer goods can be provided by home industry, and manufacturing is now about as important in terms of value as agriculture and mining combined. There is a concentration of industry in four small areas:

1. **Cape Town and West Province.** Here the second largest port, backed by a rich agricultural hinterland, provides a market and plenty of negro labour. Main products are clothing, canned and processed foods and textiles. Other industries include distilling, printing and leather working.

2. **Port Elizabeth and Uitenhage.** Again a large port with a productive hinterland provides a market and labour. The industries include the manufacture of footwear, tyres, batteries, the assembly of cars and canning of food. There are also engineering and metal industries.

3. **Durban and Pine Town.** Durban is the largest harbour in South Africa, and is the nearest to the interior gold fields. A rich agricultural hinterland, together with abundant water and coal, and Asiatic as well as Bantu labour, provide the basis for large-scale industry. Products include paper, pulp, paint, chemicals, hardware, clothing, jewellery and furniture.

4. **Southern Transvaal.** This is the most important area of South Africa's industry. The markets were created by the gold mines, which also helped the development of water, power and transport services. Skilled labour was attracted by the gold mines, and, in addition to abundant coal, steel from Pretoria provides an important raw material.

The Orange River Project. During the next thirty years South Africa's

FIG. 95. *The city of Cape Town lies at the foot of Table Mountain.*

biggest river will be harnessed in the world's largest irrigation and hydro-electric scheme which will change the face of an area twice the size of the United Kingdom. Total cost of the project will be about £225 million. 750,000 acres of arid land will be irrigated by the building of three great dams near Bloemfontein, and the diversion of water by a 51-mile tunnel through the mountains to the Eastern Cape. There will be 20 hydro-electric stations.

Problems of South Africa

Although the most prosperous of all African states, South Africa has a variety of problems. Among these are:

1. *Water.* Much of South Africa has deficient or unreliable rainfall. Many new dams are being built to extend irrigation and insure against drought.
2. *Soil Erosion.* Over-grazing of pastures and bad farming practices have led to widespread soil erosion. A Soil Conservation Act was passed in 1946 and much good work has been done since in stabilizing gullies, restoring grassland, planting trees, etc.
3. *South-West Africa.* This United Nations Trust territory has been treated by South Africa almost as annexed territory. This has aroused the criticism of the Afro-Asian group in the United Nations.
4. *Race Relations.* The dominant white minority in South Africa is determined to maintain its supremacy. Negroes, Asians and Coloureds (people of mixed descent), although a majority, have little or no say in the government of the country as a whole. The "Apartheid" (separation) policy of the Nationalist (Afrikaner) Government has increased racial tension.

Trade

In recent years the most valuable exports have been gold, wool, fruit,

FIG. 96. *Ploughing with ox teams on the South African veld.*

hides and skins, minerals, diamonds, uranium. The total value is approaching £800 million per annum (excluding gold), and the best customers are Britain, United States, Japan, Germany and Belgium. Imports, the value of which is running at about £1,117 million per year, come chiefly from Britain, the U.S.A., Germany and Japan. They consist almost entirely of manufactured goods of all kinds.

QUESTIONS

1. Draw a column graph to illustrate the racial composition of the population of South Africa.
2. Describe, in terms of variation in rainfall, vegetation, crops and land use, two journeys: one from Cape Town northwards to Bechuanaland, and another from Durban to Port Nolloth on the west coast.
3. Suggest reasons why South Africa has become the most industrialized country in Africa.

CHAPTER 24

CENTRAL AFRICA

Rhodesia, Zambia and Malawi

THE total area of these territories is 485,000 square miles—larger than France, Germany and Britain combined—yet the total population of 12½ million is less than that of London and the Home Counties.

Relief. Most of the area is occupied by high plateaux averaging 3,000 to 4,000 feet in height. They have been cut into by rivers (e.g. the Zambesi and Kafue) whose deep valleys greatly hinder communications. Some parts of the plateau rise to considerable heights, e.g. the Inyanga mountains (over 8,000 feet) and Mt. Mlanje (nearly 10,000 feet).

Climate and Vegetation. Temperatures are everywhere modified by altitude, and are not normally oppressive. Rainfall comes chiefly in summer (November-April) and is brought by the south-east Trades which blow strongly at this season. Rainfall generally falls off towards the west (compare South Africa) from a normal 40 in. (1,000 mm) along the eastern border to less than 20 in. (500 mm) in the south-west These features are demonstrated in the following statistics:

Salisbury. July average 13° C. (56° F.), December average 21° C. (70° F.); rainfall 32 in (812 mm).

Bulawayo. July average 14° C. (57° F.), December average 22° C. (72° F.); rainfall 25 in. (635 mm).

Near Lake Nyasa, the mountain areas may have as much as 56 in. (1,400 mm) of rain. The floors of the deep valleys have much higher temperatures and high humidity. Most of the country is covered with savanna: grassland with scattered trees. But where rainfall is sufficient, and along the rivers, there are evergreen forests.

Population. In 1966 the population of Zambia was 4,040,000. The population of Rhodesia was 3,849,000 (1961 census), and of Malawi 2,921,100 (1961 census).

Farming. Most of Rhodesian farming is centred on the kraal—the African village. Most of the work is done by women as the men usually work outside the reserve in order to earn money. The land is divided between the families of the village on a "strip" system, the total amount of land per family being three to four acres. The principal crop is mealies, a poor variety of maize. Sowing begins in November, and the harvest is in March. Other foods include

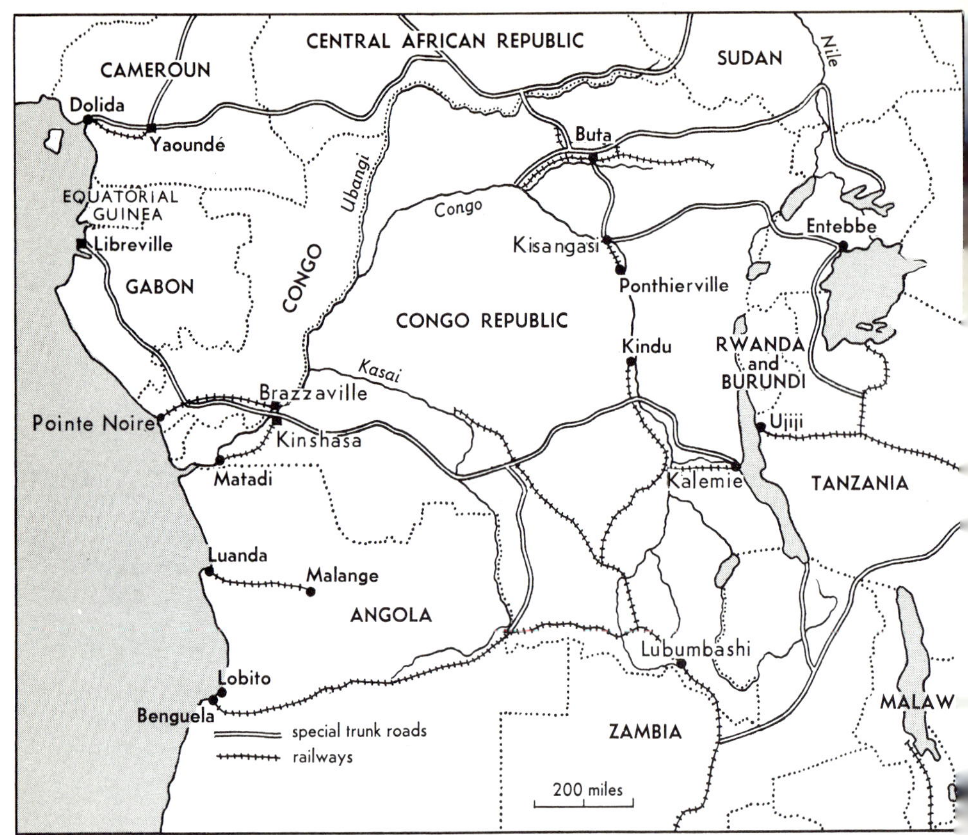

FIG. 97. *Central Africa—communications.*

melons and occasionally meat, but very little milk is drunk. Yields are only about one-sixth of what a European farmer would expect. Each kraal is allocated a certain number of cattle which it must not exceed, as overgrazing leads to soil erosion.

In Zambia shifting agriculture is still practised. Under this system clearings in the forest are made by cutting and burning. Extra timber from surrounding land is also piled up and burnt, and millet seed is sown in the ashes. Next season the soil is hoed into mounds, and beans, groundnuts, kaffir corn or cassava are planted. Meanwhile another piece of ground has been cleared and planted with millet. After four or five years the land becomes exhausted and weed-infested, so each year as one patch is cleared another is abandoned and will not be recultivated for 20 to 30 years. This is a very wasteful system, 30 to 60 acres of ground being required to keep one acre under cultivation. Crop yields are low and the constant shifting makes it difficult to improve housing or amenities.

Colonial Settlement and Farming. Although the majority of Europeans live in towns there are certain areas where white farmers have settled in fair numbers. These are chiefly areas with a favourable climate, i.e. the higher areas and areas within easy reach of a railway. The chief crops grown on the large farms (averaging 3,000 acres) in Zambia are maize, tobacco, cotton, potatoes and groundnuts. In Rhodesia maize, tobacco, citrus fruit, sugar and cotton are the main crops. Cattle are important in both countries, both for beef and dairy products, the main markets being the mining towns of the copper belt and the larger towns of Rhodesia. The Federation of Rhodesia and Nyasaland broke up in 1963 and Zambia (formerly Northern Rhodesia) and Malawi (formerly Nyasaland) are now independent. Both have large negro majorities and numbers of white colonists have left these countries for Rhodesia or South Africa. Rhodesia declared itself independent unilaterally in 1965. In Malawi, tea and cotton are the main crops grown by white plantation owners. The Africans practise subsistence farming. Independence will help to remove the gulf between the advanced colonial farms and the backward African ones. Already there has already been a considerable amount of re-allocation of land to African co-operation ownership.

Mining. Zambia and Rhodesia are important mineral-producers: copper being the most valuable, accounting for 97 per cent of the total. Zambia is the chief producer, while Rhodesia's main mineral resources are gold, asbestos, coal and chrome ore. The total value of minerals produced was £261 million (Zambia 1967) and £32 million (Rhodesia 1965).

Manufacturing. In Rhodesia, readily available sources of European and South African capital and credit, together with the local supplies of coal, power, iron and steel and agricultural products, have led to rapid development of industry. In 1964 industry formed 16·8 per cent of the total economy, and employed 80,840 people. The main products are iron and steel and metal goods, processed foods, textiles and clothing, tobacco, cement and paper. In order to supply power for these growing industries, the Zambesi has been dammed at Kariba. This great scheme was completed in May, 1960. The dam, 420 feet high, 2,025 feet long and 40 feet thick at the top, holds back the waters of the Zambesi River to form a lake 180 miles long and over 20 miles wide in parts. The water flows through underground power stations which generate power for the copper mines, for the big cities and for the new factories. The lake should eventually support a great fishing industry.

Trade. Rhodesia's chief exports are tobacco, asbestos and copper, though foreign trade since 1965 has been badly hit by trade sanctions. Zambia exports almost entirely metals, while Malawi exports mainly agricultural products—tobacco, tea, groundnuts and cotton. All three countries import manufactured goods, machinery and transport equipment. Britain is the chief trading partner except in the case of Rhodesia who now trades mainly with South Africa.

The Congo Basin

Physically, the region drained by the Congo River and its tributaries is a true basin, the centre of which is nevertheless at a high level—everywhere above 1,000 feet. Unlike much of the African plateau, the central Congo is covered by relatively young sedimentary rocks, mostly sandstones and alluvium. Surrounding this central area, the ancient basement rocks form a rim of higher land, rising to over 8,000 feet in Angola in the south, and over 10,000 feet in the volcanic highlands around the Rift Valley. In the north the surrounding uplands rise to over 4,000 feet on the borders of the Chad Basin, while in the east the Crystal mountains complete the circle, forming a barrier over 3,000 feet high in parts, which cuts off the central Congo from the sea. This vast basin and its surrounding uplands, something like 1½ million square miles in extent, and with a high average rainfall, is drained by a multitude of rivers which converge towards the centre and are finally united in the Congo River—over 3,000 miles long and one of the world's greatest rivers. The huge mass of water flows westwards, cutting through the highland rim in a series of rapids and gorges, entering its 85-mile long estuary near Matadi.

Owing to its position on the Equator, temperatures are always high in this region, and this heat generates the convection currents in the warm moist air which are the immediate cause of the heavy rainfall. The climate figures given show clearly a double maximum of rainfall immediately following the apparent northward and southward movement of the overhead sun.

Yaunde 2,461 ft.	Jan.	Feb.	Mar.	Apr.	May	Jun.	
	23·5	23·5	23·5	22	22	22	° C.
Temperature	74	74	74	72	72	71	° F.
Rainfall	1·6	2·7	5·9	9·1	8·1	4·5	in.
	40	69	150	230	206	114	mm.
	Jul.	Aug.	Sept.	Oct.	Nov.	Dec.	
	21	22	22	22	22	23	° C.
Temperature	70	71	71	71	72	73	° F.
Rainfall	2·6	3·3	7·6	8·9	5·9	2·0	62 in.
	66	84	192	225	150	50	1,575 mm

The figures show that there is practically no seasonal variation in temperature.

To the north and south of the Equator the wet seasons and the dry seasons tend to become more distinct, and the régime changes to a division of the year into two seasons—a wet season and a dry season.

Only about 55 per cent of the Congo basin is covered by dense evergreen forest. Wherever the annual rainfall is less than about 50 in. (1,270 mm.) and there is a pronounced dry season, the vegetation changes to savanna.

The average density of population in the Congo is very low—about 12 people per square mile. This is much lower than on most parts of the East African plateau or in West Africa, and is caused by unfavourable climate, dense forest, poor soils and general cultural backwardness. The backwardness results in part from the area's inadequate system of roads and railways—even though these are supplemented by river transport. (See Fig. 97.) Some of the short stretches of railway were built as "portages" to transport goods around such obstacles as waterfalls and rapids.

Shifting cultivation, similar to that already described, is carried on over most of the region. The crops grown include rice, mealies, yams, bananas, groundnuts, millet, cassava and sweet potatoes. Cattle are reared in the savanna areas (where there are no tsetse flies), especially on the peripheral high plateau regions. Additional food supplies are obtained by hunting and fishing. Cash crops are grown mostly by plantation companies, although wild palms still provide a large amount of palm oil. Plantation crops, in addition to palms, include rubber, coffee, bananas and cotton.

The mineral wealth of the Congo basin is considerable. The most valuable deposits are those of Katanga province of the Congo Republic, where the famous copper belt extends southwards into Zambia. Great efforts have been made to develop and export the mineral products which, in addition to copper, include cobalt, uranium, tin and zinc. The communications map shows the railways linking Katanga with the ports of Benguela, Matadi, Beira and Lourenço Marques. The most important mineral found in other parts of the Congo is diamonds, obtained in several places, the most important being Bushimaie on the Kasai River. The Congo basin exports more diamonds (chiefly industrial) than any other region in the world.

There are large resources of hydro-electric power in the Congo, but so far development on any considerable scale has taken place only in Katanga, where the power is needed for the mines and for refining metals. Coal is also being produced in Katanga on an increasing scale.

The industries of the Congo basin are still largely confined to the processing of primary products—e.g. the refining of minerals and the preparation of agricultural products. However, some of the larger towns are rapidly developing manufactures. Kinshasa (pop. 1,226,000) makes textiles, shoes, chemicals and jute, and has railways and ship repair yards.

QUESTIONS

1. What are the differences between a typical white settler's farm and a native African farm in Rhodesia?
2. Write short notes on the following: minerals in Zambia; the Kariba dam; shifting agriculture.
3. Account for the relatively low density of population in the Congo basin.

CHAPTER 25

EAST AFRICA

Kenya, Uganda, Tanzania

THESE three territories were all formerly under British control. At the end of the nineteenth century Kenya and Uganda were taken over by Britain and Tanganyika by Germany. Britain acquired a mandate over Tanganyika (now linked with Zanzibar to form Tanzania) after the First World War.

Relief. As with most African territories, the interior of East Africa consists of a series of plateaux at various altitudes. Unusual features, however, are the volcanic peaks, such as Kilimanjaro and Mount Kenya, and the great Rift Valleys which run from south to north. Along the coast is a plain, varying in width from 10 to 40 miles in Tanganyika, but then widening considerably at the mouth of the Tana river in Kenya. The main plateau of the interior has an average elevation of about 4,000 ft.; however, the edges are often higher than this, so the interior forms a kind of basin, partly occupied by the shallow lakes Victoria and Kyoga. Rivers—in particular the Nile and the Ruaha—have cut through the rim, so that the large lakes drain into the sea and hence do not become salt. The remarkable Rift Valleys form two great trenches, one running along the western border of this region. This trench is largely occupied by the series of deep lakes: Nyasa, Tanganyika, Kivu, Edward and Albert. The eastern branch forms a great valley across Tanganyika and Kenya, and northwards into Ethiopia. It contains a series of smaller lakes—most of them salt: such as Lake Natron—and one large lake, Lake Rudolf. Areas of higher ground in East Africa, such as the volcanic peaks and the Kenya Highlands around Nairobi, are usually areas of cool climate and good rainfall and therefore attracted European settlers.

Climate. No part of the region is far from the Equator, so that temperatures would normally be high all the year round, were it not for the modifying influence of altitudes:

Town	*Lat.*	*Alt.*	*January Av.*	*July Av.*	*Rainfall*
Nairobi	1° S.	5,495 ft.	19° C. (66° F.)	15° C. (59° F.)	39 in. (990 mm.)
Entebbe	0	3,842 ft.	22° C. (71° F.)	21° C. (69° F.)	58 in. (1,498 mm.)
Mombasa	4° S.	50 ft.	27° C. (80° F.)	24° C. (75° F.)	47 in. (1,193 mm.)

From these figures, it will be seen that the annual range in each case is small, for the sun is always high in the sky at noon. The influence of altitude is apparent in the temperature figures for Mombasa and the interior stations.

The relatively high rainfall of the interior stations is also partly explained by altitude. There is a double maximum of rainfall at each station just after the passing of the overhead sun: which is the period of greatest heat. This indicates that much of the rainfall is convectional. The wind pattern for East Africa and the adjacent Indian Ocean is shown on the maps. Note how the south-east Trades are drawn across the Equator towards the low-pressure centre in north-west India and Pakistan (known there as the south-west monsoon). In the northern winter, the normal north-east Trades re-establish themselves. They cross the Equator and are then either deflected to the left or drawn into the South African "low." The main rainfall of East Africa falls during the period of change-over of the winds in the months of March, April and May; and again in October, November and December, when the air is relatively still and convection currents can develop. The areas of heaviest rainfall are along the coast and around Lake Victoria.

Kenya

Kenya may be divided into three main areas on a basis of relief and climate. The narrow coastal plain is fertile and has ample rain, especially towards the south, and is hot throughout the year. Formerly the area was forest, with mangroves along the coast. But much clearing has taken place. Among the crops grown by the Kenyans (many of whom are of Arab blood) are sugar cane, rice, yams, cassava and maize—in other words the typical Equatorial food crops. Copra (dried coconut meat) from the coastal coconut palms is a source of oil used in the production of soap.

Mombasa (pop. 180,000) is the chief coastal town and one of the best and most important ports of East Africa. As the terminus of the railway from the interior, it handles nearly all of Kenya's foreign trade, as well as some from north-east Tanzania (Tanganyika).

The most important part of Kenya is the interior plateau and highlands. The altitude of most of these areas is between 5,000 and 8,000 ft., with ranges such as the Aberdare range and peaks like Mount Kenya rising above 14,000 ft. The highlands are divided by the great Rift Valley. To the west the land slopes down to Lake Victoria, Africa's largest lake. The Rift Valley, a very striking feature of the landscape, is 30-40 miles wide, and its floor is dotted with volcanic cones and lakes. Many of these are soda lakes, and one, Lake Magadi, provides sodium carbonate (washing soda) for export.

Owing to the altitude of the highland areas, the climate is very favourable to both European and African settlement. In fact, this is the most densely populated part of Kenya and one of the most densely populated parts of Africa. In addition to having a favourable climate, a large part of the highlands is covered with rich red volcanic soil, suitable for a variety of crops, but especially for coffee. Other commercial crops include tea, sisal, wattle bark, pyrethrum oil (used in insecticides) and cotton. Subsistence crops include maize, bananas, wheat and sugar cane. Animal products are also important—meat, bacon, hides and dairy produce. Sheep have increased greatly in importance in recent years.

Nairobi (pop. 479,000) is the capital and chief town of Kenya. It began in 1899 as a tiny depot on the Uganda Railway, and developed (during colonial days) as the centre of the important "white highlands," becoming an administrative, commercial and industrial city serving not only Kenya, but the other East African territories as well. It is a centre of air communications with all parts of the world, a railway junction and road centre. The surrounding "big game" country makes it a popular tourist resort. As an industrial town it is the most important in East Africa, being supplied with electricity by the great hydro-electric scheme at Owen Falls. The more important industries are flour milling, and the production of margarine, soap, mineral waters, cement, beer, cigarettes, furniture, footwear and clothing.

Agriculture. Kenya's economy is based on agriculture. Until recently the European settlers produced most of the export products, but in recent years African farmers have begun to make an important contribution to the production of cash crops.

The table shows the principal exports of Kenya. The main imports are machinery, transport equipment, fuel oils and lubricants.

Kenya Exports 1967

Coffee	£15,676,000	Tea	£7,396,000
Residual fuel oils	£3,141,000	Meat & meat products	£2,857,000
Pyrethrum extract	£2,423,000	Sisal, fibre and tow	£2,064,000
Jet fuel	£1,862,000	Hides and skins	£1,742,000

Tanzania

Tanganyika. The essential division of Tanganyika into a narrow coastal plain with a hot and rainy climate and an interior plateau with unevenly distributed rainfall has already been indicated. The coastal plain is densely populated except where infertile soils derived from coral are found. Rice is an important crop, and there are plantations of sisal and coconuts.

There are two main factors governing the land-use of the interior plateau. One is the amount and reliability of rainfall, the other is the tsetse fly. Away from the line of the Equator, rainfall becomes more and more concentrated into a single rainy season: the summer (November to March). However, the amount of rainfall is low and unreliable in many inland areas and, in fact, only about one third of the whole country is reasonably well watered throughout the year. The central plateau is the driest part; the relatively small semi-temperate and higher rainfall areas, on higher ground, lie around the margins, and are not concentrated as in Kenya. Vegetation over the whole interior plateau is generally rather poor savanna, but where rainfall is sufficient, as on some of the very high land, true forest is found.

Subsistence agriculture predominates over practically all the interior of Tanganyika, the only variations stemming from variations in rainfall. In the wetter areas, maize, millet, groundnuts, rice and cassava are the food crops. Where rainfall is lower, millet and groundnuts are the main crops. Stock-rearing is more important in areas free of the tsetse fly. Cash crops such as cotton, coffee and oilseeds are being cultivated on an increasing scale, and sisal, tobacco and pyrethrum oil are mainly produced in the highland areas, e.g. around Mount Kilimanjaro.

Tanganyika is an important producer of diamonds, and various other minerals, including gold, tin and salt, are worked on a small scale. Nevertheless the economy is mainly agricultural. In addition to supplying practically the whole domestic demand, agricultural products plus livestock account for four-fifths of all exports. The chief exports in 1962 are listed on page 256.

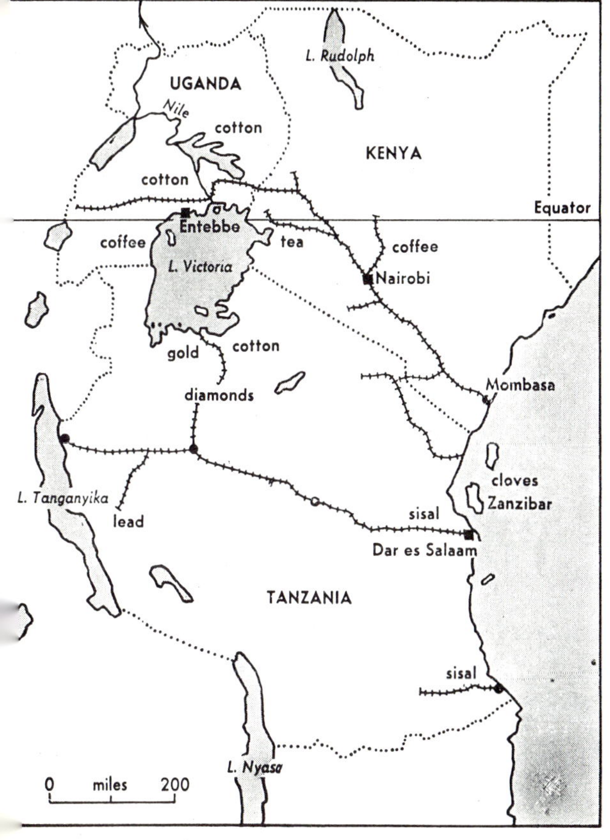

FIG. 98. *East Africa—commercial crops and minerals.*

Main exports 1967		*Main imports* 1967	
Cotton	£12,568,000	Machinery, incl. electrical ..	£11,710,000
Coffee	£11,939,000	Metals & metal manufacturers	£10,621,000
Diamonds ..	£11,146,000	Transport equipment ..	£10,536,000
Sisal	£10,046,000	Petroleum & refinery prod. ..	£5,269,000

Industry is, as yet, little developed in Tanganyika. However, it is the aim of the government to increase the production of essential manufactures. The main interest is still the processing of crops for local consumption or export, e.g. cotton ginning, tobacco curing, flour milling, sugar and tea processing. Secondary industries are beginning to grow however, and these include the manufacture of tin cans, paint, nails, rubber products, footwear and cotton textiles. Dar es Salaam (pop. 372,500), the capital and chief town, is also the main port, being sited at the seaward end of the only railway running across the whole width of the country. Tanga (pop. 50,000) is similarly situated on the railway to Arusha and Moshi, but has a smaller hinterland.

Zanzibar. The off-shore islands of **Zanzibar** and **Pemba**, detached fragments of the coastal plain, have been politically linked to Tanganyika since 1964, forming the Republic of Tanzania. The climate of the islands resembles that of the adjacent coast. Both islands are famous for the production of cloves, but copra is also important. Subsistence crops include the normal Equatorial food crops—rice, fruit, maize, yams and cassava. Both islands are very densely populated.

Uganda

Although far smaller than the other two East African countries, Uganda is more densely populated.

The average altitude of the plateau around Lake Victoria is about 3,000 ft. to 4,000 ft. To the west, the high block of Ruwenzori rises up on the Congo frontier to nearly 17,000 ft. In the east, Mount Elgon is an isolated cone rising to over 14,000 ft. As the climatic figures for Entebbe show, temperatures are always high, and the range is low, owing to the Equatorial location. Rainfall is high, partly because of the influence of the huge lake, and is well-distributed throughout the year.

Apart from the usual equatorial subsistence crops, the farmers of Uganda grow large quantities of cotton, coffee and tea, and the large numbers of cattle provide hides and skins. Uganda is the largest producer of coffee in the Commonwealth, and one of the largest producers of cotton, and these two products are the main exports.

Kampala (conurbation—pop. 170,000) is the capital and chief commercial town.

The Owen Falls dam at Jinja has stimulated industry, and manufactures include cement, bricks, tiles, soap and textiles.

Problems of East Africa

The newly independent countries of East Africa have many problems to face, some of them peculiar to themselves, some of them common to the whole region. Among the most common of these are:

(1) **Racial Diversity and Tribal Divisions.** When the Europeans partitioned Africa in the nineteenth century they paid little or no regard to existing tribal divisions when drawing up boundaries. As a result, many tribes were divided by frontiers or separated from traditional grazing grounds, e.g. the Masai. In any case the present territories are inhabited by dozens of different tribes with little or no tradition of co-operation. It is proving difficult to weld the peoples together into "nations." It must also be recorded that, in the beginning, the Europeans overcame by force all active opposition to their rule by any tribe or group of tribes. In other words, European rule started by crushing African "nationalism," and ended with trying to revive it.

(2) **Pests and Diseases.** The mosquito carries both yellow fever and malaria, two diseases which cause thousands of deaths each year. More important however is the chronic ill-health which they cause among a large proportion of the population. The tsetse fly carries human sleeping sickness and a fatal cattle disease. For this reason much good land in Africa is useless for grazing cattle, so that other areas tend to become overstocked. About half of Tanganyika is infected. The bad effect of these and many other diseases is accentuated by the malnutrition from which most Africans suffer.

(3) **Rainfall.** Over much of East Africa rainfall is both inadequate and unreliable. The dry season is a period of anxiety over food supply, for if the next rainy season is delayed, or the rainfall is below normal, famine may follow. It was this factor which contributed to the failure of the great Groundnut Scheme of 1948. Nomadic pastoralists inhabiting land which is normally rather arid, are particularly vulnerable when the rains fail, so African governments are building dams in many areas in order to reduce the risk.

(4) **Soil Erosion.** Overcropping and overgrazing by native farmers has resulted in the soil losing its structure or its vegetation cover. This leads to erosion and gully formation by the rain, which in Africa usually falls in heavy showers. Soil conservation is being carried out in some areas—for instance the Kenya Highlands—by means of contour ploughing and strip farming, i.e. the land is ploughed along the contours so that each furrow acts as a dam for water, and the crops are grown in strips which help to hold up soils washed from above.

(5) **Communications.** East Africa has a very inadequate railway system and those railways that do exist are expensive to run because of the great expanses of wilderness and semi-desert which provide no traffic. Moreover, there are many steep gradients on the railways, and as East Africa lacks coal, the engines have to be fuelled with expensive imported oil. Roads are few and

FIG. 99. *The great hydro-electric dam across the Zambesi at Kariba is vital to the economies of Zambia and Rhodesia. The lake formed by the dam will also support a large fishing industry.*

hard surfaced roads almost unknown. In the wet season most roads become muddy and frequently impassable.

QUESTIONS

1 Discuss the influence of relief on temperature with particular reference to East Africa, using the climatic figures given.
2. Discuss the factors which govern land-use in interior Tanganyika.

CHAPTER 26

WEST AFRICA

FOR centuries West Africa was exploited as a source of slaves, gold, ivory, and palm oil. Towards the end of the nineteenth century permanent colonies were established, chiefly by the British and French (who were both systematically engaged in building empires in Africa), under whose control most of the area remained until recently.

Physical Features

The coastline of West Africa is very smooth, so that good natural harbours are rare. From the coastal plain, the land rises northwards to a series of low plateaux of ancient rock. Still farther north the plateaux slope gently northwards to the interior of the Sahara. Rivers radiate from these plateaux. Some, like the Niger, the Benue and the Volta first flow northwards and then bend round to flow southwards to the sea. In the east some rivers flow inland to the Chad basin, a region of inland drainage. The rivers that flow directly southwards or westwards are shorter and more rapid. Few of these rivers are of much value for navigation because of falls and rapids, sand bars and seasonal variation in depth. However, the Niger and Benue are navigable over long distances inland, and the Gambia River is an important navigable waterway for over 300 miles from its mouth.

The northward and southward movement of the equatorial low-pressure belt causes a seasonal reversal of wind over most of West Africa, i.e. there is a monsoon effect. In January, the low-pressure zone (the inter-tropical front) lies over the coastlands of Nigeria and Ghana, and the north-east Trades blow at this season from the Sahara over most of West Africa. These winds are hot, but because of their dryness they assist evaporation and therefore have a cooling effect on the human body. They make a welcome change from the humidity of the conditions during the rest of the year.

In July we see that the low-pressure belt has moved northwards towards the Sahara, and the south-west monsoon winds blow strongly over the land at this season, bringing heavy rains. (Note that these winds originate as south-east Trades south of the Equator, and are drawn northwards by the pull of the "low." On crossing the Equator, they are deflected to the right and become south-westerlies.) As a result, the coastal areas receive heavy rainfall: up to 180 in., but this decreases northwards to under 10 in. along the desert border. Temperatures are always high in West Africa—the "normal" temperature being about 26·7° C. (80° F.). It hardly ever falls below 21° C.

Freetown (223 ft.)

	Temperature		*Rain*			*Temperature*		*Rain*	
	° C.	*° F.*	*in.* (157)	*mm.* (4,000)		*° C.*	*° F.*	*in.* (157)	*mm.* (4,000)
Jan.	27·2	81	0·4	10	July	26·1	79	35·6	904
Feb.	27·8	82	0·3	7·6	Aug.	25·5	78	36·6	929
Mar.	27·8	82	1·2	30	Sept.	26·1	79	28·5	723
Apr.	27·8	82	4·1	104	Oct.	26·7	80	13	330
May	27·2	81	11·5	292	Nov.	27·2	81	5·1	130
June	26·7	80	20	508	Dec.	27·2	81	1·4	36

Ibadan (748 ft.)

	Temperature		*Rain*			*Temperature*		*Rain*	
	° C.	*° F.*	*in.* (50)	*mm.* (1,280)		*° C.*	*° F.*	*in.* (50)	*mm.* (1,280)
Jan.	27·2	81	0·4	10	July	24·5	76	6·9	175
Feb.	29	84	0·9	23	Aug.	23·5	74	3·5	89
Mar.	29	84	3·7	94	Sept.	25	77	7·2	183
Apr.	27·8	82	5·8	147	Oct.	25	77	6	152
May	26·7	80	6	152	Nov.	26·1	79	1·7	43
June	26·1	79	8	203	Dec.	24	75	0·4	10

Timbuktu (820 ft.)

	Temperature		*Rain*			*Temperature*		*Rain*	
	° C.	*° F.*	*in.* (9)	*mm.* (230)		*° C.*	*° F.*	*in.* (9)	*mm.* (230)
Jan.	21·7	71	0	0	July	31·7	89	3·5	89
Feb.	23·5	74	0	0	Aug.	30·6	87	2·8	71
Mar.	28·3	83	0·1	2·5	Sept.	31·7	89	1·1	28
Apr.	33·5	92	0	0	Oct.	31·7	89	0·4	10
May	35	95	0·3	7·6	Nov.	27·2	81	0	0
June	34·5	94	0·9	23	Dec.	21·7	71	0	0

(70° F.). As would be expected, however, there is an increase in annual range with distance from the sea, as shown in the figures for the stations given.

Vegetation

As a result of the steady decrease in rainfall northwards, so the vegetation changes. Along the coast there is often a zone of swamp forest usually mangrove, succeeded inland by fresh-water swamp. North of this where the rainfall is sufficient (above 45 in.) there is a zone of rain forest, as already described. Farther north still, decreasing rainfall and its more marked seasonal occurrence causes a gradual change to tree savanna (Guinea savanna) and then bush savanna (Sudan savanna), and finally semi-desert (Sahel savanna). As the climate and vegetation are very similar throughout each zone, so are the ways of life of the people; but everywhere farming is the main occupation of 90 per cent of the population. The potential of the area—once there is equipment for clearing the jungle, and adequate fertilizers—is obvious.

Farming

By contrast with the forests of the Congo Basin, the West African forest is one of the most densely populated parts of Africa. Traditionally this is an area of shifting cultivation, the chief food crops being such roots as yams, cocoyams and cassava, together with maize, rice and bananas. Tree crops, including palm oil, cocoa, coffee, kola nuts and rubber provide cash income. Goats and poultry are common, but there are few cattle because of the tsetse fly.

With the increasing density of population shifting cultivation has been abandoned. The fertility of the small farms, often only three acres in extent, is maintained by crop rotation, the use of fertilizers and of household refuse for manure. The main implement is the hoe, the soil being heaped up into mounds on which the crops are grown. The land belonging to one farmer may be scattered over a wide area of forest, and each of the plots is sown with a variety of crops. Planting and harvesting continue for most of the year for there is little difference between one season and another. Oranges, bananas and vegetables are grown in gardens near the home.

There is more land under cultivation in the savanna as the labour involved is not so great. Grains become more important, especially Guinea corn, millet and maize. Groundnuts form an important part of the food supply and are also an export crop. Cotton is another export crop. Conditions are better for cattle, and great numbers are kept, chiefly by the nomadic "cow Fulani." Their animals graze on the fallow land and help to manure the ground.

Farther north, where the dry season is very long, cattle are the main interest, although grain and groundnuts are still grown. This is the area where irrigation schemes, such as those of the middle Niger, may be expected to have the greatest results. Export products of this region include hides, skins and leather, groundnuts and cotton.

Most of the Governments of West Africa are trying to modernize agriculture by teaching modern methods in schools and colleges, by consolidation of holdings and by developing new varieties of crops. A new kind of settlement is also being tried. These are large holdings of up to 1,500 acres divided into 30-acre farms which are leased to young farmers for forty-nine years. The crops and cropping system, marketing, processing, etc. are all organized and planned by the Government. Each farmer must grow two cash crops—cocoa, rubber, oil palm or coconuts, grow five acres of food crops, and keep some livestock—poultry, cattle or pigs. Loans from the Government provide a house, tools, equipment etc., and the holding is heritable by one person only.

Minerals and Communications

The region is rich in certain minerals, the most important being tin, columbite, oil, manganese, diamonds, bauxite, iron ore, phosphates, thorium and titanium. Almost the entire production is exported abroad, owing to the undeveloped state of West Africa's industries. One important result of the exploitation of these minerals has been to stimulate the development of communications, for ports and railways have been constructed or improved to serve the mineral industry.

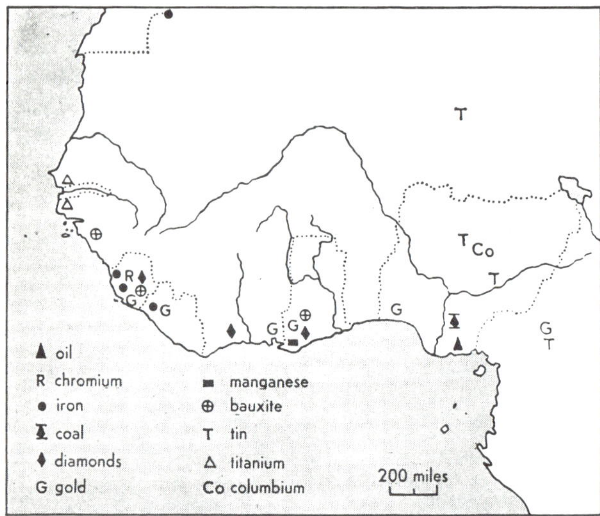

FIG. 100. *West Africa—minerals.*

Most of the railways merely lead from the coast to the interior, and there is need for interior link lines. Another disadvantage is varying gauge, for the former French colonies were constructed on the metre gauge, whereas the former British territories favoured 3 feet 6 inches. Construction of new lines is hampered by the necessity to import practically all materials and equipment, and by the ravages of termites which make it impossible to use wood for sleepers. The most ambitious construction project is the extension of the Nigerian railway to Maiduguri.

FIG. 101. *The artificial harbour at Takoradi, Ghana.*

Roads are playing a greater role nowadays, and lorry transport is well developed in most areas, but very few have hard "all weather" surfaces, and most become very difficult in the wet season.

Towns

Most of the important towns of West Africa are seaports. Owing to the unindented nature of the coast, in the past any site which offered possibilities for harbour construction became important, and many of them were developed into modern harbours, e.g. Lagos, Dakar and Freetown. Where there was no natural harbour it was necessary in many cases to unload goods into surf boats which made a perilous passage through the breakers to the shore. Such conditions existed for many years at Accra. These open roadsteads have now been, or are being, replaced by artificial harbours, such as those at Takoradi, Tema and Cotonou.

Problems of West Africa

The problems of West Africa are similar to those of other undeveloped parts of the continent as already described: poverty, malnutrition, backward farming methods, ignorance, disease, etc. All the newly independent governments of the region are trying to solve these problems by investment in education, research, communications, etc. Some of the great development schemes which are planned or being carried through involve the expenditure of huge sums, a large proportion of which must come from abroad. One of the greatest of these schemes is the Volta River Project in Ghana. This great river, over 1,000 miles long, has been dammed near Akuse. The dam, 350 feet high, holds back a lake covering 3,500 square miles and over 200 miles in length. The water can generate over half a million kW of electricity, part of which is used to make aluminium from local bauxite. The rest of the power available for other industries in Accra, Tema, Sekondi and Kumasi. About 250,000 acres can be irrigated in the dry Accra plain, and fishing in the lake is expected to yield 18,000 tons of protein a year. Other projects are the Niger Dams Project (Nigeria) and the Niger Project (Mali). The Nigerian scheme calls for the construction of several dams on the Niger and Benue. The regularizing and deepening of the rivers will make them far more useful for transport, and much-needed electric power will be generated. The Sansanding dam (completed in 1947) is part of the Mali plan. It provides irrigation water for 100,000 acres and it is planned to extend the irrigated area to 2 million acres. Rice, cotton, groundnuts, fruit and vegetables are being grown on the irrigated land.

Nigeria

Nigeria has an area of 357,000 square miles (over four times the size of the United Kingdom), and the population of over 56 million is the largest of any African state. The map shows its location in West Africa and its extent in latitude. Owing to this wide extent it includes parts of all the major zones of climate and vegetation—from the mangrove swamps of the coast to the semi-desert of the interior.

Relief. From the coastal plain, much of which is really the vast delta of the Niger, the land slopes upwards to the interior plateaux of Yorubaland and the Cameroon Highlands. These plateaux are separated from each other, and from the northern plateaux, by the wide troughs of the Niger-Benue river system.

West of the Niger delta, the coastline consists of smooth sand bars backed by a line of lagoons. These lagoons can be entered by ships where rivers entering the sea break the sandspit. One such example is the harbour of Lagos. The delta, which extends across much of the eastern part of the coast, has an area of more than 10,000 square miles, a large part being swamp. Behind the swamps, lagoons and mud flats, the low coastal plain extends inland for 150 miles. The plain becomes hillier inland, and then rises fairly sharply to the interior plateaux, where the monotony of the landscape is relieved only by the *inselberge*, or "island mountains," which rise sharply above the general level. In the north, the main plateau surface lies at about 1,000 feet, but the highest part of the Bauchi plateau averages 4,000 feet with a high point of 5,841 feet.

The vegetation and farming zones of Nigeria are typical of West Africa, but the cultivation of commercial crops is more highly developed than in most of the region. Compared with other West African states the railway system is relatively well-developed.

Commercial Crops in Nigeria. *Palm oil and palm kernels:* The oil palm grows wild in the southern rain forest zone, although there are some commercial plantations. The tree yields large clusters of fruit weighing as much as 100 lbs., each fruit consisting of a nut surrounded by a fleshy covering, which when boiled, yields palm oil; the inner kernels provide palm kernel oil when crushed. This oil is used locally for cooking and lighting, and a great deal of oil and most of the kernels are exported via Lagos, Port Harcourt or Calabar. *Cocoa:* Nigeria is the world's second greatest exporter of cocoa (after Ghana). The rather small trees grow well in the rain forests of West Africa. In Nigeria, the cocoa is grown by small peasant farmers, not in plantations. The beans grow in large pods which are harvested when ripe and cut open. The beans are extracted and allowed to ferment under banana leaves for five or six days. They are then dried, bagged and sold. Lagos is the chief port of export. *Rubber* is another rain forest plant. In Nigeria it is grown either by small farmers or in plantations. When the trees are mature, the sap is tapped by cutting a groove in the bark and fixing a cup to catch the latex which, after coagulation with acid, is smoked, dried and sent to Lagos or Port Harcourt for export. *Cotton* is grown in central and northern Nigeria, for too much rain is harmful. Small farmers produce the bulk of the crop. There are two products: lint and seed, which is pressed for oil and cattle cake. *Groundnuts,* or peanuts, grow best in sandy soil in rather dry areas. They therefore do well in central and northern Nigeria. Nigeria is the world's largest exporter. The seeds are exported to western countries and used to produce oil for margarine, and other domestic products.

Towns and Industrial Development. Nigeria produces a number of raw materials which have served as the basis for the development of local industries, e.g. cotton, palm oil and groundnuts, clay, coal, hides, tobacco and timber. Naturally, most industries tend to develop in or near towns, especially those towns which have good communications by road and rail or by water. The most important of these developing towns are given below.

Lagos is the Federal capital and chief port, and has a population of over 450,000. It owes its importance to its position on an island in a lagoon connected by a channel to the open sea. Originally a slave port, it was taken over by Britain in 1861 and became the base for the extension of British control. Today, as well as being a great port and administrative centre, it is rapidly developing industries. Manufactures include metal drums, beer, soap, mineral waters, margarine, pottery, cotton textiles and aluminium ware. Ibadan (pop. 600,000) is the capital of Western Nigeria and the third largest town in Africa. It was founded by the Yoruba people before the coming of the Europeans, and, like the medieval towns of Europe, the city still contains much farmland. It owes much of its modern importance to its position on the main railway north from Lagos. It is also a local road centre, and hence a market for cocoa, palm kernels, etc. Factories make plastics and cigarettes, and fruit is canned. The city is an educational as well as administrative centre, and has two universities. Enugu (pop. 63,000) is the coal-mining centre of Nigeria, and is the administrative capital of its region. It has railway workshops and makes pottery. Port Harcourt (pop. 72,000) is of comparatively recent origin, but is already the second port of Nigeria. Situated on deep water at the seaward terminus of the railway to Enugu and the north, its importance has been accentuated by the discovery of oil in the area. Thirteen million tons of petroleum were exported in 1965. Manufactures include cigarettes, furniture and metal building components. Kaduna (pop. 40,000) is the capital of the Northern Region and owes its importance to its central position and its location at a railway junction. In 1957 a large cotton mill was opened here, and it now supplies cloth to a wide area. Kano (pop. 130,000) is the largest city in the north and is a centre of traditional craft work in metal, leather and cloth. Local supplies of cotton and groundnuts supply weaving mills and oil mills. Other products include soap, cigarettes, shoes, enamelware, confectionery and leather.

Ghana

Ghana (92,000 square miles) is roughly the size of the United Kingdom, but the population is only 8·4 million. Nevertheless, it has the second largest population in West Africa, and is probably the most advanced economically. Ghana was also the first West African state to achieve independence.

Physical Features. There are no high mountains or plateaux in Ghana. Most of the country consists of plains and low uplands. These uplands are diversified by ridges or scarps which rise to as much as 2,000 feet in some parts. The chief river system, the Volta and its tributaries, drains a large part of the country. Ghana does not extend as far inland as Nigeria, so that only the forest and savanna zones of vegetation are included in its borders. Rainfall is greatest in the south-west and decreases northwards and eastwards. In the east, the Accra plain is a region of unusually low rainfall which supports only scrub vegetation.

Farming in Ghana. Ghana is still primarily an agricultural country, and farming, fishing and forestry occupies two-thirds of the labour force. The subsistence type of agriculture practised by the majority of the farmers is similar to that already described, but Ghana is unique in its dangerous dependence on a single export crop—cocoa. The cocoa is grown by small farmers in the forest zone, and it accounts for 53 per cent by value of Ghana's exports. The Government is trying to develop alternative cash crops, such as rubber, bananas, coffee, cotton and sugar, and is also attempting agricultural development schemes. The livestock industry is handicapped by the prevalence of tsetse fly over much of the country.

The timber industry has expanded greatly since the last war and its exports now make an important contribution to the economy. The chief timbers are obeche, sapele, mahogany and utile.

Fishing is a traditional occupation, and in recent years the Government has established a modern trawler fleet based on the new port of Tema. The Volta project is expected to give a 50 per cent increase in fish supplies.

Minerals. The chief products by value are gold, diamonds and manganese, but bauxite will probably become very important when the Volta scheme is completed. The most important mineral zone is the south-west.

Industry in Ghana. Most of the industrial establishments are found in or near the big towns. The Government is fostering the expansion of industry, especially those making use of local raw materials. The major centre is Accra (pop. 492,000), the capital and second largest port. The port was formerly merely an open roadstead, but this has been replaced by the new harbour at Tema. Takoradi (pop. 41,000) is the best and oldest seaport, but has modern facilities especially for the loading of minerals and timber. It handles 80 per cent of the imports and 90 per cent of the exports of Ghana. Tema, the new harbour, will probably develop into a great industrial centre, and 800 acres of land have been set aside for an industrial estate.

New industries in Ghana include an aluminium smelter (part of the Volta project), an oil-refinery, a steel mill, a textile factory, a sugar-refinery, a drug factory, and shoe, cutlery and cable works.

QUESTIONS

1. Describe how climate and vegetation changes between Lagos and Timbuktu.
2. Describe the railway network of West Africa, and suggest how it could be made more complete. What are the handicaps faced by railway builders in West Africa?
3. Describe the cultivation, processing and exporting of the three important commercial crops of West Africa.

CHAPTER 27

NORTHERN AFRICA

North-West Africa (the Maghrib)

NORTH-WEST Africa is the nearest part of the Arab world to Western Europe. Although divided into three independent states, Morocco, Algeria and Tunisia, the Maghrib may be regarded as a distinct region united by common elements of relief, climate, language and culture. The total population in 1967 was about 31 million distributed as shown.

Morocco	14,140,000	166,000 square miles
Algeria	12,540,000	952,000 square miles
Tunisia	4,560,000	63,000 square miles

Relief

Structurally, North-West Africa is linked with Europe rather than with the rest of Africa. The rocks are similar to those of Southern Europe and were folded and uplifted as part of the great zone of Alpine earth movements. The relief map shows that in Morocco a series of high mountain ranges (maximum altitude nearly 14,000 feet) are arranged in echelon. Between these ranges on the Atlantic coast is a series of lowlands and low plateaux, across which flow the only permanent rivers in the whole Maghrib, fed by the snows of the high mountains. The main coastal plateau is sometimes called the Moroccan meseta, because of its resemblance to the plateau of central Spain.

In Algeria the high ranges diverge, forming a mountain wall to the north and south with a plateau in the centre called the Plateau of the Shotts. The two ranges continue eastwards and reach the sea as Cape Blanc and Cape Bon, separated by the Gulf of Tunis. Along the Algerian coast there is a discontinuous coastal range called the Sahels. Inland, the "tell" rises as a series of terraces to the "Tell Atlas." Both the Tell Atlas and the Saharan Atlas are deeply scored with gorges cut by the torrents which flow during the rainy seasons. The Plateau of the Shotts (named after the shallow saline lakes which dry up during the summer) has an altitude of 3,000-4,000 feet, and is broader in the west than in the east. The Saharan Atlas is bounded in the east by a deep depression, in places below sea level, partly occupied by the Shott Melrir and the Shott el Djerid.

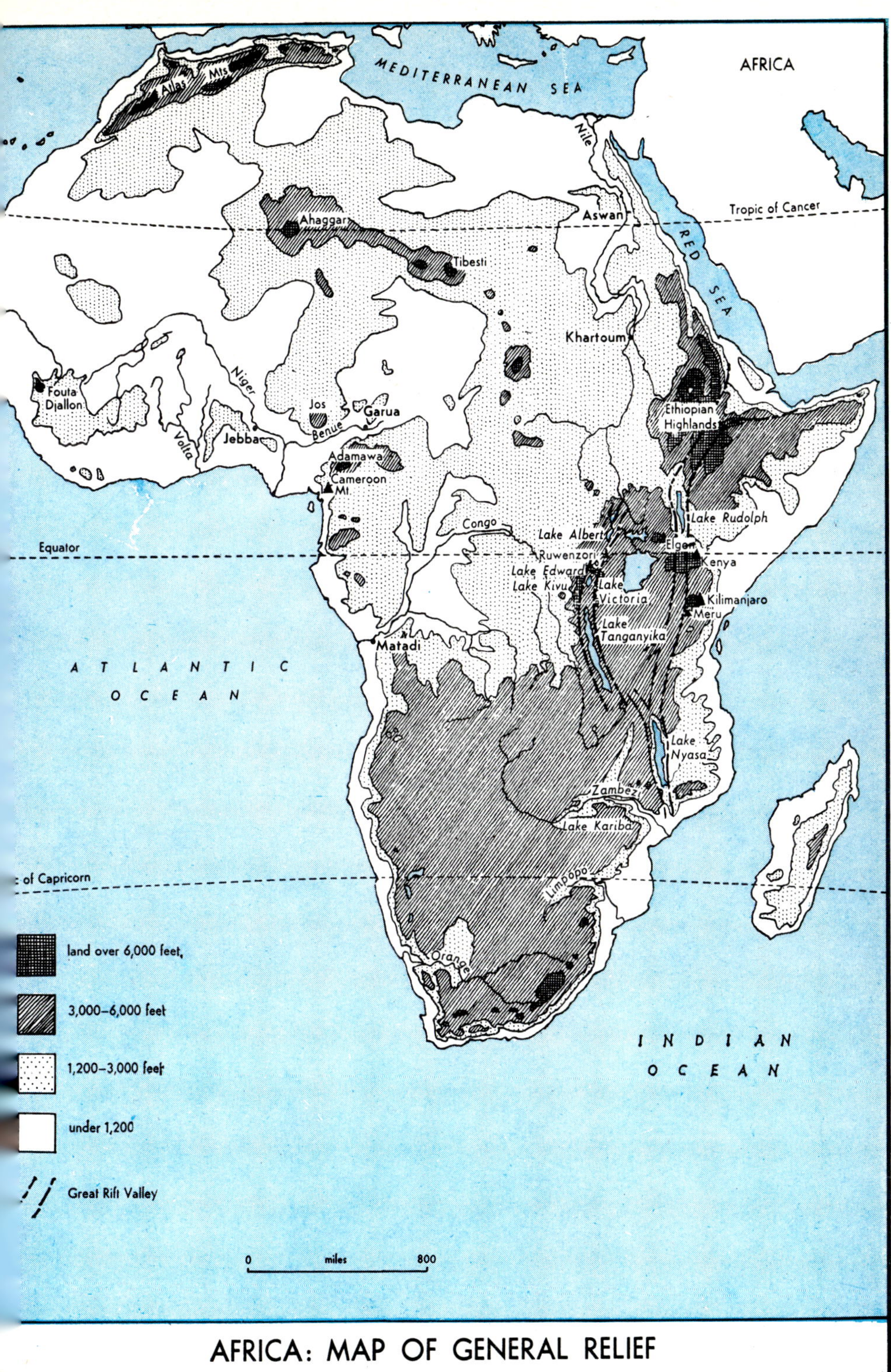

AFRICA: MAP OF GENERAL RELIEF

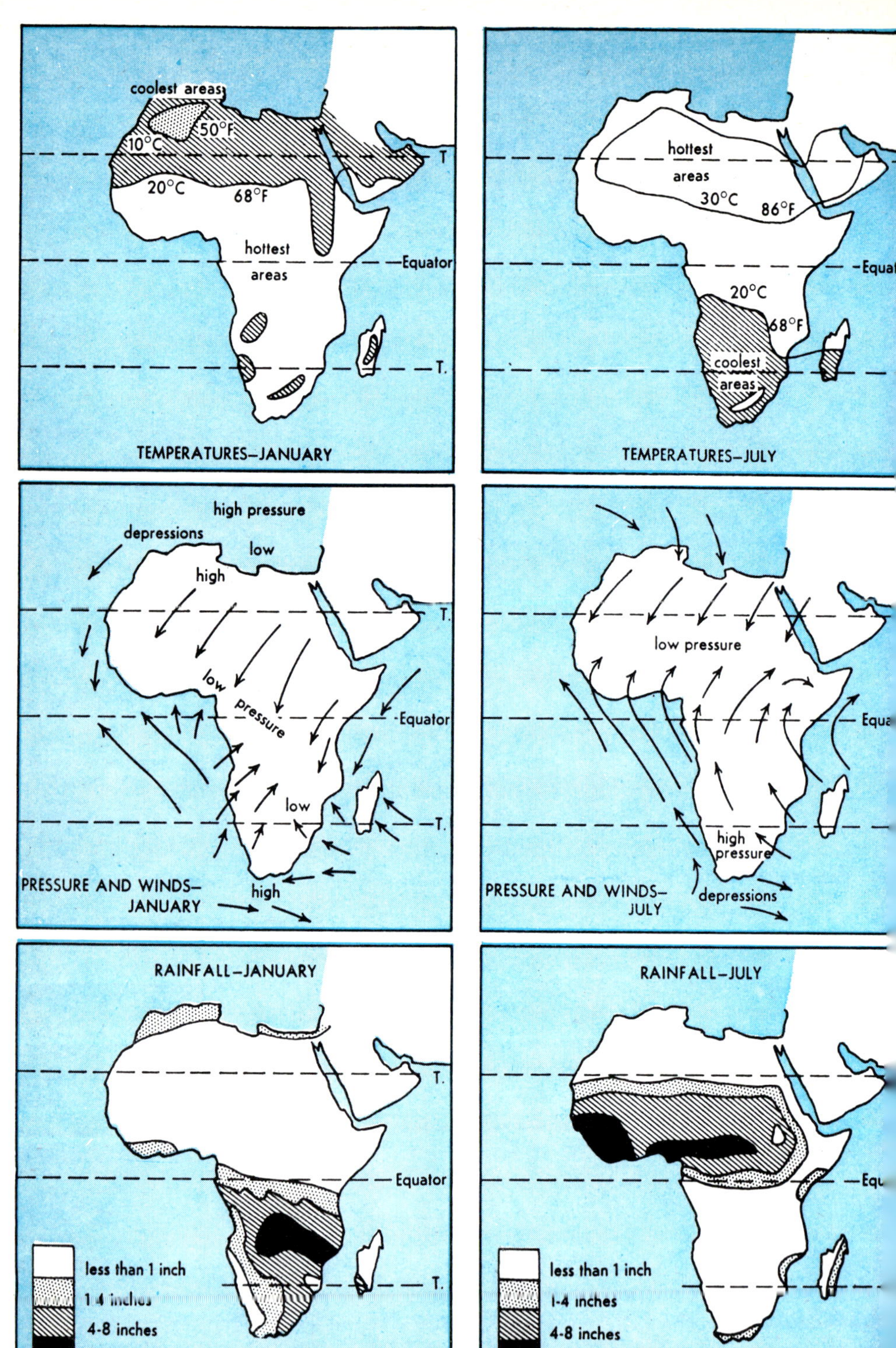

AFRICA: GENERAL CLIMATIC DETAILS

Climate

Owing to its latitude and position in relation to the land mass of Eurasia, north-west Africa comes under the influence of the westerly wind belt with its depressions that move in from the Atlantic in winter; and in summer it lies in the zone of the north-east Trades, which bring little or no rain when they pass on to the hot land. The resultant climatic type is Mediterranean, modified over large areas by relief, distance from the sea and the cold Canaries current. The main features are illustrated by the figures for the three stations Algiers, Mogador and Biskra. Note the decrease in rainfall and the increase in summer temperatures inland, and the relatively low summer maximum on the Moroccan coast.

Algiers (72 ft.)

	Temperature		*Rain*			*Temperature*		*Rain*	
	° *C.*	° *F.*	*in.* (30)	*mm.* (760)		° *C.*	° *F.*	*in.* (30)	*mm.* (760)
Jan.	12	53	4·2	106	July	25	77	0·1	2·5
Feb.	13	55	3·5	85	Aug.	26	78	0·3	7·5
Mar.	14	58	3·5	85	Sept.	24	75	1·1	32
Apr.	16	61	2·3	58	Oct.	20	68	3·1	79
May	19	66	1·3	33	Nov.	17	62	4·6	117
June	22	71	0·6	15	Dec.	14	57	5·4	137

Mogador (33 ft.)

	Temperature		*Rain*			*Temperature*		*Rain*	
	° *C.*	° *F.*	*in.* (13·2)	*mm.* (335)		° *C.*	° *F.*	*in.* (13·2)	*mm.* (335)
Jan.	14	57	2·2	56	July	20	68	0	0
Feb.	15	59	1·5	38	Aug.	20	68	0	0
Mar.	16	60	2·2	56	Sept.	21	69	0·2	5
Apr.	17	63	0·7	17	Oct.	19	67	1·3	33
May	18	65	0·6	15	Nov.	17	63	2·4	61
June	20	68	0·1	2·5	Dec.	15	59	2	51

Biskra (400 ft.)

	Temperature		*Rain*			*Temperature*		*Rain*	
	° *C.*	° *F.*	*in.* (7)	*mm.* (178)		° *C.*	° *F.*	*in.* (30)	*mm.* (178)
Jan.	11	51	0·5	13	July	32	89	0·2	5
Feb.	13	55	0·7	17	Aug.	31	88	0·1	2·5
Mar.	16	60	0·8	20	Sept.	28	82	0·6	15
Apr.	19	67	1·2	30	Oct.	21	70	0·8	20
May	24	75	0·6	15	Nov.	15	59	0·4	10
June	29	84	0·4	10	Dec.	11	52	0·6	15

Vegetation

Vegetation is almost everywhere of the Mediterranean type, i.e. the plants are able to survive the long drought of summer by such adaptations as bulbous underground stems, long deep roots or small, waxy leaves. The most characteristic vegetation is the bush, or *maquis*, found in the coastal areas and on the lower slopes inland. Where rainfall is heavy on the high mountains, oak and coniferous forests are found, and grasslands are found still higher. In the interior, and on the Moroccan meseta, the decreasing rainfall supports a poorer kind of maquis with smaller and fewer bushes. Still lower rainfall produces grass steppe, where the principal vegetation is esparto grass, which is harvested for export. Beyond the Saharan Atlas, semi-desert and desert stretch southwards as far as the savanna country.

Farming

Where rainfall is sufficient, or there is irrigation, the main agricultural products are those typical of a Mediterranean climate: wheat, barley, early vegetables, grapes, citrus fruit and olives. Much mediocre wine is produced in Algeria (and a little in Tunisia and Morocco), and on the mountain slopes cork oaks, walnuts and deciduous fruits are grown. In the dry steppe areas of the interior the population is generally nomadic, cereals being sown on the low ground in summer and left until the harvest, while the people with their flocks and herds move to the highlands.

Minerals

In the Maghrib proper the most important mineral is phosphate. All three countries produce this valuable fertilizer, but Morocco is by far the largest exporter. Iron ore is also found in all three countries, and there is a considerable export. Other minerals produced include lead, zinc and manganese.

Towns and Industries

The largest town in Morocco is Casablanca (pop. 1,177,000). It is a great port, handling over three-quarters of the overseas trade of the country, and is also a great industrial centre. It manufactures superphosphates, processed foods, soap and cement. It is an important fishing port, and cans and preserves fish. The main exports are wool, grain and phosphates.

Marrakesh (pop. 264,000) is the largest inland town, and the chief city of the south. It is, in fact, a large oasis, and a traditional market for the fruits and cereals of the north, and the dates, hides and skins of the drier south. Rabat (pop. 262,000) is the capital, with manufactures of carpets, embroidered cloth and pottery. Fez (pop. 249,000) is the ancient religious capital—the "Mecca of the West." It manufactures textiles and has a university.

In Algeria, the capital Algiers (pop. 943,000) is an important port, a

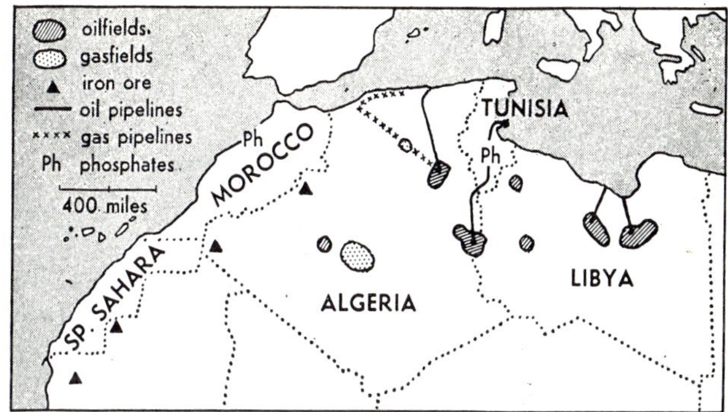

FIG. 102. *Resources of the Sahara.*

tourist centre and the chief commercial and manufacturing city. The main exports are wine, cereals and fruit. Oran (pop. 324,000) is the second port of Algeria, with manufactures of glass, cotton textiles and soap. The main exports are wine, cereals and wool. Constantine (pop. 255,000) is a railway and administrative centre, with leather and woollen manufactures. Arzew has been developed as the Algerian terminal for the export of liquefied natural gas, which is brought in pipelines from the Saharan fields.

In Tunisia, the largest town and capital is Tunis (population 642,000). It has a good position in the productive northern part of the country, and is the chief port, with exports of iron ore, phosphates and agricultural products. Sfax (pop. 250,000) is the second port, with exports of phosphates, olive oil and dates.

The Sahara

The Sahara is the greatest of the hot deserts, covering an area almost as large as Europe. The scenery is quite varied, some parts being sandy, with dunes, other parts having a stony or pebbly surface, while in the interior a series of isolated dissected plateaux reach a height of over 11,000 feet in the Tibesti Range and nearly 10,000 feet in the Ahaggar.

The climate is everywhere dry, but occasional showers allow a scattered scrub vegetation to grow in some parts. Owing to the cloudless skies high day temperatures are followed usually by low night temperatures. The incredible heat of summer is shown clearly in the figures for the oasis of In Salah (920 feet):

J.	F.	M.	A.	M.	J.	Jly.	A.	S.	O.	N.	D.	
13	15	20	24	30	35	37	36	33	27	20	14	°C. Rainfall is almost
55	59	68	76	86	94	99	97	92	80	68	58	°F. nil.

Population in the Sahara is very sparse. Only in the occasional oasis and along the Nile is sedentary agriculture possible. The desert nomads depend

FIG. 103. *Yearly variations in the flow of the Nile can cause either disastrous droughts or dangerous floods. Dams and barrages have been built along the Nile valley since the most ancient times, but none has ever had more than a marginal effect in controlling the river—which is virtually Egypt's only source of water. However, the new High Dam at Aswan will fully control the river, ensuring adequate water supply at all times, and providing* 10 *million kW of hydro-electricity. Shown under construction here is the intake section on the diversion canal.*

chiefly on their herds of camels, goats and sheep, so that their lives are dominated by the constant search for fresh pasture for their animals. Before the French conquest of the Sahara nomadic tribes like the Tuareg were often able to plunder or exact tribute from the cultivators of the oases, and from the caravans. In recent years, large deposits of oil, natural gas, copper, manganese, lead, tin and iron have been discovered, and the most accessible are being exploited. The map shows the location of the important oil and natural gas fields with the pipelines linking them to ports. The importance of these resources to the states with Saharan territories is great. It is hoped that the revenues will finance great social and economic changes.

The Nile Basin

The River Nile, about 4,200 miles long, is considered to be the longest in the world. Its sources lie on the East African Plateau, from which the stream descends by a series of falls and rapids through gorges to the flat basin of the Bahr el Jebel, or Upper Nile. From this basin the river flows slowly across the plain of central Sudan to beyond Khartoum. Three important tributaries join it in this section—the Sobat, the Blue Nile and the Atbara—all fed by the rains of the Ethiopian Highlands. North of Khartoum the river flows in a shallow gorge cut in sandstone, and is interrupted over a distance of 1,700 miles by a series of six cataracts caused by outcrops of old hard rocks which extend from the highlands on the east. Between Aswan and the Delta the valley widens out to an alluvial plain 10-15 miles broad, bounded to east and west by sandstone or limestone bluffs. Below Cairo the river has built up a large delta 8,500 square miles in area. The East African lakes even out the flow of the river, but the summer rains of the Sudan cause flooding in the plain of the Bahr el Jebel, and large quantities of water are lost by evaporation. The river is also obstructed by large masses of weeds in this area—the floating *sudd.* The rivers coming down from the Ethiopian Highlands also bring summer floods: the Blue Nile rising as much as 40 feet in September and bringing down up to fifteen times as much water as the White Nile. This flood water begins to pass down the river in June, reaches a maximum in September and subsides in November and December. This is the flood which deposits the fertile silt of Egypt, and which fills the numerous reservoirs in the desert, so that irrigation can be carried on throughout the year.

Changing climate causes changes in natural vegetation along the river. In the East African plateau, savanna is normally found, but along the river itself a modified type of rain forest is common. The Basin of the Bahr el Jebel is rather sparse savanna country, changing to pure desert farther north as rainfall decreases. The whole of Egypt may be considered to have a desert climate, for even at Alexandria the rainfall averages only 8 inches (203 mm.).

The Sudan

The Sudan is the largest country in Africa, covering an area of nearly 1 million square miles, but a large proportion of the country is desert or semi-desert, and the average population density is only about ten per square mile. The most densely populated parts are the irrigated areas near the Nile, especially the Gezira, a clay-covered plain extending over five million acres in the region between the Blue and White Niles. Here over one million acres are being scientifically irrigated by canal from a reservoir formed by a dam at Sennar on the Blue Nile. This dam, two miles long and with 800 miles of distribution channel, was opened in 1925 and has transformed the life of the district. The most important crop is cotton, but the cropping plan lain down by the irrigation board provides for rotation with wheat, millet, vegetables and groundnuts, and regular fallowing. This is one of the most prosperous farming areas in Africa.

Towns. The two chief towns are the adjoining cities of Khartoum and Omdurman. Khartoum is the capital and administrative centre, the chief commercial town, particularly for the cotton trade, and a route focus. It has a university. Omdurman, on the other side of the White Nile, is an old market centre. The combined population of Khartoum and Omdurman (including Khartoum North) is about 360,000.

The chief exports of the Sudan are cotton, groundnuts, gum arabic and sesame, which are exported mainly via Port Sudan on the Red Sea.

Egypt

Like the Sudan, Egypt is a large country (386,200 square miles), but in the past population has been almost totally confined to a narrow strip of land along the Nile where cultivation is possible. Even today, apart from this area, and a few oases, the rest of the country is mostly desert with practically no people. However, the population is rising rapidly and the total now exceeds 30 million.

Within the last ten years, the Egyptian government has made great efforts to raise the standard of living, to extend the cultivable area, and to introduce modern industry. The most spectacular project is the Aswan High Dam scheme (see caption to photograph, page 274).

Relief. The course of the Nile in Egypt has already been described. Between the river and the Red Sea lies a zone of deeply dissected highlands composed of crystalline rocks. These rise to over 7,000 feet. The southern part of the Sinai peninsula is very similar, but rises still higher to over 8,000 feet. To the west of the Nile the huge western desert is a low plateau, separated from the Libyan plateau in the north by the Qattara depression, which sinks to 440 feet below sea level.

Climate. As we have seen rainfall is everywhere low—below 10 inches (254 mm.) in the north and 3 inches (76 mm.) or less in the south. Temperatures reach 43° C. (110° F.) or even 49° C. (120° F.) in the hottest parts in summer, although along the Mediterranean coast summers are somewhat cooler. Occasionally, hot winds blow off the desert (the Khamsin). These winds may reach a velocity of 90 m.p.h. and raise the temperature by 19° C. (35° F.). Such winds can do great damage to crops.

Irrigation and Farming. Egypt has been described as the "gift of the Nile" because of its absolute dependence on the Nile waters. From time immemorial the annual flooding of the river has enabled rich crops to be grown in a desert climate. The traditional method of cultivation—the basin system—depends on the annual flood to cover the fields with water which is retained by the banks of earth around the fields. The water deposits fertile silt, and thoroughly soaks the ground. Seeds sown in this wet earth rapidly germinate and grow quickly under the hot sun. This system still serves about a quarter of the cultivated land. The basin system supports only one crop a year immediately following the flood, though there has always been a limited amount of irrigation, using primitive techniques. The most important system in use today involves damming the river to store flood water, the gradual release of this water throughout the year and its distribution by means of canals and channels. This system enables three or four crops a year to be produced, providing fertilizers are used. The chief crops grown are wheat, maize, millet, barley, rice, sugar, lucerne, beans, sesame, onions, citrus fruits, grapes, dates and cotton. The most profitable crop is cotton—as the long-stranded Egyptian variety is in great demand all over the world. Other important export crops are rice, sugar and onions. Many cattle are kept, the lucerne being used as fodder.

The need to increase the cultivated area in order to feed the increasing population and raise the standards of the whole country, led to the construction of the new High Dam at Aswan. This dam is 360 feet high and conserves 130 million cubic metres of water in a lake covering 739 square miles. This water increases by 30 per cent the cultivated area of Egypt and drives the world's largest power station, producing 2,100,000 kW, which multiplies Egypt's electricity supply by ten.

Towns and Industries. Under Egypt's first Five-Year Plan for industry (1957-60) £230 million was spent on factories for the production of textiles, motors, cement and superphosphates. A total expenditure of £423 million was involved in the second Plan (1961-66), to be spent on developments in the production of petroleum, iron, steel, chemicals and textiles. Other industries already producing are those concerned with alcohol and beer, chemicals, cigarettes, glassware, leather, paper, matches, rubber, salt and films.

Cairo (pop. 3½ million) is the capital and chief town, a great historic city of the Arab world. It has four universities and is a great cultural, commercial and industrial centre. The pyramids and other ancient monuments, and the sunny climate have made Cairo a famous tourist city.

Alexandria (pop. 1½ million) is another historic city, and is Egypt's largest port. The chief exports are cotton and cotton seeds. Others include rice, vegetables and manufactured goods. Other large towns are Suez (pop. 203,000) and Port Said (pop. 244,000) situated at the opposite ends of the Suez Canal, and Ismailia (pop. 111,000). The largest oasis town is Faiyum (pop. 102,000), situated in a depression watered by canal from the Nile.

The Suez Canal. The total length of the canal is about 100 miles. It was built by the French engineer de Lesseps and completed in 1869. There are no locks because of the flatness of the isthmus of Suez. It was used by about 18,000 ships every year, and was an important source of revenue to the Egyptian Government. The canal saves an enormous journey around Africa

FIG. 104. *Surplus natural gas being burnt at Hassi R'Mel in the Sahara.*

for ships sailing between Asia and Western Europe and was much used by tankers going to and from the Near and Middle East oilfields. But during the war with Israel in 1967, the Canal was blocked and remained closed after that.

Trade. The main exports are raw cotton and cotton yarn, and cotton cloth: raw cotton accounted 72 per cent of agricultural exports a decade ago, but now petroleum products and other secondary exports are increasing in importance. Imports are chiefly manufactured goods, fuel and mineral oils, iron and steel and wheat and flour.

QUESTIONS

1. Describe the characteristic features of a Mediterranean type of climate. How far does the climate of Algiers conform to the type? What are the changes that take place as we move inland from the Algerian coast?
2. Write notes on four of the following: the relief of the Maghrib; the Sahara Desert; minerals of North Africa; irrigation in Egypt; manufacturing in Algeria; the Aswan High Dam; the Gezira scheme.
3. Describe broadly the contrasts in climate and vegetation which would be noted in a journey down the Nile from Lake Victoria to Cairo.
4. Account for the very high density of population on the cultivated land of Egypt.

SOUTH ASIA

CHAPTER 28

SOUTH-WEST ASIA

SOUTH-WEST Asia, stretching from Persia west to Turkey and south to Arabia, forms a land bridge between the Asian and African continents. This has been, and still is a vitally important region in the world. Man's first crops were grown here and his first cities were developed in conjunction with the earliest irrigation schemes. Routes from Asia to Africa crossed with those from Europe to the East. The opening of the Suez Canal in 1869 accentuated the importance of the region. In the last fifty years the discovery and development of enormous oil resources have brought considerable foreign capital and attention. Lying along the southern borders of U.S.S.R., the area has additional strategic importance.

Relief

There are two contrasting physical zones: in the north, arcs of young fold mountains extend from the complex highland region on the remote borders of Turkey and Persia and enclose high plateaux in both countries (page 280). In the south lies the ancient stable block of the Arabian peninsula, which slopes from west to east. Between the two there curves a lower zone, largely filled with recent sediments which include oil deposits on the gently folded margins. This is the "Fertile Crescent," where winter rain in the west and the snow-fed Rivers Tigris and Euphrates in the east bring considerable water supplies. Faulting in the southern block has produced rift valleys, the largest one running from the Red Sea northwards through the Dead Sea and Jordan valley into Syria.

Climate

A great deal of the region experiences the "Mediterranean" climate pattern of winter rain and summer drought (Fig. 105). In summer mainly easterly winds bring tropical maritime air which is part of the Indian Monsoonal current, now dry after crossing land. These winds flow round a local

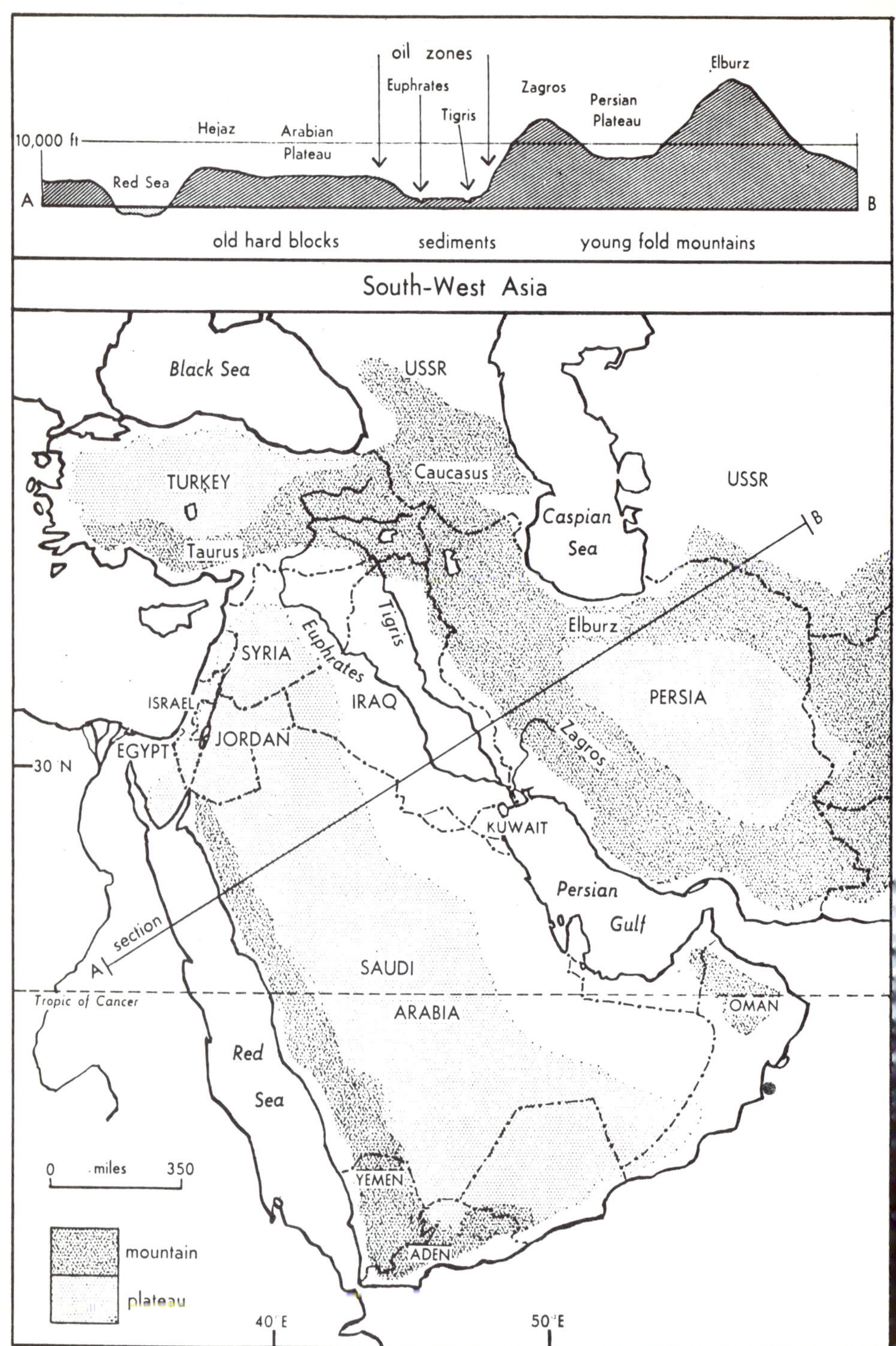
oil zones
Euphrates
Tigris
Zagros
Elburz
Persian
Plateau
Hejaz
Arabian
Plateau
10,000 ft
Red Sea
A
B
old hard blocks
sediments
young fold mountains
South-West Asia
Black Sea
USSR
Caucasus
TURKEY
Taurus
Caspian
Sea
USSR
Elburz
SYRIA
Euphrates
Tigris
IRAQ
ISRAEL
JORDAN
EGYPT
PERSIA
Zagros
30 N
KUWAIT
Persian
Gulf
section
SAUDI
ARABIA
Tropic of Cancer
OMAN
Red
Sea
0 miles 350
YEMEN
ADEN
mountain
plateau
40°E
50°E

low-pressure system and on the coasts the lower 2,000 feet becomes damp, producing intolerable conditions of heat and humidity without rainfall along the Persian Gulf and Red Sea coasts. In south-west Arabia there is heavy summer rainfall from monsoonal air drawn directly from the Indian Ocean by the low pressure of upper Egypt; and in northern Turkey depressions from the Atlantic which have skirted the high pressure prevailing in the west Mediterranean also bring summer rain.

Although in early autumn spasms of maritime air from the Mediterranean upset the stable summer conditions, the real rainy weather does not begin until late December when series of depressions move in from the west. Most of them originate in the Mediterranean. Some Atlantic depressions do penetrate this far, bringing heavy rain. The worst weather comes with the "cold front" of a Mediterranean depression which draws in semi-continental air from eastern Europe and causes raw cold rainy days. However, frequent interruptions to the westerly depressions occur. Polar continental air from Asia may sweep the area, giving clear, bright cold days; or, much worse, and especially in early spring, tropical continental air from Africa blows in hot dry winds of gale force, which can dessicate the newly sprouting crops. Temperatures can rise 17° C. (30° F.) in a few hours.

Because it is the lower atmospheric layers which produce most of the rain and these are disturbed by mountainous relief, there are considerable variations of rainfall within even a few miles. The influence of the Zagros, for instance, lying athwart the westerly winds, allows agriculture and grazing on the steppes of the Fertile Crescent, while to the west lies desert. There is enough rainfall on most of the mountains to support forest, most of which was felled long ago. The constant grazing of the hillsides particularly by the omnivorous goat has restricted re-growth with resulting soil erosion. Scrub merges into desert as the rainfall decreases, and large parts of the area are empty except for a few nomads.

Resources

Farming. The familiar cereals—wheat, oats and barley—are cultivated descendants of wild grasses of the steppe-lands of South-West Asia. These cereals are the main food crop in the north and west of the region. Here too the olive tree with its narrow leaves and long roots flourishes on the poor hillside soils over a wide area and provides oil for cooking. In the south and east rainfall decreases and temperatures increase and irrigated rice and maize, grown in the summer, gradually replace the winter-grown cereals. Often both cereals are grown, thus enabling the land to be used continuously.

The hardy date palm also provides starch in the hotter south-east, and with its high food value and keeping qualities it is a useful food supply for the nomads. Wherever cultivation is possible, vegetable gardens and orchards are found in bewildering variety. Deciduous fruits like apples and pears are grown in the cooler areas, and apricots, peaches, pomegranates and citrus fruits where it is warmer. Commercial production of oranges in Israel or dates in Iraq and sultanas in Turkey, demonstrate the importance of fruit-growing, but the crops are mainly used locally. Other cash crops include tobacco on the Turkish coast, cotton in many areas and opium.

Pastures. Around every village there are grazing lands for sheep and goats, which, although sometimes owned by the villagers, often belong to nomadic herders, who follow the flush of grass brought by the winter rain, or in the mountains by the spring snow melt.

Over three-quarters of the people depend on the land, and the village, invariably clustering around a well or spring, is the most common form of settlement. Because of poor communications most communities are largely self-supporting and it is difficult to avoid alternating periods of glut and famine.

Minerals. Since earliest times, small local supplies of metals have been manufactured with great skill into useful and ornamental articles of gold, silver or copper. However, south-west Asia has few sources of useful ores none of which is sufficiently large or accessible to be important today, except in Turkey where there are deposits of coal, iron and chrome. The major source of wealth is oil. The diagram (page 280) shows the two gently folded zones where oil deposits have been preserved; today this area is the major world producer, exceeding even the U.S.A. and reserves of oil are enormous and still expanding.

The majority of the early development was financed by foreign powers, chiefly British and American, and since both these countries have a vital interest in this source of oil, the strategic importance of the area has been heightened. The revenue from the sale of oil could play a major role in raising living standards of the countries themselves among the impoverished agricultural peoples. Oilfields necessitated the construction of pipelines to the ports: since Persian Gulf ports are shallow and ships had to pay heavy Suez dues, pipelines were laid across the desert to the Mediterranean.

Large-scale industry has developed slowly, although textiles have long been made in the villages, such as carpets from the rough wool of the local sheep, or fine lace and cottons. There are modern cotton mills in bigger centres such as Tel Aviv and engineering has been developed, largely to produce armaments in this troubled zone.

People

The 100 million people of S.W. Asia are unevenly distributed (Fig. 106). In the closely settled areas along the Mediterranean coasts and in the Tigris-Euphrates valley rulers, merchants and labourers live in towns while the impoverished country people are often only tenants or poorly paid farm

FIG. 105. *South-West Asia: rainfall distribution.*

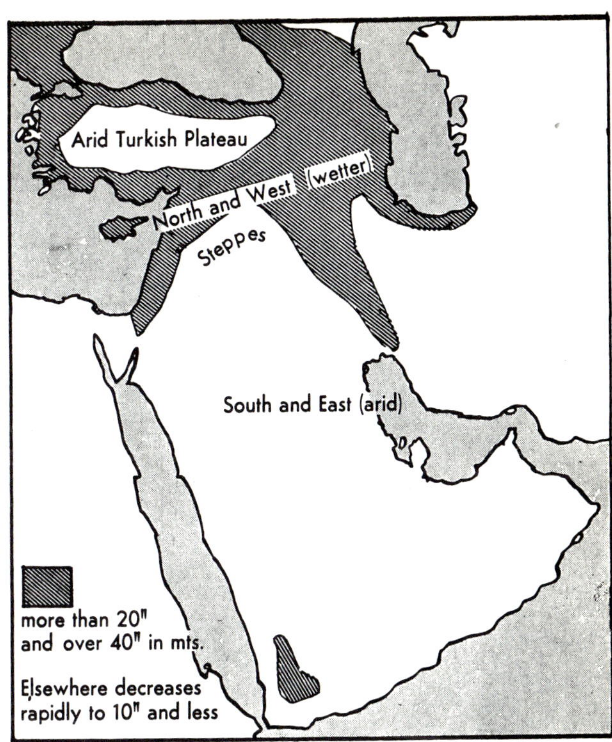

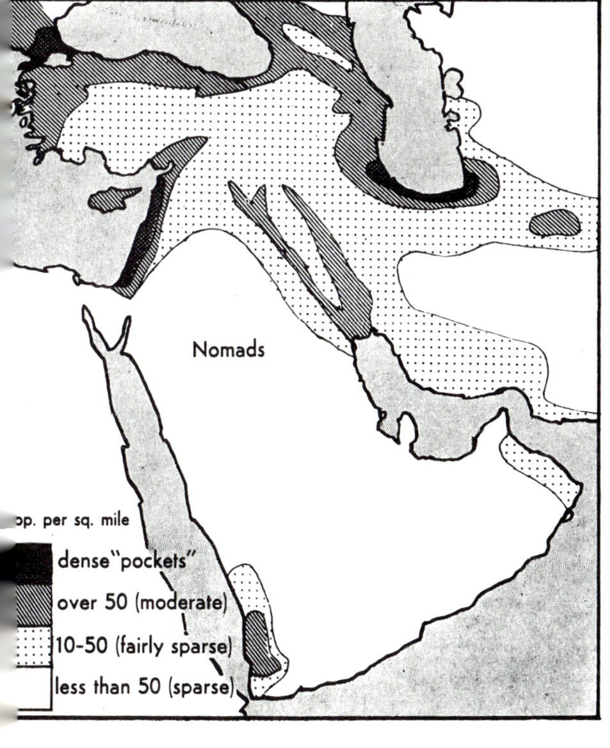

FIG. 106. *South-West Asia: population density.*

workers. Population is outstripping the food supplies in many parts and, although better farming methods, more irrigation and reclamation will help, increased industry is essential.

TURKEY

Turkey occupies the peninsula of Asia Minor and has a small European portion separated from the West by the narrow Bosporus Strait. Before the First World War Turkey had ruled a great deal of south-east Europe and the Near East. Shorn of these possessions in 1918, the country was poor, and a revolution in 1923 initiated many modern developments. Two-thirds of her 33 million people rely on the land, with considerable concentration on export crops. Industries based on coal and iron, chrome and copper, have raised living standards above those in most of S.W. Asia. There are great areas of poverty, especially on the plateau. Turkey's strategic position flanking Russia, and commanding her access to the Mediterranean, has brought her Western aid and influence in recent years.

There are four distinct regions of Turkey. The Black Sea coast with its high rainfall, steep slopes and lack of harbours is fairly thinly peopled, except in the coalfield of Eregli, and round the iron and steel town of Karabuk behind it. In the east the ancient town of Trebizon commands one of the few openings into the interior. The Sea of Marmara in the west is a drowned basin bordered by steep faults and a series of such inlets, continued inland by river basins, comprises the second region, the Aegean coast, which is much more open and habitable; it is the richest part of Turkey, with a large production of sultanas and other fruits which are dried and are exported from Izmir. *Istanbul* in the small European section of Turkey commands the route linking the Black Sea and the Mediterranean.

FIG. 107. *Turkey: mineral deposits, pipelines and main towns.*

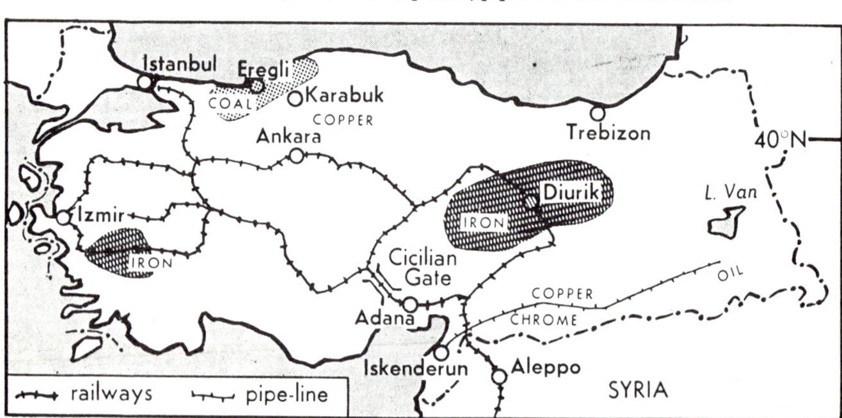

The Mediterranean coastlands are divided by ranges of fold mountains into isolated river basins; it is drier and warmer than the Aegean area, and with suitable irrigation, cotton and rice are cultivated.

The bulk of the country lies on the high inland plateau rimmed by mountains, which restrict the influence of the sea; rolling hills separate dried-up lake basins where cereals are cultivated; from these some wheat and barley may find their way into world markets. The plateau merges eastwards into the little-known tangle of mountains on the borders of Russia, Persia and Syria, where winter snows lie long, and the summer heat is intense; people become more scarce and poorer farther east.

Ankara replaced Istanbul as the capital after the revolution. Although an ancient centre, it is in a rather poor steppe region, reached by routes which for centuries have followed a trough lying behind the coastal mountains.

IRAN

Iran (Persia) is second in size to Saudi Arabia in S.W. Asia and its 26 million people compared with 7 million reflects the greater proportion of useful land. It is ruled by a Shah who is very conscious of the influence of Russia, and in recent years has introduced reforms in ownership of land and modern industry, in an attempt to raise the low standard of life of his people. However, Iran largely remains a poor agricultural country.

The Zagros Mountains sweep down the western half in parallel ranges of folded rocks which enclose river and lake basins with no outlet to the sea. In the north where damp Mediterranean air brings appreciable amounts of winter rainfall, slopes are forested and in basins such as those of the saline lake Urmia, wheat, tobacco and opium are grown. *Tabriz*, lying back from the variable lake shore, has an ancient market where nomads who, with their sheep, follow the mountain pastures as the snow clears in the spring exchange their wool for food. The wool is used in the local factories.

As the rainfall decreases southwards, forest is replaced by grass and shrubs and eventually bare rock. There are important agricultural basins centred on such ancient market towns and route centres as Isfahan and Shiraz. Mountain streams have deposited rich alluvium and also provide irrigation water for orchards of figs, apricots and peaches; cotton and wheat are also grown. In the southern Zagros great new dams are being built in Zuzistan.

Oil. Most of Iran's large oil reserves are found in the south-western folds of the mountains and crude oil is sent by pipeline to Abadan on the Persian Gulf for refining or export. In 1961 oil was also discovered offshore. Almost nine-tenths of the country's revenue comes from the sale of oil. Flanking the eastern edge of the Zagros is a line of towns such as Yezd and Kirman, which obtain water from the mountains through ancient and skilfully constructed subterranean channels called *qanats*, and are thus able to grow cereals and cotton. They lie where routes follow the mountain foot, but avoid the treacherous salt-encrusted marshes which cover large tracts of central Persia. This arid, uninhabited zone has possibly the highest temperatures in the world in summer, but bitter Siberian winds in winter. *Teheran*, the capital, lies between this "dead heart" and the Elburz Mountains to the north on a belt of more fertile alluvial soil. With its modern buildings and chemical and engineering factories, Teheran is like a Western city. Beyond the forested mountains a narrow coastal plain lies along the dune-fringed Caspian Sea, which is shrinking by eight inches in depth each year, leaving the ports almost useless. The high summer rainfall and temperatures make this a natural "greenhouse" area, which supports a fifth of the country's population; rice, sugar cane and tropical fruits are grown.

The Elburz broaden eastwards into a grassy plateau area with pastoral nomads rearing sheep and goats and on into the Kopet Dagh Ranges on the Russian border. These turn north-south to form the barren hill country bordering Afghanistan. Here again, wide basins support a little irrigated agriculture centred on market towns such as Mershed and Sistan; it is poor, remote, sparsely-peopled country. The southern coastal fringe of Persia has little rain and very few people.

FIG. 108. *Persia: Elburz Mts. covered with a mantle of snow.*

INDIA AND PAKISTAN

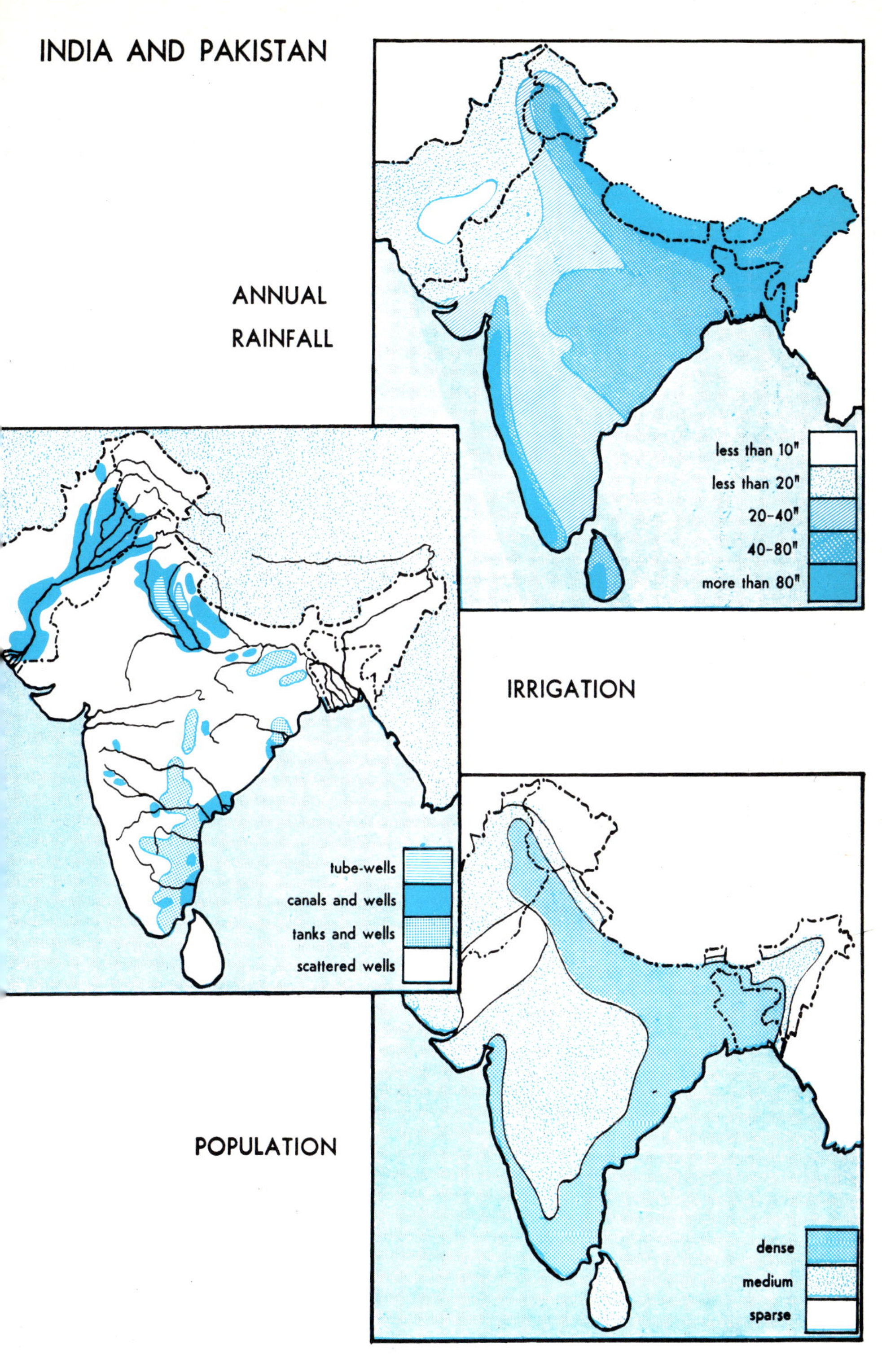

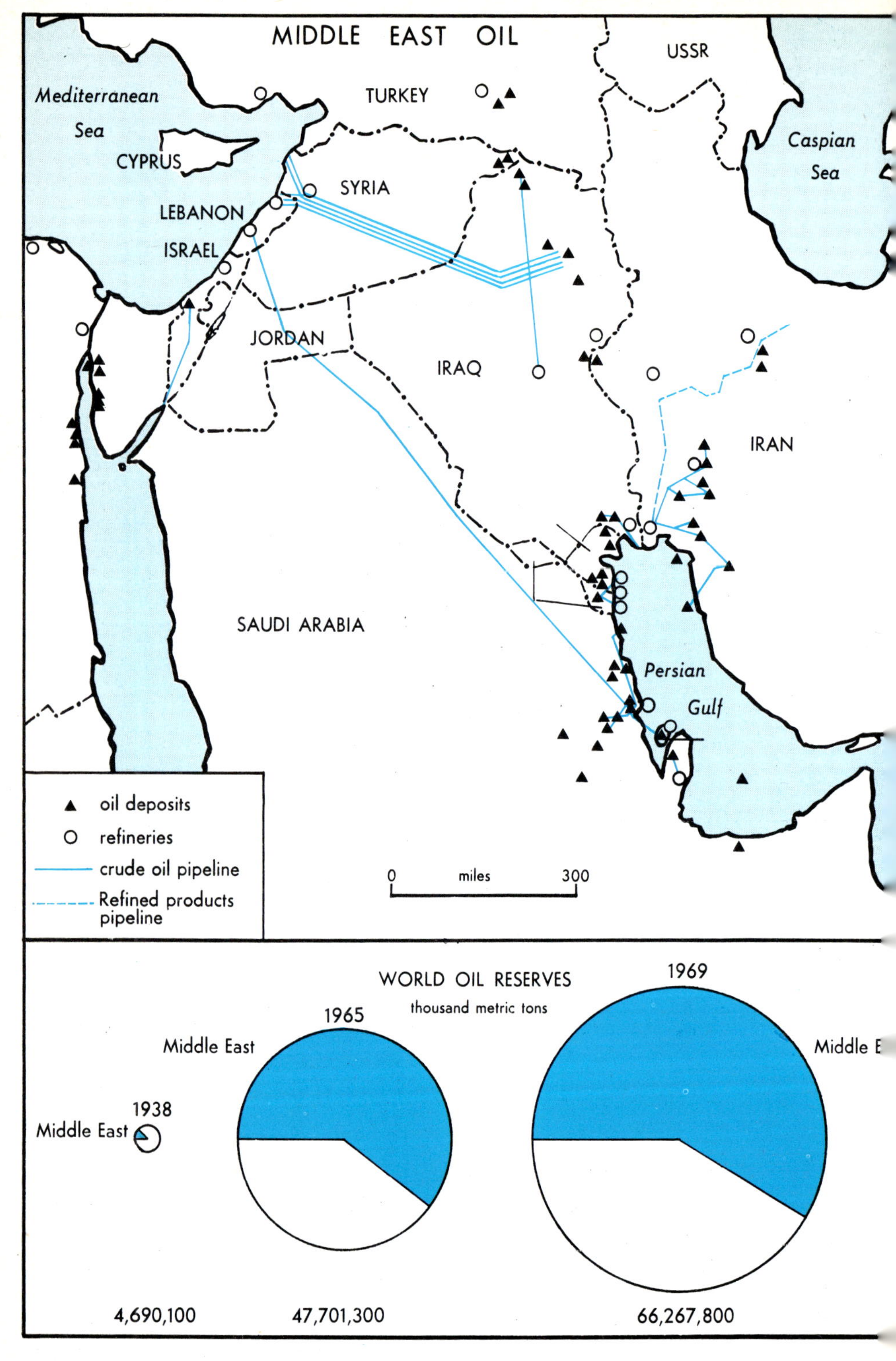
MIDDLE EAST OIL
Mediterranean Sea
CYPRUS
LEBANON
ISRAEL
SYRIA
TURKEY
USSR
Caspian Sea
JORDAN
IRAQ
IRAN
SAUDI ARABIA
Persian Gulf
oil deposits
refineries
crude oil pipeline
Refined products pipeline
0 miles 300
WORLD OIL RESERVES
thousand metric tons
1938
Middle East
4,690,100
1965
Middle East
47,701,300
1969
Middle E
66,267,800

IRAQ

Iraq (Fig. 109) occupies much of the lowland of the Rivers Tigris and Euphrates; its 9 million people are part of the Arab world. The name, "iraq," from the Arabian word for "cliff," refers to the low edge of the Arabian plateau to the west, a region of stony desert cut by the valleys of occasional *wadis*, temporary streams which flow after the rare desert storm. Bedouin graze sheep and goats on the sparse vegetation. On the east Iraq borders Persia along the Zagros Mountains, from which swift streams flow into a fertile, irrigated foothill belt. In the north-east lies the rolling steppe-land of Assyria, the chief cereal-producing area, with adequate winter rains for barley and wheat. The Kurdish people of north-east Iraq have much in

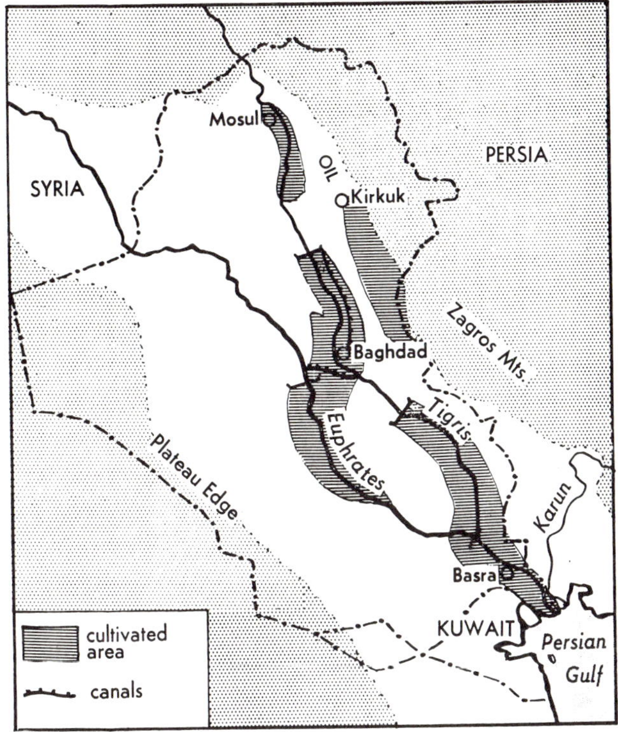

FIG. 109. *General map of Iraq.*

common with their kinsmen across the Persian border and have demanded the creation of a separate Kurdistan. *Mosul* on the Tigris is an old market centre with cotton- and wool-making. *Kirkuk* is the centre of a 60-miles long oilfield, from which a pipeline leads 400 miles across the desert to the Syrian port of *Banias*; a great deal of oil revenue (about half the total revenue) is allocated to economic development. This mainly means improved irrigation, for in spite of its importance since ancient times, Iraqi agriculture is very backward and the people very poor.

The rivers flood with the snow melt in spring, decrease during the long dry summer and rise slightly with the winter rains. In the north they have cut into the older alluvium and water must be lifted out; south of Baghdad they are above the general level of the country, which is built of recent alluvium, and simple canals can lead the water to the fields, helped by the storage of flood waters in barrages. *Baghdad* lies on the Tigris in the "waist" of land where the two great rivers come within 30 miles of each other, a zone where several ancient cities including Babylon have arisen. It combines squalid slums housing country Arabs who have migrated from poorer villages, with modern factories and offices, as well as ancient craft quarters and bazaars.

The rivers join to form the Shatt el Arab, a region of marshes separating intensively cultivated patches; 90 per cent of the world's commercial supply of dates grows here, irrigated naturally when the river waters are regularly ponded back by the tide; they are exported from *Basra*, which although 70 miles from the sea, is Iraq's only port.

SYRIA

The Turkish rule in Syria ended after the First World War, when the country came under French control. Syria finally became independent in 1950.

From a narrow coastal plain backed by mountains and stretching for less than 100 miles along the eastern Mediterranean, Syria widens abruptly east of the Orontes valley and then tapers north-east for 400 miles across the steppes of the upper Euphrates. There is a small population of 5½ million, but as only a fifth of the total area is cultivated, they are concentrated in the more favoured areas, chiefly the coastlands. Here, there is rain almost all year, since the prevailing south-west winds travel over a long stretch of sea and temperatures are never below those necessary for plant growth. Fertile land is confined to small flat pockets between rocky peninsulas and limited inland by limestone hills. Syria has only one rather poor port, *Latakia*, which has no rail to it, hence a great deal of her trade goes via Lebanese ports to the south.

The River Orontes meanders through the flat floor of a rift valley which continues the line of the Dead Sea and Jordan. Stretches of swampy salt marsh separate cultivated areas growing wheat, vegetables and fruit and irrigated by the ancient *noria* or wheel lifting the water into channels; Home, Harmar and Idlib are all market centres in these irrigated zones and have craft industries like metals and textiles. In the south, however, the Anti-Lebanon Mountains reduce the rainfall even further, so that rivers are fewer and desert prevails except in "oases" such as *Damascus*, which there is

FIG. 110. *Syria: water-wheel near Hama.*

intensive cultivation of Mediterranean fruits and vegetables over 400 square miles around the capital. Along the whole hillfoot zone *qanats* (see page 286) bring water to village lands.

Rainfall decreases eastwards but in the north there is sufficient winter rainfall to support sparse grasslands, the home of nomadic herders. *Aleppo*, the largest Syrian city, lies where these steppes meet the foothills of the Turkish mountains and accessible routes can penetrate them. Artesian wells have recently been tapped and cotton-growing increased. Like Damascus, Aleppo has a considerable industry making cottons, woollen and silk cloths. Population increase can possibly be absorbed in Syria for a time by further development of artesian wells and of the further 20 per cent of the country which is thought to be cultivable.

LEBANON

Lebanon became a state completely separate from Syria in 1950. It is the smallest country in S.W. Asia and with its population of about 2½ millions, comprising equal numbers of Christians and Moslems, is the most densely settled. The climate and country change quickly in the 60 miles Lebanon stretches from the coast. From the narrow plain, rarely over five miles wide, intricate stone-faced terraces climb the slopes of the Lebanon Mountains. Felling and grazing have left little trace of the famous cedars which once clothed these hills and many places suffer from severe erosion. In the Beq lower rainfall and more extreme temperatures necessitate irrigation.

Springs along the line of junction of alluvial clays and the limestone of the hills provide a water supply. Eastwards again the higher Anti-Lebanon Mountains reach 7,000 feet; their winter snows and pine-covered slopes attract increasing numbers of tourists. Also numerous relatives who have emigrated to America or western Europe send home money, so that on the whole the country enjoys a higher living standard than some of its neighbours.

ISRAEL

Palestine is the ancient homeland of the Jews and the birthplace of Christianity, and the larger part of it became the modern Jewish state of Israel in 1948; bitter enmity between Jews and Arabs make Israel hated and unrecognized as a state by the predominantly Moslem countries which surround her. After the Arab-Israeli war of 1967, Israel remained in possession of areas of adjoining Arab countries.

With nearly 3 million inhabitants Israel's population density is second to that of the Lebanon; although many Arabs left the country as refugees after 1950, thousands of Jews from Europe came to settle, bringing with them skills and an enthusiasm for their new homeland. Despite the poverty of much of the soil, and the lack of minerals or power resources, Israel is forging ahead as a modern nation in industry and agriculture.

The north-south pattern of coastal lowland, hills and then interior plain, which runs through Syria and Lebanon, continues into Israel and Jordan. However, the belts become wider and more complex and Israel has an additional region in the barren Negev Plateau in the south, sweeping round to the Sinai peninsula and the Egyptian border.

The coastal plain is fringed in part with sand dunes, which make harbours scarce and encroach on the farmland; the sandy soil behind them, watered by wells reaching down to impervious clays below makes ideal conditions for the citrus fruit grown in large orchards. Most of this is exported from the modern port of *Tel Aviv*, which is also a holiday resort and industrial town making textiles and hardware. In the north, *Haifa*, in a rocky sheltered inlet, has lost much of its former importance for oil-refining since the pipeline feeding it from Kirkuk in Iraq, was cut by the hostile Arabs of Jordan; however, it imports crude oil by tanker to refine. Throughout the plain new well-planned Israeli villages surrounded by productive farmland contrast with ancient mud-walled settlements.

The Lebanon Range decreases in height over the Israeli border and merges into a basalt plateau yielding rather better soil in the region round Nazareth. Southward the fertile plain of Esdraelon cuts through the hill zone and beyond it lie the barren hills of Judea. The wilderness sweeps up to the city gates and the city owes its growth entirely to its religious significance in different ways for Jews, Christians and Moslems.

FIG. 111. *Jordan Valley. The meandering pattern of the river on its course from Lake Tiberias to the Dead Sea. Note the deep trough-like valley with its steep slopes and the enormous loops of the meanders. The valley is thought to be part of a general rift structure which continues southwards through the Red Sea and Ethiopia to Central Africa.*

Precipitous slopes drop to the Dead Sea shores; its bed lies 2,598 feet below sea level and potash evaporated from the waters is Israel's only mineral wealth. The Negev is a region of nomads, except where the Israelis have established new settlements in regions with deposits of loess and, by the skilful use of the heavy dew fall of the desert, they grow wheat, maize and barley.

JORDAN

Trans-Jordan was the name given to the area east of the Jordan which became a country under British Mandate after the First World War; it became the independent country of Jordan in 1945, and later extended its frontier westwards into Palestine. Jordan is the poorest country in S.W. Asia, with very little lowland, uncertain water supply and no oil resources yet discovered; indeed three-quarters of the country is useless desert. Assistance from the world powers interested in her strategic position plays a large part in the country's economy.

In the far north-west Lake Tiberias lies behind a sill of basalt extending from the Nazareth region of Israel and, in cutting through this, the River Jordan falls swiftly enough to be used in generating electricity. The river then meanders through desert, its banks fringed with willows and tamarisk, or by saline swamps. At intervals small irrigation schemes and land-reclamation give cultivated tracts, where the high temperatures all year round enable tropical fruits such as bananas to be grown. The frontier follows the river south of the Dead Sea through mainly desert country to the Gulf of Aqaba and Jordan's only seaport, *Aqaba*, which handles the country's meagre exports. The principal exports comprise tropical fruits from the valley floor, phosphate rocks from the Dead Sea region, olive oil from the hill slopes and leather from the grazing lands east of the hills.

Under the rule of King Hussein most of the nomadic tribesmen have been settled in agricultural villages as part of efforts to improve food production; peasants have been granted land ownership, and co-operative efforts in buying machinery encouraged. However, the problems of drought, coupled with rising population, make it difficult for Jordan to become self-supporting.

Arabian Peninsula

The Arabian peninsula is similar in size to India and Pakistan, yet the populations of the two areas are respectively 15 million and 618 million. Although Saudi Arabia dominates the peninsula, around the edges are a series of states varying from the tiny Bahrein to the large Aden; they are mainly sheikdoms under some degree of western influence, except for the Yemen.

From the narrow, barren coastal plain along the Red Sea, where *Jidda* is the only port for 2,000 miles, the Arabian plateau rises sharply in the Hejaz (meaning "barrier"); where there is sufficient rainfall, steppe-lands develop, and the great Nejd plateau beyond the mountains is still the home of many Bedouin tribes whose diet consists mainly of dates, bread and milk, and who drive their herds of sheep and goats between the pastures. The Hejaz rises southwards through high pasture land of Asir to a maximum of 12,000 feet in the mountains of Yemen, a remote and hostile little country.

The Tuwaiq Mountains run from north to south through the centre of Arabia; rain falls on their limestone rocks and percolates along the beds eastwards where it accumulates in enormous pits in oases like Hofuf and Wadi Kharj and is pumped into canals which irrigate fields of melons, tomatoes, onions and other vegetables. Extension and improvement of these and other aspects of farming are being carried out with some of the vast oil revenues received by the king, for the great Near-East oil zone extends down the eastern flanks of Arabia. The tiny sheikhdom of Kuwait, otherwise a barren desert, and the small island of Bahrein are exporters of oil.

Aden contains much of the Rub'al Khali, or "empty quarter," a largely unexplored waste of sand-dunes and uninhabited save for a few little-known primitive tribes of hunters and gatherers; towards the coasts of the Arabian Sea, oases occur where water percolating from the Yemen can be tapped by wells, and there is enough rainfall for sparse pasture. Along the south and east of the peninsula the Arabs have long been fishermen and traders; their large sailing dhows, a common and, not so long ago, a feared sight for their slave raids on the east of Africa. In the remote country of Oman, where the land rises to 10,000 feet, there are inland communities still with Negro slaves, who were brought in at the port of Muscat, which lies in a torrid, sheltered and deep inlet.

QUESTIONS

1. Discuss the extent to which South-West Asia has been a world cross-roads in the past, and remains one today.
2. Describe and account for the difficulties of the Arabian environment.
3. Write an essay on oil in South-West Asia.

CHAPTER 29

INDIA, PAKISTAN AND NEIGHBOURING COUNTRIES

THIS vast area contains widely contrasting regions, from the equatorial regions of Ceylon (6° N.) to the continental regions of northern Afghanistan (40° N.); and from the deserts and bare ridges of Baluchistan to the thick tropical forests clothing the mountain arcs along the Burmese frontier. It also rises from sea level to the highest peaks of the world in the Himalayas.

Like Egypt on the Nile and Mesopotamia between the Tigris and Euphrates, the Indus plain in West Pakistan was the home of ancient civilizations based on planned irrigation, agriculture and large complex cities, rich in activities, trade and culture. Throughout history there have been repeated invasions and movements of peoples from central Asia through Afghanistan and across the harsh ranges by the Khyber and other passes into the Indo-Gangetic plains and south into the Deccan peninsula. Many strands have become woven together into a complex tapestry of race, language and religions. The dominant racial type has dark straight hair; skin and eye colour are lighter in the north and west than in the south, and people there are often taller. In remote jungle-covered hills, there remain primitive tribal groups of dark frizzy-haired Negritoes, speaking a variety of ancient languages which have only recently been written down and worshipping local gods, perhaps a rocky hill, tree or stream.

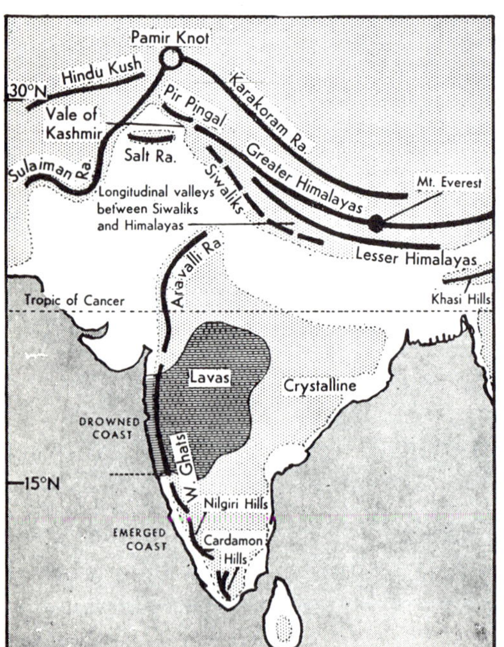

FIG. 112. *India: relief.*

There are 400-500 million Hindus; their religion is characterized by sharp division into castes of which there are four main groups, the priests, warriors, farmers and craftsmen and the "out-castes," formerly "untouchables," who are scavengers, leather-workers and washermen. There are many large and increasing numbers of sub-castes, although strict observation of the links with particular occupations and the rules barring inter-marriage or even eating with another caste, are weakening and are no longer officially recognized. Untouchability is outlawed, following the teaching of modern India's great moral and national leader, Mahatma Gandhi, although it will take a long time for all trace of it to disappear.

Relief

There are three major regions, the northern mountain arcs, the southern plateau and the intervening plain. The Himalayan Ranges average 300 miles wide and over 2,000 miles long; even the major passes lie over 18,000 feet above sea level and there are many peaks well over 20,000 feet. Ranges fan out from the complex Tertiary folds of the Pamirs mainly in U.S.S.R. (Fig. 112). They decrease in height where they swing southwards at each end; in the west in the barren jagged ranges of Afghanistan linking with those of Persia and in the east in the jungle-covered parallel ranges bordering Burma.

Peninsular India, the Deccan, is thought to be a fragment of the great southern continent of Gondwanaland which in the course of drifting apart was largely responsible for pushing and folding the Himalayas out of the deposits of the ancient Tethys Sea which bordered it to the north. The plateau, including Ceylon, has a base of ancient Archean gneiss, with low ranges of even older schists standing above the general level. Faulting has caused rift valleys and depressions now occupied by rivers mainly following the general slope from west to east, and building deltas into the Bay of Bengal. The Eastern Ghats are a lower and less continuous rampart than the Western Ghats, which drop steeply, cut by short swift rivers to the narrow plain bordering the Indian Ocean. The north-west corner is covered for 25,000 square miles by sheets of lava poured out through long vents in Tertiary times and the Ghats here are carved along the beds into a series of steps.

The Indo-Gangetic Plain, 2,500 miles from end to end, has been built from layer upon layer of alluvium brought by flooding, by wandering rivers from the mountains to the north and to a lesser extent from the Deccan. In detail there are distinct variations in levels and soil types. Along the foothills, a zone of coarse waterless gravels has been laid down by torrents from the outer ranges. Each of the major rivers has formed its own cone-shaped area of deposits, which become finer away from the river and from the mountains until the zone between the cones is one of fine clays. In the arid west the rivers have cut deeply into the cones, leaving them "high and dry," and flow in narrow flood-plains; further east they become less deeply incised, until in Assam the Kosi River wanders freely over its cone, bringing frequent disaster.

The Indus, fed by the five great rivers of the Punjab, has built a delta of shifting channels and sandbanks into the Indian Ocean; in the east the combined Ganges-Brahmaputra Delta covers an area the size of England. Here the main distributaries now flow through the eastern half and the western half has been left with stagnant water channels with no regular renewal of silt.

Climate

The Monsoon regime dominates the climate with its division into three seasons. The cool dry season from November to February has clear sunny days, warm in the south but with cool winds in the north. Westerly depressions bring 1-3 in. of rain a month to the north-west, allowing the winter growth of wheat; these depressions either come from the Mediterranean or originate in troughs in the upper air and also account for snowfalls in the mountains. Cold spells follow sometimes bringing frost well down the Ganges valley. Meantime south-east India and north-west Ceylon are experiencing their main rains from the equatorial maritime air involved in depressions over the Bay of Bengal from the retreating summer monsoon; along with the high southern temperatures they encourage the growth of tropical crops.

April, May and June are the months of the hot, dry season over most of the subcontinent. Even in the shade day temperatures often exceed 37·8° C. (100° F.) and remain above 32·2° C. (90° F.) at night. By May the lower layers of the atmosphere are humid and conditions even more trying. Bengal has some squally storms called "Northwesters" derived from troughs in the upper air, Assam shares the early monsoon with Burma and South India also has some early rains; however, elsewhere upper air conditions prevent the humid air from rising enough to cause condensation.

In mid-June comes the "burst of the Monsoon." A high, dry-season air-stream moves suddenly with the general summer movement of pressure and wind belts, from south to north of the Himalayas, the humid lower air is able to rise; heavy rainfall begins as a series of storms and then prolonged wet periods varying in amount with relief and exposure to the prevailing winds. Breaks occur in the rains, longer and more disastrous in the regions of lower rainfall; as they tail off, intolerable conditions of heat and humidity in October merge into the next cool dry season.

Resources

Minerals. The main exploited coal and iron deposits are in the north-east corner of the Deccan and there are iron deposits and lignite in the south. The oldest oilfield is in Assam and there is some in the salt ranges of West Pakistan. India may have important oil deposits in Gujerat, while Pakistan obtains natural gas from Sui, north of Karachi. Although long famous for opals and rubies, Ceylon has little wealth suitable for modern industry.

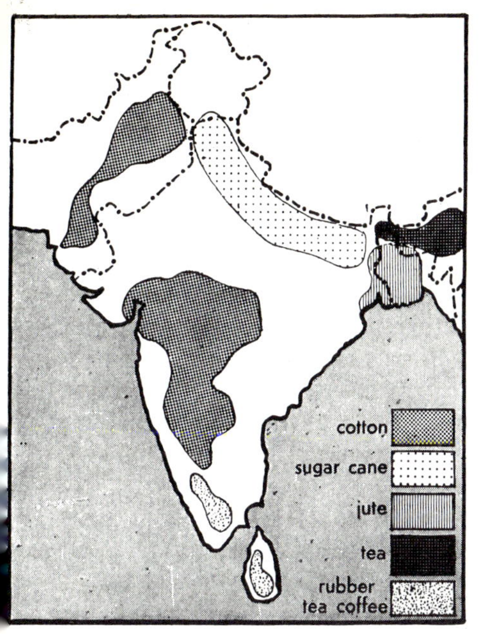

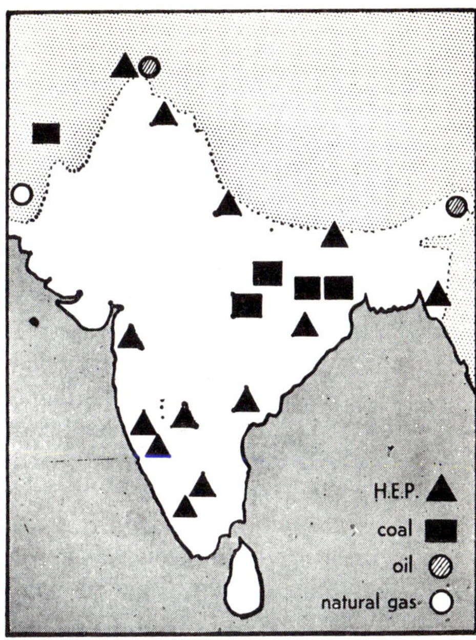

FIG. 113. *India: chief cash crops (left); power resources (right).*

FIG. 114. *Paddy fields between Jammu and Kashmir (left); Kerala, one of India's many hydro-electric dams.*

Increasing supplies of hydro-electric power are also being developed along the Himalayan edge and in the Western Ghats and India may well obtain power from atomic reactors in the near future.

Industry. Since Indian independence in 1947, the amount and range of industry has increased tremendously; before then there was the heavy iron and steel industry associated with the coal and iron deposits, textiles, mainly cotton, in the cotton-growing areas and jute in the delta region of Bengal, and a multitude of craft industries such as gold and silver work. Now there is development of the manufacture of machinery, electrical equipment, machine tools and fertilizers, in addition to the great new steel plants which have been built. A number of industrial areas are being established in the predominantly rural landscape.

Farming. Over four-fifths of the people live in villages and are engaged in agriculture either as owners, tenants or labourers or as village craftsmen. Village types vary from the loose linear pattern along the sandbars of Kerala or the levees of Bengal where water supply is no problem, to the compact villages of the arid Punjab and northern Deccan. In between is the looser village of the "tank country." The water supply is the predominant factor which determines the character of village and the crops grown, and irrigation has been practised for thousands of years. Wells are most common in the alluviums of the Indo-Gangetic plain, but are found everywhere; the *tank* is an artificial lake holding the monsoon floods for use during the dry season in southern India. Modern canal-irrigation was started by the British in the clay soil of the Punjab and is associated with low barrages downstream and higher dams in the foothills. There are important new schemes and extensions of old ones along the Himalayan edge, in Afghanistan, in the Deccan and in northern Ceylon.

Apart from the areas of plantation crops started by Europeans, for tea in Assam, and coffee and tea in south India and Ceylon, the methods of farming are simple though not unskilled and mechanization is quite rare. Holdings are small and each may consist of many patches or strips. Efforts are being made to increase efficiency and production by

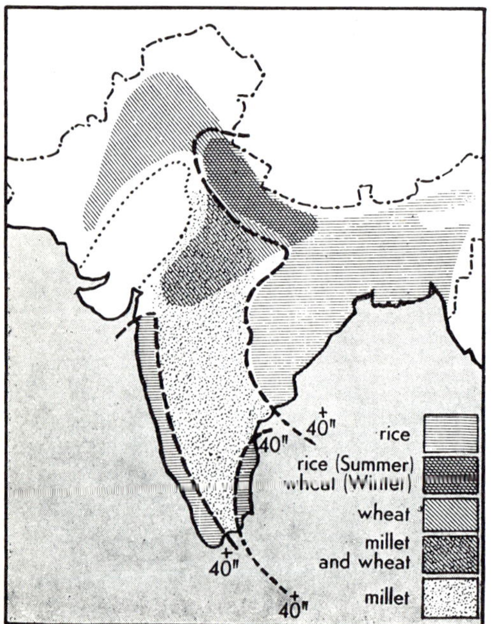

FIG. 115. *India: chief grain crops.*

schemes of rural education, often by example on one farm in a district. The chief crops are shown on the maps. Although the basic cereal, rice, grows where soil and water conditions are suitable, in many drier areas it is a prized rarity and coarser hardier grains are the staple; in the north-west wheat becomes increasingly important. Vegetables, especially those used in the preparation of spicy curries, are grown in tiny patches. Coconuts and bananas become important in the south.

Population densities are about as high as the resources of an area can support and sometimes higher, leading to migration and even hunger; there is little room for the rapid increase taking place without advances in farming techniques and industrialization. Thus each country in the subcontinent has adopted some form of planned development with aid from international bodies like the Colombo Plan or the World Bank. Ceylon has one of the few areas where new land can be opened up for farming; it lies in the drier north, where an ancient irrigation scheme had decayed. The settlement of thousands of families here is failing to keep pace with an increase of 3 per cent in the total population each year. India is developing heavy industry in order to make her own machinery for more factories. In the country "Community Development" areas, each covering 100-200 square miles, encourage the people to help themselves in crop improvement, and road, school and hospital building. Pakistan imports machinery for new factories.

AFGHANISTAN

Afghanistan lies among mountain belts and desert basins which stretch south-west from the Pamirs; this whole western end of the mountain zone is arid, most of the bare rocky surfaces yielding only a little poor grazing. The higher Hindu Kush have enough water for hydro-electricity and are flanked by coal deposits. Agriculture is mainly confined to alluvial fans along the hillfoot and the major river basins of the Helmand, Kabul and Zhob; here cereals and fruit (especially apricots) are grown, while nomads use the arid outer areas. This type of country continues south and east into Baluchistan, part of West Pakistan, where *Quetta* is the administrative centre.

Northern States

South-east from the Pamirs lies Kashmir, centred on the beautiful Vale of Kashmir, once a lake-bed along the Jhelum River. Although the former Hindu Maharajah acceded to India in 1947, most of the people are Moslem. Villages lie on alluvial fans, or terraces of the former lake, and intricate rice terraces are watered by mountain streams. Herds of sheep and cattle move up to the spring pastures in the mountains. *Srinagar* is the centre of an important tourist industry.

As the great mountain arcs stretch eastwards, they become more humid, and the southern edge becomes flanked by swampy jungle, while forests stretch farther up the slopes. Between Kashmir and Nepal lies a 300-mile stretch of mountains belonging to India's northern states, and containing hill-stations such as Simla and Naini Tal, and the beautiful Kulu Valley.

Nepal, strongly influenced by India and bordered to the north by the ranges which include Everest, is 500 miles long. Swampy "terai" on its southern edge can now be developed, especially since the advent of malaria control. All but the highest mountains are used, either for stock grazing, including the yak, or for cultivation of potatoes and barley, while the great interior valleys are rich rice producers. The Indian-controlled country of Sikkim, lying on the ancient pack routes to Tibet, has similar country, while south of it a portion of West Bengal has many high quality hill tea-gardens round Darjeeling and Kalimpong. East of this Bhutan slopes from the snow-capped mountains of the Tibetan borders southwards to the Brahmaputra plains; here too roads from India are bringing influence from the south into the forested remote hills and rice-growing valleys. Buddhism survives strongly in Bhutan and Nepal. The wild forested country of Bhutan extends eastwards to that great tangle of mountains where India, China and Burma meet. The Brahmaputra gorge cuts through them to the plain of Assam, which is shut in to the east by north-south ranges. All these mountains are tribal territory, often with hilltop villages and rice fields in the valleys; being remote they are subject to famine, which only air drops of supplies can help. The Nagas are one group which has achieved some degree of self-rule within the Indian Republic. Where there is a larger valley or basin, as in Manipur, larger settlements are found.

The Assam plain is bounded on the south by the crystalline horst of the Khasi hills which support shifting cultivators. Around the hill station of Shillong there are moorlands and pinewoods, while Cherrapunji on the south-facing slopes receives the full impact of the summer monsoon and records one of the highest annual rainfall totals in the world (428 in.).

PAKISTAN

Pakistan lies in two contrasting parts separated by 2,000 miles of India (Fig. 117). Although more favoured in soils and rainfall, East Pakistan has more pressing problems, since its dense population is isolated from former outlets in Assam and the factories processing its jute were left over the border in Calcutta after the partition of India. Since then many new factories have been built, hydro-electric power developed at Karnafuli to supply *Dacca*, the growing capital, and the port of *Chittagong* improved to handle the export of raw and woven jute. West Pakistan is larger and more varied (Fig. 116). The bare ranges of Baluchistan and the North-West Frontier Province slope down through scattered hillfoot oases to the desert banks of the Indus, irrigated seasonally in parts by inundation canals and perennially by the

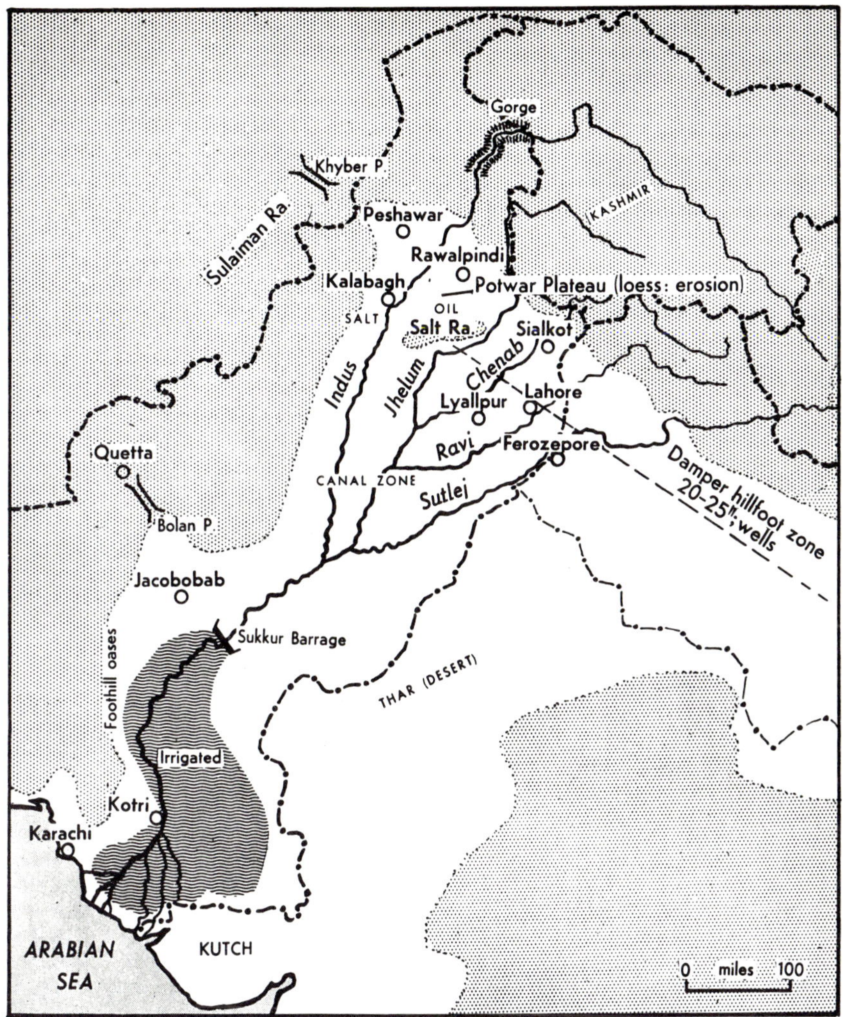

FIG. 116. *Indus valley.*

Sukkur Barrage. *Karachi* to the west is a modern sea- and air-port. The new capital, *Islamabad*, not yet completed, lies near Rawalpindi on the loess plateau of Potwar in the far north. East of it ancient well-irrigation along the hillfoot and more modern canal-irrigation further out on to the plain, encourages the growth of wheat in winter and rice in summer. Numerous old towns with craft industries lie in the hillfoot zone. The south and east become progressively drier until the scrub and thorn of the Thar Desert is reached.

INDIA

The Indo-Pakistan border cuts through the Canal country of the Punjab, and this led to bitter disputes over the sharing of water. The great Bhakra-Nangal project, built since partition, now leads water south of the Sutlej to the deserts of northern Rajasthan and provides power for several cities including Delhi. Ludhiana, with cycle and sewing-machine factories and workshops making car parts, largely employs refugees from West Pakistan. In the canal and well-irrigated zones south of the hills and where rainfall is more abundant wheat gives place to rice as a food crop, and to sugar-growing which gives cash to the farmers and industry to the many towns. Delhi stretches from the banks of the Jumna through the crowded slums and bazaars of the old town to the spacious capital of *New Delhi* built in 1912.

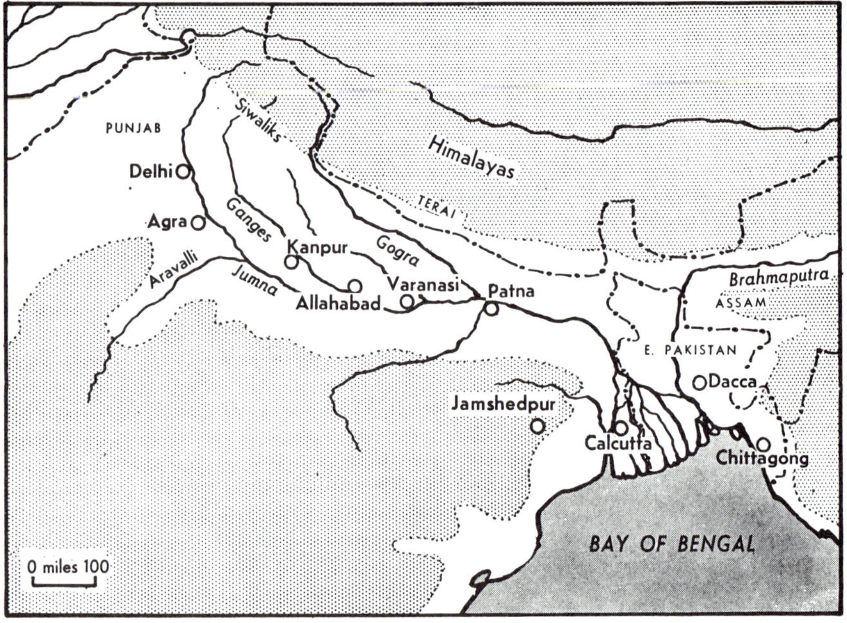

FIG. 117. *Ganges valley.*

The forested Terai along the Nepal border is being developed for farming since the coming of malaria-control. Further east the long-settled alluvial cone of the Gandak contrasts with that of the frequently flooding Kosi. The broad alluvial plain of the Ganges occupies the southern third of the vast lowland stretching from the Himalayas to the Deccan; it is a gigantic "rice bowl" watered by the heavy summer rains and supporting both dense rural populations and many cities such as *Varanasi* (Benares) and *Kanpur*.

FIG. 118. *India: iron and steel works at Jamshedpur.*

In the wet north of Bengal and again in the Assam valley terai jungle was cleared largely by European capital to provide land for medium-quality tea-gardens. Eastern Assam has coal and oil and new farmland is being won from swamp and jungle; industry and trade are increasing in the river ports in spite of flood and earthquake hazards. Farther south in the very densely populated western half of the Bengal delta, deserted by the main distributaries, rice and increasing amounts of jute are produced. The produce is sent to the factories which are strung along the River Hooghly.

Calcutta, with its chemical, paper and machinery making, has strong links with the industrial towns of the north-east corner of the Deccan, for which it is a commercial centre and outlet.

Peninsular India

The humid north-east part of the Deccan has a landscape of flat-topped plateau surfaces of overgrazed grass and scrub with forested hills rising above them. Rice terraces flank the valleys and rainfall is sometimes supplemented by irrigation. Lac (for lacquer) is collected from insects on certain trees; most of the forest is leafless in the hot dry season. New irrigation and power projects like Hirakud, serving the Chattisgarh area of the middle Mahanadi basin, are increasing the productivity of farm and factory. The Mahanadi delta is a rice-growing area subject to hurricanes from the Bay of Bengal in spring and autumn. South of the delta, on the narrow coastal plain, lies Visakhapatnam, India's main shipbuilding port.

Rainfall decreases westwards into Madhya Pradesh, where the valleys and basins begin to contain black soil derived from the lavas which, being retentive of moisture, encourage the growth of cotton. Sorghum is the main food cereal. The remoter hills still have tribal groups including the Gonds, from whom Gondwanaland is named; the busy airport of Nagpur lies at the junction of trans-Indian airlines and railways. The nearby coalfield provides power here and for the engineering and electrical factories of *Jubbulpore.*

The two chief cotton-growing areas centre on the textile towns round Ahmedabad and on the plateau inland from Bombay where the wide rather dreary landscape of semi-arid lavas is interrupted by trees lining the roads or surrounding villages. Sorghum, wheat, groundnut and linseed are also grown. From *Poona*, a military and engineering town, the electric train to Bombay descends the steep Ghats, with their basalt cliffs carved into bizarre shapes, to a totally different landscape. Rainfall of 80-100 in. combined with the alluvial soils derived from the lavas makes this a vivid green landscape of rice in the rainy season. *Bombay*, India's leading port, is cramped on to a long island. Hydro-electric power generated in the Ghats feeds cotton mills, car and engineering factories. South of the city the drowned coastline has fishing ports sheltered by lava headlands. Beyond Goa the coastal landscape changes to one of lagoons and sandbars, lined with coconut palms. Consequently fishing is supplemented by coir and copra making in *Mysore* and *Kerala*. Hydro-electric power harnessed from the steeply falling rivers is carried eastwards to Mysore and Madras and west to coastal towns, such as Alwaye, where it provides power for aluminium-smelting. The rich alluvium behind the sandbars enables two or even three rice crops a year and supports a rural population reaching 3,000 people per square mile, one of the highest in the world. Between the coastal fringe and the Ghats lies a lateritic plateau where tapioca is the main foodstuff; although a high yield is obtained it is protein deficient. The lower slopes of the Ghats have rubber plantations, while higher up tea and coffee are grown on the slopes. Elephants are used in the transport of hardwoods from the rain forest on the wetter western slopes and, over on the drier eastern slopes, the deciduous teak.

In the rain-shadow area east of the Western Ghats and south of the lavas lies the tank-irrigation country of Mysore, Andhra and Madras; rice and sugar are grown below the tanks, and eleusine millet, groundnut and sesamum on the dry fields. The rich gardens of areca, betel vine, spices, fruit and vegetables are irrigated from wells where possible. *Bangalore*, sprawling over the plateau, is the largest town of south India; its textile, aircraft, electronic and machine tool factories use hydro-electricity. Between the Cardamon and Nilgiris hills the low corridor of the Palghat has black soil which encourages cotton-growing, supplying textile towns like *Coimbatore*.

The Eastern Ghats and inland hill ranges have a foothill belt of intensive well-irrigated rice and vegetable lands. The great Cauvery delta, which has had canal irrigation for at least a thousand years, has an intensive pattern of rice, banana and sugar patches, and several ancient temple towns. The south-east tip of India is dry and tank-irrigation comes right to the sea.

Madras, in the centre of a lagoon-fringed coastline, has an artificial harbour; it has an increasing number of industries including cars and textiles.

Northwards the joint deltas of Krishna and Godavari are a major tobacco region. The inland areas in the heart of the Deccan have been opened up by the irrigation and power from the Tungabhadra Dam.

Fig. 119. *Ceylon.*

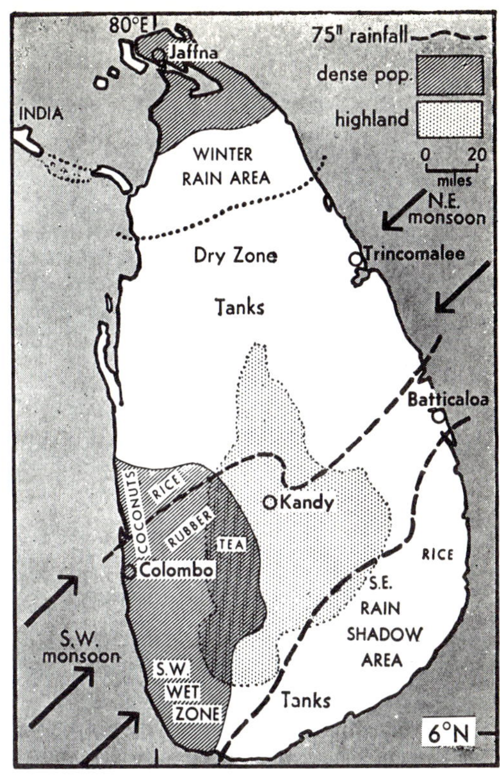

CEYLON

In recent centuries population in Ceylon (which is Buddhist) has been centred on the Wet Zone (Fig. 119), especially in the coconut and rice-growing coastal plains of the south and in the mountain valleys. The old Arab trading ports and later the influences of the Europeans, Dutch, Portuguese and British resulted in commercial towns on the coast, like *Colombo* or *Galle*. Rubber plantations on the lower slopes, and tea and coffee higher up opened up the forested highlands and attracted Tamil workers from the Indian mainland, who are a substantial minority in the independent Buddhist state. Again in the limestone Jaffna peninsula there are the Tamil people, who date from a much earlier invasion and who grow rice with irrigation from ground water in the limestone. Between Jaffna and the Wet Zone lies the thinly inhabited Dry Zone, where although rainfall is about 50 in. evaporation is high, and dry jungle abounds. Once the centre of an ancient canal and tank system irrigation supporting a much higher population, it is now mainly dry scrub with an occasional village irrigating rice fields from a tank. Malaria and invasions have resulted in the decline of its population and great cities like Anuradhapura. Today redevelopment is in progress, using the northward flowing streams of the wet highlands, in an effort to supply farming land. Also, a complete aerial survey of the island has been made to help the planned development of this small country where, since the advent of malaria control, population increase has been very rapid.

QUESTIONS

1. Describe and account for the variations in climate and crops in the Indo-Gangetic Plain.
2. Write an essay on irrigation in India and Pakistan.
3. Compare Calcutta, Bombay, Karachi and Colombo.

CHAPTER 30

SOUTH-EAST ASIA

SOUTH-EAST Asia lies astride the Equator between 10° S. and the Tropic of Cancer. On the Asian mainland are Burma, Thailand, Malaya, Cambodia and Vietnam, while offshore are the island groups of the Philippines and Indonesia.

Relief

The relief of the area may be simplified by considering three significant features. A core of rocks, among the oldest in the world, stretches from eastern Burma, through Thailand, Malaya, northern Java and Sumatra and finally to Borneo. These rocks are the roots of an ancient mountain system once higher than the Himalayas, and, although some mountains reach 10,000 feet, they have been smoothed and rounded by erosion, and contain minerals such as the gold of Burma. The grain of the mountains is north to south, producing a similar trend to the mainland rivers and the Malayan peninsula. The Malacca Straits, the Java Sea and the South China Sea are the shallow drowned fringes of this ancient land. Secondly there are two great arcs of much younger fold mountains which sweep round the core, giving higher, jagged relief. One runs from the Arakan Yoma in western Burma, through small island groups in the Indian Ocean, and reappears as the high backbones of Java and Sumatra; the other stretches south from Formosa through the Philippines to Celebes and the eastern islands of Indonesia. The intersection of these two mountain arcs gives rise to the bizarre shapes of Celebes and the Moluccas. Here deep ocean basins, contrasting with the shallow seas further west, lie amid mountainous coral-fringed islands.

South-East Asia is a region of considerable earth movement with many volcanoes and frequent earthquakes. The third main feature is the alluvial lowlands, the most important economic region built up by numerous short rivers fed by the heavy rainfall and by distant snows in the heart of Asia. On the mainland the Irrawaddy, the Chao Praya, the Mekong and the Red Rivers all form the fertile heart of a country; north-eastern Sumatra is a swampy lowland fringed by mangroves and all the islands have coastal lowlands of greater or lesser width.

Climate

Since South-East Asia lies within the Tropics, temperatures are high all the year except where modified by altitude. The amount and seasonal distribution of rainfall vary with altitude, position and exposure to the prevailing winds (Fig. 120). In a broad zone bordering the Equator the warm moist air of the North-East and South-East Trade Winds converge, resulting in the formation of weak depressions often causing rain. In addition, strong upward convection currents build up in this very unstable air, producing frequent violent

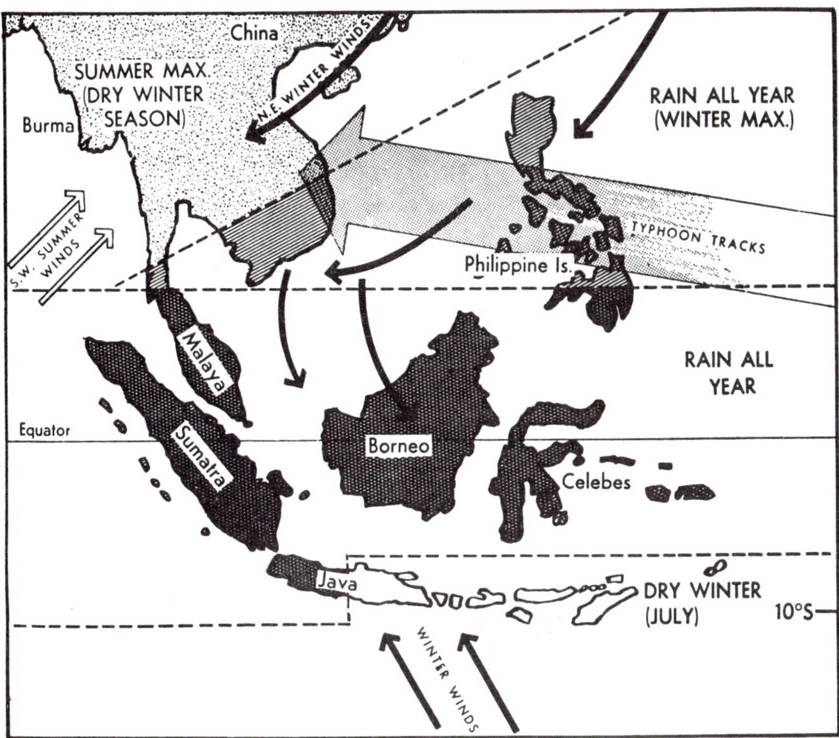

FIG. 120. *South-East Asia: principal areas and climate.*

thunder and electrical storms with torrential rain. Conditions vary little throughout the year. Northwards, however, a distinct dry season begins which increases in length with distance from the Equator. The winter winds are north-easterly; for part of the time they may be an extension of the great outflow of air from Asia which has now crossed the China Sea (Fig. 120) and, as in south China, bring cool cloudy weather but little rain. At other times they bring Pacific air, which yields heavy rainfall, especially in early winter before the winter monsoon is in full swing. The Philippines, Vietnam and north-east Malaya experience their maximum rainfall at this time. The rest of the mainland is swept by the wet Indian Summer Monsoon from the south-west, producing very heavy rainfall on the mountain ranges, which, as already noted, extend from north to south directly in its path. Between them there are "rain-shadow" zones, particularly in upper Burma and in Thailand. The south-west monsoon is dry before it reaches the eastern parts of the mainland. Away from the equatorial zone, there is more sunshine, less cloud and summer temperatures are much higher.

Finally, south from the Equator, there is a belt stretching eastwards from Java which also has a dry season, but in July and August, the winter months of the Southern Hemisphere, winds from the heart of Australia bring warm dry air to this region. About one hundred typhoons—tropical revolving storms—cause moisture-laden high winds which blow across Vietnam and the Philippines every year between July and November.

Resources

Minerals are rather scarce, the chief ones being the alluvial tin washed from the ancient rocks of Malaya, Thailand, Burma and Sumatra, and oil deposits in the gentler folds edging the high mountains of Sumatra and on the fringes of Borneo. Although only two per cent of the world's total is produced, the oil is important for shipping and for the Asian and Australian markets. There is little coal and hydro-electric power is still in its infancy. Although iron-ore deposits are widespread they are difficult of access and poor in quality; the lateritic soils yield valuable bauxite deposits.

Timber is an important export and source of fuel and raw materials for building, which uses three-quarters of the total cut each year. Most of the trees are broad-leaved hard-woods, comprising the tall evergreens of the true equatorial forests with valuable but scattered rosewood and mahogany trees and the deciduous forests of the drier areas, which drop their leaves in the dry season. This group includes the valuable teak forests of Burma and Thailand. Mangrove swamps fringing the coasts provide timber for fuel and building and are cultivated in Malaya. Most widespread and important of all, however, is the bamboo, really a tall grass and used for all manner of constructions and for paper-making. Over-exploitation of the forests has led to soil erosion in many parts, aggravated by the many farmers who make clearings for their crops and then move on when the soil is exhausted. Most of the countries now have a policy of forest and soil conservation.

South-East Asia is a major crossroads of the world; the islands have acted as stepping stones between Asia and Australia, crossed by the trade routes from China to Europe and India. The Malacca Straits compares with Suez and Panama as a routeway, except that the Malacca Straits is a natural one. The Hindu religion—and also Buddhism (still the religion of Burma and Thailand)—was brought by traders and conquerors from India over two thousand years ago; later the Arabs imposed their Moslem faith, which remains in Malaya and Indonesia. Western countries seeking oils and spices gradually usurped the whole area in the last three hundred years and exploited the mineral and agricultural resources for their own benefit. They brought in both Indian and Chinese labourers who today form an integral part of the population of the area. Export crops such as tea and rubber were cultivated on organized plantations and cities and railways were built to serve the export trade. After 1945 independent nations emerged with enormous problems of destroyed cities, growing populations and a vital strategic position in world politics.

Almost every racial type and every stage of civilization is found in South-East Asia, from the primitive food gathering Sakai tribe of the Malayan jungles to the settled rice farmers on the plains and the shopkeepers and factory workers in the coastal cities. The most densely populated areas outside the cities are the fertile plains, especially in the volcanic soils of Java and the rich alluvium of the Red River, Mekong and Chao Praya valleys. People have migrated from these areas, but so far South-East Asia, except for Java, does not have the extreme population pressures of India or Japan. If the present rate of increase goes on, however, there will be difficulty, and the economic development must keep pace. Industry employs only a tenth of the people, the rest being dependent on the land. Industries must be developed both to provide employment and to make goods which are at present imported.

Although there are many small-scale traditional industries, such as rice milling, cotton and silk weaving, there are not as yet many heavy industries.

Farming. There are three types of farming in South-East Asia. First, there are the shifting cultivators who grow yams, vegetables and hill rice (which grows without being covered with water) for domestic use. Then there are the more numerous rice farmers of the plains who use 90 per cent of their land for this one vital food crop, supplemented by spices and highly flavoured vegetables, and very often by fish. The fish are caught in ponds, streams and flooded paddy fields and round the coast from innumerable small vessels, often still under sail. Finally, there are the big plantations of rubber, tea and oil palms established by the Europeans, using Chinese and Indian labour. In Thailand and Burma and South Vietnam, rice for export is grown.

BURMA

In 1948 Burma became independent and left the Commonwealth. Since then attempts at democracy have alternated with dictatorships. High, forest-clad mountains extend along the western side of Burma (Fig. 121) separating it from India by a 200-mile belt, thinly peopled by shifting cultivators. The

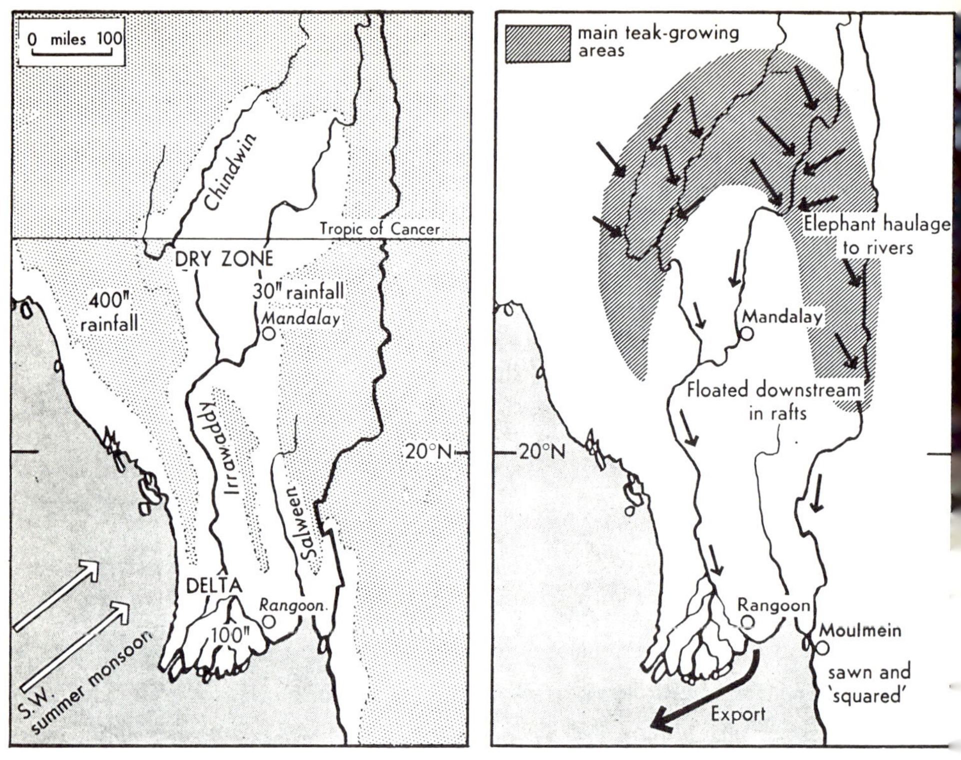

FIG. 121. *Burma: general features (left); teak (right).*

mountains sweep round the northern edge of Burma, into the almost unexplored confusion of mountains and gorges on the Chinese borders where the Yangtse, Hwang Ho, Salween, Mekong and Red Rivers originate within 100 miles of each other. Although most of it, like western Burma, is a sparsely populated jungle, on the southern edge there is an important teak-growing region. Along the south-east of the country Burma's mountain rim decreases in height until it merges into limestone hills and the old rocks of the Shan plateau, dissected by the Salween into remarkable gorges. The Shan people, who differ racially from the Burmese, are settled farmers who grow tea on their hill slopes to sell to the lowlands, and extract tung oil from the forests. The once-famous lead, silver and ruby deposits are now much depleted.

The real heart of Burma, the Dry Zone, was peopled long ago by northern tribes who were repelled by the swampy disease-ridden delta. A wide variety of crops including rice, vegetables, cotton, are grown, mainly for the people themselves, who go out to their fields from compact villages surrounded by trees, near water supplies. Some irrigation is practised from artificial ponds similar to the "tanks" of south India. Cotton, the only export, is also grown. Some oil deposits are exploited, the oil being piped from Prome.

The Irrawaddy Delta has been settled only during the last hundred years;

before then it was a swamp similar to the present north coast of Sumatra. The opening-up of the delta for commercial rice-growing took place at the same time as the wheatlands of North America were being created from the natural grasslands; mechanization has been of little help to the Burmese, and even today the rice is still transplanted and harvested by hand. However, rice accounts for the bulk of Burma's exports carried by small boats through the delta to the wharves of Rangoon.

South-east from the delta a narrow coastal plain has been built out by the turbulent Salween and other shorter rivers; tin is dredged from the alluvium and sent to Singapore for smelting.

South-East Corner of Asia

All of the south-east corner of mainland Asia except Thailand was a French colonial possession only given up in 1954 after very bitter fighting. The area now contains the Communist state of North Vietnam; and the states of South Vietnam, Cambodia, Laos and Thailand, all of which receive military and economic aid from the West, particularly the U.S.A. It has become an area of strife.

There are three main rivers, each pushing a delta out to sea by many hundreds of yards each year. The Red River and the Mekong, each over two thousand miles long, rise in the far north-west in China; much shorter is the Chao Praya with its three main tributaries which have built up the plain of central Thailand. The thinly peopled mountain ranges separating the valleys consist of old rock and are not so high or steep as those of western Burma or Java. The ranges are deeply dissected by gorges and narrow valleys, and are clothed in forest and scrub. Summer rainfall, decreasing towards the east and influenced greatly by relief, affects the whole area, but the north-eastern corner also has beneficial winter rains (Fig. 120). The climate is always hot, except high in the mountains.

Thailand, the largest country in the region, is in some ways a smaller edition of Burma. There is a ring of mountains enclosing and sheltering a plain, which has less than 40 in. rainfall. This country also has a similar source of revenue in exports of rice, rubber, teak and tin. The

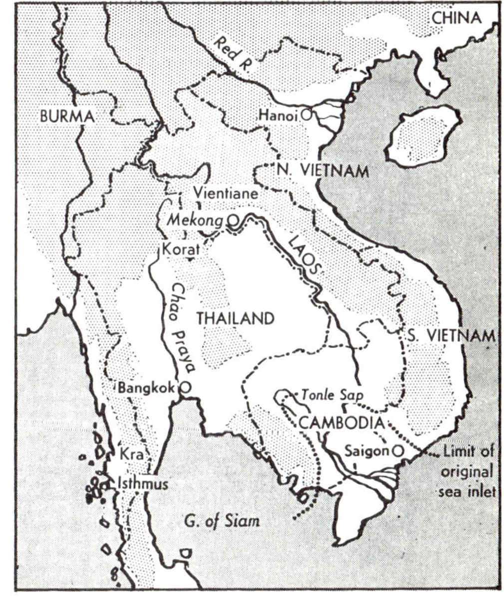

FIG. 122. *South-East corner of Asia.*

river delta has been cleared for rice-production and is dominated by the capital and chief port of Bangkok. Since the Kra Isthmus causes a rain-shadow over the Thailand rice lands, river water is essential for irrigation. In the north-east of Thailand, the Korat plateau leads down to the border along the Mekong; this is a region of flat-topped sandstone hills, covered with grassland on which cattle are reared and sold to pull the lowland ploughs of the rice-cultivators.

Thailand's few factories are concerned with refining sugar and making jute bags and cement. There are surveys and plans to utilize the main rivers for power and irrigation, and a new highway connects Bangkok to the north and east of the country.

North-east of Korat lies the wild country of *Laos*, consisting of the western slopes of the Annamite mountains down to the Mekong Valley. The primitive tribes living in the hills gather food or practise shifting cultivation, while settled rice farmers occupy the valley floors. There is a string of small settlements including the capital *Vientiane* along the Mekong which is the main highway in this landlocked underdeveloped little state.

Again to the north-east the Red River basin forms the nucleus of North Vietnam; on the delta, which is very similar to that of the Si Kiang in China farther north, population density exceeds 2,000 per square mile. Although this too is a rice-growing area like the other deltas, the better distribution of rainfall enables two crops a year. None of the rice is exported, for there is such a huge domestic requirement. The shallow, sandy distributaries of the Red River make it difficult to reach *Hanoi*, the capital, 80 miles inland. Chinese aid and influence come from the north along the railways from Kunming and a new one from Canton.

South of the Red River, the Annamite Mountains reach the sea in rocky peninsulas enclosing small but very densely peopled pockets of lowland. Although these areas are subject to flooding from the short turbulent rivers fed by the rains in the hills behind, they lack reliable rainfall themselves.

The lower Mekong basin is divided between **South Vietnam** and **Cambodia.** The basin is a flat plain, 200 square miles, and once an arm of the sea now silted up to yield productive ricelands. Tonle Sap is a remnant of this inlet. Like the lakes of the Yangtse, it widens greatly in the summer floods, protecting the downstream areas; the Mekong flows into it in summer when its width increases from 20 to 60 miles.

The main part of the Mekong delta, which is crossed by five main distributaries and a host of linking waterways, is another rice granary, similar to those of Burma and Thailand. A variety of floating rice was developed here; with its stems up to 18 feet long it can grow in the rising flood waters. The rice is milled at Cholon, twin town to *Saigon*, the capital. The rice is exported to the plantations and towns of Malaya down the Saigon River.

MALAYA

The Malayan peninsula points like a long finger south-east from the mainland of South-East Asia. The population contains many Chinese and Indians as well as Malays. The backbone of parallel mountain ranges down the centre of the country reach 6,000 feet and are gently rounded and jungle covered. On the western flanks, tin is dredged from the alluvium.

The sultry equatorial climate prevails over the whole peninsula except in the north-east where Pacific winds bring a winter rainfall maximum.

Although Malaya is more developed than any other Asian country except Japan, in 1800 it was a thinly peopled swamp-fringed jungle. The great development was stimulated by the rubber plantations conceived by European companies, and the tin mines opened by Europeans and Chinese. Rubber, introduced from Brazil, thrives in the lateritic soils and humid climate. The plantations lie along the western slopes of the mountains, where rainfall is greater. Local farmers have copied the methods and now almost half the production comes from small holdings. The large estates with their factories and research laboratories are more efficient, but the small farmer can return to rice farming if the price of rubber drops. Because of price fluctuations owners have a greater variety of crops; palm oil is of increasing importance in Malaya. Although rice is the main food crop, grown wherever possible and irrigated from ponds, rivers and canals to give more yield, Malaya must import rice from Thailand and Burma to feed the plantation workers. Since she is forced to buy most of her manufactured goods and machinery, more industrial development is necessary. *Penang*, the major port from the northern half of the country, is situated on an island with sheltered waters facing the mainland. *Kuala Lumpur*, the capital, lies farther south and inland with Port Swettenham as its port. Population is dense on the west coast plains and sparse in the mountains; the rice-farming plains of the north-east, like Kelantan, are the most Malayan part of the country. Population pressure is not yet great in Malaya.

Singapore is a small crowded island less than 20 miles across. In 1819 Sir Stamford Raffles, realizing the vital strategic and trading position

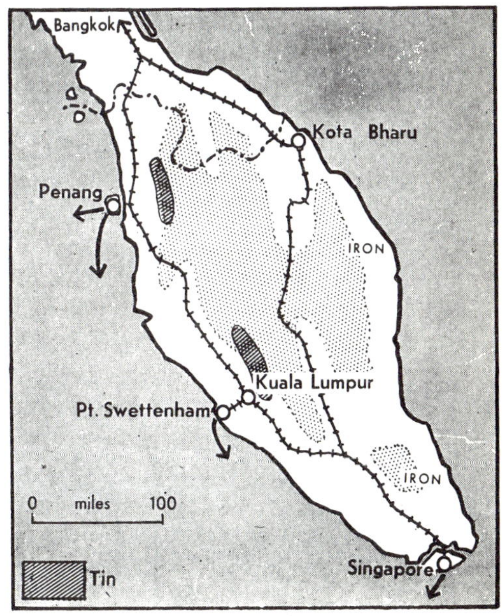

FIG. 123. *Malaya and Singapore (arrows indicate direction of trade).*

of Singapore, acquired for Britain the then uninhabited island. Now the busy harbour where ships from all over South-East Asia bring goods for re-export, the big office blocks and warehouses and the teeming streets, are evidence of his forethought (Fig. 123).

In 1957 the Federation of Malaya became independent within the Commonwealth, to be followed in 1958 by the small island of Singapore. In 1963 the Federation of Malaysia came into being, consisting of the Federation of Malaya, Singapore and the colonies of North Borneo (renamed Sabah) and Sarawak; this meant that the U.K. relinquished sovereignity over Singapore, North Borneo and Sarawak. However in 1965 Singapore broke away from Malaysia.

THE PHILIPPINES

The Philippines are a group of over 7,000 islands stretching for a thousand miles between Formosa and Borneo. Over 90 per cent of the land area is in eleven of the islands. They fall into three groups: two large islands, Luzon in the north and Mindanao in the south, are separated by the cluster of the Visayas from which the long thin finger of Palawan points away to Borneo.

After a period of American control, the Philippines became fully independent in 1946. The Spanish influence is shown in the language, Roman Catholic religion, the architecture and the large *haciendas*, or estates; the Americans encouraged the development of export crops and mineral wealth such as the gold of Luzon.

The mountain ranges of eastern Formosa continue through the Philippines, reaching heights of 9,000 feet, and containing both old, mineral-bearing rocks, and newer volcanic peaks, some of which are still active; Mt. Manay in the Visayas has erupted thirty times in the last 300 years. Earthquakes and typhoons occur sometimes resulting in the loss of many lives. One typhoon, "Freda," made 80,000 people homeless in 1956. Although the area is well suited to hydro-electric power with its short swift rivers and heavy rainfall, very little has been developed. The islands are fringed with alluvial plains which stretch right across the south of Luzon; where the original rock is volcanic, they have very fertile soils.

Temperatures are high except in the higher hills and although rainfall varies in amount and according to season there are no really dry areas. The natural forest on the mountains yields some timber, much of which has been burned by the shifting cultivators who live there and is replaced by poor "secondary" growth. Two-thirds of the people are agrarian, most of them being small-scale farmers who own or rent only an acre or two, which they go out to cultivate each day from the long straggling villages. They grow cash crops as well as rice and vegetables. Even the one-time head-hunting *Ifuagos* of the Cagayan valley in north-east Luzon, who built the

FIG. 124. *Luzon, Philippine Islands: for many centuries various Filipino tribes have been building these mountain rice terraces.*

famous rice terraces, now grow export tobacco on the valley floor. Their houses, like those of the Malayans, are built on stilts and made of bamboo and thatch. There are some big estates which employ people to work on them—growing sugar in south Luzon or abaca in Mindanao. Abaca, a plant related to the banana, produces a dozen or so sword-shaped leaves which are cut to make the fibre hemp used in rope-making. Although coconuts cover more ground than any other commercial crop, sugar is much more valuable. The nuts are gathered as they fall and are dried in the sun or smoked to make "copra", one of the sources of oil for soap and margarine.

The diet of the Filipinos consists of rice or maize, supplemented by bananas and beans, and many strongly flavoured vegetables; fish is important too as in the whole of South-East Asia. The people live mainly on the plains of Luzon and the Visayas, while many of the smaller islands are almost uninhabited. As in Java and Japan, it is difficult to induce people to live permanently on the small islands, although there is a lot of seasonal migration within the islands to work at the different harvests. Manila, the dominant island of the Philippines, has an old Spanish sector behind the busy waterfront where produce from all the islands is collected for export.

INDONESIA

Indonesia, lying astride the Equator (between 10° S. and 13° N.), is a nation of mountainous islands amid seas which have been the thoroughfare of traders and conquerors from earliest times. The Dutch, who for two hundred years controlled and developed the region, left in 1948, and conceded West New Guinea in 1962: it is now called West Irian.

The pattern of the islands can best be seen if it is divided into four regions (Fig. 120); there are the three large islands of Borneo, Sumatra and Java and the group of smaller islands to the east, the largest being Celebes. While all have more mountain than plain, and experience an equatorial climate (although the south-east has a marked dry season), the four areas show considerable contrasts in their present and future geography which makes a convenient way to study them.

Borneo, the largest, has a smoother coastline and is more compact than any of the others; the mountains which cover most of the islands are of ancient, smoothly worn sandstones which yield poor soils and are thickly covered with jungle. Java and Sumatra are long and relatively thin. Sumatra, being much the larger, has the distinctive feature of a belt of peat more than 100 miles wide, bordering the east coast which faces Malaya. With its volcanic peaks up to 12,000 feet the mountains of Java and Sumatra are much higher than those of Borneo; and, even more important, the soils derived from them,

FIG. 125. *River scene in North Borneo. Note the houses built on piles.*

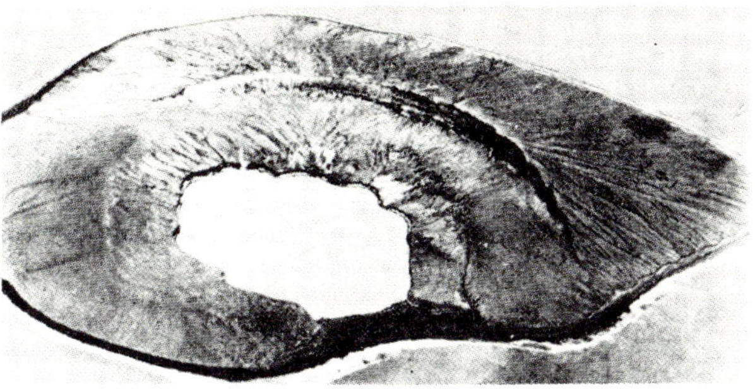

FIG. 126. *Krakatao, a small volcanic island in the Sunda Strait, made perhaps the loudest noise ever, when in* 1883 *the volcano blew its head off. The noise was heard* 3,000 *miles away.*

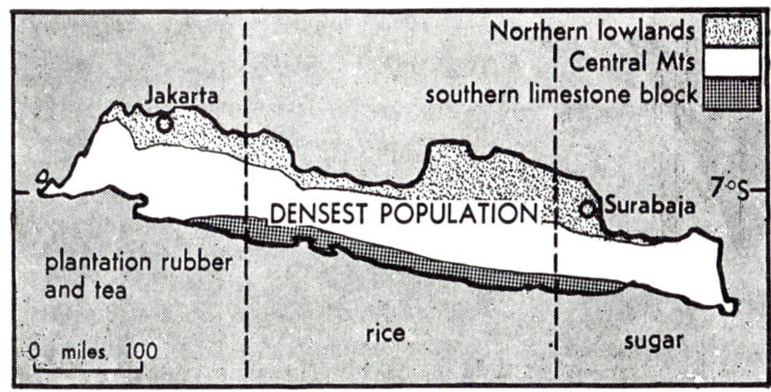

FIG. 127. *Simplified map of Java.*

especially in Java, are much more fertile. The eastern islands have coral-fringed and highly indented coasts, with little or no plain, and forested mountain interiors; the Great Sunda Barrier of hazardous coral reefs, stretching 200 miles south-east from Borneo, compel the shipping routes to go south between Bali and Lombok.

The greatest contrasts occur in the man-made landscapes. In this, Java stands alone; with 73 million people and densities reaching 6,000 to the square mile, it is one of the most intensively settled and cultivated places in the world. Farms are so close that the countryside looks more like a suburb. The two main reasons for Java's huge population are the Dutch influence in developing agriculture to produce export crops of rubber, tea, spices and sugar and the frequent renewal of soil fertility by volcanic eruptions. Terraced paddy fields, as remarkable as those of Luzon, climb the slopes often dangerously close to the craters; a modern warning system has reduced the loss of life from eruptions. In the eastern sunnier half of the island, sugar cane is an important cash crop. In Sumatra, however, there are few terraces, and the only small areas of dense population are plantations which have been started round Medan and near the oilfields in the south. Two-thirds of the people of Sumatra are still shifting cultivators or primitive hunters, although the interior basins have some groups of settled farmers such as the Christian Batak. Borneo has even less agricultural development; most of the tribes of

FIG. 128. *N. Borneo: a Dyak youth blowing a poisonous dart used in hunting.*

the interior do not grow crops at all; some rice cultivation occurs round the oilfield of North Borneo and around these in Balikpapan. In the eastern islands the once-important quantities of spices and copra which brought traders here are now declining and large-scale plantations are developing in the more accessible areas of the main islands.

The major towns are all ports, with only a few industries based on processing local raw materials of sugar, rubber and copra. Farming is the main occupation. Indonesia lacks coal and iron—essential ingredients of heavy industry; oil and tin are her main sources of mineral wealth.

The chief problem is to encourage people to move from the crowded farms of Java to develop the empty jungles of Sumatra and Borneo; but people are reluctant to move into these difficult areas which present so many obstacles to settlement.

QUESTIONS

1. Write an essay on the rice-exporting regions of South-East Asia.
2. Account for the contrasts between Java and the rest of Indonesia.

CHAPTER 31

CHINA

CHINA lies along the east of Asia between 20° N. and 50° N. One of the oldest empires in the world, it became a republic in 1912. The old Chinese Empire consisted of the densely populated regions of China proper, Manchuria and the less important regions of Turkestan, Tibet and Mongolia. In 1931 Manchuria was taken over by the Japanese, who developed its natural resources. Japan invaded China in 1937, and, by the end of the Second World War when the Japanese were expelled, China was disunited and weak, and the Government quite incapable of restoring the country. Strong, disciplined Communist forces with Russian aid quickly gained control, and the Chinese People's Republic was declared in October, 1949.

Despite setbacks, China is now a major modern power and anxious to be recognized as such.

Relief

From the Pamir Mountains in the far south-west, mountain ranges over 20,000 feet high fan out to the north, east and south-east (Fig. 129). They are built of crystalline and sedimentary rocks folded in Tertiary times and eroded by rivers and ice into jagged ranges. In the south they enclose the high Tibetan plateau which is over 16,000 feet above sea level and the lower Koko Nor Basin, and culminate eastwards in the little-known Kam or "Region of Great Corrosion" and the Great Snowy Mountains which tower over Szechwan. In this region four great rivers rise, the Hwang Ho, the Yangtse, the Mekong and Salween; their mouths are as much as 2,000 miles apart. Farther north the mountain ranges are lapped by a great peneplain which covers over a million square miles from Manchuria to Sinkiang. Warping has formed lower basins, now filled with sediments, eroded by wind and forming endless desert landscapes like the Gobi.

The eastern edge of the peneplain has been tilted to form a low rim with a faulted eastern edge; thus the Khingan Mountains seem low from the west and drop sharply to the rolling Manchurian plain on the east.

The alluvial plains of the lower Hwang Ho and Yangtse Rivers, which coalesce along the small indeterminate Hwai River, represent most of China's lowland which has sufficient rain for crops.

The gently sloping mature mountains of eastern Manchuria continue

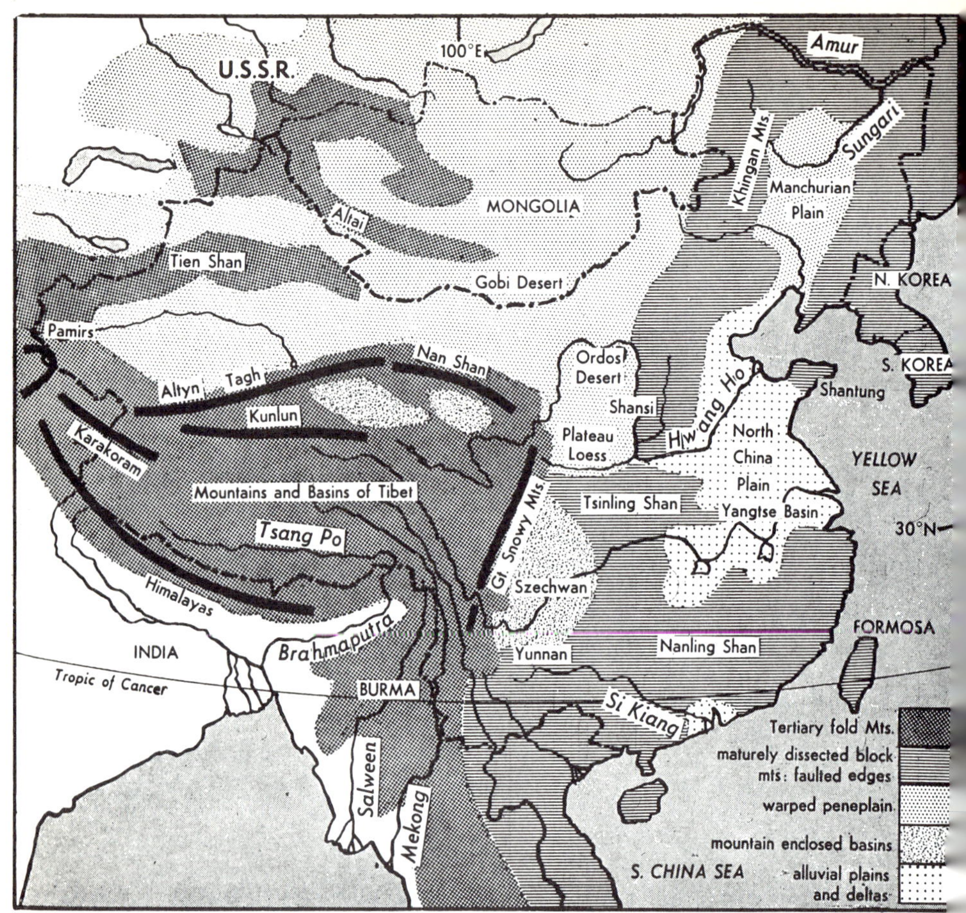

FIG. 129. *Simplified relief map of China.*

southwards into Korea and reappear across the shallow China Sea in Shantung. They are similar in structure to the great mass of hilly land which fills southern China; some of the earth's most ancient (Pre-Cambrian) rocks are found, containing minerals such as tungsten and antimony in the south, and later sediments including coal measures in Shansi and Shantung (literally "eastland" and "westland"). Although the Tsinling Shan are composed of similar rocks, they are higher and much more dissected with canyons over 1,000 feet deep.

There are smaller alluvial plains at the delta of the Si Kiang, on former lake beds in the Yunnan plateau and around Chengtu in Szechwan. Most of Szechwan (the Red Basin) consists of rolling sandstone hills, in which coal measures are also found.

Although China's main rivers follow the general slope eastwards, the western third of the country drains into interior deserts or salt lakes. The Hwang Ho,

2,600 miles long, swings in a great U northwards to enclose the Ordos Desert. Wind-blown dust, or loess, from the interior deserts, has been deposited to the south-east, in the fantastically sculptured Loess Hills region of Shansi. The river carries loess alluvium down to its flood plain, and dykes to contain it have caused deposition on the river-bed, which is now higher than the surrounding country. The taming of the Hwang Ho by a series of dams in the middle and upper course is a major modern project. The Yangtse, the sixth longest river in the world (3,200 miles), has a flood-plain 1,000 miles long between the Ichang gorges by which it leaves Szechwan and the sea. Natural lakes such as Poyang alleviate the flood danger by acting as reservoirs. Tributary valleys of the Hwang Ho, such as the Fei and the Wen and the Yangtse tributary the Han, have long been important centres of settlement. The Si Kiang valley is little more than a trench in the southern mountains.

Climate

Although the monsoonal climate with its marked seasonal contrasts prevails over China, there are great differences in rainfall and temperature between south and north, and between mountain and lowland (Fig. 130). In winter a widespread but shallow high pressure area forms over Siberia which may extend over the whole of China. Strong, intensely cold winds blow from the north-west, but the weather is often sunny; padded clothing is worn in northern China and freezing temperatures may reach south as far as the Yangtse. As these winds swing round over the China Sea they become warmer and more humid and the air rises to form a continuous layer of cloud over south China, resulting in overcast, cool weather. The latitude also produces higher temperatures than in the north. Interruptions in this prevailing winter weather are brought by stormy depressions from the west which may dispel the anticyclone for a few days, and before it returns there is often an unsettled spell. The Yangtse is noted for these winter depressions. In the north there are frequent dust-storms in winter, for the ground is devoid of vegetation at this time; there may also be light snowfalls.

In spring temperatures rise rapidly everywhere and the interruptions in the winter monsoon increase in length. Warm, humid air spreading from the south mingles with the cold retreating anticyclone to form many depressions which bring a belt of very heavy rain moving steadily northwards over the country during May and June. This is the "Mai U" or "mouldy period," when high temperatures and humidity stimulate the growth of fungus on shoes and books. After this rain belt comes the full summer monsoon, not as wet as the Mai U but hot and sunny with thunderstorms. Typhoons bring heavy rainfall to the east and south from June to November and they may also cause much damage. Gradually the winter pattern becomes established as temperatures fall, and autumn is a sunny season.

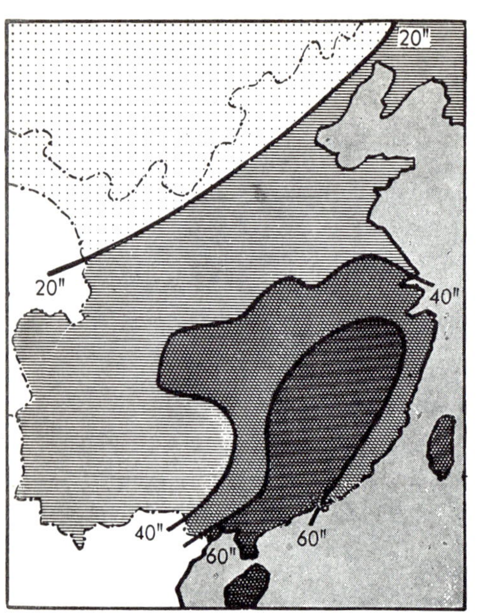

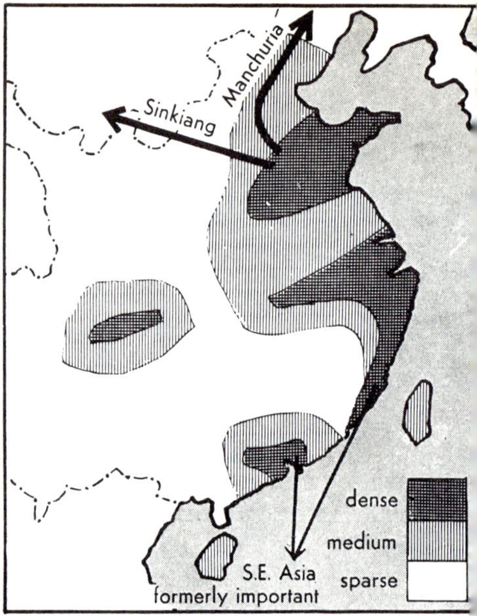

FIG. 130. *China: rainfall distribution (left); population density and principal directions of migration (right).*

Resources of China

In her effort to become industrialized, China has the advantages of her large and widespread resources of the basic raw materials—coal and iron. In her production of both of these materials she ranks as high as those of the U.S.A. and U.S.S.R. There is not much oil compared with that of the U.S.A. or U.S.S.R., though new fields are being opened up in Sinkiang; the hydro-electricity developed in central Asia on the Hwang Ho and Yangtse valleys provides a quarter of the country's power. Many minor minerals essential in steel production, such as tungsten and antimony, are found in the ancient crystalline rocks of the southern mountains; the coal and iron resources are mainly in Szechwan, Shansi and Manchuria.

The third Five-Year Plan, begun in 1966, gave priority to agriculture; earlier plans emphasized the development of heavy industry to produce iron and steel to be used in making the machinery necessary to manufacture other goods; rails, bridges, locomotives, lorries were also important for transporting raw materials. Then the expansion of the consumer industries was planned. Existing industrial areas in Manchuria, Szechwan and the lower Yangtse valley have been expanded, and big new ones developed as in the arid loess area of the north-west, where now Paotow and Lanchow are important steel and textiles towns. *Lanchow* is also the hub of a new network of routes that includes the China-Soviet Highway stretching westward through the remote province of Sinkiang (also a new industrial area with oilfields) up to the Russian frontier.

New railways are under construction along this route and southwards to Tibet. The Szechwan Basin is now also linked with northern China by rail through the Tsinling Mountains. A five-mile bridge over the Yangtse at Hankow completes the line along the 1,500 miles which separate Canton from Peking. Water transport is important in the south and along the Yangtse which ocean vessels can ascend for 500 miles to Hankow. The dreaded Ichang gorges have been blasted to allow smaller steamers up to Chungking. Locally, farmers use junks to carry their products to the towns; and, although motor roads and transport by bus and lorry are increasing, the narrow flagged paths of south China, and the dusty cart tracks of the north are still the main means of communication for the country people.

Farming. There are over 120 million peasant families dependent on agriculture in China, comprising five-sixths of the population and concentrated on the alluvial plains (Fig. 130). There is a considerable contrast between the wheat-growers in the north and the rice-growers of the south, between the brown, treeless landscapes of the north and the lush green bamboos of the south. In addition the mountains have their own pattern of forests and terraces in the wetter south, contrasting with the bare or grassy slopes of the north and west, where cattle become important. Drought and flood, pests and diseases have drastically reduced the population of the crowded plains in the past; since 1949 it has been a major aim of the Chinese to eliminate these problems and at the same time to increase food-production to keep pace with the rapidly increasing rate of population growth.

Unlike other Asian nations the Chinese have regarded their vast and increasing population as an asset rather than a problem and use human labour on a massive scale to build dams, roads, bridges and railways. Even so, congestion on the eastern plains has resulted in emigration; there are 20 million Chinese in South-East Asia, where they are highly successful traders. Many others have gone north of Manchuria and westwards to the steppelands, and established farming communities. China has been more fortunate than Japan in finding many outlets for her people.

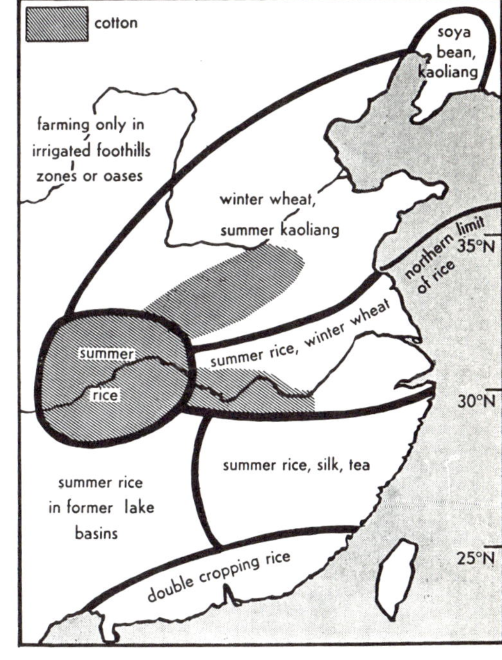

FIG. 131. *Distribution of crops.*

South China

Between the narrow hilly valley of the Si Kiang and the North Vietnam border there is a sparsely peopled upland of sandstone which has been weathered into strange shapes. It contrasts strongly with the fertile plain built out by the Si Kiang and its two main tributaries (Fig. 132). Here 9 million people live on 3,000 square miles of delta, one of the most intensively cultivated patches of the earth's surface. High rainfall (70 in.) and warm temperatures which never drop below 10° C. (50° F.) allow crops to be grown throughout the year on the alluvium between the tree-lined villages strung along the raised banks bordering the numerous waterways. Three rice crops a year and mulberries and fruits such as lychees are grown. In Canton, the leading town of south China, many of the people live in junks on the river. Tung oil from the hills to the north is exported from Canton, but this city is 80 miles from the sea and not a major port.

Hong Kong, an island (Fig. 132) at the mouth of the Si Kiang, is a British possession along with some 400 square miles of leased territory on the adjoining mainland; it is a busy, thriving, commercial and industrial port, while Macao is a decaying Portuguese town in the south-west corner of the delta.

West of the Si Kiang lies the high, cool plateau of Yunnan, a name meaning "beyond the clouds." Amid rocky hills there are numerous silted-up lake basins forming good farmland suitable for the growth of rice and vegetables. *Kunming* is a growing communications centre of China's contacts with South-East Asia; a railway built 40 years ago by the French links it with North Vietnam, and the famous Burma Road runs westwards and is to be

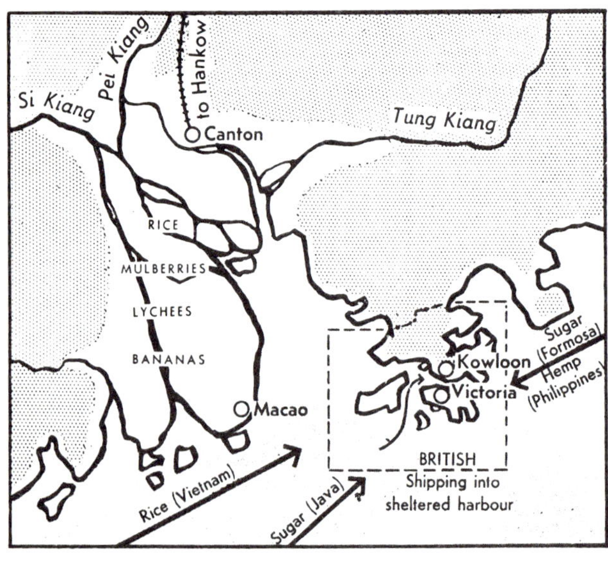

FIG. 132. *Canton and Hong Kong.*

followed by a railway. A new line is being built north to Chungking. North of the Si Kiang, the Nanling Shan rise over 3,000 feet; though their forested slopes are much denuded and eroded, they still yield tung and camphor and contain much mineral wealth only now being exploited. Along the coast, isolated plains such as those round Amoy and Swatow form densely-peopled pockets which have fostered seafarers and emigrants.

Central China

The river basin of the Yangtse, which is a transitional zone in many ways between the south and north (Fig. 129), can be divided into three regions (Fig. 133). Little is known of the first 1,500 miles of the river's course through high forested mountains. Emerging into Szechwan, the Yangtse and its tributaries have built out alluvial fans, which give the fertile basis for the

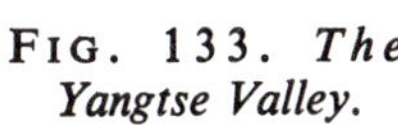
FIG. 133. *The Yangtse Valley.*

intensive irrigated farming of rice and vegetables; the Chengtu plain, for instance, supports 5 million people. Although fairly high and far from the sea, the winter temperature in Szechwan is kept above freezing point most of the time by winds like the Föhn of the Alps which sweep down from the high mountains to the west. The region is notorious for its foggy winters and cloudy summers, when there is a reliable 30 in. of rainfall. *Chungking* was the capital of China during the war with Japan. A great deal of the bombed areas have been replaced by wide streets and steel and textile factories.

Below Ichang where the gorges through which the Yangtse leaves Szechwan end, there is a wide flood plain stretching 1,000 miles to the sea. From the north the Han River flows through a basin not unlike a smaller Szechwan and from the south two tributaries enter through lakes Tungting and Poyang, which, although expanding greatly in size with the summer rains, protect the lower Yangtse plain from bad floods. A 90-feet high dyke protects *Wuhan*

FIG. 134. *Shanghai: the city commands the trade of the Yangtse and has developed as a commercial and manufacturing centre.*

(a triple town comprising Hankow, Hanyang and Wuchow), an important cotton-manufacturing and route centre at the Han confluence. Fifty miles downstream new steel mills have been built to use local iron ore at Tayeh; at the head of the delta is Nanking, a former Chinese capital, while south of the Yangtse mouth lies *Shanghai*, China's second port.

Waterways intersect the landscape of the Yangtse valley and are the main routeways for the villagers who with the help of the summer rains grow rice, mulberries along the stream banks and a wide variety of vegetables and fruit. The Chinese plan one day to divert the surplus waters of the Yangtse northwards through the Hwai basin (noted for its floods and the scene of a major project in water control) to irrigate the drought-stricken plain of the Hwang Ho.

North China

The only really navigable section of the Hwang Ho is a short stretch near the coast, the upper stretches being shallow and erratic (Fig. 136). In the valley of the Wei Ho, its chief tributary, the fertile loess-alluvium and sheltered climate provided the "cradle" of Chinese civilization, and today this is an expanding industrial region, controlling routes westwards which follow ancient caravan routes through the "Jade Gate" to Mongolia. North of the Wei Ho valley is the remarkable area of the loess plateau with its cave-dwellings cut into the 200-feet high vertical cliffs, which are liable to collapse with slight

FIG. 135. *On the loess plateau millions of people live in bizarre cave dwellings cut into the vertical cliffs, sometimes over 200 feet high.*

earth movements. Farmland is poor and dry, and millets and wheat are grown, with sheep and cattle more important than elsewhere in China. *Paotow* and *Lanchow* are rapidly developing textile towns on the Hwang Ho and lie on important routes between China and Russia.

East of the U-bend of the Hwang Ho lies the rocky plateau of Shansi, cut by the Fen Ho which links a series of fertile basins like a string of beads. Shansi has China's most important coalfields and there is iron in the north of the province. Shantung, a rocky peninsula jutting into the China Sea, has fishing villages round the coast and coalfields at the inland edge. The maritime influence on the climate encourages vine-growing and a little rice is also grown.

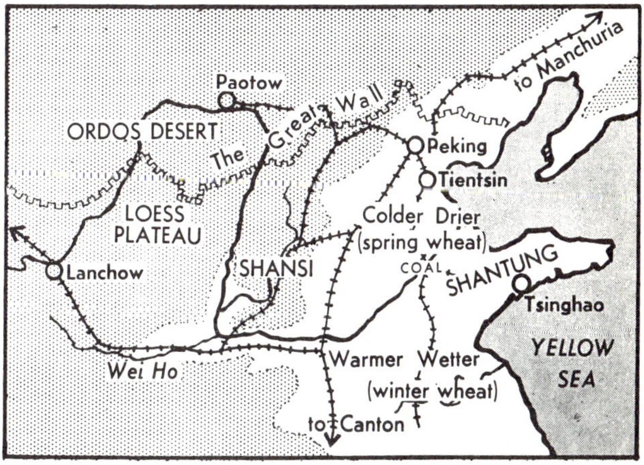

FIG. 136. *Simplified map of Sinkiang.*

The North China plain, really the gigantic delta of the Hwang Ho, is composed of yellow alluvium washed down from the loess plateau. Villages are built well back from the raised bed of the river which has often burst through and changed its course; a hundred years ago it actually reached the sea south of the Shantung peninsula. It is a land of bitter, dusty winter winds and scorching summer sun with most of its 25 in. of rainfall coming in spring and early summer. Although the renewal of silt by flooding has made it a fertile land growing wheat and millets, it is overcrowded and people leave for Manchuria or the western steppes. The plain was guarded in the past from marauding tribesmen by the fortified city of *Peking* in the narrow coastal plain to the north. Peking is now the Chinese capital and has flourishing industries, such as textiles and machinery. *Tientsin* is now 50 miles inland, because of the silting of the Yangtse, which must be continually dredged, and has an outport, Tangke; nevertheless it is the leading Chinese port and has three million inhabitants.

Manchuria

Beyond Peking in the far north lies Manchuria, restored to China in 1945 by the Japanese who had controlled it for 40 years; to Japan it was an outlet for her crowded peoples and a source of raw materials for her industry. The country is 800 miles wide and 1,000 miles from south to north, much of it being rolling lowland with fertile black soils underlying the natural grassland. Winters are colder and longer than in north China and the summers not so warm. The empty plains had attracted many Chinese farmers even during the Japanese period, and even more move there today to grow wheat, millets, kaoliang and the soya bean for export. Manchuria is even more important to modern China as an industrial area with coal and iron deposits and heavy industry developed at *Anshan* and *Fushan* by the Japanese and now greatly expanded to produce half of the country's steel output.

The rocky peninsula of Korea matches that of Shantung over the China Sea. Only the northern half is under Chinese control; it is a continuation of the industrial zone of southern Manchuria. On the east coast is the ice-free port of Hanyang, rivalling the Russian Vladivostock further north which is frozen in winter. South Korea is gentler and more like a continuation of southern Japan, with tiny rice fields and forested mountains; the 38th parallel the line delimiting North and South Korea, was finally recognized after the Korean War which involved both the western nations and Russia in 1950-52.

Western Extremities

The main part of China occupies the basins of the three main rivers and is less than half of her total area. From earliest times there were long struggles between the settled farmers of the plains and the nomadic pastoralists of the deserts and grasslands to the north-west. Now the Chinese are penetrating and farmers are settling in these lands. The surviving nomads are being forced to settle in villages and to grow crops. Sinkiang, lying between the Tibetan plateau and the Tien Shan (the "heavenly mountains"), drops below sea level in places. Although there is less than 5 in. of rain a year, the snow-fed streams from the mountain rim have long been used to irrigate wheat and cotton

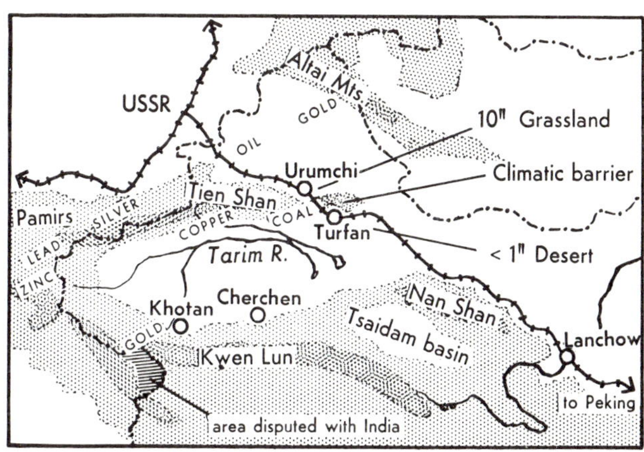

FIG. 137. *Map of the Tarim Basin.*

fields, and melon gardens round *Turfan* and the other towns of the basin. The oil at Yumen and coalfields are stimulating industrial development in this remote part of central Asia. Tibet too is now under firm Chinese control; in contrast to Sinkiang, it lies above 12,000 feet, a country of mountains and basins. *Lhasa* lies in the valley of the Tsang Po (Brahmaputra) and is now reached by a new road from Sinkiang and a railway is planned. The sparsely peopled pastures of Tibet now also have many Chinese settlers. Mongolia is an independent, largely pastoral and desert country which looks more to Russia than to China. The former exchanges and co-operation between China and Russia have given way to some hostility in recent years.

QUESTIONS

1. Contrast north and south China.
2. Compare economic development in China and India.

CHAPTER 32

JAPAN

JAPAN consists of an arc of mountainous islands off the east coast of Asia, stretching for 1,200 miles from 30° N. to 45° N. The islands are in part the crests of a submerged mountain range, which rises over 10,000 feet above sea level and plunges 35,000 feet below it east of the country. Instability in the earth's crust results in a considerable number of earth tremors every year and there are 500 volcanic peaks including Fujiyama.

Climate

Although Japan lies within "Monsoon Asia," there are wide differences in climate; temperatures drop sharply from south to north and from lowland to highland. It is never too dry for crops to grow and the mountains have excessive rainfall. In summer moist tropical air sweeps in from the Pacific, resulting in sultry weather and thunderstorms with the heaviest rainfall in the south and east. In winter, cold monsoon air from Asia may result in dry, clear spells; or it gathers moisture to yield heavy snow falls in north and west; or it swirls with ocean air in widespread depressions. In the spring these cause the "plum rains" when summer monsoon air becomes involved. Autumn is a changeable period often with a few typhoons.

FIG. 138. *Japan: climate (left); land cover (right).*

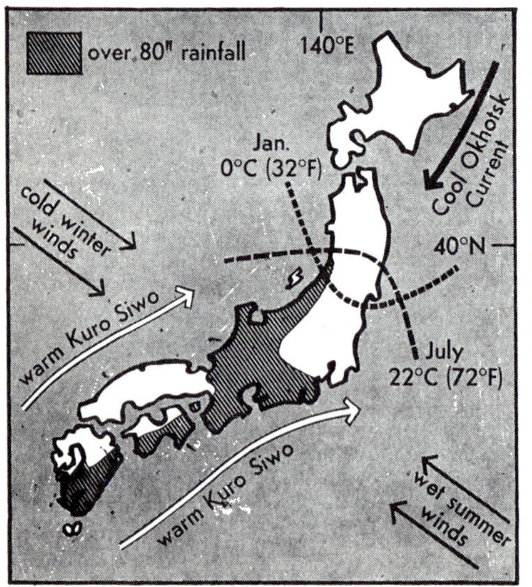

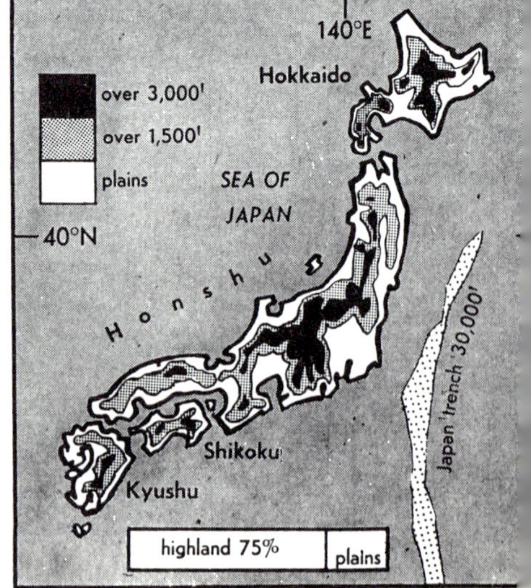

Fig. 139. *Inside a silk factory in Osaka.*

Fig. 140. *Pearl industry: women pearl divers at work* (*left*)*; pearls being carefully selected and graded* (*right*).

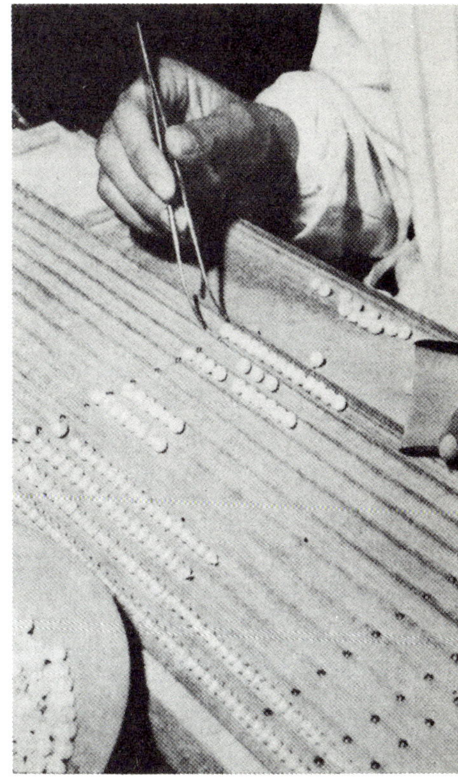

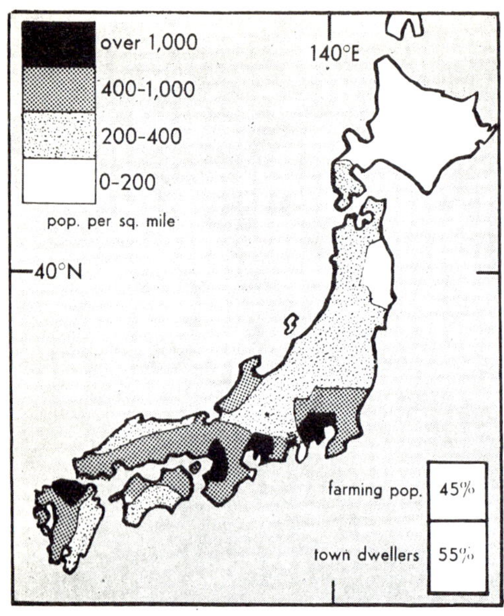

FIG. 141. *Japan: population density.*

Resources

Japan has a shortage of natural resources; her two coalfields have thin, steeply inclined seams with little anthracite or steam coal; ten per cent of her oil requirement is home-produced. Hydro-electricity is abundant, from the steep, swift-flowing rivers in the high mountains of Central Honshu.

In 1853 the West forced Japan to leave her centuries-old feudal isolation. Since then she has become a leading industrial nation, following Western methods. Population increased rapidly and to find markets, sources of raw materials and living space, Japan conquered first Manchuria, then much of China and South-East Asia. Forced to surrender all this in 1945, the 100 million Japanese are more crowded than ever on to their limited plains where population density reaches 3,000 per square mile.

Since farms average only two acres in size, one member of the family often works part-time in a factory and many farmers are also fishermen. The precious alluvium of the plains is utilized to the utmost, irrigated by ditches, canals and ponds to prolong the growing season. There is no space for fodder crops and with little good natural pasture (the dwarf bamboo harms animals), livestock are rare except in Hokkaido. Chemical, animal and human fertilizers are used liberally and each plot of land is cultivated to produce the highest rice yield in Asia.

Over half the people are factory workers in the industrial belt of southern Japan; the four main areas are round Tokyo, Nagoya, Osaka and in northern Kyushu. Big modern factories are important as are the innumerable small workshops and domestic industry. Living standards are rising more rapidly than in other parts of Asia. Japan concentrates on producing goods of a high standard many of which now compete with those from Europe.

Southern Japan

The "nucleus" of Japan lies round the warm, sheltered waters of the Inland Sea. Agriculture is most intensive on the pockets of alluvial plain between the forested mountains; roads and villages hug the foothill edges. There are four main plains, the largest of which, Kwanto, although only 5,000 square

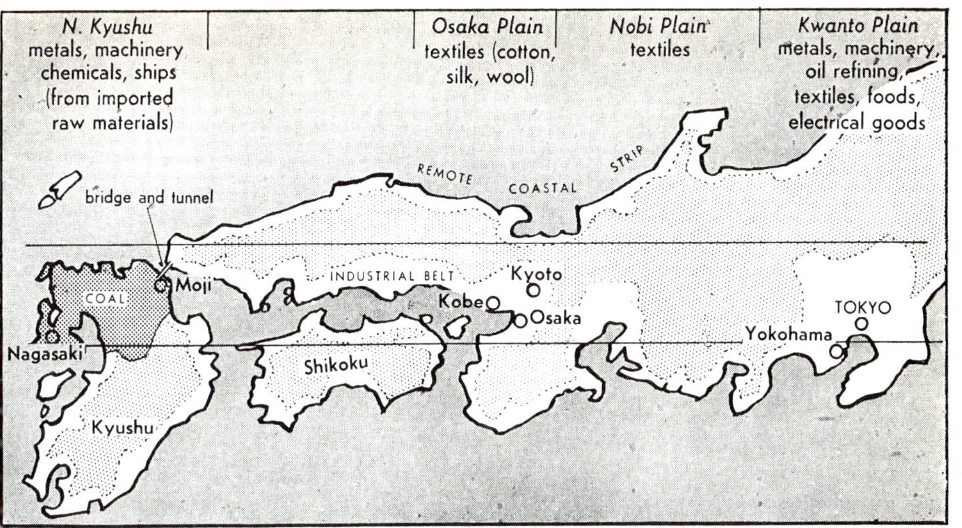

FIG. 142. *General map of South Japan.*

miles contains 30 towns and one-fifth of the population. The plains consist of mainly poor gravel and are difficult to irrigate or cultivate. **Tokyo,** the capital, lies at the head of a shallow bay, with Yokohama, its deep-water port, 12 miles down the coast. Both cities have a western appearance with their office blocks and factories.

The Osaka Lowland also contains several cities; *Kobe*, the chief Japanese port, imports raw cotton and wool for the textile industries of *Osaka*, a smoky industrial town spanned by canals. Inland, *Kyoto*, the ancient capital, still has traditional Japanese industries, such as lacquer and brassware. Temperatures are higher than in the Kwanto Plain. Oranges for export and bamboo for paper are important hillside crops, while three crops of rice a year may be grown on the lowland.

The Chikulu coalfield in Kyushu provides power for the Osaka area and for an industrial region along the north coast of Kyushu, where heavy raw materials such as oil, cement, iron and chemicals are imported and processed. In spite of the dropping of the atomic bomb in 1945, the isolated port of Nagasaki is now the site of a huge industrial combine of shipyards, steel mills and electric plants. Japan has been the leading shipbuilding nation since 1955. The remainder of Kyushu and the whole of Shikoku are hilly lands of farm and forest where productive land is scarce and people migrate to the towns of the industrial belt.

Northern Honshu

Northwards from Tokyo subtropical conditions diminish. Forested mountain ranges including clusters of volcanic cones run north-south, enclosing small basins sheltered from heavy rainfall and having higher ranges of temperature than the coasts. Mulberries grown on the slopes provide food for silkworms, the basis of the traditional silk industry as in the Suwa Basin. However this industry is being superseded by the making of synthetic fibres with its associated chemical industries. One of Japan's poorest regions lies east of the mountains. This area is cooled by winds blowing from the Okhotsk Current and lined by sand dunes ponding back lagoons and marshes. There is a low population density; most of the people are farmers growing one crop of rice a year and small amounts of millet. Iron at Kamashi provides the only industry.

West of the mountains the full force of the snow-laden winter storms lashes the fishing and farming villages in the isolated plains, so roofs are weighted down with boulders. The more sheltered areas however are able to export some of their rice to the industrial towns of the south and the Iwaki Basin is famous for its apples. Japan's oil wells are found east of Akita.

Hokkaido

Hokkaido is the only part of Japan capable of supporting more people than it already does—a marked contrast with the rest of the country. Because of the disagreeable bleak climate, the southern Japanese are reluctant to settle there. The timber from the rich coniferous forest of the central mountains feeds huge paper mills at Sapporo, and there are deposits of coal and iron. Although a quick-maturing rice (growing in 90 days) is grown on the Ishikari plain, potatoes and oats are more common. Sheep and dairy cattle graze on the slopes, and this is the only part of the country where milk is important in the diet. The landscape resembles that of the American Mid-West with its rectangular pattern of roads and settlements and its similar storage barns. There are plans to build a tunnel to link Hokkaido to Honshu.

QUESTION

1. Describe the major problems facing Japan.

REVISION SUMMARY

EUROPE AND THE U.S.S.R.

Page

FRAMEWORK OF EUROPE

Europe is a continent **without extremes** of climate or physical con- 137
ditions. It contains about one-seventh of world population, but contributes a greater share of manufactured goods and agricultural produce. **Surrounded by shallow seas** swept by warm ocean waters: unusually warm for its latitude. Advantages in belonging to same land-mass as Asia.

A. RELIEF

Earthstorms in Tertiary times formed the present major European 138
relief features. The landscape was finally moulded in the Quaternary Ice Age. Great ice caps covered Britain, Scandinavia and the North European Plain as well as the Alps and Pyrenees. On melting, the ice left on the plains dumps of material carried from the mountains. It had also carved U-shaped valleys and mountain peaks. Changes of sea-level formed the Baltic and drowned low plains to form the North Sea.

B. ANCIENT LANDS OF NORTHERN EUROPE

Ancient crystalline rocks form the *Baltic Shield.* It is a low plateau 138
scraped bare by ice; covered with forest, swamp and lakes. From Spitsbergen through western Norway to north-west Britain ancient *Caledonian* fold mountains were later eroded into plains and again uplifted. They have been deeply dissected but contain remains of ancient level surfaces.

C. HERCYNIAN EUROPE AND PLAINS

Much of Central Europe and France is formed by remains of 141
Hercynian fold mountains from Carboniferous times, later eroded away and broken in Tertiary times. In several places they are covered by younger sediments that form plains, or have been etched into scarps and vales. Hercynian Europe has wooded hills, deep gorges, broad valleys and basins that form rich farmland.

Plains left by retreating ice. Loess and limon.

D. EUROPEAN RANGES

The alpine mountains comprise giant folds of weak sediments amid 146
which lie older crystalline rocks. Quaternary glaciation modified the earlier relief into pyramidal peaks and U-shaped valleys. Along the flanks, broad depressions have been filled by material eroded from the mountains to form low plateaux.

Page

(1) **The Alps** proper, extending in a great curve from the Mediterranean to Vienna, comprise parallel mountain chains. A crystalline core is surrounded by schists and shales: the outermost belt is low sandstone hills or limestone plateaux. Entrances to the U-shaped valleys are frequently marked by moraine-dammed lakes. In the Jura, upfolds mark ridges and downfolds, valleys; but the northern edge is a limestone plateau. 148

Between the Splügen Pass and the Dora Riparia river are the most majestic peaks, often pyramidal with sharp rocky ridges. Intense frost results in splitting of the rock. Large snowfields feed glaciers. Main valleys are U-shaped with hanging tributary valleys. Farther east the work of ice is less impressive.

(2) **The Balkan** Mountains and the Stara Planina form a southern continuation of the Transylvanian Alps. The flat-topped Rhodopi is an ancient uplifted block. The Danube is deflected north by the similar but lower Dobrogea. Numerous mountain basins are good farmland. 149

(3) **The Carpathians** form a great curve from Bratislava to Iron Gates. On the north are rounded forested hills, but the High Tatra has jagged crystalline peaks. On the south, limestone, crystalline and volcanic rocks form the Slovak Ore Mountains, rich in minerals. On lower slopes loess forms rich soils. The eastern Carpathians are rolling forested mountains, highest in the Transylvanian Alps. They enclose the fertile basin of Transylvania. 150

(4) **The Dinaric Mountains and Greece.** Part of the alpine system, the Dinaric ranges lie parallel to the Adriatic coast. Some have been partly drowned to form offshore islands. Limestones, common in these mountains, give barren country known as *Karst.* The eastern flanks descend in forested hills to the warm, cultivated Sava-Drava valleys. The highest parts are the Albanian-Macedonian Highlands. The Morava-Vardar valleys form a vital routeway from the Aegean Sea to the Danube Plains. The Greek mountains are similar, with earthquakes still common. The islands of the Aegean and Ionian Seas are drowned ranges. 150

EUROPEAN CLIMATE

A. The North Atlantic Drift makes Europe unusually warm in winter for its latitude. Depressions are winter rain-producers in the 157

Page

Mediterranean basin, which is dominated in summer by dry continental air.

B. The plains of north-west Europe generally allow easy penetration by Atlantic air, but may also allow cold Siberian air to bring hard winter conditions. The Norwegian coast is warmed by Atlantic waters in winter, but the shelter of the high mountains limits the penetration of this winter warmth. 158

C. The Mediterranean region has a strong seasonal rhythm, with cyclonic rain in winter, and hot dry summers, with coastal heat tempered by sea breezes. Persistent north winds blow in summer in the eastern Mediterranean. Iberia has moist Atlantic conditions on the north and west, but the Meseta has hot dry summers and cold dry winters. The North Italian Plain has a modified Mediterranean climate. Mountains along the Adriatic and Aegean coasts prevent deep penetration inland of Mediterranean conditions: the Balkans have summer rain with cold snowy winters. 158

D. Central European climate is transitional between maritime and continental conditions of the Russian plain. Cold snowy winters follow hot summers with thunderstorms. The Hercynian uplands have colder winters, and sheltered basins are unusually dry. Drought is a serious problem. 159

E. The Hungarian Plain has cold winters and hot dry summers. The Alps, Carpathians and Dinaric mountains experience cold winters with much snow and clear skies, while summers are warm or hot, with thunderstorms. 159

F. The Balkan mountains and plateaux prevent inland penetration of Mediterranean climate, but allow deep southward penetration of more northerly features. Northern slopes are mostly forest. Cold air draining into valley floors and basins and movement of the warm **föhn** winds influences the siting of settlements. 160

SOILS AND VEGETATION IN EUROPE

A. Age-long cultivation has greatly influenced soils and vegetation in Europe. **Tundra**—poor moor and bog, has been least modified by man. **Coniferous forest** is mainly in glaciated country and contains the major timber reserves. Small clearings are farmed on acid and infertile soils. **Deciduous forest** has been cleared for farming. Poor areas have been planted with conifers. 161

Dark soils are found on limestone rocks, e.g. the **black earths** which have formed over loess. These dark soils are the most fertile. These

Page

are important in the Paris Basin, central Europe, southern Poland and the Ukraine. They are used for arable farming.

B. Plants in the **Mediterranean** climate must withstand the long, hot, dry summer. Forests are mostly conifers and there is much scrub. Olives and cork oak are found. The destruction of forests has caused rapid devegetation and soil erosion. Sedimentation and waterlogging in lowlands and development of malarial swamps. Soils are generally limey and poor in humus. Irrigation. 162

PEOPLE OF EUROPE

Population about 435 million. Racially very mixed. Population growth was particularly rapid in the nineteenth and twentieth centuries as industrialization spread from west to east. 164

TRANSPORT AND ROUTEWAYS IN EUROPE

Roads remained bad until modern times. Railways. Influence of the World Wars. Importance of Alpine tunnels. Roads and railways better developed in west than east. New motor roads. Airways and pipelines. Waterways for freight. 165

EUROPE'S MINERAL WEALTH

(1) Most minerals needed in Europe are **imported,** because local deposits are inadequate or absent. Coal is, however, plentiful on the northern edge of the Hercynian massifs. Brown coal is used chiefly to generate electricity and as the basis of the chemical industry. 169

(2) **Coal** now supplies less than half of Europe's energy requirements. Petroleum consumption is rapidly expanding, though there are relatively few deposits in Europe, and the bulk is imported from the Middle East. Exploration of the North Sea. The main supplier to Eastern Europe is Russia. Natural gas in the Low Countries. Bauxite and iron ore are worked on a large scale. Uranium, mercury and mineral salts. 169

(3) Thermal electricity is still the most important, but hydroelectricity supplies in Scandinavia and the Alpine lands. Nuclear power is being developed. 170

LAND USE IN EUROPE

(1) **Farming** varies widely in regional character. Despite a large production of foodstuffs within Europe, much food has to be imported. 171

(2) The proportion of people engaged in farming increases eastwards and southwards. 172

(3) Differences between the agriculture of the Communist countries 172

Page

of Eastern Europe and the capitalist democracies of the west. **State farms, collectives and co-operatives.**

(4) The Atlantic coastlands are noted for dairy and beef cattle. 172
Arable farming is more important inland—in the north cereals are grown and in the south wheat and sugar beet in the loessic soils. Importance of the potato. The eastern loess belt in Polish Galicia and the Russian black earth lands are important grain producers. Maize, tobacco, limes and sunflowers are grown in the south-east Europe. Sheep and goats are common in the Mediterranean and Balkan regions. The Mediterrannean region grows citrus fruit, olives, wine and vegetables.

(6) The vine is widely grown south of 50° N. France and Italy 174
produce almost half the world's wine.

(7) **Fishing:** In the northern waters the main fish caught are 174
herring, cod and flat fish. In southern waters pilchards, sardines and tunny. Shellfish in the inshore waters.

COUNTRIES OF EUROPE

FRANCE

Paris: only city with over a million people. Farming standards low. 175
Wine; cattle. **Industry** in north. **Chemicals;** Lacq oilfields; bauxite. **Hydro-electricity.**

NETHERLANDS

Two-fifths below sea level. **Delta project.** Densely populated. 177
Dairying; vegetables, bulbs and flowers. **Natural gas. Europoort.**

BELGIUM

Two languages. Much industry. Intense agriculture. 179

NORWAY

Half population live in **Oslo. Fishing** and **farming. Hydro-electricity.** 180

SWEDEN

Hydro-electricity (95 per cent of power). Forest industries. **Iron.** 180
High quality manufactures. Crops in south, livestock in north; Lapps breed reindeer.

DENMARK

Most densely-populated Scandinavian country. **Dairy farming,** 181
high yields. Shipyards. **Electricity imported.**

FINLAND

Only 9 per cent cultivated. Rest—swamp, lake, river, **forests.** 182
Farms in clearings; **forest products. Copper.**

Page

GERMANY

(1) **West Germany:** One of the greatest **industrial powers** in the world. Coal, iron, lignite; some oil. Ruhr. Agriculture also important wine. Modern plant—severe war damage quickly repaired. 183

(2) **East Germany:** Deficient in mineral resources but possesses lignite, mineral salts, non-ferrous metals and uranium. Effects of war damage more lasting than in West Germany.

AUSTRIA

Remnant of former great state. **Forest industries. Tourism.** 187

SWITZERLAND

National groups. **Dairying. Precision industries;** hydro-electricity. **Tourism.** Rout centre. **Tunnels.** 187

POLAND

Important food exporter. 189

CZECHOSLOVAKIA

Steel engineering; chemicals. Arable farming. 189

HUNGARY

Farming. Wine. Industry growing rapidly. Lignite; metallurgy; chemicals. 190

BULGARIA

Farming—half the country arable land. 191

ROMANIA

Farming. Natural gas, oil (less important than formerly). 191

ITALY

Agricultural. **Wine;** wheat; olives. **Industry** mostly in north. **Tourism.** Hydro-electricity. 192

JUGOSLAVIA

Agricultural; wine. **Industry** increasing rapidly. Hydro-electricity. **Tourism.** 194

GREECE

Agricultural. Poor. **Tourism.** 195

SPAIN

Agricultural. **Wine;** olives; **vegetables; citrus fruit.** Some chemicals and steel. 197

PORTUGAL

Agricultural. Poor. Population concentrated Lisbon, Oporto, Douro Valley. **Wine.** Fishing. 197

Page

tayga forest, and this becomes drier towards the east.

B. PERMAFROST 209

Short summers and long, deep frost penetration under little winter snow causes most Siberian ground to be permanently frozen. Nearly all the tundra is permanently frozen; islands of unfrozen ground in northern tayga. Water welling up from beneath the frozen layer causes surface deformation.

C. THE FARMING LANDS OF THE MIXED FORESTS AND WOODED STEPPE 210

European Russia has a belt of deciduous forest tapering eastwards, and a great deal of ground has been cleared for farming. Potatoes, rye and flax are grown. Large forests remain, though this tends to thin towards the southern wooded steppe—which passes southwards into true steppe. Fertile soils have encouraged farming in this belt. Black soils are suitable for wheat and maize, etc.

D. THE STEPPE 210

In European Russia and in Siberia, most of the best steppe has been taken into cultivation. The steppe is Russia's most important farming country, but bad farming has caused gullying and erosion. Drought is also a major problem, the rainfall being meagre (mostly in Spring). The remaining natural steppe is suited only to pastoralism and includes much salty ground.

E. DESERTS OF THE SOVIET UNION 211

South of the steppe, around the Caspian and Aral Seas, desert occurs. It can be made into rich country by irrigation, like the oases along the rivers and on the mountain footslopes and basins. Many oases have been greatly improved and new irrigation schemes have been built.

F. SMALL AREAS OF SPECIAL NOTE 212

The south-west of the Caucasian isthmus has a warm humid climate and tea and fruit are grown. In the Amur basin climate is monsoonal and rice and soya beans are cultivated.

THE SOVIET PEOPLE

Slavonic Russians (Great Russians, Ukrainians and Byelorussians) form over three-quarters of the total population. The fifteen largest national groups have own republics, forming the Union of Soviet Socialist Republics. 213

South of Leningrad and west of the Volga, two-thirds of the people

Page

live in one-sixth of the country's area. Central Asia and Caucasia have many well settled pockets of country. Growth of urban population. Large-scale building programme since 1945.

RESOURCES, INDUSTRY AND OCCUPATIONS IN THE SOVIET UNION

A. The Soviet Union has large coal and petroleum reserves, besides iron and its alloy metals and non-ferrous metals. Minerals for the chemical industry are also plentiful. Main coalfields: Donbass, Kuzbass, Karaganda, Vorkuta. Petroleum comes from the Volga basin, Caucasia and the Caspian Sea. Construction of pipelines and refineries. Natural gas from the oilfields and new sources. Although minerals are widely scattered the main areas are in the Urals, eastern Ukraine, western Siberia and the Altay country. 218

B. ELECTRICITY 219

Most electricity is still generated by thermal plants, but large hydro-electric stations have been built. Electricity is widely used for railway traction. Atomic generators at Obrinsk (Moscow).

C. INDUSTRY 221

(1) Three-quarters of all Soviet industrial capacity is in European Russia and the Urals, notably in the Ukraine and around Moscow. Industry also in Eastern Siberia, the Far East and Central Asia. The discovery of oil has been important in industrial growth in the Volga basin. Northern European Russia has been developed for iron ore, coal and petroleum. Plans exist for industrial development in Southern Siberia.

(2) Three-quarters of Soviet iron and steel output comes from the Ukraine and the Urals. Iron ore (Krivoy Rog) and coal (Donbass) are the raw materials in the Ukraine. The Urals have local ore but coal comes from the Kuzbass and Kazakhstan. Western Siberia and the Moscow region are also significant steel-producing areas. Nearly all large towns have some engineering, but Moscow, Leningrad, Kharkov and Kiev are especially important. Chemical production is widespread.

D. FARMING 223

(1) Mixed farming predominates. Dairying, rye and flax in the north; wheat and arable crops farther south; sugar beet in the western Ukraine; sunflowers in the south. Favourable areas cultivate the vine. Siberia is noted for dairying on rank river meadows, while in the Far East, Russian settlers cultivate "Chinese" crops (e.g. soya beans, perilla).

Page

(2) Large areas of unused steppe from Northern Caucasia across Kazakhstan and western Siberia as far as the Altay foothills have been cultivated. Low and unreliable rainfall and soil erosion restrict cultivation. To reduce the burden on transport of food to Arctic settlements, special northern farms in small, climatically favoured areas have been established. 226

Small areas of unusually favourable soil and climatic conditions produce crops which make the Soviet Union largely independent of outside supplies. The oasis lands of central Asia are especially important, where the irrigated area has been considerably expanded. Crops include cotton, mulberry, rice and fruits. Sticky red soils in humid Western Transcaucasia are also noted for crops of citrus fruits, tea and vegetable oils.

E. PASTORALISM 228

In the drier steppe pastoralism is common. Cattle are raised near the settled farming districts. Sheep are raised in the poorer steppe. Horse raising was also once important; camels and yaks are bred. In northern forest and tundra reindeer are herded. Fur-bearing animals bred on farms are now more important than those trapped.

TRANSPORT AND COMMUNICATIONS

A. Rivers and the portages between them were the traditional Russian routeways. Now railways are the most important means of transport. Railways designed primarily to move heavy, bulky freights. Problem of Permafrost in Siberia. 230

B. GREAT RIVERS OF THE U.S.S.R. 233

Rivers are frozen during the long winter period and in spring the thaw causes great floods. In late summer low water hinders navigation. Rivers flow generally northwards or southwards, whereas most movement in modern Russia is east-west. The most used river system is the Volga and its tributaries, which canals link to the Baltic and Black Sea. Another important river is the Dnieper. Cement, oil and timber are typical goods on river boats and there are great log rafts as well.

C. RUSSIAN ROADS 234

Few well-surfaced major highways, mostly in European Russia.

D. SEAS AROUND THE SOVIET UNION 235

All suffer from being frozen in winter, though ports may be kept open by icebreakers. Fog is also a serious summer hazard.

AFRICA

Page

GENERAL

Africa is a large, compact continent lying athwart the Equator. It has a relatively even coastline. The continent consists of: 237

A. RELIEF

(1) Small narrow coastal plains; 237

(2) Extensive interior plateaux with great Rift valleys in the east; 237

(3) Fold mountains in extreme north and south; 237

(4) High isolated volcanic peaks, such as Mount Kilimanjaro. 237

Drained by great rivers which have much hydro-electric potential but little value for transport because of waterfalls and rapids. 238

B. CLIMATE

"Hottest" continent, with hardly any "cold" lands. 238

(1) Movement of "Heat Equator" with overhead sun; 238

(2) This causes corresponding movements of pressure belts, wind belts and areas of heavy rainfall. 238

(3) *Equatorial zone* (heat, and heavy rain) dense forest. 239

(4) *Tropical zone* (summer rain, winter drought) savanna. 239

(5) Desert and semi-desert. 239

(6) Two zones of "Mediterranean" climate. 239

(7) "*China*" zone (heavy summer rain). 240

SOUTH AFRICA

Mainly a tilted plateau, higher in the east. Main rivers drain westwards. A large country (500,000 square miles); population 16 millions (3 million whites). Most advanced agriculture and industry in Africa. 241

A. RELIEF AND CLIMATE REGIONS 241

(1) **Cape Ranges:** Mild winters with rain. Hot, sunny summers. Fruit growing (deciduous, citrus and vines), wheat and sheep. 241

(2) **East coast hill region:** Area of grassy, hummocky hills and escarpments. Summers hot and wet. Winters dry and mild. Most rivers flow only during the wet season. Large Bantu population have many cattle and grow maize. Nearer coast plantations of sugar cane and citrus and other fruit. 241

(3) **South-west Africa:** Interior highlands of west and coastal desert. Sparse population. Extensive cattle ranching. 242

(4) **Kalahari and Orange River basins:** A plateau surrounded by higher land. Rainfall low, sparse vegetation. Sheep and cattle. 242

(5) **The Veld:** A high plateau—3,000-6,000 feet. Climate temperate —rainfall declines westwards and is variable. Main crops grown 242

Pag.

in tsetse-free areas. Plantation crops include rubber, coffee, bananas, cotton.

(6) **Minerals and industry:** Copper in **Katanga,** also cobalt, uranium, tin, zinc, diamonds. Much hydro-electric potential. Some manufacturing in larger towns: textiles; shoes; chemicals; shoes. 251

EAST AFRICA

A. GENERAL 252

(1) **Relief:** Series of plateaux with high volcanic peaks and narrow coastal plain. Great Rift valleys in interior, many lakes. 252

(2) **Climate:** Temperatures lowered by altitude but little seasonal variation. Rainfall fairly heavy along coast, decreasing inland. Double maximum. 252

B. KENYA 253

(1) Narrow coastal plain hot and wet—crops grown include sugar cane, rice, yams, cassava, maize, copra. Interior plateaux and highlands rise to 5,000-8,000 feet with higher peaks. Rift valley 30-40 miles wide. Many lakes. Cool climate, suitable for European settlement. Good soil. Cash crops: tea, sisal, pyrethrum, cotton, coffee. African subsistence farming; meat, hides, dairy produce. 253

(2) **Industries:** Hydro-electric power from Owen Falls. Flour milling, margarine, soap, cement, consumer goods. 254

C. TANZANIA 254

(1) The coastal plain is hot and rainy, with dense population. Rice, sisal and coconuts important. Zanzibar and Pemba grow cloves. Vegetation mostly savanna. Subsistence agriculture predominates. Crops include maize, millet, ground-nuts. Cattle reared in tsetse-free areas. Cash crops include cotton, coffee, oilseeds, sisal, tea, tobacco and pyrethrum. 254

(2) **Minerals:** Diamonds very important, some gold, tin, salt. 255

(3) **Industry:** Not yet highly developed. 256

D. UGANDA 256

Densely populated. Fertile plateau at 3,000-4,000 feet. Equatorial climate. Subsistence farming and cash crops—cotton, coffee, tea. Many cattle.

E. PROBLEMS OF EAST AFRICA 257

Racial diversity and tribal divisions; pests and diseases; unreliable or low rainfall; soil erosion; poor communications.

WEST AFRICA

		Page

Page

NORTH-WEST AFRICA (MAGHRIB)

A region of Arab culture with total population of 24 millions separated from rest of Africa by desert.

A. RELIEF 268

Interior folded ranges—Atlas Mts., reaching nearly 14,000 feet—with lowlands and plateaux along coast.

B. CLIMATE AND VEGETATION 271

Mediterranean modified by relief. Total rainfall decreases inland until desert reached. Scrub, oak and coniferous forest on higher ground; grassland above. Grass steppe (esparto) in interior.

C. RESOURCES 272

(1) **Farming:** Typical Mediterranean crops—wheat, barley, early vegetables, grapes, citrus, olives. Some nomadism. 272

(2) **Minerals and industries:** Iron ore and phosphates exported on a big scale. Others include lead, zinc and manganese. Oil and gas deposits in Sahara. Industries—glass, processed foods, clothing, textiles, soap, phosphates, cement. 272

D. REGIONS 273

(1) **The Sahara:** Greatest of the hot deserts. Three types surfaces—sandy, stony, rocky. Not entirely without rain. Great mineral wealth—oil, gas, manganese, copper, lead, tin, iron. 273

(2) **Nile Basin:** Nile longest river in world. Between Aswan and delta, valley widens to alluvial plain, 10-15 miles broad. Importance of irrigation (Nile). 275

(3) **The Sudan:** Largest country in Africa—nearly one million square miles. Much desert or semi-desert. Irrigated areas near Nile, densely populated, especially Gezira. Main cash crop cotton. Food crops: wheat, millet, vegetables, groundnuts. 275

(4) **Egypt:** About 386,200 square miles with population of over 30 million. Population confined to delta and narrow strip along the Nile valley. Climate very dry and hot, irrigation necessary for crop growing. Main crops wheat, maize, millet, barley, rice, sugar, lucerne, vegetables, citrus fruits, grapes, dates and cotton. Main export crops cotton, rice, sugar and onions. 276

Industries: Many new factories set up under Five-Year Plans to produce textiles, motors, cement and superphosphate. Other factories built or planned will produce petroleum products, iron and steel, chemicals and textiles. 277

SOUTH ASIA

SOUTH-WEST ASIA

Page

(2) **Northern States** include:

(i) **Kashmir** (disputed between India and Pakistan). 301

(ii) **Nepal, Sikkim, Bhutan** (influenced by India). 302

(3) **Pakistan** 302

(i) **East Pakistan,** comprising the "active" eastern half of the Bengal delta; rice and jute are grown. *Dacca* is the capital and *Chittagong* the port. 302

(ii) **West Pakistan:** larger and less densely peopled; it includes a mountain rim, a plain irrigated by canals and desert. 302

(iii) **Punjab:** cotton and wheat and textile manufacturing. 304

(4) **India** 304

(i) The **Gangetic Plain** which in the west is often irrigated by canals and has wheat and cotton cultivation; becomes wetter to the east, with sugar, rice and jute cultivation. Tea in Assam. *Delhi* (capital) and *Calcutta*, the port. 304

(ii) The **Peninsula** which varies from: (*a*) the humid north-east with jungles and minerals; (*b*) the lavas of the north-west with cotton growing and manufacturing; (*c*) the *West Coast* plain with rice, coconuts and fishing; *Bombay*; (*d*) the tank country beyond the Ghats—millets; the industrial town of *Bangalore*; (*e*) the *East Coast* Plain—a series of fertile irrigated deltas; *Madras*. 305

(5) **Ceylon** has four regions: 307

(i) The wet south-west plain with rice, coconuts and the port of *Colombo*. 307

(ii) The hilly centre with rubber, tea and coffee plantations. 307

(iii) The *Dry Zone* now being redeveloped. 307

(iv) The limestone *Jaffna* peninsula. 307

SOUTH-EAST ASIA

A. RELIEF 308

(1) An ancient core of rocks stretches from eastern Burma, through Thailand, Malaya, Java and Sumatra to Borneo. 308

(2) Two intersecting arcs of Tertiary fold mountains, with volcanoes and deep sea trenches. 308

(3) Alluvial plains along the rivers and round the coasts. 308

B. CLIMATE 309

(1) Temperatures high all year; rainfall is more seasonal and sunshine more prolonged away from the Equator. 309

(2) Rain all year near the Equator. 309

(3) Winter rain in the Philippines, Vietnam and north-east Malaya. 309

	Page
(4) The remainder of the mainland has the summer monsoon with rain shadows between the mountain chains.	309
(5) South of the Equator the Australian summer monsoon produces a dry season to eastern Java.	310
C. RESOURCES	310
(1) **Alluvial tin:** Malaya, Thailand, Burma, Sumatra.	310
(2) **Oil:** Sumatra, Borneo.	310
(3) **Hydro-electricity:** A vast undeveloped source of power.	310
(4) **Timber:** Tropical hardwood (Borneo); teak (Burma, Thailand); bamboo (throughout); mangroves (coastal swamps).	310
D. CULTURE	310
(1) Early Hindu, Arab and Muslim influences on language, race and religion.	310
(2) Western conquest and development of plantation crops, cities and communications, from the seventeenth century onwards.	310
(3) Since 1945, independent nations have been trying to develop their economies to absorb population increase. Little industry.	311
E. FARMING	311
(i) **Shifting cultivators** in the forested hills of the interior.	311
(ii) The **rice farmers** of the plains.	311
(iii) **Plantations.**	311
F. REGIONAL GEOGRAPHY	311
(1) **Burma**	311
(i) The forest-clad mountains on the western side of Burma which extend round the northern edge of Burma on the Chinese borders; sparsely populated; teak forests.	311
(ii) The **Dry Zone** of the Middle Irrawaddy, the heart of Burma; well irrigated, rice and cotton growing. *Mandalay*.	312
(iii) The delta of the **Irrawaddy**, reclaimed in the last 100 years; rice is grown for export from *Rangoon*, the capital.	312
(iv) The **Arakan Coast**, an alluvial lowland; tin dredging.	313
(2) **The south-east corner**	313
(i) Although complex, it is shaped like a hand, the fingers representing the mountain chains. Each country has a core of alluvial lowland.	313
(ii) Basins of Red River, Mekong, Chao Praya.	313
The countries are **Thailand, Laos, North Vietnam, South Vietnam** and **Cambodia.**	313
(3) **Malaya**	315
After independence (1957) Malaya stayed in Commonwealth. Now main portion of Malaysia.	

CHINA

Page

(2) China became a republic in 1912. 321

(3) The Japanese occupied part of China from 1937-45. 321

(4) A Communist government took control in 1949. 321

B. RELIEF 321

(1) Tertiary mountain chains fan out from the Pamirs and enclose basins (Tibet, Szechwan). 321

(2) A vast peneplain lies to the north, tilted westwards to form the Khingan Mts. 321

(3) The low hills of Manchuria, Korea and south China consisting of ancient mineral-bearing rocks. 321

(4) Lowlands are scarce and confined to the coasts and the great river valleys: (i) Hwang Ho; (ii) Yangtse. 321

C. CLIMATE—The monsoon pattern prevails: 323

(1) Winter: cold dry winds from the Siberian high pressure region curve over the sea and bring a continuous layer of cloud to the south. There are also westerly depressions. 323

(2) Spring is warm and rainy as the air masses intermingle. 323

(3) Summer is hot and rainy; typhoons. 323

(4) Autumn is calm, sunny and cool. 323

D. NATURAL RESOURCES AND DEVELOPMENT 324

(1) **Mining and power** 324

Coal—Vast deposits (Shansi, Szechwan, Manchuria). 324

Iron—Vast deposits (Shansi, Manchuria). 324

Oil—Very little discovered (Sinkiang). 324

Tungsten, Antimony }—Good supplies (South China hills). 324

Hydro-electricity—Vast potential (Western Mts.). 324

(2) **Industry**

(i) Rapid development since 1950 under three **Plans** starting with basic machinery, leading to consumer goods. 324

(ii) Former industrial centres developed in Manchuria and Szechwan, and new ones created, e.g. *Lanchow*. 324

(3) **Communication** 324

Transport has been greatly developed.

(i) New railways in the west. 324

(ii) Roads. 324

(iii) Waterways in the east. 324

(4) **Agriculture** 324

(i) Four-fifths of the people are dependent on agriculture; Communes. 324

JAPAN

NORTH AMERICA

CHAPTER 33

THE CONTINENT AS A WHOLE

WHATEVER the true objective of Columbus, it is certain that he visited many of the islands of the West Indies and sighted the Orinoco delta; it is equally certain that Amerigo Vespucci made at least two voyages to Mexico and the Brazilian coast. It is not at all certain that the name "America" owes its origin to Vespucci. At about the same time (1498), John Cabot returned to Bristol after journeys to Newfoundland and Eastern Canada, and along the shores of the Carolinas and Florida. The customs roll of the port shows that he received two payments of £20 each from his friend Richard Ameryk, the senior collector of customs (and later Sheriff of Bristol). Cabot probably named the new territory after him for it appears as America on the first Bristol maps produced before any comparable European maps.

From the sagas of Norse history we find that the first Europeans to discover the continent were Vikings. Probably Bjani, leaving Iceland in A.D. 986, reached the Labrador coast but its ice-bound condition prevented him from landing there; but more likely still, Leif the Lucky and Karlsejni in A.D. 1002 went ashore either here or farther south and named the area Vinland (Woodland). Perhaps "Wineland", another name given to it by them, was suggested by the presence of the wild currants. Although very little settlement took place, there must have been much voyaging and fishing, and Columbus and Cabot obviously had some encouragement to make their journeys from the rumours which filtered through to them.

Nowadays, America is not thought of as meaning the whole of the New World; and to most people, any reference to "America" or an "American" implies association with the U.S.A. only. On the other hand, Canada means a specific part of North America; its name is derived from "Kanata," an Indian word for "settlements."

Position

North America, stretching from the Isthmus of Tehuantepec to 70° N., lies entirely north of the Equator and covers an area about two and a half times that of Europe. Canada and the U.S.A. are approximately the same size.

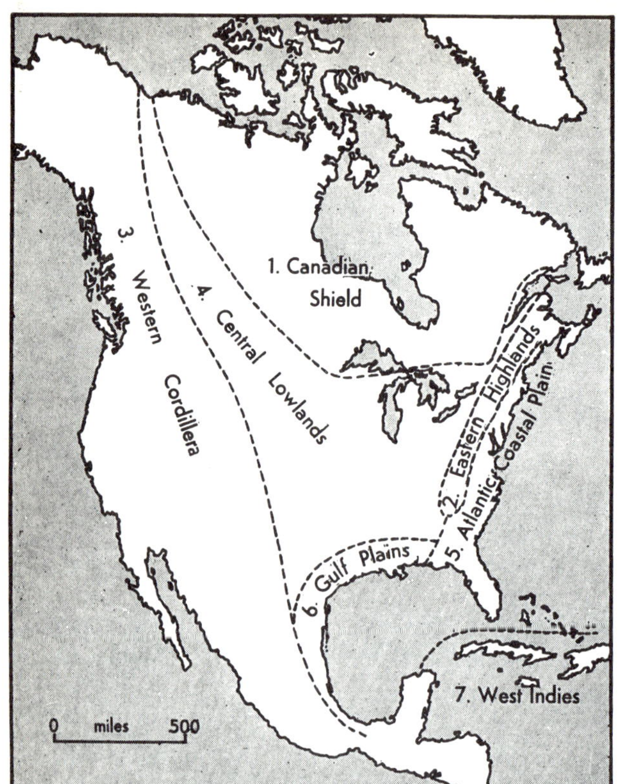

FIG. 143. *North America divided into seven major relief regions.*

Although not to the same extent, North America is, like Asia, a continent of contrasts with its wide area of tundra of the north, its deserts of the south-west, its high mountain ranges and its plains; and there is the great temperate area at the continent's widest part, bounded by the latitude lines 26° N. and 55° N. with rich resources. Southern Canada, the U.S.A. and the plateau region of Mexico have conditions of alternating cold and warm seasons favouring hard work and the settlement of Europeans, and this area of North America has four times as many people of pure European stock as are to be found in the rest of the world outside Europe and the U.S.S.R. See map on page 410.

Relief and Structure

The continent can be divided into seven relief regions as shown in Fig. 143.

(1) The Canadian Shield, covering more than half of Canada, is generally of low, hummocky relief with hills and ridges rising about 200 feet above the adjacent lakes and valleys. The Shield is lowest around Hudson Bay but locally it may rise to 2,000 feet and in Labrador the plateau reaches 5,000 feet. The southern edge in Quebec drops quite steeply to the St. Lawrence river.

The rock structure is essentially Pre-Cambrian. The Shield has undergone

FIG. 144. *The Eastern Highlands with a section drawn along the line AB.*

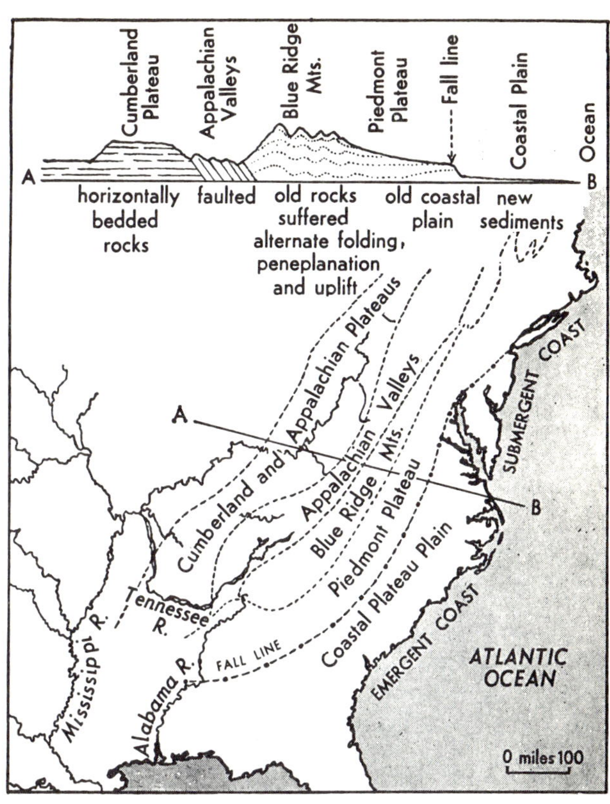

FIG. 145. *Hamilton Inlet, Labrador, part of the Canadian Shield.*

complex vertical movement, and sediments have been laid down during successive periods of subsidence and then removed by later erosion processes. Uplift during this period was accompanied by volcanic action and igneous intrusions. Rich ores and varied deposits have resulted from the solidification of igneous rock masses during the later stages of intrusion and cooling.

The many lakes of the Shield have been caused mainly through the action of ice.

(2) The Eastern Highlands stretch from Labrador through Newfoundland and New England to Alabama in a general direction from north-east to south-west ("the Caledonian trend"). They can be divided into physical regions trending in the same direction, as shown in Fig. 144. The present structural form has resulted from geological history, the main stages of which can be summarized briefly as follows: (*a*) In very early times the present position of the Western Piedmont marked the western edge of a continent, and farther west still the floor of the ocean was covered with successive horizontal layers of sediment derived from the erosion of this continent to a peneplain. Coal measures were formed in the shallow marginal waters. (*b*) Amorican folding contorted the peneplain to give the ridges and lifted the deep sea deposits to form the horizontally-bedded Cumberland Plateau. The Appalachian valleys with their coalfields are situated between these

FIG. 146. *Athabasca Glacier, Jasper National Park in the Western Cordillera.*

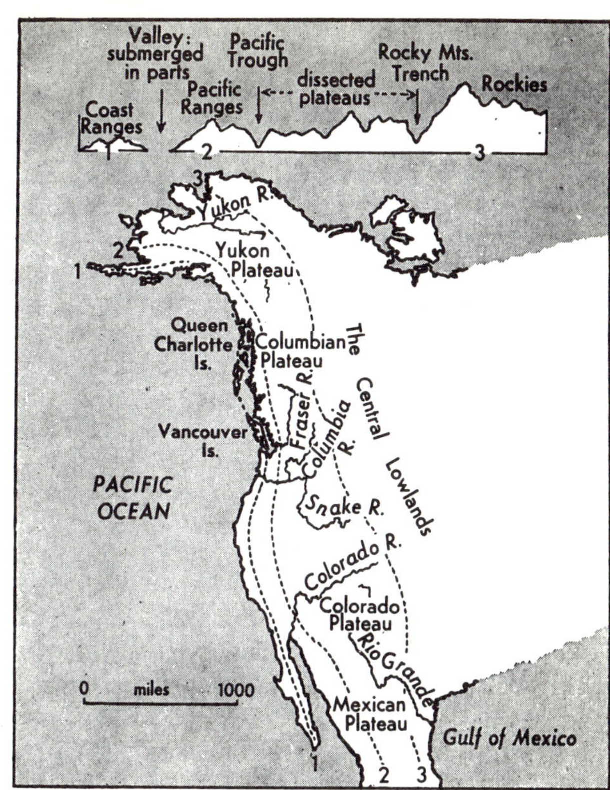

FIG. 147. *Western Cordillera with a generalized section across British Columbia.*

two formations. (*c*) After a period of erosion and deposition giving the eastern coastal plain, a further uplift caused the present Piedmont Plateau and the consequent rejuvenation enabled the rivers to cut deep valleys through the ridges and to give a Fall Line at the junction of the old and new coastal plains. The present character is the result of a final Tertiary uplift.

(3) The Western Cordillera, also complex in character, consists of three major ranges with intermontane plains and plateaux, which have been folded comparatively recently and are consequently much higher and more extensive than the other highlands of the continent.

Since the main period of folding, there has been much erosion; and local movements accompanied by volcanic action have given the Fraser, Columbia and Snake Rivers complicated courses where they have had to cut through rising barriers and lava flows. The Rockies are high, sharp and rugged, their tops consisting of bare rock still carrying remnants of glaciers. The coastal plains are narrow and are covered by great thicknesses of varied material, the result of intense erosion. The Rocky Mountains and the Pacific ranges form the great mineral areas of the Cordillera.

(4) The Central Lowlands form a broad belt stretching from the Arctic to the Gulf of Mexico. For the most part, they are vast plains composed of almost horizontally-bedded sedimentary rocks of Cretaceous and Tertiary formations lying on old rocks which stand out occasionally as plateaux and generally form infertile areas.

(5) **The Atlantic Coastal Plain** spreads from the edge of the Piedmont, which is marked by the Fall Line and consists of sands, clays and limestones in a series of outcropping edges, lying parallel to the coast. The more resistant materials form scarps which face inland, while the gentle dip slopes face towards the sea. In the north-east there is a rugged indented coast and the narrow plain has a glacial drift cover. From the Hudson River to Chesapeake Bay, there are irregular peninsulas close to the Fall Line providing the sites of the U.S.A. Atlantic ports. Southwards the Fall Line recedes and a line of long, low sandy reefs cuts off fresh- and salt-water marshes from the open sea. The rivers have wide, open valleys with bottom lands of silt deposited in flood time. The neighbourhood of Cape Hatteras has been a kind of fulcrum about which tilting has taken place to give a coast of submergence in the north and one of emergence in the south.

FIG. 148. *Flood plain of the Mississippi. The* 2,450-*miles long river has its source in Lake Itasca, Minnesota, and annually pours* 675,000 *cubic feet per second of water into the Gulf of Mexico. Vast quantities of alluvium are carried to its mouth, creating a productive delta. In its lower reaches the Mississippi flows through an alluvial valley which in parts is lower than its flood level. Navigation is difficult for the river meanders, changes its course, and forms enormous loops.*

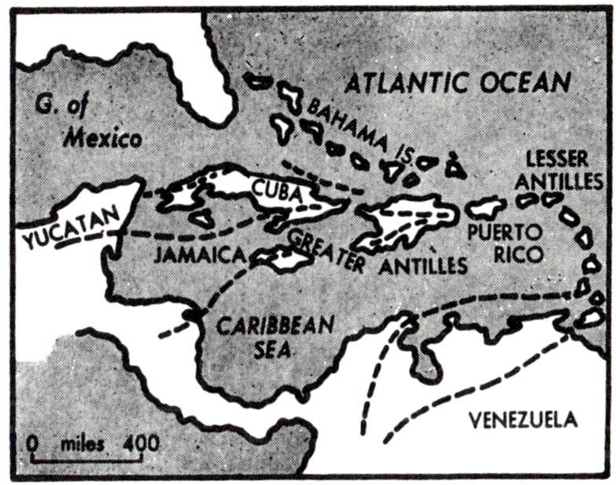

FIG. 149. *Fold systems of the highlands of the West Indies. Note how they meet at Puerto Rico. The Greater Antilles are connected to the W. Cordillera; the Lesser Antilles are linked with the Andes.*

(6) **The Gulf Coast** is a lowland of barrier beaches, lagoons, mud flats and the bird's-foot delta of the Mississippi. On the eastern side of this river, sandy flats replace the mud.

(7) **The islands of the West Indies** form a great curve from Yucatan to eastern Venezuela separating the Gulf of Mexico from the Caribbean.

QUESTION

1. Select *one* coastal area and *one* inland area of Canada showing extensive results of glaciation. Describe these results and indicate how far they affect man's work in those areas.

CHAPTER 34

GENERAL FEATURES OF CLIMATE AND NATURAL VEGETATION

THE maps in Figs. 150-154 show the broad outline of the climate of the continent and demonstrate:

(i) the large bends in the isotherms which indicate a greater degree of cold over the interior than over the coastal regions in January, and the reverse conditions in July;
(ii) the vast area of the continent which is below freezing point in winter;
(iii) the greater degree of winter warmth on the west coast compared with the east coast in the same latitudes;
(iv) the seasonal migration of the isotherms of 10° C. (50° F.) and 18° C. (64° F.);
(v) rainfall total diminishes from the coastlands to the interior, and from the south towards the north over the Central Lowlands;
(vi) most of the continent has a summer maximum of rainfall.

Temperatures. The latitudinal spread of the continent causes temperature conditions to vary from those of the Arctic in the far north, through the cool and warm temperate zones, to the subtropical coastlands of the Gulf of Mexico; the long coastlines are affected by varying winds and ocean currents and therefore show widely differing rainfall conditions. The large longitudinal spread gives an extensive interior with its particular continental conditions of high temperature ranges and low rainfall totals with a summer maximum. Great differences in altitude result in corresponding variations in temperature.

Air Masses. Over recent years, investigation of the upper air conditions has revealed the importance of air masses and Fig. 151 shows those which particularly affect the North American continent. There are two sets of similar atmospheric pressure systems operating east and west of the continent respectively, and a continental pressure system which is reversible with the seasons (Fig. 153).

The polar continental air mass, predominant in winter, results primarily from the uniformity of land surface with an extensive snow and ice cover and relatively stagnant air conditions. Tongues of the mass penetrate southwards giving rise to the north winds usually known as "cold waves."

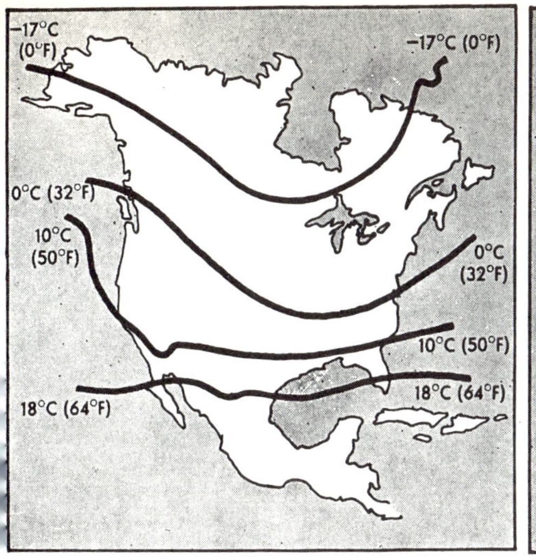

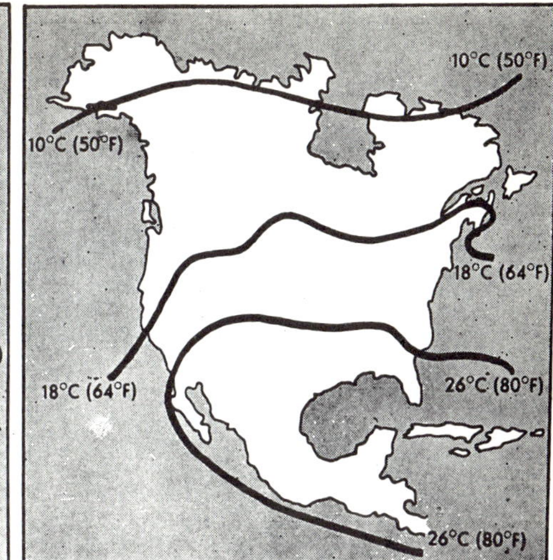

FIG. 150. *Isotherms: January (left); July (right).*

In summer, the air from it moves southwards as streams, which differ considerably from each other according to whether they are moving over land (e.g. the Prairies) or water (e.g. the Great Lakes).

Tongues from the polar maritime air over the North Pacific and North Atlantic Oceans bring cool, moist air to the coastal areas throughout the year; but in more southerly latitudes, tropical maritime air moves northwards through the Central Lowlands carrying thunderstorm conditions from the warm Gulf of Mexico and the Gulf Stream. The same effect is not found in the south-west coastlands because there the waters of the Californian current are cool and the mountain barrier prevents any modifying influence entering from the east. On the other hand, tropical continental, hot, dry air is found to a limited extent over the interior.

The meeting of the polar continental air with either the polar maritime or the tropical maritime air causes the development of depressions which follow regular paths along the Central Lowlands and the St. Lawrence Valley; and it is the passage of these temperate and tropical cyclones together with the relief of the continent which governs the precipitation. In the summer months, the temperate cyclones are

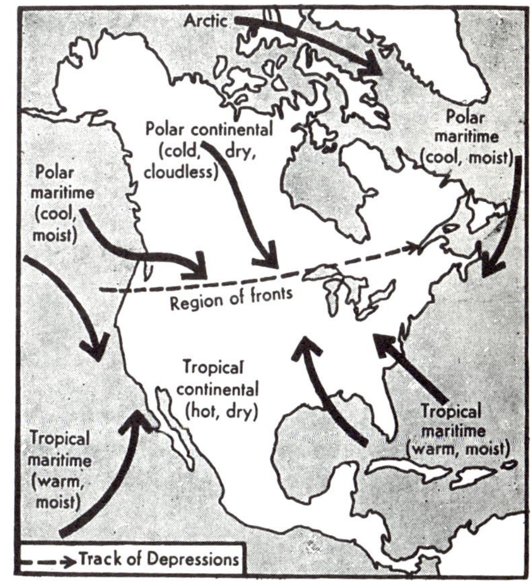

FIG. 151. *Climate: air masses.*

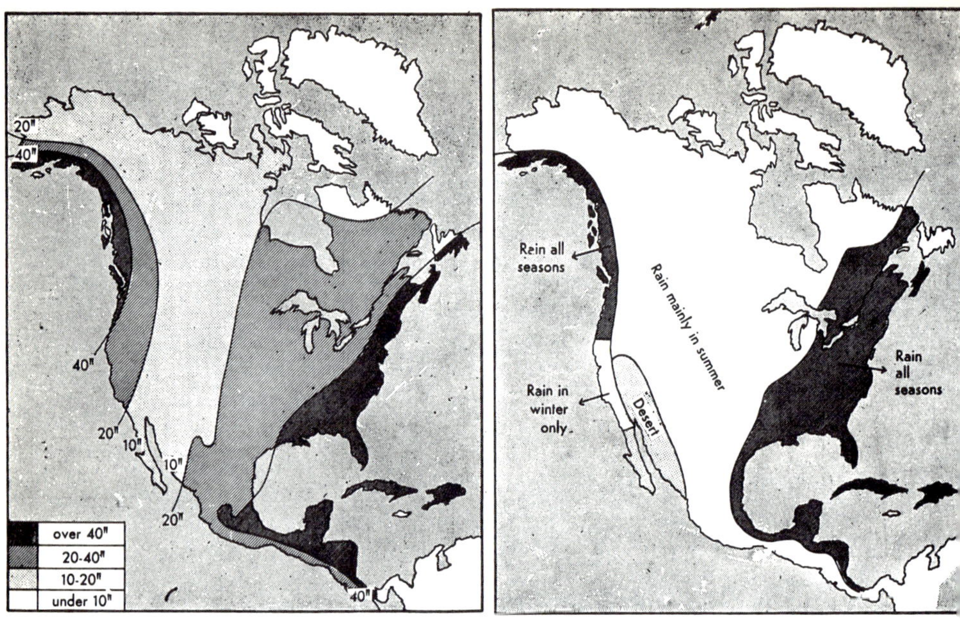

FIG. 152. *Rainfall: annual distribution (left); seasonal distribution (right).*

confined to a west-east line from British Columbia to the Gulf of St. Lawrence; but in the winter, the cyclones enter via Puget Sound, bend southwards almost to the Gulf of Mexico and leave in the neighbourhood of Cape Hatteras. The tropical cyclones enter from the Gulf and leave as temperate cyclones through the Gulf of St. Lawrence. Winter rainfall, therefore, is confined to coastal margins and the adjacent highlands from onshore winds; the interior receives most of its rain in the summer months. The Californian area experiences most of its rain in the winter and during one period of two months there is no rain. The intermontane plateaux of the south-west lie in the region of the tropical continental air mass and, because they are shut off from the maritime influences, are very arid.

Ocean currents exert the influence normally to be expected. The Westerlies over the warm Pacific Drift and the North-east Trade Winds over the Gulf Stream raise the temperatures of places on the coastlines of British Columbia and Florida respectively and increase the rainfall. The cold Labrador current plays a minor part in causing the freeze-up of the Gulf of St. Lawrence from November to March; but its most important influence is in the formation of fogs on the Grand Banks of Newfoundland, where it meets the Gulf Stream. The cold Californian current is principally responsible for the low mean summer temperatures of the coast and for persistent summer fogs. San Francisco, for example, has a very low temperature range of 6° C. (11° F.) and its maximum mean monthly temperature of 15° C. (60° F.) occurs in September.

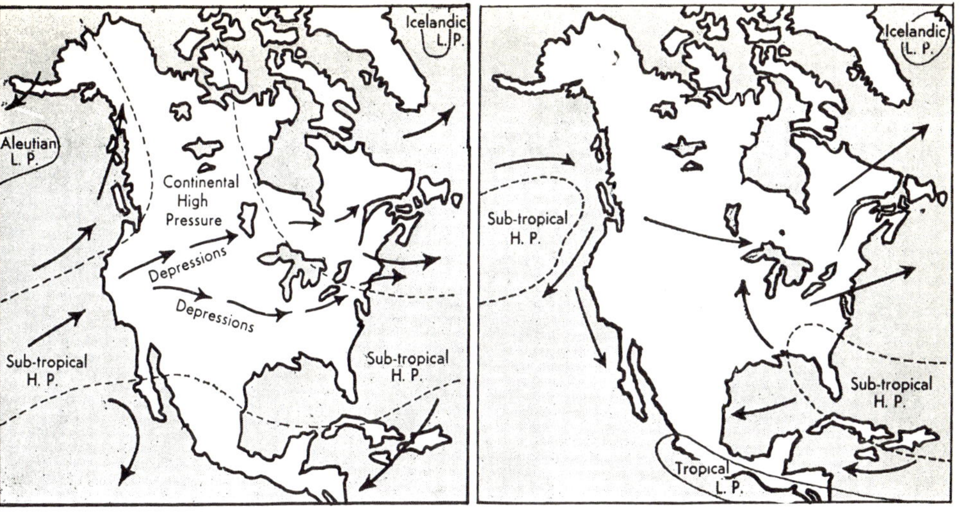

FIG. 153. *Pressure zones: winter* (*left*); *summer* (*right*).

The Chinook. One other special feature is the Föhn character of the Chinook. This is a warm, drying winter wind blowing generally over Alberta. A depression moving across the Central Lowlands draws humid air from the west across the Western Cordillera. As this descends on the eastern slopes, it is warmed by compression and it is thus able to raise the temperatures of the land over which it blows and absorb water vapour. The main benefit of this wind is that it enables the cattle-grazing lands of Alberta to be used in the winter in spite of the altitude.

Climatic Regions (Fig. 154)

(1) *Arctic coastlands.* Very cold winters; no temperature above 10° C. (50° F.); little precipitation, mainly snow.

(2) *Cold interior.* Short warm summers; very cold winters; light rainfall.

(3) *Interior.* Wide temperature fluctuations and ranges; summer maximum of rain; winter cyclonic activity with snow; annual rainfall and mean annual temperatures increase from north to south.

(4) *Warm temperate.* Hot summers and mild winters; high humidity at all seasons; summer maximum of rainfall; affected by the warm air mass over the Gulf of Mexico; rainfall decreases westwards.

(5) *Cool temperate western margin.* Moderate temperatures all year, with low range; rain all seasons with cyclonic influences in winter; variation in temperature and rainfall between the mountain slopes and the valley floors.

(6) *Mediterranean.* Winter rain and summer drought; temperature modified by cold current; local variations are owing to mountain and valley aspects.

(7) *Desert.* Large temperature range; cold winters on the plateaux.

(8) *Cool temperate eastern margin.* Considerable range of temperature

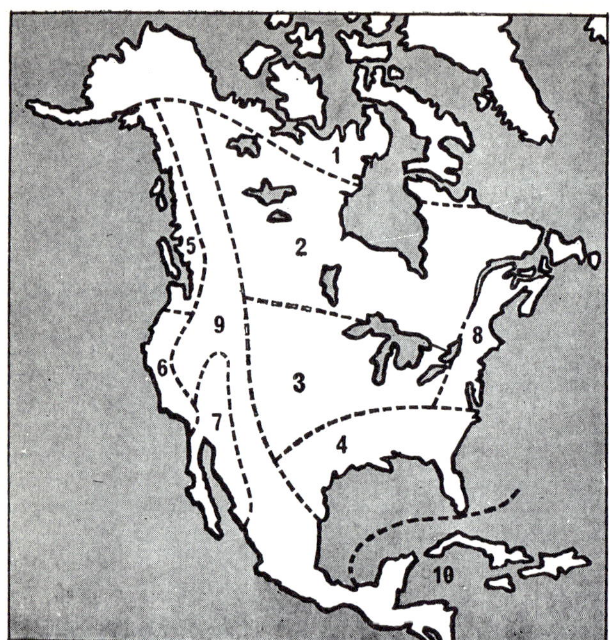

FIG. 154. *Climatic regions of North America.*

owing to cold winters; winter conditions bleak because of cyclonic influence, the cold waves from the north and cold moist air from the Atlantic.

(9) *Rocky Mountains.* Altitude reduces all temperatures below the average for the latitude; cyclonic effects in autumn and summer; marked rain shadows in the valleys.

(10) *Tropical.* High temperatures; small temperature range because of maritime influences; rainfall is high on the windward slopes of the hill regions; a summer maximum of rainfall; hurricanes.

Natural Vegetation

North America has a wide variety of vegetation.

(*a*) *Forest:*

(i) The very thick, rich coniferous forests along the western mountain slopes where rainfall is high, winters mild with no long frost period and summers warm; Douglas fir, pines.

(ii) The *taiga* of the northern part of the continent with its long, cold, snowy winters and short, warm summers.

(iii) Hardwood deciduous forest of oak and maple of the Mississippi, St. Lawrence lowlands and the Appalachian slopes.

(iv) The evergreen cypress and yellow pines of the warmer south-east.

(*b*) *Grasslands of the interior.* In the east, the higher rainfall stimulates the growth of tall grasses; true prairie lies west of 100° W., where the rainfall is reduced to about 20 in.

(*c*) *Tundra* is found in the far north of the continent, where the short summer, long, cold winter and low rainfall stimulates the growth of small plants and some grass but no trees.

(*d*) *Semi-desert* of the dry intermontane plateaux and the peninsula of Lower California, where the vegetation is coarse and leathery.

(*e*) *Swamp conditions* exist in the delta of the Mississippi and in the sub-tropical wet southern tip of Florida where they are known as the Everglades.

(*f*) *The tropical regions* of Mexico and the West Indies have natural vegetation which varies from a form of tropical forest in the lowlands to grasslands on the plateaux.

(*g*) *The eastern end of Cuba and the central American coastlands* have a cover of equatorial rain forest.

QUESTIONS

1. Describe the variations of climate to be found on the West Coast of the continent between the latitudes of 60° N. and 30° N. and explain their occurrence.
2. Explain: (*a*) the fact that Montreal is frozen up in winter while Vancouver is ice-free and (*b*) that the rainfall total over the Mississippi Basin varies with latitude and longitude.
3. Describe and explain the distribution of semi-desert over the continent.

CHAPTER 35

DISTRIBUTION OF POPULATION

THE wide variations in the physical character of the continent lead naturally to corresponding variations of population density. The north of the continent is cold and the south warm, while the 20 in. isohyet runs north to south through the middle of the U.S.A. to give the relatively damp east and the dry west. Most of the people of North America, therefore, live in the south-eastern part, from the Mississippi eastwards and from the St. Lawrence valley southwards, where conditions are favourable for farming and there are fuel and raw materials. Population is least dense in the west where the highlands and its intermontane semi-deserts limit the occupations to those of a pastoral character, with the occasional more-densely-occupied regions of mining and of irrigation farming. In the far north the low temperatures, thin soils and tundra and taiga conditions which follow them, hinder settlement, unless there is mineral wealth.

The area of moderate density in the agricultural southern prairies of Canada and the Middle West of the U.S.A. has patches of higher density superimposed on it where trading and industrial centres have developed. On the West Coast, especially in U.S.A., the intensively irrigated farmland and the new industrial zones have attracted more people.

Three-quarters of Canada's 21 million people (1968) live in the southern part of Quebec and Ontario, particularly in the St. Lawrence Lowlands and the Lake Peninsula of Ontario, where the advantages of hydro-electric power and water communications through the Great Lakes and the St. Lawrence have created the industrial heart. In the Maritime Provinces, people live mainly along the coast but the relatively sparse population (ten per square mile) of the southern prairies is distributed evenly over the agricultural area. Scattered populated centres are found in the southern parts of the Canadian Shield and in British Columbia. In the narrow band of the country in the south, there is an average density of about thirty people per square mile, varying from the 100 per square mile in the Great Lakes-St. Lawrence area to under five per square mile on the High Plains of Alberta. In 1901 the population was largely rural; 37 per cent was urban with only two cities, Montreal and Toronto having more than 100,000 people. In 1968 65 per cent was urban and the big cities had 48 per cent of the people of the country.

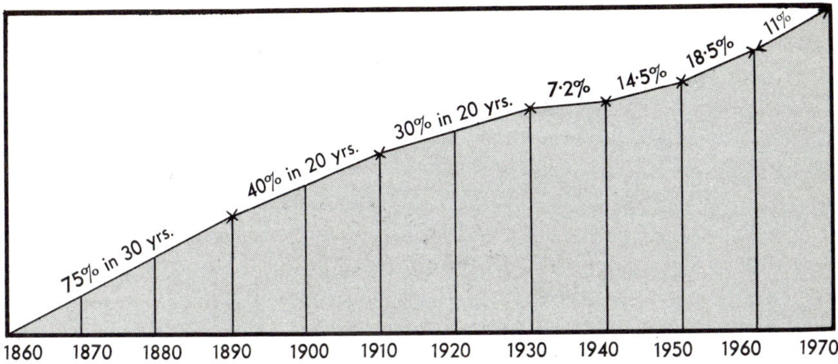

FIG. 155. *Percentage increases in the U.S. population over varying periods.*

Prince Edward Island and Saskatchewan are the only largely rural provinces, but even here the rural character is changing.

The total population of the U.S.A. in 1970 was over 202 million, thus exceeding the expectation of 200 million by 1975. The north-east of the country with its healthy climate and its vast natural resources of timber, fuel and power is the most densely populated. Towards the Mississippi and the south, eastern U.S.A. has evenly populated farmland, and, as in Canada, there are many industrial patches. In the western interior there are numerous areas of high density representing the commercial and mining centres and the same features also appear on the West Coast. In 1960 approximately 70 per cent of the people were classified as urban, but again the rural population is not necessarily identical with the farm population.

Mexico has a population of approximately 49 million with about 24 per cent living in cities, over half of the people being urban. The south is more densely populated than the north where mining and pastoral farming in the drier regions provide the only occupations, *Monterrey* being the only large settlement. In the east the presence of oil and gas has promoted the development of the chief industrial centres and so is much more densely populated.

The population in the West Indies is distributed unevenly, and although most of the islands have an overall density of 250-300 per square mile, the densities for the Bahamas and Barbados are respectively 26 and 1,400.

Population Growth

Once the New World had been discovered, it was natural that, through the increased commercial possibilities, movement from the Old World would occur and that these people would settle on the East Coast. However, the Spaniards, having crossed the narrow Panama neck, settled on the West Coast. After this there were movements westwards along the natural routeways from the newly established centres on the East Coast. In Canada, the St. Lawrence

FIG. 156. *Night view of steel mills in Pittsburg, Pennsylvania, an industrial centre in the densely populated north-east of the U.S.A.*

lowlands were exploited for farming, fishing and lumbering in the period 1845-95 so that the east grew rapidly. Between 1901 and 1913 this area showed no great changes but expansion took place in the west, again with the growth of farming and forestry. Conditions were then reversed between 1914 and 1929 with the industrial growth of the east. More recently, development has occurred on the margins of what has been regarded as the habitable belt. New varieties of wheat, oats and potatoes specially suited to their environments have extended agriculture into the Peace River district and the Clay Belt round Hudson Bay. The exploitation of mineral wealth in North Quebec, Labrador and the northern Cordilleras, and the proliferation of airways, roadways and tractor trails into North-West Territory have stimulated further movements of people as well as the use of new methods of cultivation and pastoral farming. This northward expansion has not been accompanied by any diminution in the sizes of the metropolitan cities of the south. On the contrary, Montreal, Toronto, Hamilton and London are being developed to supply the industrial needs of the north and Edmonton and Vancouver are the bases from which the opening up of the Athabasca, Great Bear and Great Slave regions is taking place.

The rate of increase has averaged 2·1 per cent per annum, greater than that for the U.K. (0·44 per cent), U.S.S.R. (1·14 per cent), U.S.A. (1·36

FIG. 157. *Small township in the farmlands of the Connecticut Valley.*

per cent) and even for the Asian countries, e.g. India (1·25 per cent) or Korea (1·99 per cent). It is estimated that by A.D. 2000 the population of Canada will be 40 million and that Ontario and Quebec will each have 10 million and British Columbia about 4 million with at least half of her population in Vancouver. It is not suggested that the expansion in the north of Canada will necessarily affect the overall distribution of population. It is thought, however, that the total mining and smelting population may reach 150,000 and that for lumbering and farming another 100,000. The new Kitimat in British Columbia might have 50,000 and Venture in the Yukon, 20,000.

In the U.S.A., development was rapid and much more even in direction. Movement along the river valleys preceded the filling up of the land between them. The discovery of gold in California and of the routes to it through the Cordillera opened up the West and eventually the dry High Plains on the eastern side of the mountains. From 1860-90 the source of immigration was mainly N. and W. Europe and their descendants moved westwards. Since then most have come from the interior and southern countries of Europe. By 1850 the movement had reached the states immediately west of the Mississippi and the consolidation was complete by 1880. From 1900-10, only the three Pacific states and Oklahoma and Pennsylvania showed any real increase. From 1920-30, however, the migration was in reverse, Michigan, Ohio, New York, New Jersey, Florida and Illinois showing the advances.

The increase in California was continuous throughout the whole period. There have been similar recent changes in the detailed distribution of population. The greatest numerical increases have taken place in the West, in the Pacific States, and Colorado is growing twice as fast as the nation as a whole. Other significant increases have occurred in Florida, Michigan, Texas, Maryland and Virginia. Corresponding decreases have been recorded in Arkansas, Mississippi, North Dakota and Oklahama, where there has been the decline of the rural functions.

It was very natural that the U.S.A. rather than Canada would receive more immigrants. The development was in terms of farming and the richer soils and the warmer climate of the southerly lowlands attracted the people from the farming interior of Europe. Conditions have changed and nowadays hydro-electric power, timber products and minerals exert a greater influence. The idea that Canada is merely an extension of the U.S.A. is ceasing to hold the truth it once had. British Columbia is no longer a continuation of Washington and Oregon, nor the St. Lawrence Lowlands part of the Central Plains. Canada was quite prepared to go ahead with the St. Lawrence Seaway scheme on her own, and has spent more than twice as much as the U.S.A. on the project.

Native Peoples of North America

Prior to the coming of the Europeans, there were three major groups of native peoples (the *Amerinds*), each probably reaching the continent by way of the Bering Strait. In the extreme north, the *Eskimos* have adapted themselves to the environment through the food available and their ingenuity in the use of tools and methods of movement. They number about 11,500 and are no longer living in scattered communities or as isolated groups of nomads. The introduction of reindeer herds has given them another occupation and source of food. Their inherent mechanical ability is being used on the airfields, and sealing and fishing are carried on from large co-operatively owned motor boats. The kayak is now lashed to the decks of these and the only igloos seen are used by sportsmen from the south and are made of plastic.

The forest belt separates the Eskimos from the second of the native groups, the *Red Indian* who belonged to the plains where the freedom of movement allowed them to hunt the bison. They also inhabited the forest regions and some lived as fishermen along the northern Pacific coasts. The two groups meet on the shore of Hudson Bay and on the Mackenzie Delta at Aklavik. The continent lacked animals which could be domesticated and the only grain was maize east of the Mississippi. The Indian had no knowledge of the plough or of the wheel; they were neither herdsmen nor cultivators. Their lives were controlled by the movement of the bison which gave them food and hides until the advance of the railways and the white hunter's rifle reduced the bison and prepared the land for cattle. Indian reservations are now widespread in the U.S.A. In Canada they are no longer a dying race; indeed they are increasing rapidly and they are seeking employment away from the reserves. Many are working on the new Canadian nuclear power station at Douglas Point on Lake Huron.

The third native group, the *Aztec* of the Mexican Plateau, are sedentary people who knew how to cultivate and irrigate their land before the Spaniards arrived. Interbreeding of Indian and European has created a mestizo population. In the West Indies, the original Caribs have almost disappeared, the bulk of the population being either Negro or mulatto (i.e. mixed Negro and European) and is derived from the descendants of the slaves who were brought to work there as the plantation agriculture of the islands was developed.

QUESTIONS

1. Compare the population densities of the lands immediately surrounding: (*a*) Hudson Bay and (*b*) the Gulf of Mexico. Account for the differences you mention.
2. Find out what you can of the lives of the *Eskimo* and *Aztec* and show how far the differences between them are related to their geographical backgrounds.

CHAPTER 36

THE ST. LAWRENCE APPROACH

CARTIER, Cabot and Champlain made their entries into the continent by the St. Lawrence approach and the Gulf and its surrounding lowlands formed the heart of the new Canada. The comparative friendliness of the Huron tribe of Indians encouraged the setting up of new European settlements on old native sites such as Quebec and Montreal, but the physical character of the river banks offered little opportunity for the growth of big towns. On the north side, highland comes close to the river between Anticosti Is. and Quebec; slopes are steep and unbroken so that the few settlements are on the coastal terraces, or at the mouths of turbulent streams such as Tadousac on the Saguenay River. On the southern shore, the Notre Dame Mts. also rise up sharply leaving only a narrow terrace. There is a high degree of isolation along the banks of the St. Lawrence and although the river is a commercial highway, there are few harbours. The shallow water inshore and the cover of winter ice for five or six months are further difficulties, weighing heavily against the establishment of settlements.

FIG. 158. *Regions:* 1. *Newfoundland and Maritime Provinces.* 2. *St. Lawrence Lowlands:* (*a*) *Gaspé Peninsula;* (*b*) *South shore;* (*c*) *North shore;* (*d*) *Eastern Townships;* (*e*) *Upper St. Lawrence.* 3. *Lake peninsula of Ontario.*

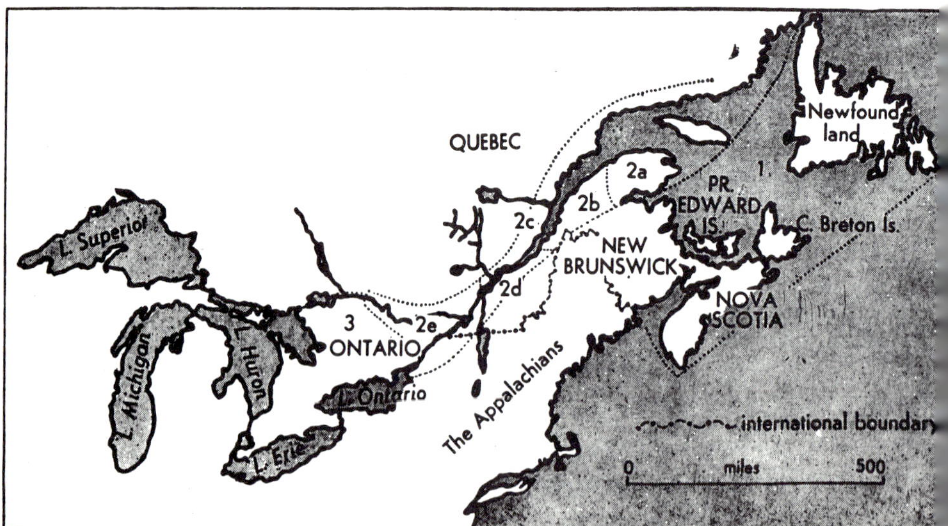

Relief

The region (Figs. 158 and 159) is an extensive depression containing the Great Lakes and the St. Lawrence which connects them to the Atlantic Ocean. The Maritime Provinces and Newfoundland are a northward extension of the Appalachians, and although generally of low elevation form a hilly, flat-topped, dissected plateau which rises into the Notre Dame Mts. and Shick-shock Mts. (4,200 feet) of the almost uninhabited Gaspé Peninsula.

The lowlands of New Brunswick and Cape Breton Island are underlain by little-disturbed Carboniferous rocks to give the Minto and Sydney coal-fields. Triassic Red Sandstòne is found along the coast of the Bay of Fundy and intrusive materials have been responsible for the gold in quartz veins in Nova Scotia. During the Pleistocene period of glaciation, central New Brunswick and the highlands of the Gaspé Peninsula were local gathering grounds of ice, and tongues of the Labrador sheet greatly affected Newfound-land and the eastern Maritime Provinces.

Broadening westwards, the extensive lowland of Quebec stretches along both sides of the river. The south side is often referred to as the "Eastern Townships" region and has much cultivable glacial soil; on the north side the land takes the form of a ridge parallel to the river and rises to a height of about 1,000 feet which marks the edge of the Canadian Shield. Most of the tributaries, e.g. the Saguenay and the St. Maurice, of the St. Lawrence flow south from the Shield through lakes which can serve as reservoirs and are potential sources of water power, the importance of which is difficult to assess. Fingers of good land spread away from the estuarine zone up the valleys of such rivers as the Ottawa and the Richelieu.

FIG. 159. *St. Lawrence approach: structure and minerals.* (*Higher land is shaded.*)

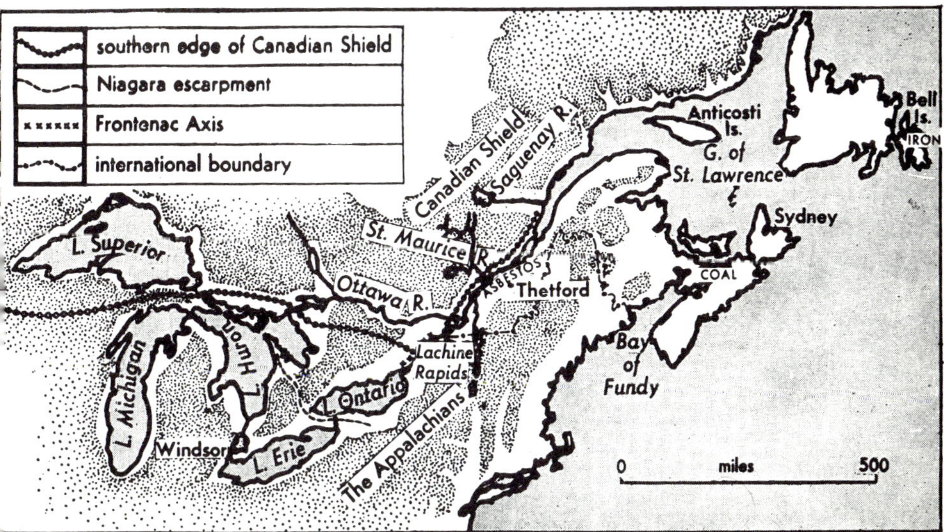

Fifty miles west of the Ottawa, a projection of old rocks called the Frontenac Axis extends from the Canadian Shield through Brockville to Kingston, situated at the eastern end of Lake Ontario, and has caused the very broken river floor marked by the Thousand Isles and by the Lachine Rapids from Lake Ontario to Montreal. It also delimits the extent of marine conditions in the geological history of the area, and thus acts as a boundary between the St. Lawrence Lowlands and the Lake Peninsula of Ontario. The latter is a triangular-shaped area of glacial drift lying between Lakes Eire, Huron and Ontario on Palaeozoic rocks which are almost horizontal. It is crossed by the abrupt, east-facing Niagara escarpment of Silurian limestone and in the south-west, the older rocks have been folded into low domes giving the salt deposits under Windsor and the recently exploited gas and petroleum. The escarpment is cut by the Niagara River to give the falls of the same name.

The lake area was the site of a pre-glacial river basin, draining to the Atlantic and separated by a low watershed from the Mississippi Basin. The Great Lakes occupy hollows caused by the warping of the earth's crust at the close of the Ice Age and which, in the final retreat of the ice, were filled with melt water trapped between the fronts of the lobes of ice and the hills. The process of glacial retreat was irregular; the cutting of overflow channels across the watershed (e.g. the Mohawk depression) alternated with the erosion of the Niagara escarpment, and ended with the cutting of the channel from Lake Champlain. The release of water (in the form of ice) from the land was balanced by the increase of water in the oceans; thus the sea level rose to fill the St. Lawrence depression and left the marine sediments of the lowlands. The present landscape has resulted from a much later uplift.

Climate

The climate of the whole region is a mixture of maritime and continental, for the prevailing westerly and northerly winds carry the continental conditions (and particularly the winter cold waves) to the coast, and south-east winds blowing in front of the depressions bring in sea influences. The influence of the Great Lakes is an additional climatic control and causes, particularly, a marked northward deviation of the winter isotherms, although the three western lakes are never warmed to any great depth. As a group they tend to moderate the conditions of the area and to raise the humidity. In spring the warm south-west winds are cooled by the lake waters giving cloudy, foggy weather and the length of the growing season along the lake shores is also increased. Everywhere, temperature ranges are high.

In the interior there is a marked summer maximum of rainfall, but further east, the Atlantic provides a source of winter rain, rainfall being fairly evenly distributed. Winter precipitation is in the form of snow, much heavier in this region than in the far north. Mean annual totals increase eastwards.

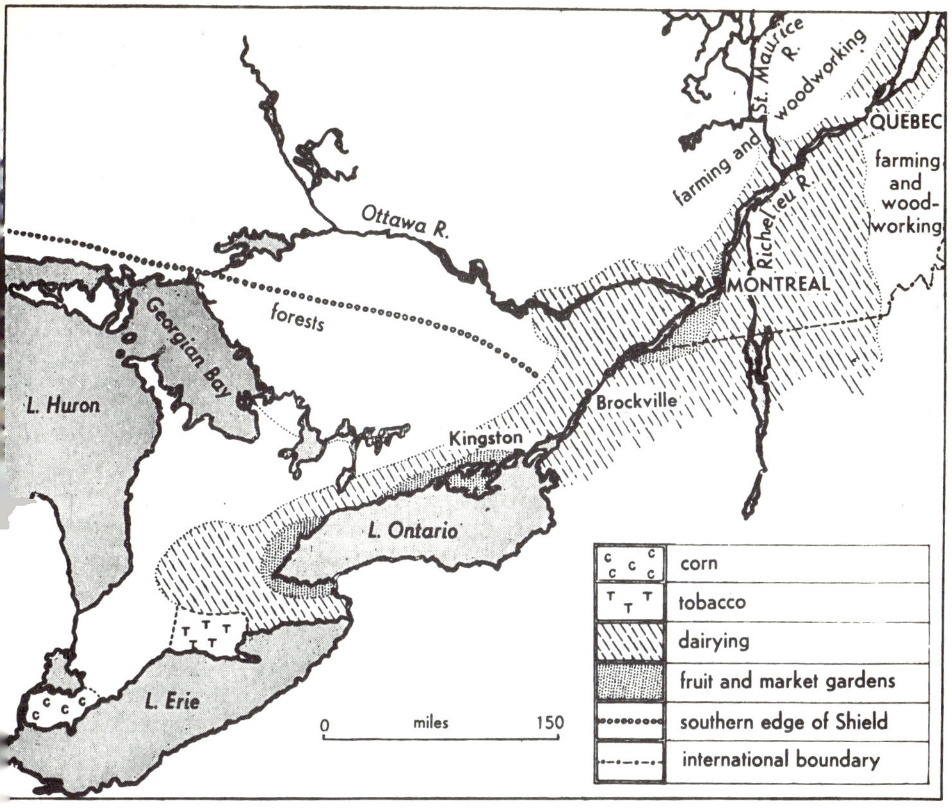

FIG. 160. *Farming zones of the Upper St. Lawrence and Lake Peninsula.*

The long winter is a handicap in the mid-St. Lawrence area; ice from November to March impedes navigation on the river and nearly a quarter of Canada's grain shipments must leave by U.S. ports. Fogs in June cause more difficulties, but to the east the all-seasons rainfall helps to give the rivers of the Maritime Provinces a uniform regime.

Forests

The variation in forest type reflects the rainfall distribution. In Newfoundland and Arcadia the mean annual rainfall (40 in.) is suited to the conifers (spruce and pine) of Nova Scotia, Prince Edward Island and most of New Brunswick; forest of a more mixed character with hemlock, birch, maple and cedar is found in south Nova Scotia and New Brunswick. With the decrease in rainfall westwards the forest character up the St. Lawrence changes slightly, although the conifer predominates; some hardwoods (oak and ash) are found in the east, and south of the river the sugar maple is important industrially. A small northerly intrusion of deciduous forest from the U.S. is present in the Ontario Peninsula, a more fertile area which has encouraged many settlements.

Wooded areas are small and are found only on the lighter soils, with beech and maple, conifers, and hardwoods such as hickory and walnut. In recent years large-scale insect infestation (particularly in the areas of spruce and jackpine) has caused changes in the composition of the forest. The useful trees have had to be replaced by the less valuable, a process which normally takes a minimum of fifty years, although newer methods of reafforestation have reduced this period.

Newfoundland

Newfoundland, an island lying across the mouth of the St. Lawrence, is a low plateau with an extremely irregular coastline consisting of bays, fiords and cliffs reaching 200-400 feet in height. One third of the island is covered by water which has filled up the rock basins. Glaciation has also stripped the surface of soil and only along the lake margins and in the river valleys is there any good farmland. The fast-flowing rivers give a vast supply of water power and the slopes are forested mainly with spruce and fir. Offshore, the broad continental shelf of the Grand Banks provides the province with the site of its very important fishing industry, but the mixing of the air over the cold Labrador current and the Gulf Stream produces frequent coastal fog.

Farming is largely pastoral, and the few crops are hay, potatoes and oats. However, farming and fishing are not carried on together as in crofting communities. The chief products are those from the mines and forests. Iron ore, mined in Bell Island from beneath the sea at Wabana, goes to Sydney in Cape Breton, or Philadelphia for smelting, and lead, zinc, copper and asbestos

FIG. 161. *Oyster fishing in Malpeque Bay, Prince Edward Island.*

FIG. 162. *Rolling farmland in the province of Quebec.*

are obtained from the interior. The pulp mill is the most important new industrial feature; there are two of them, one at Grand Falls and the other, the largest integrated pulp mill in the world, at Corner Brook. American, Canadian and British capital has made use of the extensive spruce and fir forests and the abundant water power.

Maritime Provinces

The Maritime Provinces is a region of farms and factories, separated from the rich lowland and industrial Ontario by much uninhabited land and the St. Lawrence which is frozen for half the year. An all-Canadian railway route from Halifax to Montreal has to make a long detour through southern Quebec and New Brunswick, so that most rail-users travel more directly through Maine.

This northern extension of the Appalachians consists of deeply dissected Palaeozoic rocks with a south-east—north-west trend of valleys, bays and peninsulas. The broad glaciated infertile plateaux are covered with lakes, swamps and forests, and submergence has created Prince Edward and Cape Breton Islands.

Prince Edward Island

Most of Prince Edward Island (the "Garden of the Gulf") is dairy farmland with hay, oats and potatoes as the main crops. The island is separated from New Brunswick by the Northumberland Strait and has some shelter from the Atlantic storms through its position in the great southwards curve of the Gulf. The annual average of 210 frost-free days (compared with

155 of the Quebec coast) and the rainfall of 40 in. have encouraged intensive production of hay. Glaciation here was less intense, so that soils were not removed as elsewhere. The silver fox fur farming has declined considerably since the mid 1930s. The island has important lobster fisheries and oysters are cultivated in Malpeque Bay. Other industries are cheese and butter making and the manufacture of fish-meal and fish-oil by-products.

Nova Scotia

Nova Scotia has advantages in some excellent soils and in her mineral wealth, but the province is not as important as one would expect. With its small harbours and its position on the continental shelf it shares in fisheries (from Halifax, Lunenburg and Digby) and its valley floors provide good farmlands for field crops, cattle, pigs and poultry. River and marine muds at the head of the Bay of Fundy have been dyked and drained for fodder crops and dairying, and pig rearing has developed.

The minerals suggest that the area has all the advantages for integration into one of the important world areas for steel manufacture. Cape Breton has 200 square miles of good coal on tidal water behind Sydney, and thick seams outcrop on the cliffs; more coal is mined in the Minto, Cumberland and Pictou districts. The excellent Newfoundland iron ore, also on tidal water, is near, steel being made at Sydney and Trenton, but the freezing up of the St. Lawrence always prevented the products of the area from being transported into industrial Canada and stock-piling during the winter was necessary. The obvious market for the steel is the Atlantic coast of the U.S.A. but this is not completely accessible because of tariff policies. Some of the steel is exported to the U.K. Other industries include food and beverage preparation, wood and paper products and fish canning. *Halifax*, the largest town in Nova Scotia, is a railway terminal (C.N.R.) and, with its magnificent harbour, is Canada's winter port; it has shipyards and trades with the West Indies, South Africa and Europe.

New Brunswick

New Brunswick is compact and regular in shape and is of low undulating relief. A large lake area provides the power for its saw mills and pulp mills. Although some 3° C. (5° F.) colder in January, *St. John*, its chief port, is like Halifax, ice-free and shares in Canada's winter traffic with that port. The harbour of St. John is situated on the St. John River above a narrow rocky gap with its "reversible falls." The mouth of the river is about 12 feet above low-tide level so at that time it enters the sea by means of falls; at high tide the sea is higher than the river mouth so that the sea falls inland. Only on four occasions a day is the water at the same level on both sides of the gap and therefore smooth enough to allow ships to enter and leave the port.

Population

Prince Edward Island, the only almost completely and uniformly settled area in the Maritime Provinces, is predominantly rural with an agricultural economy, and has only one city, *Charlottetown.* In Cape Breton, the people are essentially town-dwelling with 70 per cent living in Sydney. As one travels southwards into Nova Scotia the coastal population is grouped into small fishing communities with the one large metropolitan centre of Halifax. The interior of this province is sparsely populated and has no urban centres, but towards the isthmus and into New Brunswick, there are a number of small towns acting as market centres for the area. Most of the people live in the major cities of *Fredericton* (the capital), *St. John* and *Moncton.*

The southern shores of the St. Lawrence have a lowland piedmont increasing in width towards the south-west and backed by the Appalachian slopes. Subsistence agriculture is general and there are large saw mills obtaining timber from the plateau where there are scattered forest settlements. Population increases and becomes more urban towards the south-west. In the region known as the Eastern Townships where the hill belt is deeply entrenched by streams and soils are very stony, 40 per cent of the world's asbestos is produced by open-cast mining methods from veins in serpentine. The chief centres are at *Asbestos* and *Thetford Mines* and the bulk of the product goes to the U.S.A. and Great Britain.

Along the river, English-speaking settlers from the U.S.A. follow their traditional farming methods. The marine or glacial soils and old lake sediments found in the valleys have encouraged mixed farming and dairying. Canada's largest sheep-rearing industry is in the northern part of the area; sugar beet is the main crop of the Richelieu Valley, while the maple sugar industry is dominant on the shorelands. *Sherbrooke,* situated in the hills at the meeting place of plateau valleys, is the regional capital of the Eastern Townships and has its own textile and machinery industries. The other industries of the region are concerned with rubber, furniture and tobacco at Granby and Victoriaville.

Similar conditions are found on the other side of the St. Lawrence, although the glacial cover contains more gravel and boulders. Here, the French settler carries out his own traditional methods of farming. Beef cattle and pig rearing are important in the mixed farming of the area, particularly in the Ottawa Valley. Because the Canadian Shield comes very close to the St. Lawrence River, the important forestry industry tends to keep to the bank and to the valleys of the Saguenay, St. Maurice and Ottawa, with *Three Rivers*, *Shawinigan*, *Gatineau* and *Hull* as the leading centres of production. The most important factor is the ease with which hydro-electricity can be generated where rivers have steep gradients as they drop over the faulted edge of the Shield; Quebec and Ontario have no coal.

The Upper St. Lawrence region lies mainly between the Ottawa and the St. Lawrence in eastern Ontario with the remainder in New York State. The Ottawa River has limited navigability but it is important as a source of power and in the transport of logs. Eastern Ontario has an important dairy belt supplying Ottawa and Montreal with cheese manufacture being dominant. The regional capital, Ottawa, was at one time the most important centre of lumbering in Canada, developing largely as a result of the cutting of the Rideau Canal. With its twin town of Hull, it still has a great output. The same activities are found on the American side of the river.

The St. Lawrence

From Newfoundland to Lake Ontario the connecting link of all the regions is the St. Lawrence, the course of which has four sections:

(i) upstream to Quebec, the first bridge being ten miles above the city;

(ii) Quebec to Three Rivers, the tidal limit of the St. Lawrence and a port as well as an industrial town;

(iii) through the shallow Lake St. Peter and the delta of the Sorel to Montreal;

(iv) Montreal to Lake Ontario, the section of the Thousand Isles, the Lachine Rapids and the new Seaway.

Quebec City, a regional capital, is the commercial centre for the estuary; it has some value as a port, with modern shipyards and dry docks. **Montreal,** also a regional capital, stands at the head of navigation on the St. Lawrence, where the Lachine Rapids begin. Here the river branches to enclose islands and the city developed on the left bank of the main stream. Montreal has the immediate hinterland of the St. Lawrence, Ottawa and Richelieu valleys. It handles traffic from the Lakes Peninsula and the Prairies. Although Montreal's

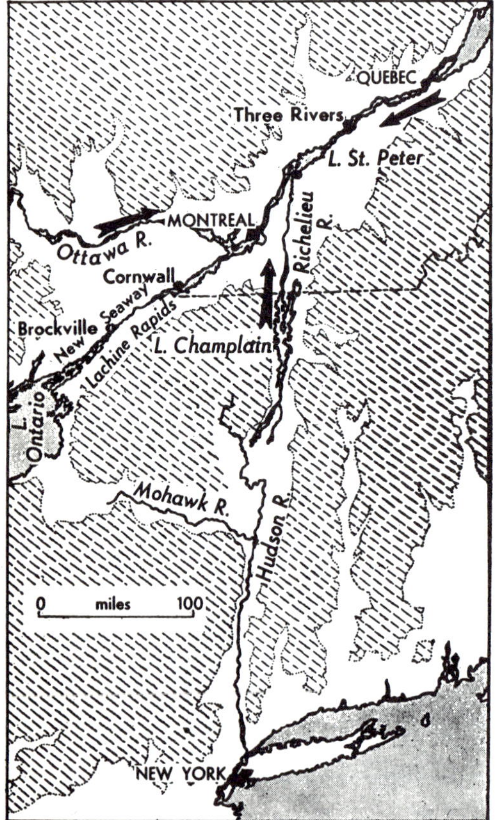

FIG. 163. *Position of Montreal. The higher land is shaded.*

FIG. 164. *The Iroquois Dam, a key structure in the St. Lawrence Seaway.*

exports are bulky, e.g. newsprint, wood pulp and grain, it has the advantage of being 300 miles nearer to Liverpool than New York. The four to five months freeze-up every year is its greatest handicap.

The St. Lawrence Seaway scheme was an attempt to solve the combination of two problems, the need for more power and the need for improved navigation. The towns of the low flat plains of Eastern Ontario formerly imported coal in order to produce their electricity; the freighters of the great Lakes drawing 25 feet could not use the Lachine Canals with a depth of 14 feet. The Seaway consists of new canals above Montreal while it has associated dams and power stations at *Cornwall* and *Beauharnois.* Although the scheme has increased the traffic in and out of Lake Ontario and thus benefited the Lake Peninsula, the one-way locks on the Welland Canal still impose a limitation and the channels west of Lake Erie will have to be deepened if Chicago and Detroit are to derive the benefit forecasted for them. Furthermore the season for lake traffic can last only eight months because of ice. The great value of the Seaway will probably be found in the transhipment westwards of iron ore from the new sources in Labrador and Venezuela to the sites of the Canadian and American steel industries, and in general cargoes eastwards. This suggests an increase of commercial traffic between U.S.A. and Canada rather than in world trade from the interior of North America.

The Lake Peninsula. This is bisected by the Niagara Escarpment, most of the area having a cover of varied glacial debris. Flat plains of clays, sands and silts line the shores of the lakes, where materials were deposited in the flooded depressions as the ice retreated. This is an important agricultural region. Market gardening is to be found everywhere close to the shores, and tobacco, fruit, flax, sugar beet, dairy pasture and maize give valuable yields.

Manufacturing Industries. Not only the presence of hydro-electric power, but also the fact that it borders the Great Lakes and therefore lies on a major trade route, has made Southern Ontario an important manufacturing region producing probably half the Canadian consumer goods. These include motor vehicles, iron and steel, rubber goods, implements and machinery of all kinds, furniture and food products (particularly meat, dairy and bakery products). The iron and steel industry of *Hamilton* is centrally situated to use the ore from the Lake Superior ports, coal from the Appalachian fields and limestone quarried from the escarpment, while the products are easily distributed throughout the region. Motor vehicles are made at *Windsor* opposite *Detroit*, *Oshawa* (originally important for making buggies) and Hamilton. Furniture and woodworking are widespread, but brickworks are close to the large cities. With the discovery of oilfields in Alberta, the production of petroleum has become a major Canadian industry, most of the crude oil now being refined at *Sarnia*.

Toronto. The most important town in the Lake Peninsula is Toronto, the capital of Ontario, with a harbour, a water front on Lake Ontario, railway facilities and many factories. The city, situated where the Don River enters the lake, has been built and developed on an old lake bed. The harbour is protected by the sandy hook of Toronto Island, the clays, sands and gravels providing the building materials. This is the best example of urban development in Canada.

QUESTIONS

1. Until the 1920s, Newfoundland was called "an empty shell with a fringe of shacks." Comment on the validity of this statement and consider the changes in its economy which have taken place since then.
2. Contrast the sites, positions and functions of *Quebec* and *Montreal.*
3. Explain the present industrial importance of Ontario and Quebec provinces within Canada as shown by: "Ontario has one-half and Quebec one-third of the total value of goods manufactured in Canada."

CHAPTER 37

THE CANADIAN SHIELD

NORTHWARDS from the St. Lawrence Lowlands, the Shield, covering half of Canada is a completely different type of physical and economic region. The fertile plains of the lowlands with their prosperous agriculture and their advantages for manufacturing give way to a plateau between 1,000 feet and 2,000 feet high surmounted by innumerable small hills rising from 200 feet to 300 feet above the plateau level and covered by irregularly shaped lakes and swamps and a very indefinite drainage. The cities of the lowlands are replaced by tiny scattered mining townships, separated from each other by vast forest areas.

The Shield is lowest around Hudson Bay, where the lowlands extend some 350 miles inland over a flat, poorly drained area. It then rises away from the Bay in all directions before ending in a steep edge overlooking the St. Lawrence Lowlands. Few signs of outcropping Pre-Cambrian rock are apparent but the very ancient ranges of mountains show much intrusive volcanic activity which accompanied extensive folding and contortion, before being eroded into a peneplain. Later the Shield was covered with Pleistocene ice and scraped bare of soil. When the ice retreated, the lowlands were smothered in drift and many of the valleys filled in; the numerous hollows which were left today form lakes. Thus the Shield is not level and appears as a succession of steep-sided rock ridges with lakes and swamps, some of which are being slowly filled with peat. Towards Hudson Bay, there is a large area of undulating, or gently rolling, moraine or boulder clay plains called the Clay Belt.

One of the most important effects of glaciation was the disorganization of the surface drainage. Generally the rivers, such as the Nelson and Albany, drain towards the Hudson and James Bays and almost all of them have steep banks; but in Labrador, drainage is very indefinite. Although at first sight streams appear to flow from coast to coast, their sources are usually in lakes occupying hollows in a comparatively flat plateau. Most of those which follow the general slope of the Shield flow from lake to lake in a series of spillways, but on reaching the lowland their gentle gradient is broken by outcropping rock ledges, which give rise to rapids and falls and hence to potential hydro-electric power. The water divide is low and is close to the Great Lakes-

St. Lawrence Lowlands. Rivers such as the Saguenay and St. Maurice are comparatively short, flowing from the lake area in the interior, over many falls and rapids, into long, deep fiords.

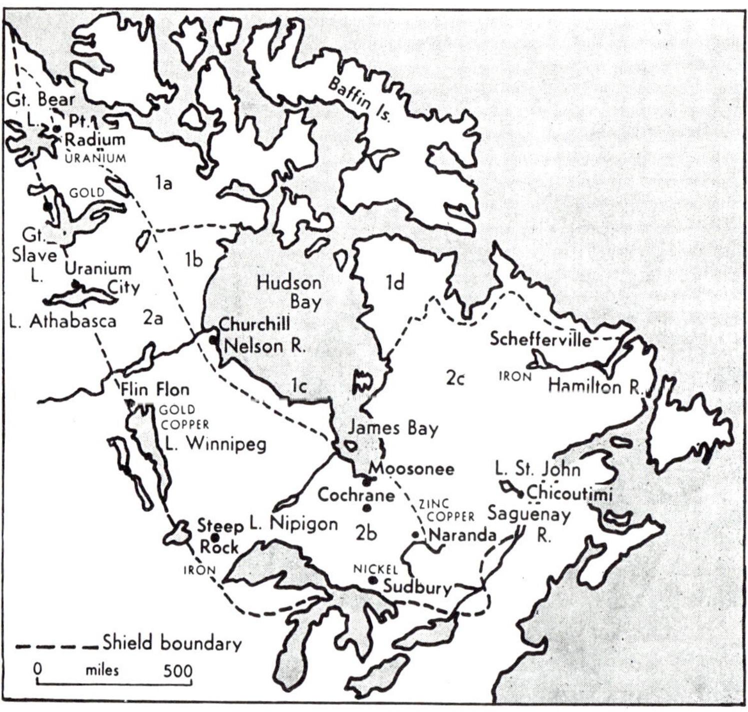

FIG. 165. *Canadian Shield. Sub-divisions:* 1*a. Arctic mainland.* 1*b. Keewatin.* 1*c. Hudson Bay Lowland.* 1*d. Ungava* (*New Quebec*). 2*a. North and marginal lake area.* 2*b. N. Ontario.* 2*c. Quebec.*

Climate

The interior is cold with short warm summers followed by very cold winters; the temperatures drop uniformly to the Hudson Bay Lowland where the average January temperature is −26° C. (−15° F.), and for July 13° C. (56° F.). At Cochrane in the Clay Belt, the July temperature of 19° C. (66° F.) is as high as that of Montreal, but the frost-free period is only half as long. The Arctic coastlands have a very cold winter and temperatures never rise above 10° C. (50° F.). The surface soils freeze in winter and thaw out in the summer, but below these there is *permafrost*, sometimes as much as a foot deep. Rainfall, everywhere with a summer maximum, decreases from south to north, with approximately 30 in. as the mean annual rainfall for Sudbury, 26 in. at Cochrane and under 10 in. on the Arctic coast. In the south, depressions are responsible for the rain and the snow; in the north, the precipitation is mainly snow and the polar air is so dry that there is very little even of that.

Natural Vegetation

The natural vegetation follows the normal pattern. In the far north, low temperatures and slight rainfall prevents the growth of trees; the Arctic lowlands is, therefore, a belt of tundra extending round the Hudson Bay and into the Ungava coastlands as far as the Hamilton River. The rest of the Shield is covered by coniferous forest suited to the cool temperate eastern margin climate and the grey infertile podsols.

Occupations

Development of the Shield has been influenced by climate and geology. The former has fixed the limits of the forest and helped to instal fur-trapping and trading. The fur trade of the Hudson Bay Company and its settlement at Moose Factory had stimulated the north-south movement across the Shield as early as 1642, for the pelts were transported along the Ottawa River to Montreal. Forests provide an ideal environment for mice, lemmings, rabbits and birds—the prey of many fur-bearing animals; beaver and muskrat feed on tree bark and roots. Although the cutting back of the forest for pulpwood has forced the animals further northward there has been no serious decline in their population. Lumbering is most important in the south of the Shield where the forest thins out towards Hudson Bay, and the pulp industries are based on new plantations. Here, too, there has been a rapid increase in the generation and use of hydro-electric power.

Farming. On the other hand, the climate is of little help to agriculture and the future is rather uncertain. There is little soil of any depth except in some of the old lake beds and the clays are cold and not well suited to plant life. Farming is also limited by the short growing season: the many hours of sunshine in the summer are offset by the short frost-free period. In Southern Ontario the cutting back of forests has released land and after the soils have been improved by open drains or by lightening the clay, crops of hay, barley and roots are grown as fodder for dairying, and vegetable growing is also important. On the Clay Belt and around Lake St. John, farming has been influenced by the areas of fertile lake muds and is mixed and dairy in character. However, the difficulties here are so great that very few are full-time farmers and logging is more important.

Mining is much more important than farming, the Shield being one of the richest areas in the world for hard rock minerals. However there is no coal, except for some lignite south of James Bay, thus restricting industrial development. The minerals are found in the forest belt in three subregions:

FIG. 166. *Schefferville, the chief mineral centre of Labrador.*

(*a*) Along the western edge of the Shield at its contact with the Mackenzie Valley. The settlements are mainly scattered small mining and fur-trading "towns" of 3,000-4,000 people, the towns are situated on rivers and lakes where there is some means of transport available as well as the minerals. Recent expansion has been made possible by improved air transport.

(*b*) The southern region has an extensive mineral belt which, although handicapped by early inaccessibility, has greatly developed with the construction of railways.

(*c*) Labrador has high-grade haematite ores lying below a very thin glacial drift cover in a north to south trough from Ungava Bay to Hamilton Creek. The chief centre is *Schefferville* on Knob Lake, linked by rail to Sept Iles on the St. Lawrence for summer transport of the ore. Freer transport conditions of the new Seaway have made it as cheap to move this ore to the Lake ports from Labrador as from the longer established Mesabi region.

Hydro-electric Power

The surface features of the Shield provide ideal conditions for abundant water power resources, which offset the lack of coal and petroleum. There are potentialities in the rivers flowing into Hudson Bay and the station at Abitibi Canyon supplies power to Cochrane, Porcupine and Sudbury. The rivers flowing over the southern edge of the Shield into the Great Lakes and the St. Lawrence have provided power locally for a long time, but the Otto Holden station on the Ottawa River now supplies places as far away as Toronto. In Quebec the Bersimis River supplies power to the Gaspé Peninsula by underwater cable, while the St. Maurice and Saguenay Rivers carry many stations.

Growth of Settlements

The Lake St. John and the Saguenay region is a small occupied area of 1,500 square miles, with nearly a quarter of a million people, in the midst of 41,000 square miles of rugged forested plateau country. A flat-floored clay plain surrounds Lake St. John and is drained south-eastwards to the St. Lawrence by the Saguenay River.

There were sporadic logging operations in the region as early as 1725. Farming settlements were first established around Lake St. John in 1849, largely by people of Norman descent, but they were very isolated and except for the forest workers there was no market for their produce. The land is at present given over in equal proportions to pasture, hay and oats, but the region has to import butter and eggs. The saw-mill industry thrived till the end of the nineteenth century when it was supplanted by the mechanical pulp industry and by metallurgical works utilizing the cheap hydro-electric power.

Arvida is a city with all modern amenities; railways have been constructed and Port Alfred has been built as a deep sea port so that bauxite from

FIG. 167. *Gold mines at Yellowknife, North-West Territories.*

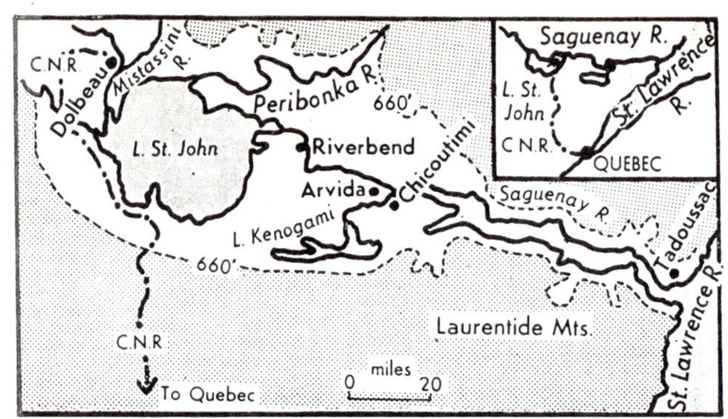

FIG. 168. *Saguenay River region.*

Guyana can be handled easily for refining in an area where power is very cheap. The many lakes provide natural reservoirs collecting water from the large catchment basins. The second harbour is *Chicoutimi* with its furniture and woollen factories and saw mills. Together with its neighbouring townships it forms a compact industrial area (Fig. 168).

Tundra

The tundra lowland shores of the Arctic, Hudson Bay and Ungava are, as we have already seen, sparsely populated. Along the Arctic coast, Eskimoes and Indians live in small scattered groups in addition to the few white prospectors, traders and social workers. In the west, the contact of the natives with the whites has influenced their lives to a marked degree but in the centre and particularly in the Boothia Peninsula they are still very primitive, existing on fish and caribou. In Keewatin and Ungava the natives are specialist hunters and trappers relying upon imported foods. The people of Ungava, mostly Indian and Eskimo, were generally nomadic, but, as in the west, their isolation has been ended by the iron ore discoveries on Knob Lake.

Churchill, on Hudson Bay, and Moosonee, on James Bay, have railways to the more important areas further south. The Hudson Bay railway from The Pas on the Saskatchewan River to Churchill passes through part of the valley of Nelson River and was intended to provide a shorter and cheaper route from the Prairies to Britain. The port is ice-free during August and September and it has a very large grain elevator.

QUESTIONS

1. Discuss the extent to which the opening up of the new Seaway from Montreal to Lake Ontario is likely to benefit the development of the Canadian Shield.
2. Describe the importance of the Canadian Shield as a mining area and indicate the ways in which its geographical conditions make exploitation of the minerals difficult. Add a brief note on the methods by which these difficulties are being overcome.

CHAPTER 38

THE CENTRAL LOWLANDS

THE Central Lowlands, stretching from the Arctic Ocean to the Gulf of Mexico, consist of gently folded sedimentary rocks covered with a varied mantle of glacial drift, river mud and wind deposits. They include the drainage areas of the Mackenzie to the Arctic, the Saskatchewan system to Lake Winnipeg and thence to Hudson Bay, the Great Lakes and the St. Lawrence, and finally the Mississippi.

Mackenzie Lowlands

This northern, rather narrow, projecting subregion of the Central Lowlands is by no means a uniform zone; the Franklin Mts. split it into two parallel sections with the main river flowing in the western branch. It is an undulating area of glacial sands, gravels and clays, the deposits becoming coarser in the north. The delta is an extensive area of shallow water, interlaced channels and sand bars and it is slowly building northwards, but the river can be considered as being continuously navigable upstream to Fort Smith, as far as the rapids of Good Hope. The regime of the river is irregular; in summer it is very shallow, but because it flows northwards, the delta freezes up earlier and thaws later than the upper reaches resulting in extensive flooding along the course in autumn and spring. The coastal fringes are Arctic in character and most of the region has a sub-arctic climate. The delta is a region of tundra, but much of the upper Mackenzie Valley is timber-covered. However, only in the south-west, where jackpine is obtained, is there any real commercial lumbering. The forests are retaining their fur-trapping value. Reindeer (originally herded from Alaska by Lapps in 1929) feed on the Arctic pastures east of the delta and their presence has encouraged some settled communities of Eskimoes. Areas of cultivation are few and widely scattered in forest clearings on the river terraces.

The soils are deeper than those on the Shield and the daylight is long enough in the summer to give a rapid growth of crops. Large yields of potatoes are obtained and vegetables and fodder for dairy cattle (hay, clover, alfalfa and oats cut green) are grown. Permafrost retards farming and prevents surface drainage. Water-filled depressions become mosquito-infested thus preventing the full use of the pastures by caribou and reindeer. Although there is some farming in the Peace River district (a pioneer fringe) and an experimental agricultural station at Fort Simpson, there appears to be little

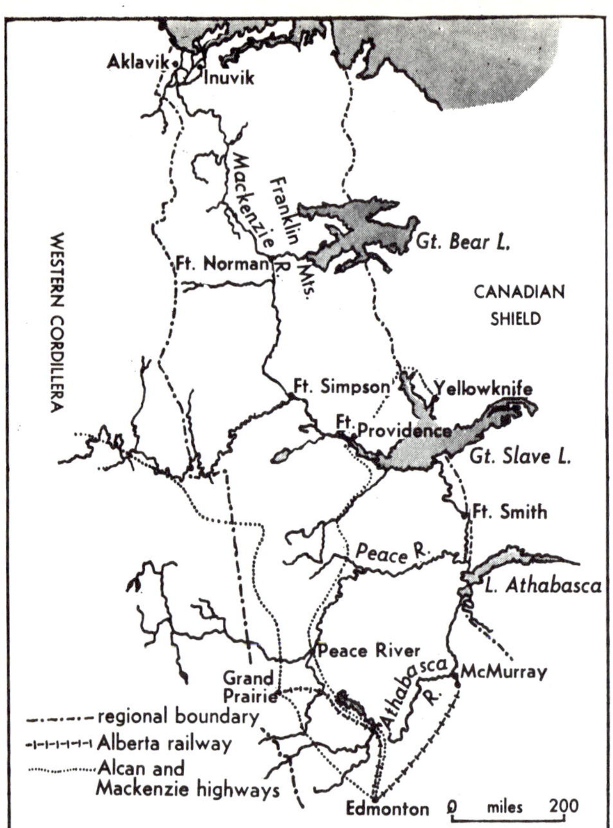

FIG. 169. *Mackenzie Lowlands.*

prospect of the region being developed as an agricultural area. The length of the growing period is too short but the main reason is the persistence of excellent harvests on the Prairies. Farmers have little incentive to go further north where transport is difficult and there is no convenient market for their produce.

Population growth has been slow but steady and settlements are found along the rivers and lakes where water transport links them. Despite the shallows and the rapids, the river is used in the summer by diesel-powered tugs, and when it is frozen, from September to May, its valley is followed by pack train. Mining equipment, precious metals, furs and fish are either tractor-hauled or carried out by air. The traditional dog sledge is still used. The region is not completely isolated, for the Mackenzie Highway starts from Edmonton and reaches Yellowknife via the Peace and Hay Rivers.

Aklavik, the original delta settlement, was built on low ground in the zone of deep permafrost. Subsidence caused by the thawing out of the silt through domestic heat, together with increased lateral erosion by the Mackenzie has necessitated the construction of a new town, *Inuvik*. This has been built on stilts on a higher bluff overlooking a more navigable distributary.

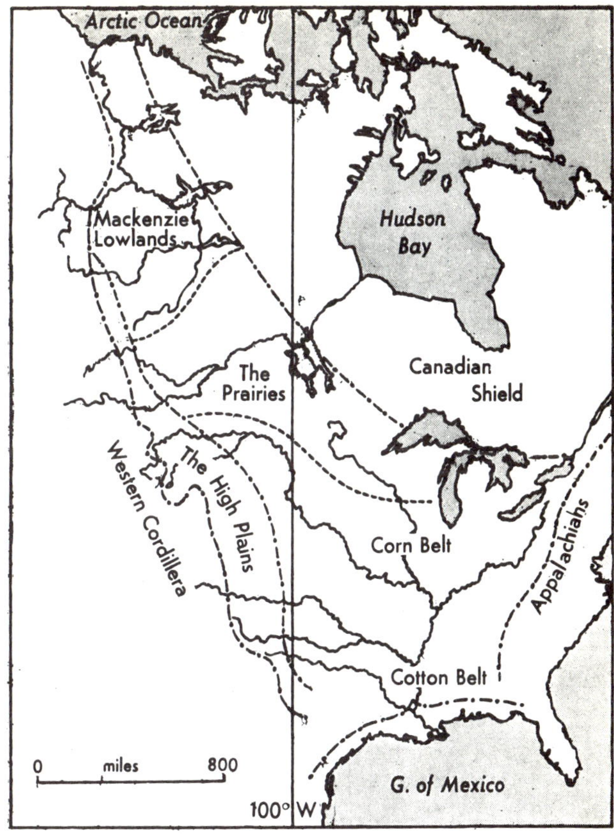

FIG. 170. *Central Lowlands.*

Prairies

The Prairies, a major part of the Central Lowlands, are reached from the north by crossing a series of low divides separating the Athabasca, Upper Churchill (Beaver) and North Saskatchewan Rivers (Fig. 173). They are undulating plains crossed by two escarpments, the Manitoba Scarp and the Missouri Coteau. These divide the plains longitudinally into three separate terraces or steps increasing in height from east to west.

(*a*) *The eastern step* is the South Manitoba Lowland and extends by way of the Red River Valley into North Dakota and Minnesota. Fertile alluvial clay deposits were left as large lakes (e.g. Lake Agassiz) in front of the melting ice sheet leaving the present Lakes Winnipeg, Winnipegosis and Manitoba as remnants.

(*b*) *The Central Prairie Plain* is mainly the dip slope of the Manitoba Scarp and is a terrace of undulating land rising gently westward.

(*c*) *The High Plains* of the west consist mainly of unconsolidated materials brought down by the Saskatchewan rivers and the tributaries of the Missouri. The plains are deeply dissected producing "Badlands" topography best seen in South Dakota and south Alberta.

FIG. 171. *Lake Louise, Alberta. Note the U-shaped valley.*

The core of the Prairies in the southern parts of Manitoba and Saskatchewan and extending across the border is true temperate grassland. The soil is deep, dark brown in colour and rich in humus. Winters are cold but summers are warm with a July mean of 19° C. (66° F.) and a growing season of 110 days. Rainfall, everywhere between 15 and 20 in. a year, is cyclonic, occurring mainly in summer. Summer thunderstorms occur when warm moist air from the south is drawn into the low-pressure centre over the heated plains. The temperatures are higher in the south and the growing season increases to 120-140 days. In the "Badlands" of both Canada and U.S.A., rainfall is much lower (12 in.) and vegetation is mainly semi-desert.

On the east, north and west of the grasslands, rainfall is slightly heavier and the soils are deep, dark brown and more nitrogenous than the typical Prairie soils. Here the natural vegetation is parkland with a mixture of tall grass and deciduous trees. Further north still, less grass and more trees, first deciduous and then high-grade coniferous, grow on the poorer soils.

Colonization. The first settlers in the Canadian Prairies approached the region from Hudson Bay and reached the Red River Valley. Early adverse reports on the possibilities of using the land to the west because of rainfall conditions restricted colonization, but later a new direct railway route across the grasslands with branch lines into the Park Land encouraged rapid settlement. The introduction of modern farming machinery, the soil character, the summer rainfall and the fact that there was no need for forest clearance contributed to the expansion of spring wheat production.

FIG. 172. *Grain truck being loaded from a combine in Saskatchewan.*

Farming. Wheat occupies about half of the total cropland in the grassland zone, but in the Park Land belt round it oats and barley may be more important than wheat. In the wheatlands, change from the early monoculture to the present mixed system has been forced on to the farmer through overploughing, soil erosion, drought, rust, grasshopper plagues and world economic conditions. Under the Prairie Farm Rehabilitation Act of 1935 a proper rotation of crops was introduced and farm practices improved. Disease-resistant wheats were developed, soil erosion reduced by tree planting and light land unsuitable for crops became pasture. Small water-conservation schemes and irrigation works have also been established where the lack of rain made them necessary. Limited irrigation has enabled green vegetables, soft fruits, soya bean, alfalfa, tobacco and maize to be grown. Mixed farming in the Park Lands covers the rearing of dairy and beef cattle, pigs and poultry, and the growing of root crops, lucerne and hay for winter fodder. Sugar beet is also important. Here rainfall is up to 20 in. a year and the belt supports a relatively dense population.

Spring Wheat Belt. The Canadian Prairies are continued into U.S.A. The farms are again large and there is a high degree of mechanization. The same subsidiary crops and dairying with the same methods of mixed farming are replacing the wheat and fallow monoculture and the same battle against drought and soil erosion is repeated here. East of the Spring Wheat belt is a region of hay production and the most concentrated area of dairying in U.S.A. Creameries, and butter and cheese factories take

OCL/GEO 2/Z

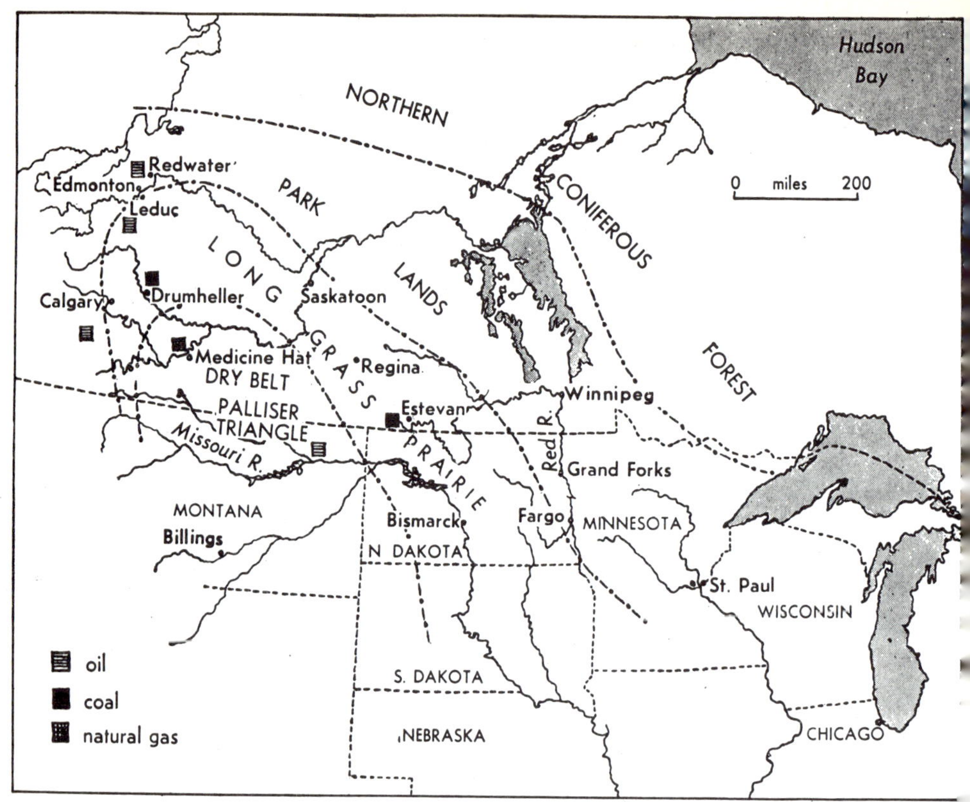

FIG. 173. *Simplified map of the Prairies.*

the place of grain elevators and very large quantities of liquid milk are supplied to surrounding areas.

While the wheat farmers of Canada were settling in the true grasslands, cattle ranchers occupied the drier parts of the High Plains of Alberta, where the brown soils give rise to the "short grass" Prairie, and early clearance of snow by the Chinook helps to reduce the length of the winter feed period. The land is generally unsuited to crop cultivation. The short grass plains continue into western Dakota and Nebraska, and extensive beef cattle ranching is to be found there also. It was from here that the first beef cattle were driven into Alberta, but whilst agriculture and ranching were spreading in Canada together, agriculture followed ranching in the U.S.A.

Minerals (Fig. 173). The Prairies are undergoing another phase of development through the exploitation of the coal, oil and natural gas fields. The coal mined at *Lethbridge* and *Drumheller* in Alberta is mainly bituminous, while some Tertiary lignite is obtained at *Estevan* in Saskatchewan. The latter is used in power stations but with the change from coal to diesel oil to run both these and the railways, there seems little future for the coal of the area. The oil has a much greater potential value for Canada, for in spite of the fact that production is only a comparatively small fraction of the total North American output, reserves appear to be large. Extraction of petroleum from the very considerable deposits of oil-shale is still in the experimental stage.

Economic Development. This Prairies region has no definite boundary except where it abuts on to the Canadian Shield. In the north and west it blurs into the foothills of the Rockies and the High Plains; in the south it is an extension of the western part of the Middle West of the U.S.A. During the first half of this century it had been completely settled by farming communities of all kinds, but now urbanization is taking place. Earlier processing industries of the farm products are being expanded and further industrialization is using the coal, petroleum and other mineral resources; the small outposts and fur-trading centres have become regional capitals. In Canada, *Winnipeg*, *Regina* and *Edmonton* (provincial capitals), *Calgary*, and the smaller urban settlements of *Saskatoon*, *Moose Jaw* and *Brandon* are centres of trade.

Winnipeg is the chief distributing centre for the wheat of Canadian Prairies, the commercial centre of the Park Land mixed farming. The city is still a fur-trading centre and is the headquarters of the Hudson Bay Company. The town has a large number of light industries. Winnipeg began as a canoe junction of the Red and Assiniboine Rivers and as a meeting place of a north to south waterway and an east to west land route. Its original site as Fort Garry on a low hill overlooking the river confluence was good but it has mainly spread on to the valley plains which are easily flooded. The modern character of Winnipeg resulted from the way in which it developed as a railway focus.

Edmonton is now larger than Winnipeg and is the terminus of the new route to the north as well as controlling the route across the Rockies through the Yellowhead Pass. There are also petroleum and natural gas deposits.

Across the border there is an absence of large cities except for the twin cities *Minneapolis* and *St. Paul.* At the former, flour milling was important utilizing the water power of St. Anthony's Falls on the Mississippi; St. Paul was head of the river navigation.

Population. The people, although mixed, are mainly of northern European stock. In Canada, most came from the west and were largely British with some French. Later, Slav immigration occurred and small minority groups were established with their own villages and religions. In the American part of the region, the core again came from the west of Europe but there was a stronger Scandinavian and Teutonic background.

Corn Belt

There is no well-defined boundary to any subregion of the Central Lowlands. The Spring Wheat zone of the Canadian Prairies and Minnesota merges gradually eastward into the hay and dairying belt of Wisconsin and the shore of Lake Michigan. Southwards to St. Louis and, stretching from the Missouri to the Ohio, is the Corn Belt, covering the eight states of the Mid-West with Iowa, Illinois, Indiana and Ohio as its heart. This, the most important agricultural part of the U.S.A., produces half of the maize of the country and large quantities of oats, hay, soya beans, wheat and vegetables. Here, there is a growing season of about 150 days, a mean summer temperature of 23°-25° C. (74°-79° F.) with warm nights, and an annual rainfall of 30-40 in. with a slight summer maximum.

The brown glacial soils, very sticky when wet, and the loess have lost much of their original fertility, but there has been intensive use of artificial fertilizers and natural manures, and the adoption of various crop rotations, particularly maize, grass and grain. The farmers do not rely so much on the marketing of their crops as on the livestock products. Most of the maize, because it is so valuable as a fattener, is fed to the pigs, cattle and sheep reared on the farms within the Corn Belt as well as those brought in from the High Plains.

Winter wheat is also an important crop in this belt, but in Kansas in the south-west, where rainfall is lower, it becomes the dominant crop. In the southern part of the State there is some use of irrigation for vegetables, alfalfa and fruit. Along the Missouri and Ohio margins, the glacial soils with their associated fertility disappear, as does the intensity of cultivation; general farming appears in a broad zone. Across the Ohio into the Bluegrass region of Kentucky, with its centre at Lexington, and the basin of the Tennessee, particularly round Nashville, the soils which result from the weathering of limestone of the Cumberland Plateau are richer; horses and dairy cattle are reared and tobacco is a very important cash crop.

Population Growth. As in the north, the settlement unit in the Corn Belt and the surrounding area is the farmstead. As the westward movement of people and the opening up of land which accompanied it took place, the distance between families became greater. There is thus no "village" in the Mid-West, but there are many towns and small metropolitan areas, such as St. Louis, Chicago and Kansas City. The region had no waterways except the Mississippi, and roads (and later railways) were able to cross the level plains without having to pass through any one particular focus. The towns were the markets and social centres of the farm groups, and were only manufacturing centres in that they satisfied the local requirements, e.g. bricks, pottery and farm implements.

Minerals. High-grade bituminous coal is mined from the eastern interior coalfield of Illinois, Indiana and across the Ohio into Kentucky, and the western interior coalfield of Iowa and Missouri into Oklahoma. In the same regions there are oilfields; more electric power is being made available contributing towards increased manufacture in some of the urban areas.

FIG. 174. *Corn and cotton in the Mississippi Basin.* 1. *The black waxy soils of Texas.* 2. *S. Texas.* 3. *Yazoo and Red River Valleys.* 4. *Alabama.* 5. *High Plains of Texas and Oklahoma.*

Chicago is more than the economic capital of the Corn Belt; its position on the shore of Lake Michigan makes it the natural capital of the whole of the Mid-West. It is the market centre for every kind of produce from the rural areas and a manufacturing centre. Like Winnipeg, it was at first a fur-trading centre at the natural junction of the road from east to west and the waterway of Lake Michigan and the Mississippi from north to south. There was also the advantage of easy portage across the morainic ridge south of the lake. Chicago developed with the coming of the steamboats to the Great Lakes, which carried settlers westwards and grain and animals eastward; it grew still more when the railways followed the old roads round the southern end of Lake Michigan. A canal in a meltwater channel carries barges from Lake Michigan to the Illinois and thence via the Mississippi to New Orleans. Chicago manufactures machinery, clothing, foodstuffs and is the largest meat-packing and canning centre in the world.

St. Louis started as a fur-trading post near the junction of the Mississippi and Missouri, but on high ground away from the effects of flooding if either of the rivers should be ponded back by the other. The city developed as a port and trading centre because of its river location, and as a road, and later as a railway, centre because it was the point from which the settlers fanned out, especially to the south-west after crossing the Mississippi. St. Louis is the collecting centre of Illinois coal and Lake Superior iron ore and has an iron and steel industry in addition to the manufacture of hardware, chemicals and leather goods, as well as stockyards and meat-packing plant.

Cotton Belt

Relief. The main physical feature of this region is the Mississippi flood plain with a cover of alluvium laid down by the river. The Mississippi and its tributaries flow through the region in a series of wide meanders, with oxbow lakes still visible in places, and with artificially reinforced levees. The edges are marked by bluffs occasionally covered with loess and beyond them are the Black Belts of Texas on the west and Alabama on the east.

Historical Development. This is the area often referred to as "the South" which, because of its own special geographical conditions, has gone through character changes different from any other part of the continent. All other regions so far considered have progressed normally through the stages of simple pioneer settlement, an intense agricultural economy and a present-day increase in urbanization. The south, however, has much warmer conditions, and everywhere there is a period of at least 200 frost-free days. Rainfall is about 50 in. decreasing westwards, coming mainly in the summer. Although there is a likelihood of harmful frosts from the north, the most common air stream is the warm and damp one from the Gulf of Mexico. This climate, humid, warm, temperate and almost subtropical in the south, was not considered suited to the white settlers. Thus a plantation economy using slave negro labour was the natural outcome, and southern agriculture became dominated by cotton and tobacco as cash crops and maize as a subsistence crop. Compared with that of the north, agriculture in this region was neither as intensive nor as mechanized. Further, there was little industry and thus the standard of living remained very low.

The American Civil War and the abolition of slavery made no apparent change. Many of the freed negroes, with no real knowledge of farm methods, stayed on the land to compete with the white men as crop sharers and cash tenants. Monoculture was retained because no alternative was known and the lack of capital halted the progress of mechanization. Towards the end of the nineteenth century, the difficulties were aggravated by the appearance of the cotton boll weevil.

The Cotton Belt Today. Cotton still remains the chief cash crop but less than 10 per cent of the area is under cotton. More efficient methods of production include the extensive use of fertilizers such as phosphates from Florida and residues from cattle cake made in the south, deep ploughing, spraying with insecticides against boll weevil and machine sowing and picking on the large and prosperous farms, particularly in the west. Two very important changes have taken place in the process called "diversification." The first is the introduction of new crops so as to keep a balanced agriculture and to overcome the problems of soil erosion. This has in turn been accompanied by an increase in the size of farms and in the use of mechanization. The "cotton belt" as such no longer exists; it is broken up into separate regions with few of the major producing areas being found in what is known as the "old south." Half of the total crop comes from the black waxy prairie soils of Texas and from southern Texas; other areas are the alluvial lands of the Yazoo and Red River tributaries of the Mississippi, and the Tennessee Valley in northern Alabama.

The second change is the slow but gradual growth of manufacturing industries. The minerals include the coal of the Alabama field and the oil and gas of the mid-continental field. Ninety per cent of the U.S. bauxite supplies come from the Ozarks at *Little Rock*, in Arkansas, and iron-ore mined at *Birmingham* is used in the steel industry. Food processing, cotton manufacture and the chemical industry are to be found everywhere, aided by increased supplies of electrical power.

The most significant result is the gradual drift of the population from the farms to the towns. Negroes are steadily migrating northwards and into California, and natural increase in Alabama, Arkansas and Mississippi is not making up for the loss in those states.

QUESTIONS

1. Describe the main physical features of the course and basin of the Mackenzie and estimate the extent to which these features have been responsible for the slow economic development of the area.
2. "While in Canada, agriculture and ranching developed together, agriculture followed ranching in the U.S.A." Discuss the validity of this statement with particular reference to the grassland belts of North America.
3. Describe carefully the position of the chief wheat-growing region of North America. State the advantages of the region for large-scale wheat production. How does the wheat exported from this region reach the ports?

CHAPTER 39

THE HIGH PLAINS AND THE DUST BOWL

THE High Plains are somewhat undefined in extent, for although the western border is fixed by the position of the Rocky Mountains, the eastern border merges into the Central Lowlands. The zone between 98° W. and 100° W. is the nearest approximation to a boundary—by reason of the changes in climate which take place across it. In the U.S.A., these High Plains tend to be plateau-like, stretching from the Missouri Plateau in the north through the Sand Hills of Nebraska, the Colorado Piedmont and the Staked Plains and Edwards Plateau of Texas. The general level of the land surface has been deeply cut by the tributaries of the Missouri, Mississippi and Rio Grande, and by the Texan rivers which flow independently into the Gulf of Mexico.

Climate. From the Central Lowlands to the Rockies, the climatic transition is gentle across the line of 100° W. The annual rainfall total declines westwards from 20 in. to under 15 in. a year and there is a distinct spring or summer maximum. Average temperatures vary according to latitude and are affected by altitude. In winter, however, the effect of the Chinook is shown by the higher temperatures of the western edge.

Natural vegetation. A corresponding variation of soils and vegetation follows the climatic transition. The soils in the east are dark with a deep alkaline horizon and once had the typical cover of the tall grass prairie, but in the west, soils are brown and the alkaline layer is closer to the surface, and the original vegetation was short-grass prairie. These were rolling grasslands and almost treeless except along the watercourses, a characteristic which discouraged settlers who needed timber for building and fuel. In the dry parts, bunch grass grew in tufts but the area was crossed quickly by the settlers on their way to the more favourable Pacific coastlands.

Cultivation. In the early 1830s, ranching was recognized as the obvious use of uncultivatable grassland and "open-range stock rearing" originated in Texas and later spread northwards; by 1875 the whole area of the Indian buffalo plains country had become cattle country. New methods of cultivation (particularly the development of new grain varieties and the introduction of dry farming) encouraged homesteaders to begin a fresh movement westwards, and by the turn of the century the land had been subdivided for agriculture.

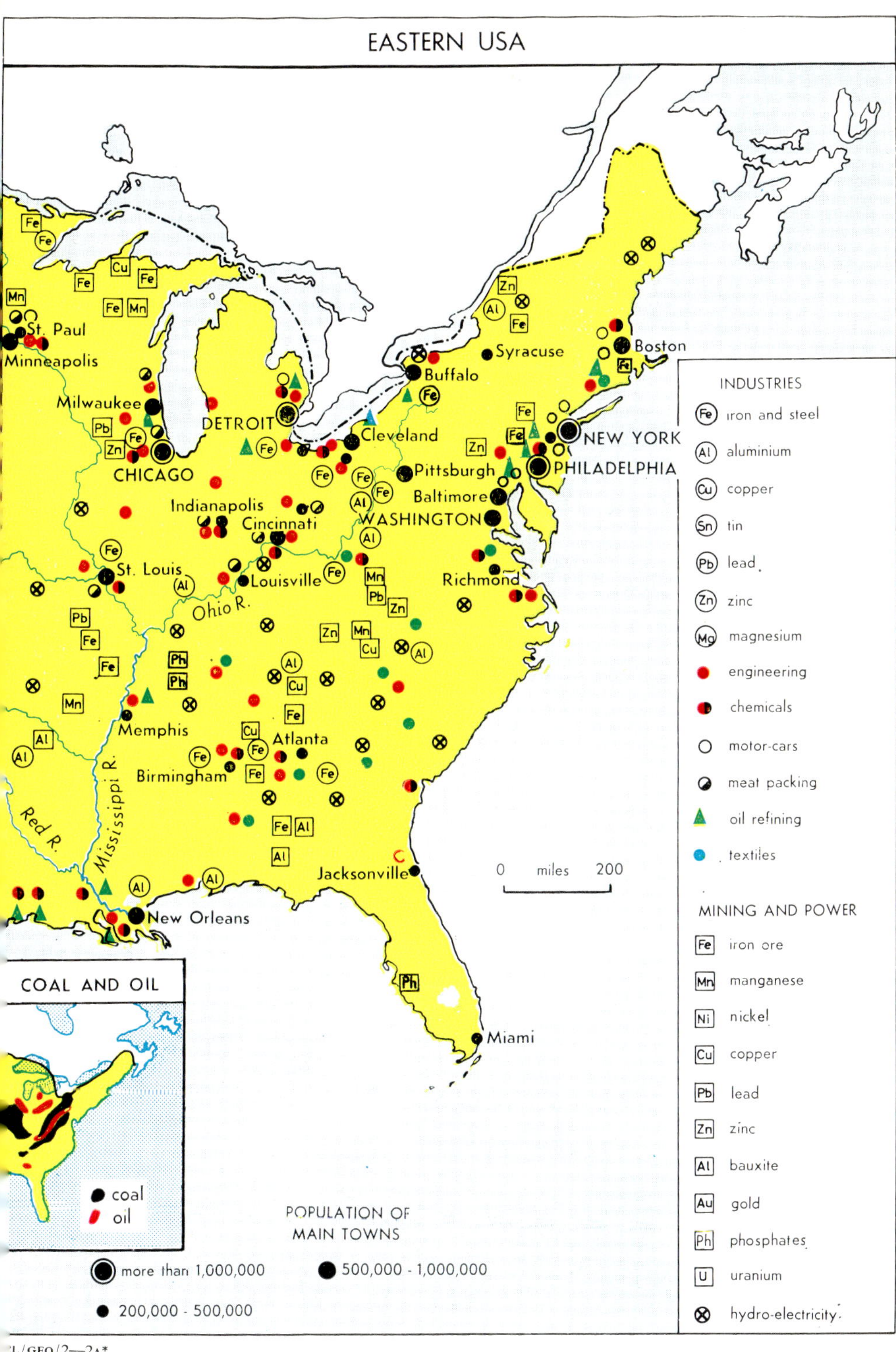
EASTERN USA
Minneapolis
St. Paul
Milwaukee
CHICAGO
DETROIT
Indianapolis
Cincinnati
St. Louis
Louisville
Ohio R.
Memphis
Birmingham
Atlanta
Mississippi R.
Red R.
New Orleans
Jacksonville
Miami
Cleveland
Buffalo
Syracuse
Boston
NEW YORK
PHILADELPHIA
Pittsburgh
Baltimore
WASHINGTON
Richmond
0 miles 200
INDUSTRIES
iron and steel
aluminium
copper
tin
lead
zinc
magnesium
engineering
chemicals
motor-cars
meat packing
oil refining
textiles
MINING AND POWER
iron ore
manganese
nickel
copper
lead
zinc
bauxite
gold
phosphates
uranium
hydro-electricity
COAL AND OIL
coal
oil
POPULATION OF MAIN TOWNS
more than 1,000,000
500,000 - 1,000,000
200,000 - 500,000

GREENLAND
Arctic Circle
Baffin Island
Alaska
36
Yukon
Hudson Bay
Churchill
32
32
33
Coast Ra.
Rocky Mts.
Labrador
Calgary
Vancouver
Winnipeg
33
Yokohama 4280
Seattle
Cascade Mts.
31
Quebec
Montreal
Toronto
31
Newfoundland
Liverpool 2872
London 3240
Honolulu 2919
Minneapolis
St. Paul
Snake
Nova Scotia
Halifax
Coast Ra.
Buffalo
40
Boston
Salt Lake City
Chicago
Omaha
Detroit
35
Denver
34
St. Louis
Ohio
Cincinnati
New York
Washington
San Francisco
37
Colorado
33
ATLANTIC
OCEAN
Memphis
Appalachian Mts.
40
Los Angeles
Dallas
35
Mississippi
Birmingham
Charleston
London 4740
Rio Grande
38
New Orleans
Florida
37
Honolulu 4741
Sa. Madre
Gulf of Mexico
Bahamas
Cape Town 6795
Tropic of Cancer
Havana
CUBA
MEXICO
Mexico City
DOMINICAN REP.
JAMAICA
HAITI
PUERTO RICO
BR HONDURAS
Kingston
HONDURAS
Caribbean Sea
Liverpool 3875
GUATEMALA
EL SALVADOR
NICARAGUA
Maracaibo
PACIFIC
Trinidad
Caracas
50°
40°
30°
20°
10°
50°
40°
30°
20°
10°

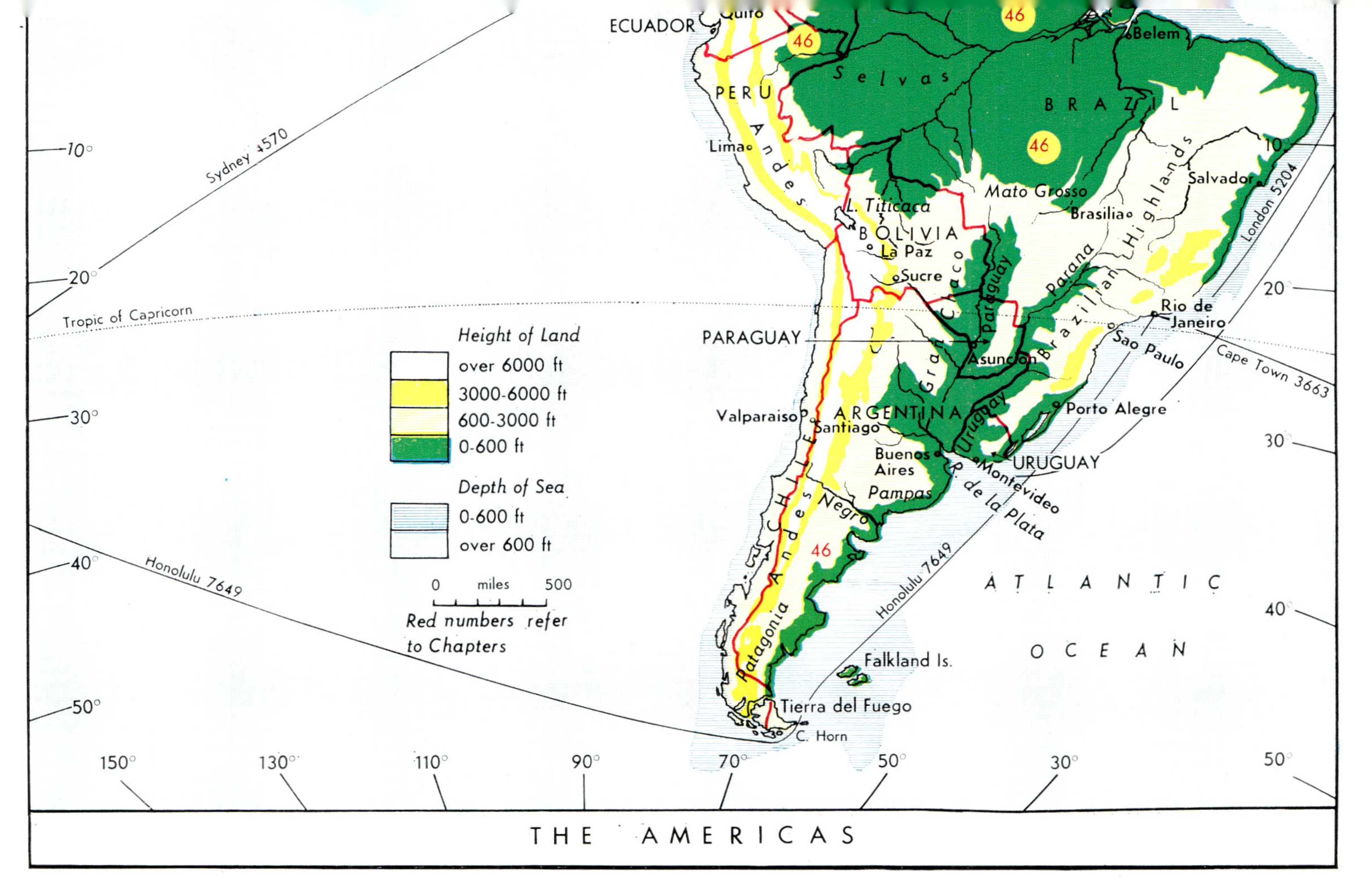

THE AMERICAS

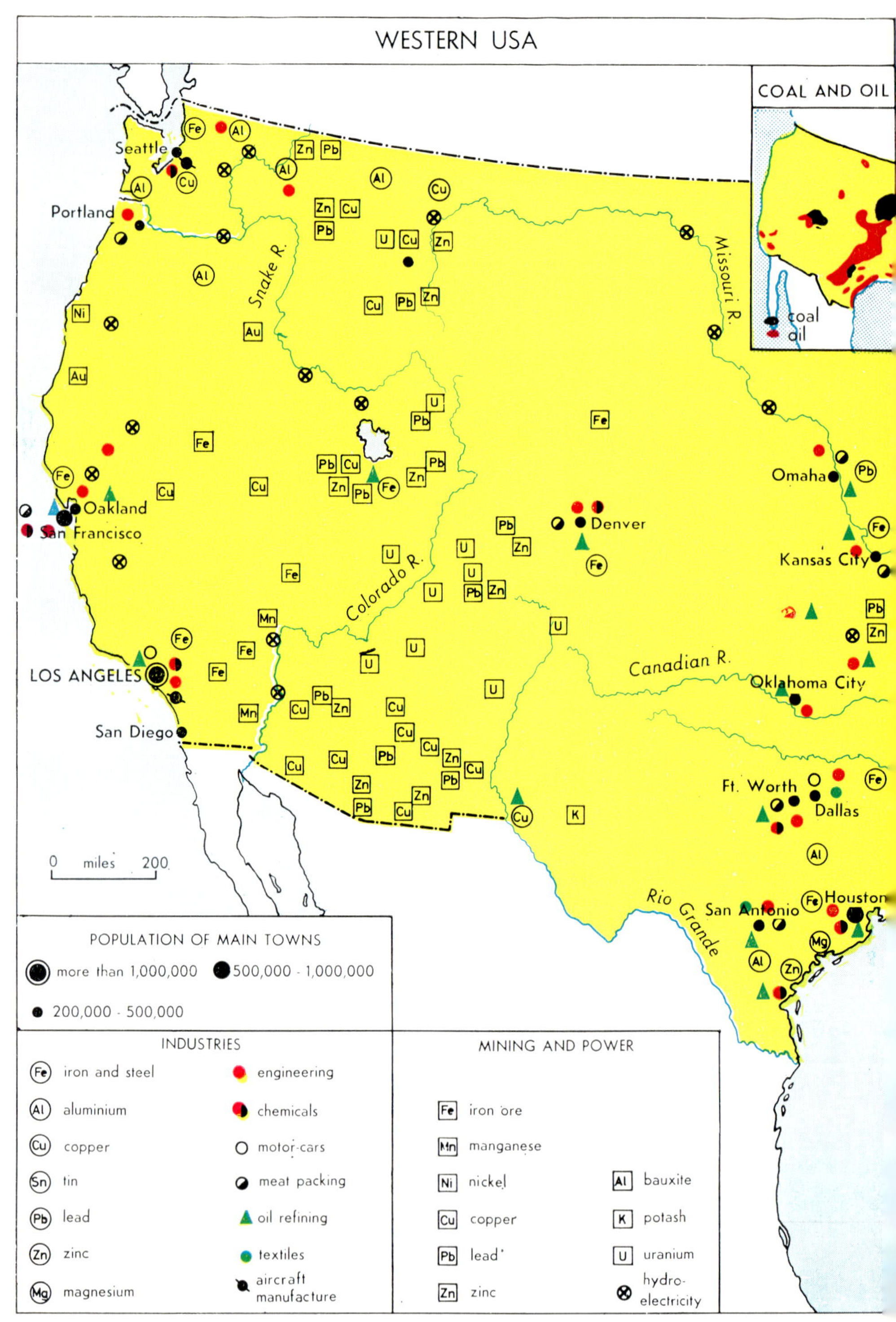
WESTERN USA
COAL AND OIL
coal
oil
Seattle
Portland
Oakland
San Francisco
LOS ANGELES
San Diego
Denver
Omaha
Kansas City
Oklahoma City
Ft. Worth
Dallas
San Antonio
Houston
Snake R.
Missouri R.
Colorado R.
Canadian R.
Rio Grande
0 miles 200
POPULATION OF MAIN TOWNS
more than 1,000,000
500,000 - 1,000,000
200,000 - 500,000
INDUSTRIES
Fe iron and steel
Al aluminium
Cu copper
Sn tin
Pb lead
Zn zinc
Mg magnesium
engineering
chemicals
motor-cars
meat packing
oil refining
textiles
aircraft manufacture
MINING AND POWER
Fe iron ore
Mn manganese
Ni nickel
Cu copper
Pb lead
Zn zinc
Al bauxite
K potash
U uranium
hydro-electricity

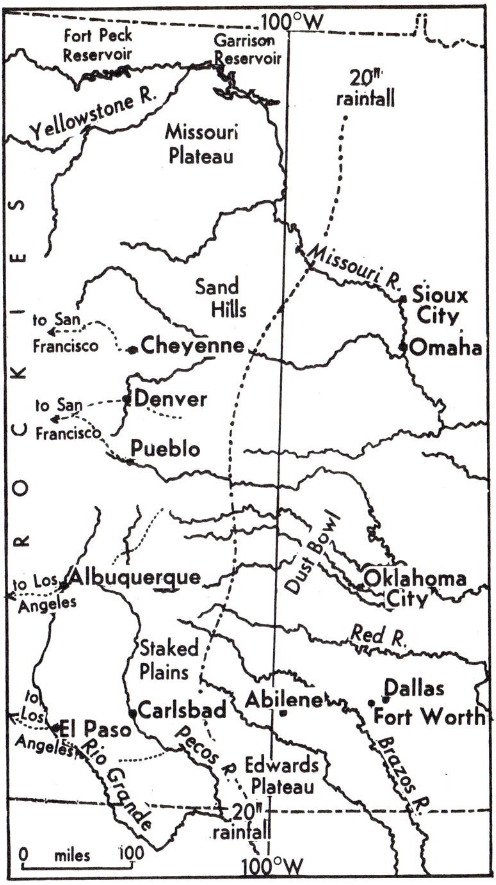

FIG. 175. *High Plains and Dust Bowl.*

In the short grass prairie today, extensive beef cattle ranching is the chief occupation. The cattle are cross-bred from English Herefords in the north and from Indian Brahmin cattle in the south. Texas, where it is warm enough for the animals to remain outside all the year has the most cattle; but large numbers are to be found in Nebraska and Kansas, where hay is grown as winter feed. Although young cattle may be sent to the lowlands for fattening, the modern tendency is to feed the animals on alfalfa and sugar beet tops on their home ranches before sending them to the stockyards and slaughter-houses of the East. Sheep flocks are also numerous and are transferred in the summer to higher alpine pastures in the Rockies to avoid the hot, dusty plain.

The wetter eastern margin is cultivated for wheat, sorghums, sugar beet, potatoes and alfalfa, but the most important feature is the extension of cotton growing even on to the Staked Plain and Edwards Plateau. Cultivation of these crops is assisted by the spring and summer maxima of rain, the excellent soils and particularly by the methods of dry farming and irrigation either from deep wells, or channels.

Since the High Plains have from the first been an area to be crossed, there are few urban settlements. The markets for the cattle are outside the region at *Omaha*, *Sioux City*, *Dallas* and *Fort Worth*, while another north-south line of towns lies at the foot of the Rockies, each controlling some sort of gap leading through them to the Pacific coast. Their relative importance varies with the local environment of farming or manufacture. *Denver* is the most important city and is situated at an old crossing point on the South Platte River, where the river leaves the mountain foot, and is close to the best developed irrigation area of the Plains; the city produces mining machinery. *Pueblo* has a large iron and steel manufacturing plant, metal smelting works and glass works and cotton mills.

FIG. 176. *Oklahoma: a farmer and his wife working on a cotton farm.*

FARMING IN OKLAHOMA AND KANSAS

FIG. 177. *Kansas: harvesting wheat. Note the broad, flat expanse of farmland.*

Dust Bowl

The Dust Bowl is the name given to one part of these High Plains extending through Kansas, Oklahoma and North Texas where soil erosion occurred as a result of the spread of cultivation into the drier west. The high prices obtainable in the late 1920s and a succession of rainfalls above the mean, encouraged the monocultivation of wheat. The natural grass cover was removed and, in the farming rotation, land was left fallow. The summer heat and the winter frost pulverized the soil and a succession of drought years produced more dust to be removed by the wind. Overgrazing by cattle added to the destruction. Farms were deserted when the occupants failed to make a living and more soil was removed by wind. When rain did occur, usually in heavy thunderstorms, it gullied the land and still more soil was lost by transport to the rivers. The issue of a government report in 1936 began the series of attempts to reduce the hazard of drought and the cause of soil erosion. The number of stock is now limited and farmers produce fodder crops and improve pastures with drought-resistant grasses within a crop rotation. Small-scale irrigation schemes and stock reservoirs, many based on sites originally intended to supply mines, have been constructed and have encouraged dairying. Strip farming and ploughing have been introduced into methods of cultivation. In the Missouri Basin particularly, irrigation and flood control schemes (e.g. the Pick-Sloan Plan) have been instituted and in some cases carried through.

QUESTIONS

1. Comment on and explain the climatic figures for Abilene and Denver as given below:

   Abilene: Rain in inches for the year 24·3 with 7·7 in spring, 7·1 in summer, 6·5 in autumn and 3·0 in winter; the mean monthly temperatures for January and July are 7·2° C. (45° F.) and 28° C. (82° F.) respectively.

   Denver: Total rainfall 14·3 in. (5·5, 4·6, 2·6, 1·6 respectively in the four seasons); January and July mean monthly temperatures −1° C. (30° F.) and 22° C. (72° F.).
2. Describe the location and physical character of the area known as the Dust Bowl of U.S.A. Explain the changes which have taken place in the farming activities of the area in the last twenty years.

CHAPTER 40

ACROSS THE WESTERN CORDILLERA

THE High Plains provided the kind of terrain which suited the early settlers on their westward journeys once they had crossed the Mississippi. However, the complex Western Cordillera formed a barrier to wagon-trail and railway construction alike. There are only two places where this barrier is really broken. The first is in the far north between the Brooks Range and Mackenzie Mts. through Fort Yukon and Tamana. The second is across the relatively low watershed between the Green and Snake Rivers. The former, being isolated, is still of little value, but the latter formed the old Oregon Trail (Fig. 179) between what are now two populous parts of U.S.A. Along the eastern front of the barrier, Edmonton, Calgary, Cheyenne, Denver, Pueblo, Albuquerque and Monterrey were sited at the beginning of trans-Cordilleran routes; the railway routes across the Cordillera also cut as directly as possible across the barrier using available passes with junctions far away from the

FIG. 178. *Journey across the Rockies.*

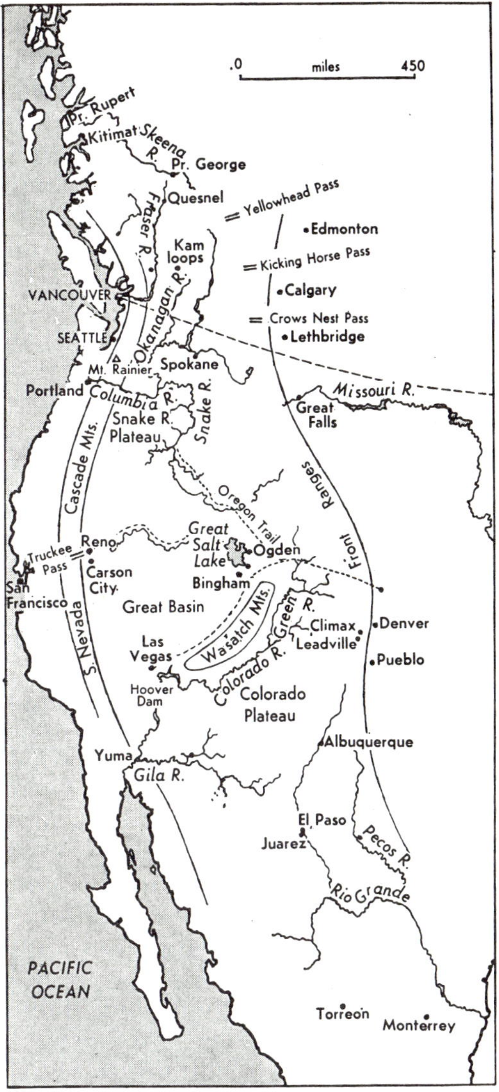

FIG. 179. *Western Cordillera.*

passes themselves. In the Canadian system, for example, where the region is narrower, the roughly equidistant Yellowhead, Kicking Horse and Crows Nest Passes take the lines from Edmonton, Calgary and Lethbridge, the junctions being as far apart as Winnipeg and Vancouver. In the last thirty years tunnels have been cut to reduce the distances through the passes. The main effect of this barrier farther south is that much of the produce of Utah, Idaho and western Montana intended for eastern markets still leaves via the Columbia River and the Panama Canal.

Relief. The Western Cordillera is a system of great complexity, stretching from Alaska southwards into Mexico. The journey from the High Plains to the Pacific coast along any line of latitude first crosses the mountain chains of the Rocky Mountains, two roughly parallel ranges which have been denuded leaving outcrops of the cores of crystalline rocks. Behind these ranges is an area of scattered mountains, plateaux and basins. Much of this inner region resulted from crustal faulting, volcanic action and vertical erosion by rivers. Weathered lava and alluvial deposits of all kinds mask the original relief and structure and recent elevations have rejuvenated the rivers which have cut enormous gorges (e.g. the Grand Canyon).

Finally to the west of this "basin and range" country, and separated from it by troughs which resulted from intense faulting, is the highland edge of the Western Cordillera. This is a mountain chain of old crystalline rocks containing many volcanic cones such as Mt. Whitney and Mt. Ranier. The most marked of the troughs are the Pacific Trough with the Middle Fraser River, Death Valley and the Salton Sink of California.

During the Quaternary period, ice from the Cordillera covered the whole of the intermontane area, the meltwater from it forming lakes in the Great Basin which either drained naturally southwards or created spillways, e.g. the Grand Coulee on the Columbia River.

Climate. The varied altitude and the enormous latitudinal extent of the region results in a great diversity of climate. In the valleys and on the lower plateaux, rainfall seldom exceeds 10 in. a year, while in south-west Arizona, the average is little more than 4 in. The western ranges are wetter and have a greater snowfall than the eastern, as the North Pacific depressions bring much relief rain and snow in the winter, but as they weaken in the summer, there is less relief rainfall on the highlands and more convectional rainfall in the interior. Over the intermontane area there is little influence from the Pacific Ocean or from the maritime air of the Gulf of Mexico. In the north there is a tendency towards a winter or spring maximum of precipitation.

Seasonal ranges of temperature are extreme everywhere, but while latitude influences the average conditions, summer temperatures over the enclosed basins and valleys are higher than those of the Central Lowlands in comparable latitudes. Sunshine figures are also high but, although the growing season is long enough (varying from 100 days on the British Columbia plateau to 300 days in Arizona), conditions are generally too dry for crop cultivation.

Natural Vegetation. The natural vegetation similarly shows zonal variations: from alpine meadows, through forests, where it is wetter or cooler (or both), woodland and grass, to scrub. The forest is dense at high altitudes and contains Douglas fir, spruce and pine, but with a decrease in rainfall it becomes more open with grass interspersed to form valuable grazing land. The woodland of the Rockies consists mainly of patchy growths of pine and oak and, at a height of about 5,000 feet, comparatively useless juniper trees, the grass being used as pasture. As rainfall decreases at lower levels, these woodlands are replaced by a mixture of bunch grass and drought-resistant sage-brush with cotton wood and willow along the watercourses. The "desert" of the south is covered with cacti, mesquite and greasewood.

Eastern Ranges

The eastern Rockies have few settlements because of their great height and many steep slopes. Early discoveries of gold did not encourage permanent settlements, but more recently the mining of copper, lead and zinc and ferro-alloy metals, coal, oil shales, and phosphate have attracted the associated smelting and refining industries. Recent hydro-electric schemes have invigorated industry at Trail, Great Falls and Salt Lake City, although most of the refining processes in U.S.A. are still completed on the eastern seaboard.

While mining provided the incentive of movement through the region, lumbering and ranching have been mainly responsible for the limited permanent settlement. Where conditions are wetter and cooler, as in the Columbia Valley of British Columbia, forestry is important. Where woodlands are more open, grass farming and dairying are complementary to the forestry industry. On the U.S. side of the border, where rainfall is lower and temperatures higher, care has to be taken to conserve water, so timber cutting is controlled. The output is small because the timber is so far from markets. From Montana to Texas cattle ranching and sheep herding are of greater importance than crop cultivation and nomadic herding takes place because natural pasture is either scattered or of poor quality. Cattle are stall-fed in winter or sold in autumn for fattening on the Central Plains.

The many valleys and small basins are usually filled with river muds and some are old lake floors. These good soils permit the cultivation of alfalfa, wheat, and barley with the help of irrigation from the mountain streams; potatoes, lettuce and fruit from Colorado are profitably grown and sent to the eastern cities.

Intermontane Plateaux

Although the vast intermontane region is very diverse in landforms and climate, the overriding feature is aridity. Agriculture must therefore be assisted by irrigation or dry-farming methods, and animals are rough-grazed.

The sub-Arctic conditions of the Yukon Territory do not favour commercial lumbering of its white spruce forests, but fur-trapping is important and there is a profitable summer tourist traffic along the Alaskan Highway. Southwards, the interior plateau of British Columbia and northern Washington is mostly volcanic and heavily glaciated. The valley of the Okanagan (a tributary of the Columbia River), one of the most important agricultural areas, has steep walls rising to over 4,000 feet; the valley floor and old lake terraces alongside have a variety of black and brown soils. In the north, farming is usually mixed, but southwards (and extending into Washington), the valley produces orchard fruits, peaches, apricots and apples. Although dry conditions necessitate irrigation there is the advantage of low humidity and therefore little danger of fungal infection.

Wheat is a major crop south of the international frontier. Near the Grand Coulee, dry farming produces a crop of wheat every two years. South of Spokane, the area between the Snake and Colombia Rivers has very fertile soils weathered from basalt flows, while in the valley of the Palouse,

FIG. 180. *The 535-foot high Grand Coulee Dam.*

tributary of the Snake, wind-blown lava dust has built dunes. This area has a typical Chinook climate with good spring soaking rains and an average summer temperature of 19-21° C. (67-70° F.). Seven-eighths of the total crop land of Washington State is given over to wheat, the most important feature being the breeding of special wheat with frost-resistant and grain-holding ears. Use of machinery is widespread and production is intensive; "grain hay" is even grown for cattle feeding in the neighbouring states.

Grand Coulee Dam. The outstanding irrigation scheme of the area is based on the construction of the Grand Coulee Dam. The Grand Coulee was a dry gorge over 200 miles long carved by the Columbia River as an overflow channel when its normal flow was dammed by ice, and then abandoned by the river when it returned to its original course. The dam has been built to pond back water to form Lake Roosevelt; from here, water is syphoned up into Grand Coulee reservoir and is then distributed. Power supplies, flood control and irrigation are covered by the complete scheme. Production is diverse including vegetables, sugar beet and alfalfa. Other dams have been constructed employing the regular flow of the Columbia, and the scheme for power supply is the result of co-

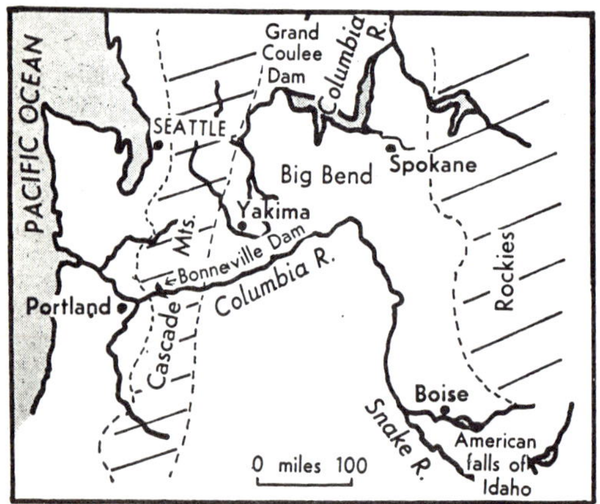

FIG. 181. *Columbia River Project.*

operation between the U.S.A. and Canadian governments. Most of the power is used in the industries of the coastal towns outside this area.

The Great Basin lies at a height of 3,000-7,000 feet and is, in the centre and south, a region of inland drainage, with saline playas lying on flat floors at high altitudes. The rocks are mineral-bearing: gold and silver are worked spasmodically in Nevada, and copper is mined by open-cast methods at Bingham at the margin of the area in Utah. Smelting of all minerals takes place at Salt Lake City and at Garfield nearby.

Because of its greater altitude, there is more rain than in the Columbia area, and there are pastures of bunch grass and sage-brush. Since the area is ringed with mountains, farming has a transitional character, using upland pasture in summer and lower plains in winter, with feed or alfalfa grown under irrigation. Mining camps provide the markets.

The Colorado Plateaux are even drier and although thin forest covers the higher rims, the bottom is almost true desert extending into the Death Valley and the Salton Sink in California and the Gila Desert in Arizona. In the last of these, all the characteristics of desert landscape, climate and vegetation are present.

The area, however, has the *Colorado Canyon*—the basis of a large tourist industry, it has the remains of Indian forms of land utilization and it has extensive modern irrigation schemes. The Navajo Indian country, now a reservation, is to be found above the Canyon, along the Colorado and its tributaries—particularly the San Juan River. This is an area of flat-topped mesas with a cover of gravel wash and volcanic remains; it is deeply dissected by streams with floors of quicksand. Packhorse is still a major means of transport. The pastoral Navajos graze sheep, goats and cattle but there has been overstocking and this has led to soil erosion.

The *Hopi* is a specialist dry farmer of the desert area, growing maize, beans, chillies and other Indian crops and rearing sheep. He will travel miles to find underground water, sows seeds and plants corn in dry sand before the summer rains and then carefully tends the plants during the rains. He obtains high yields from dwarf varieties. The Hopi live in settled communities, selling their produce to tourists or in the Indian markets, and some work in white settlements.

Irrigation. The construction of large dams and consequent irrigation schemes are comparable to those of the Columbia Basin and are dependent on the Colorado and Gila Rivers. The Colorado rises in the neighbourhood of Longs Peak in Wyoming and brings down considerable quantities of silt and sand to build a huge deltaic fan where it flows into the Gulf of California. At one time its water occupied a depression (called the Salton Sink) to the north-west, but this has dried up leaving a small salt lake below sea level. The exit from the Sink to the Gulf is by way of the Imperial

FIG. 182. *Colorado River Project.*

Valley, an area of good soil and farming conditions but subject to the hazards of flood, drought and silt deposition. In order to irrigate these dry areas of Arizona and California, three important dams have been constructed: the Boulder (or Hoover) dam at the bend of the Colorado near Las Vegas in Nevada; the Imperial Dam above Yuma to replace an older Laguna Dam; and the Roosevelt Dam in the Salt River Valley. Irrigation is used in the Coachella Valley where dates are grown and in the Imperial Valley where early lettuce, cantaloupe (a species of melon) and grapes, alfalfa, barley, sorghum and Mexicali cotton are grown.

The Mexican Plateau again shows the great diversity of this region. There is no apparent physical alteration from New Mexico across the Rio Grande nor is there any change in racial type. The plateau rises towards the south-east with a number of volcanic extrusions and islands of old rock (which contained the minerals on which its wealth was based) standing out above the gravels washed down from the mountains. The Rio Grande has a course which is somewhat similar to that of the Colorado, in that it has its source in the Colorado Front Ranges and a number of gorges and incised bends, but it is more a river of the High Plains.

Some parts of the Plateau have only 4·5 in. of rain a year and nowhere does the rainfall rise above 20 in. Large ranches (haciendas) are found here as well as mines. As social changes took place, the land was redistributed so that it now belongs to the village community, thus reverting to the original Indian system before the coming of the Spanish. Agriculture is not possible on the old rocks and the gravels, but new irrigation schemes and the water-borne or volcanic soils of the valleys permit the production of wheat, beans, alfalfa, and cotton, while ranching is also important.

Western Ranges

The western wall of the Western Cordillera is for most of its length an inner mountain range. In parts of British Columbia it comes down to the sea as the Coast Range, but in north-west U.S.A., as the Cascade Range, it develops into a true inner range, which is continued into California as the Sierra Nevada, and acts as a barrier to rain-bearing winds and depressions from the Pacific. Hence its western slopes are wet and timber covered, and provide the raw material for the enormous lumbering industries of the Pacific states and provinces. Douglas firs and many varieties of pine are important everywhere and on the Sierra Nevada there is the sequoia with its spongy bark which holds water and makes the tree fire-proof. This tree is also insect-proof and fungus-proof and since it thrives best in sheltered valleys which are not windswept, it grows to enormous heights.

Settlements are few, for naturally the people prefer the coast, but there are ribbon settlements up the valley routeways. In the north, the mountains have permanent ice-fields and on reaching the coast are penetrated by fiords. The Sierra Nevada has heavy winter snows which hinder railway operation but the water from these is available for irrigation in the south and for power generation in the north. The agriculture of the California Valley and the irrigation schemes of the Truckee and Carson City areas also rely on the waters of the Sierra Nevada. In the north, British Columbia has the Kemano-Kitimat scheme of power production. Here a dam across the Nechako River has impounded a large lake which overflows by means of a tunnel through the Coast Range to the Kemano power station. The power is used for aluminium ingot production at Kitimat, the bauxite coming from Jamaica. (See p. 395 for the reference to Arvida in Quebec.)

QUESTIONS

1. How were the pioneer trails across the Western Cordillera of North America influenced by the relief features? Describe the present pattern of the railway routes.
2. Consider the physical and economic importance of irrigation schemes in one of the following areas: the Columbia Basin, the Colorado and Gila Valleys, the Central Valley of California, Mexico.
3. Account for the specialized branch of farming on the Staked Plains, the Okanagan Valley, the Salt Lake Basin, the Palouse, the Indian Reservations, Coachella Valley.

CHAPTER 41

ALASKA TO OREGON

THE Pacific coastland has, naturally, the youngest states and provinces of the continent, for while the first surveys were carried out by sea, the settling of this region took place from the east. Juan de Fuca and Cook were looking for the North-West Passage from the Pacific side and respectively reached Washington and Vancouver; and the Pacific seal fur-trade provided the incentive for Spaniards, Russians and the Boston Fur Company to explore the coast still further towards the end of the eighteenth century. Gray, Vancouver and Mackenzie were others who mapped the inlets and islands. The incentive of furs drew the Hudson Bay Company to the mouth of the Fraser River, and Lewis and Clark who arrived at the mouth of the Columbia in 1805 from St. Louis. Astoria, and later Spokane (1812), were established as fur-trading posts in what was then the Oregon country.

Relief

The area provides one of the finest examples of a longitudinal coast. West of the Cordilleran zone is a line of depressions containing the coastal channels of Alaska and British Columbia, the Puget Sound, the Willamette Valley and the Central Valley of California. From Alaska to Oregon these depressions have been deepened by ice and then submerged; but still further south the Willamette and Californian valleys are filled with a considerable thickness and variety of alluvial material. A western mountain chain bordering the coast and extending as far as the Strait of Juan de Fuca can be recognized in the north by discontinuous islands and peninsulas; from here to the mouth of the Columbia River, the chain is high and rugged but further south in Oregon it becomes a series of low rolling hills.

Climate

The coastlands belong to the same cool temperate marine type as N.W. Europe, but again there are variations—particularly of rainfall total. The weather is controlled by the Aleutian Low Pressure system, but the westerlies blowing over the North Pacific Drift are not as warm or as humid as those over the North Atlantic and January temperatures are 3° C. (5° F.) lower than those over the European coast. Rainfall comes in all seasons, with a winter maximum, the total varying with relief.

FIG. 183. *Fishing in British Columbia: a hand brailer being used in transferring salmon from a seine net to the ship's hold.*

Occupations

Fishing. The Pacific Ocean has greatly influenced the economy of this coastal region. The advantage of the penetration of the fiords into the coastlands is offset by the cold climatic conditions and there are no major ports except for those of Puget Sound and Vancouver. The fiords, however, do provide the headquarters of fishing fleets catching halibut, cod, salmon and crab. Ketchekan is the salmon capital of Alaska and Prince Rupert has the world's largest cold storage plant for fish; Puget Sound and the long sandy beaches of Washington teem with shellfish and the Columbia River is second only to the Alaskan area in salmon. However, river-damming and over-fishing have reduced the salmon catch, and a Fisheries Commission has been set up jointly by Canada and the U.S.A. to regulate the quanties of fish caught. Washington State is the chief area involved in the fishing industry and most of the fish companies have their headquarters in Seattle. The Aleutian Islands are the main seal-breeding grounds.

Forestry. The wet slopes are forest clad and there are many pulp and saw mills. In Alaska, hemlock and spruce feed the mills of Sitka and Juneau; logging railroads were constructed on Vancouver Island, Vancouver City becoming the main lumber port of Canada. In the U.S.A., Oregon and Washington are the chief producers of forest products, the Douglas fir reaching its best development. Washington is known as the "Evergreen State" and after 100 years of logging operations half of the state is still covered with forest.

FIG. 184. *Alaska to Oregon.*

Farming. From Alaska to the Puget Sound, the mountains come so close to the sea that most of the streams of snow-melt or from the glaciers have only torrent tracks and level land is very scarce. Agriculture exists only on the floors of the narrow valleys or along the fiord strands. Alluvial land on the shorelands of the Puget Sound has been reclaimed by dyking (as in the Bay of Fundy), and there has been slow clearance of fern and tree stumps on the leached soils in the interior. Farther south still, the flat-floored Williamette Valley, less humid than around the Puget Sound, is the richest of these depression lands.

Alaska produces only 10 per cent of its food, which is grown on the only two good farmland areas: Matanuska near Anchorage, and the Tanana Valley near Fairbanks. Dairy produce, poultry, potatoes, lettuce, vegetables, barley and oats are the major products but there is a specialist production of strawberries at Haines. Homesteading is difficult at the best of times and often impossible because of the lack of roads and the vast distances from any possible markets. Settlement in the Matanuska area was entirely sponsored by the Federal Government during the depression of 1935. The greatest future value seems to be in the meat industry, beef, mutton and reindeer meat; the difficulty lies not in looking after the animals but in organizing and establishing markets. There are many areas of good grass, but the heavy summer rains make it difficult to harvest sufficient hay for winter feeding.

Liquid milk production has become important, and the principal areas are in the lower Fraser Valley and in western Washington. Around the Puget Sound, with its English type of climate, there is intensive dairying for cheese and butter, and deciduous and small berry fruits are grown for jam manufacture. Kent, just south of Seattle, specializes in lettuce production and Puyallup near Tacoma has a bulb and flower industry. The Willamette Valley, "the

New England of the West," is important for its general grain, grass, hay and livestock farming, and is now competing with California in prune plums and cherries. Throughout these Pacific coastlands the early stimulus to farming was the gold rushes. Some of the needs of the miners were supplied from the adjacent farms, but the farmers of the Willamette Valley thrived because they were nearer the Californian mines than the Atlantic farmers.

Mineral wealth is mainly confined to the carbonaceous materials, although between 1950 and 1961, 31 per cent of the total mineral production of Alaska was of gold, with approximately the same amount for coal, petroleum and natural gas. The Kennicott copper mines are now exhausted, but placer gold-mining in the area between Seward and Fairbanks is still important although its production is declining. Coal, mainly sub-bituminous and lignitic, is obtained from the Healy River area near Fairbanks and from Matanuska, and reserves are probably extensive. Both of these areas have the advantage of being close to the Alaska Railroad and to military installations which provide a market. Negotiations for the export of coal to Japan are in hand but climatic conditions make mining costs prohibitive so that little may come of the project. Petroleum, drilled at the Swanson River field on Kenai, has the advantage of the adjacent tideway; some is used locally and the remainder goes to Seattle.

Although in 1969 this provided the whole of the 10 million tons produced by Alaska, more significant finds have occurred on the North Slope of Northern Alaska, where most of the major world petroleum companies are exploring. Climate and accessibility to markets are difficulties to be overcome. Fairbanks gets its oil from the Gubik field in this area and a pipe line is to be laid to the ice-free

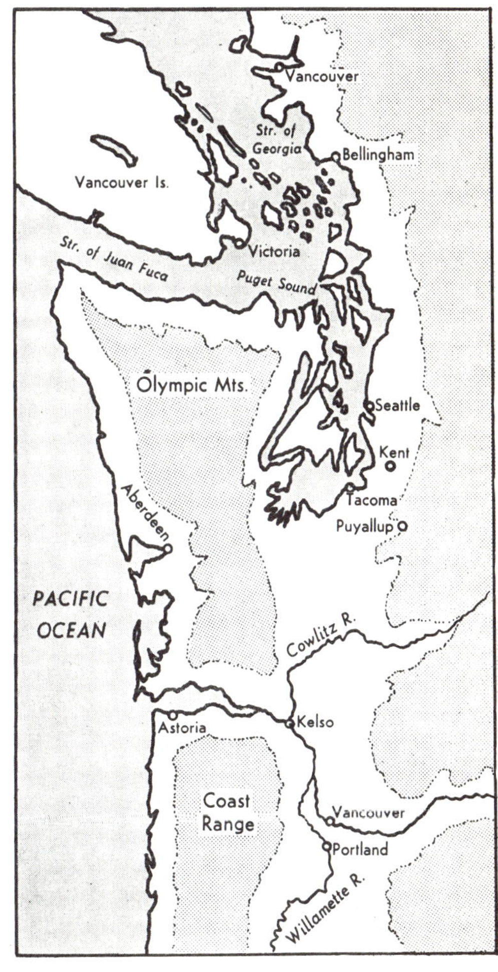

FIG. 185. *Position of Seattle and Vancouver.*

port of Valdez; experiments are being conducted with specially built ships with a view to using the North West Passage to the east coast of North America. A refinery at Anchorage is now in operation.

Natural gas from Kenai is piped to and used in Anchorage and is also being sent in liquefied form by giant tankers to Japan. The natural gas potential of the North Slope is also considerable and the economics of piping gas 2500 miles are under consideration.

From the same geological formation, coal is mined from thick seams at Nanaimo in Vancouver Island. The market, however, is small and is decreasing rapidly, since Vancouver is now the terminal of the oil pipeline from Alberta and of a natural gas pipeline from the Peace River District. On the other side of the strait, there is a small output of coal from Bellingham in Washington.

North of Vancouver, the main industries are the canning of fish and the production of logs and pulp, but there are factors working against the increase of the importance of manufacturing in Alaska. High transport costs, insufficient capital, inadequate public facilities and long distances to established markets make rapid expansion in the future improbable. By contrast, in the south there is greater development of the same industries as well as aircraft, steel, shipbuilding and aluminium refining industries. Portland and Seattle also have textile industries.

Communications

The physical nature of the coastlands demands that the separate sections must be linked by sea. Seattle was the port of embarkation for the Yukon gold miners, and it is still the terminus of Alaskan trade. Recently, a state ferry service between Prince Rupert and Skagway has been opened to encourage a tourist industry and to make transport easier along the Panhandle of the State.

Settlements

The area around the Strait of Georgia and Puget Sound and on to the mouth of the Columbia River is one of expanding population, with a correspondingly large urban growth. It is a region of cities—Vancouver, Seattle, Tacoma and Portland. The railway made Vancouver a great port and the opening of the Panama Canal increased its value to one of world importance. It has a share in both the Pacific and Atlantic trade. In contrast, Prince Rupert failed to attract the promised trade between Canada and the Orient, and thus has grown very slowly and still has less than 10,000 people. Because they are converging points of transcontinental routes, both Seattle and Portland are large industrial cities, commercial ports and regional capitals; Seattle for Puget Sound and Portland for the lower Columbia Valley and the Willamette Basin.

Native Population

One other important change occurs as one travels southwards along the coast—the distribution of the native population. South-west Alaska and the Aleutian Islands form a volcanic chain, treeless and valueless because of its stormy conditions, occupied by the native *Aleuts* who are related to the Eskimoes but have their own customs and language. The Seward Peninsula is Eskimo country. The winter is a time of strong winds and high humidity while summers are cool, rainy and foggy. The Eskimos here live in igloos of earth, sod, driftwood and whalebone. The Indian population also tends to spread into the interior, where some are nomadic herdsmen, though many are gradually adopting a western way of life, becoming mechanics at the ports and airfields. Farther south the *Indian Klingits* work in the fish canneries and at the salmon traps; while in the Annette Islands, the *Metlakatla* Indian colony is one of the prosperous settlements utilizing all available natural resources. Indians are found in all kinds of settlements in British Columbia, while in Washington, Indian reservations attract the tourists.

Summary

These coastlands have vast potential. In climate they are comparable to the British Isles, with Kodiak in the north having a summer temperature the same as that of western Scotland and a winter temperature the same as St. Louis. Such countries in other continents have farms and pasture lands as well as fish and forests. Washington and Oregon have vast resources of farm and forest land still untouched, and the Fraser Lowlands and Vancouver must grow with the westward spread of Canadian Prairie importance. On the coastlands of British Columbia, forestry will remain the leading industry for some time.

Alaska is often erroneously thought of as a state completely devoid of potential, relying only upon fish, fur and forest products to provide a small return. It is larger than the five states Washington, Oregon, California, Nevada and Arizona together, and is the only state in the U.S.A. to cover four time zones. Internally it has only 4,000 miles of highway and two railways; but it lies on the main polar air routes.

QUESTIONS

1. What is the importance of Alaska to the economy of the U.S.A.?
2. Compare and contrast the sites and importance of Vancouver and Prince Rupert.

CHAPTER 42

CALIFORNIA AND FLORIDA

LIKE its northern neighbours, California cannot be treated as one region, and its eastern sections have already been described as part of the Western Cordillera. Its western subregions, however, are relatively more important and are showing rapid increases of population, industry and agriculture. In the south-east of U.S.A., Florida is showing the same rapid increases and, indeed, the two states are strong competitors for markets for their products in the north-east, although they differ so markedly in relief.

California

Position and Relief. The Great Central Valley is an immense trough floored with a deep alluvial cover and flanked on the west by the Coast Range stretching from the Klamath Mts. southwards into lower California. This range is broken only in the centre where the subsidence of a rift valley has created the Golden Gate and San Francisco harbour. The Coast Ranges swing eastwards to become the San Bernadino Mts., which lie at right angles to the general trend; south of these are the Los Angeles lowlands. The longitudinal coastline with many small harbours stimulated the fishing industry of the Pacific Ocean, but while Washington State looks north, the fleets of San Francisco, San Diego and San Pedro go to Peru and Hawaii.

Climate. The Valley has mild winters and hot summers, resulting in a long growing season (a Mediterranean-type climate). It is drier and more extreme than the coastlands but has more rain than the south. As is the case on the coastlands farther north, California is subject to the Pacific depressions which give a high rainfall to the Coast Range (over 40 in. a year north of San Francisco) and a high snowfall to the Sierra Nevada, but these move northwards in the summer when the sun is over the Tropic of Cancer. In southern California winds are mainly offshore and very dry conditions prevail. Around Los Angeles, the "Santa Anna," a hot, dry wind descending from the mountains, necessitates a tree shelter belt to protect crops. Coastal temperatures are below the average for the latitude, and there is much fog, partly because of the presence of the cool Californian current, and also because of a north wind blowing from the high pressure cell. This ocean current also encourages fish breeding, as fish prefer cool water, and provides a second reason for the presence of the Californian fishing industry.

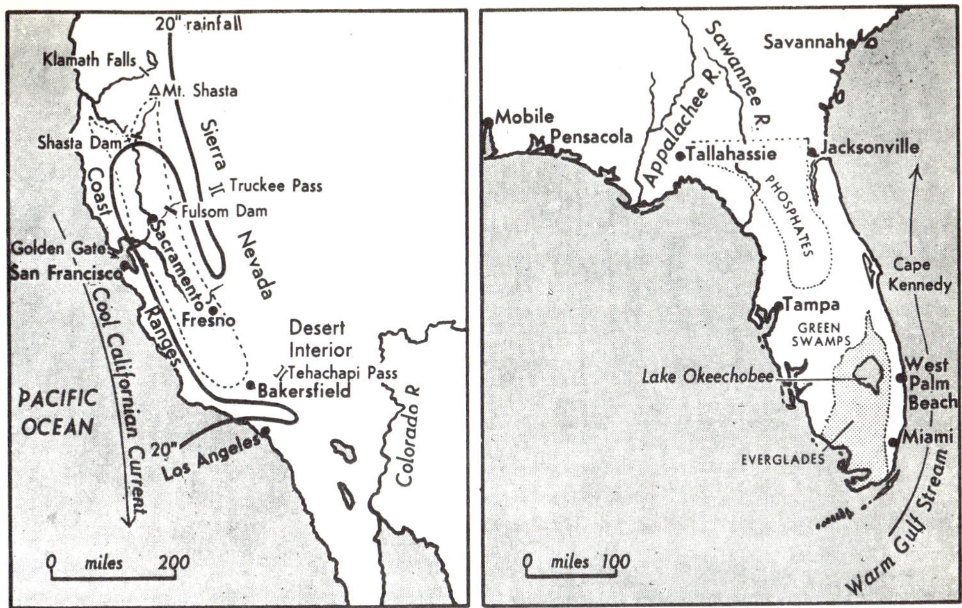

FIG. 186. *California* (*left*); *Florida* (*right*).

Farming. California is the most valuable agricultural state in the U.S.A. and Los Angeles county the richest farming county, but three-quarters of the Californian crops come from the Central Valley. Cotton, barley, truck crops (such as tomatoes and soft fruits, which are highly perishable, so are brought daily, when in season, to the urban centres), sugar beet and fruits are produced in greater quantities in California than in any other state.

The most powerful factor in the increase of agriculture is the long growing season and the accompanying sunshine value. Additional factors have been extensive irrigation schemes, a well-developed transport system (with the use of refrigeration cars particularly), co-operative marketing methods and a demand for high quality produce by an affluent home market.

Irrigation. The vast irrigation works and flood control systems were based on the fact that the Sacramento, flowing down the centre of the northern half of the valley, carried a surplus of water, while the much larger San Joaquin basin, the southern half of the Central Valley, was receiving but a third of its needs. The Central Valley Project involved the construction of the Shasta and Keswick Dams in the north to store the Sacramento water derived from the winter rain and melting snow, and the Delta Cross Canal to carry that water to a pumping plant at Tracy and to hold back the sea water. It is then lifted to continue its flow southwards up the San Joaquin Valley. The course of the San Joaquin has also been diverted southwards and the Friant Dam built to store water from the Sierra Nevada for the irrigation channels.

Around Los Angeles, the alluvial fans have been irrigated to the great benefit of dairy farming and the growing of oranges, lemons and vegetables.

FIG. 187. *California—first home of the motion picture.*

Fruit. Because of the great length of the Californian Valley there is a wide variety of fruit grown including apples in the north, prune plums in the Santa Clara Valley and grapes at Fresno in the San Joaquin Valley. They ripen at different times thus making it possible to use seasonal (mainly Mexican) labour.

Minerals and manufacturing industries. California is still a leading gold-mining state, but except for petroleum, the total mineral production is not large and there is no coal. The south-western corner of the state is rich in petroleum and natural gas which power important manufacturing industries. There are refineries at Los Angeles, Bakersfield and Richmond near San Francisco. Hydro-electricity is harnessed mainly from outside the area but some is coming from the new Keswick and Fulsom Dams. In addition to the canning of fish and fruit and the processing of foods, *San Francisco* has textile, electronics and shipbuilding industries, while Los Angeles has aircraft and motor assembly industries (developed since 1943). A declining industry is that of film-making, centred on Los Angeles. As a port, San Francisco serves the Central Valley and some parts of the mountain states, and is America's main gateway to the Pacific and the Far East.

People. Over four-fifths of the Californian population lives in urban settlements, a very high proportion considering that there is an apparent predominance of agriculture, but the two large conurbations of San Francisco and Los Angeles are great manufacturing centres. From 1950-60 the state population increased by 48·5 per cent; San Francisco grew by only 24 per cent, while Los Angeles increased by 67 per cent.

Florida

Florida has a very different geographical make-up from that of California, but they have a common characteristic in their highly profitable citrus fruit industries. Both are referred to as "sunshine" states. The citrus industry in California is confined mainly to the production of oranges and lemons in the region around Los Angeles, and the growing of apricots. Peaches and grapes, grapefruit, oranges and limes are more important in Florida than any other fruit.

Relief. Florida is a low-lying peninsula formed of an uplifted platform of recent limestone, and partially covered with sand. Its coast is one of sand bars, spits and lagoons, and in the south there is a string of coral islands called "keys." Local subsidence of underground caves has left a number of sink holes, and there are many solution lakes. Black salt-water muds give rise to low-lying swamplands such as the Everglades and Big Cypress swamp.

Climate. Florida's greatest asset is the climate, which is very different from that of the western zone of California. Everywhere in Florida, the summer temperature is about 27° C. (80° F.); in winter, even in north Florida, the temperature does not fall below 13° C. (55° F.) as an average, although the low relief has sometimes allowed a rapid penetration of cold air and killing frosts from the north. Although Florida lies mainly in subtropical latitudes, the whole state benefits from its insular location and from onshore air blowing

FIG. 188. *Florida: piling-up sponges (left); a highway, lined by cypresses, a common sight (right).*

over the Gulf Stream. Winter sunshine hours are longer than anywhere else in the U.S.A. It is, however, one of the wettest states with an overall average of about 50 in. a year. Indrawn, warm, humid, tropical air comes from the north-east to the east coast (where the mean annual rainfall is about 60 in. a year) and from the southerly quarter to Pensacola and Tallahassie. Wide ranges between minimum and maximum amounts of rainfall have been recorded everywhere, e.g. Pensacola 28·7-90·4 and Miami 28·7-84·4 in, a year. About 60 per cent of the rainfall comes between June and September with spring as the driest season.

These are excellent rainfall conditions which are missing in California where irrigation must be used. Conversely, most of the northern and central parts of Florida have porous surface rocks and soils and there is so much loss by either evaporation or percolation that 22 per cent of the productive land for vegetables and citrus fruits is under well or spring irrigation. The size of the water supply is becoming increasingly more important because population, industry, crop cultivation and tourism are growing very rapidly.

Water Supply. In the northern half of the state there is plenty of water stored close to the surface in the very thick limestones, and this is obtained by tapping major springs. In the extreme south-east region, the permeable limestones and sands lie on impermeable clays. Here the water-bearing rocks vary in thickness but there is good usable water, fluctuating in quantity with the seasonal fall of rain. Efforts are being made to control flooding in the Green Swamp area and to reclaim this and some of the Everglades for new citrus groves, but the possibility of water shortages might arise. It is in connexion with the tourist industry, now the most important occupation in Florida, that the problem of water supply is becoming serious. A fall in the level of water, perhaps because of increased evaporation, may result in the lowering of lake level and the curtailment of recreational facilities, particularly the fishing camps. Along the coast, the fall of the water-table causes the salt water to rise in the wells when the pressure of the fresh water is decreased, and holiday resorts consequently suffer. The problem of water supply is so pressing that pilot plants for converting sea water to fresh water have been set up at Daytona and St. Petersburg.

Farming. Although commercial production of oranges began in Florida, a freak freeze-up towards the end of the nineteenth century annihilated the industry in the north and forced it to the slopes above the lakes in the centre of the state. In 1960-61, 78 per cent of the oranges and 73 per cent of the grapefruit of the U.S.A. were produced in Florida, the result of scientific production and marketing over the last twenty years. Truck crops, mostly winter vegetables and flowers, are also important and these two branches of farming account for just over half of the value of the agriculture of the state. The growing of sugar cane and tung-oil nuts, and more recently the rearing of beef cattle, are the other agricultural activities. Important phosphate deposits worked in the north-west are processed at Tampa and Jacksonville for export to the Cotton Belt of the South.

Industry. There has been a substantial industrial growth too, although no one industry dominates the economy. Pulp and paper mills, food processing, furniture, chemicals and the manufacture of electronic equipment are to be found in the larger cities. The abundance of water is an important factor, and it was the exceptional quality of the water in the Pensacola area which led to the establishing of a large nylon plant. The increased supplies of natural gas piped from Texas and Louisiana have also helped in the expansion of industry.

Tourism is the main reason for Florida's growing wealth. The warm winters, the sandy beaches along a warm-water coast and the comparatively close link with the north by air have attracted visitors. These have also roused more agricultural and industrial activities, through the obviously expanded market. Miami, on the east coast, is the most important resort and it is closely followed by West Palm Beach and St. Petersburg on the west.

Population. Florida is growing much more rapidly than California. In 1940, Florida was the least populated of the twelve south-eastern states, but in recent years has had the greatest population increase. The reasons for this have been the industrialization, the tourist trade and the fact that it has become the home of many of the retired wealthy northern business men. Its chief town, Miami, shows the same outstanding increase; it has been the quickest-growing town in the country, slightly more than San Diego. The resorts with Tampa and Jacksonville, provide the high urban character, but with the exception of the Everglades and swamp regions, there is otherwise a fairly even distribution of the population. A special nucleus of scientists is established at Cape Kennedy, the rocket base.

QUESTIONS

1. Examine the reasons for the importance of fruit growing and irrigation in both California and Florida.
2. Compare and contrast the physical and human geography of California and Florida.

CHAPTER 43

THE GULF COASTLANDS

FROM Yucatan to Florida, both peninsular platforms of Tertiary limestone, the coastlands of eastern Mexico and southern U.S.A. surround the Gulf of Mexico with a ring of offshore bars, sand spits, lagoons and coastal swamps, broken only by the constantly growing delta of the Mississippi. Immediately behind are the low-lying plains of recent, unconsolidated marine deposits of sands and clays with the occasional band of mud left by rivers such as the Rio Grande, the Sabine and the Alabama. The inner boundary of this region is difficult to delimit for there is a gradual transition to the main Central Lowlands; in Mexico, from Vera Cruz to Brownsville, it is narrow and then widens into southern Texas. Baton Rouge might be thought of as its limit on the Mississippi.

Relief and Climate

Physically, the area was not amenable to settlement and the pine forests on the sandy soils and the river swamp forests of cypress were difficult to traverse; furthermore, hurricanes have always been an autumn hazard. This is a region of warm winters and hot summers and, because of the small chance of frost, has a long growing season. The rainfall total varies; in the east it is about 60 in. a year, decreasing westwards to the Mexican border where it is about 20 in.; farther south the rainfall total again rises till at Vera Cruz it is 68 in. but the low Yucatan peninsula only has about 30 in. The area belongs to the Eastern Warm Temperate type but there is a slight monsoonal complication; although there is no complete seasonal wind reversal, the prevailing winds bring a strong summer maximum of rain.

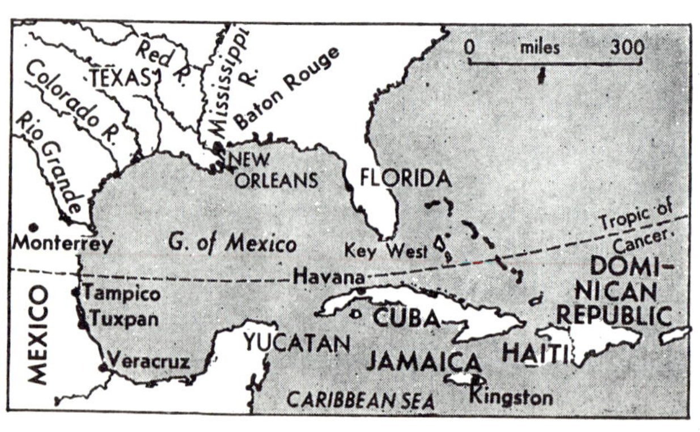

FIG. 189. *Gulf Coastlands.*

FIG. 190. *Mechanical cotton picker.*

Occupations

Farming. Where the advantages of this climate combine with those of soil, as in southern Texas and western Louisiana, agriculture is important; and where the swamplands are drained, the black soil is cultivated for rice and sugar cane. In the area east of the delta, agriculture is not important except for some vegetables and fruit, the only extensive sugar cane area in the country being near the delta. At Covington, on the edge of Lake Pontchartrain, early spring strawberries comprise a valuable crop. On the western side of the delta and extending into the more humid part of Texas is a large rice-growing area. The soils are deep and water is pumped from shallow wells and streams. The labour force is mainly negro and most of the work is mechanized; cattle-raising is often combined with rice-growing. Although farther west, with the somewhat drier conditions, cotton and vegetables replace rice, cattle-raising still remains important and farming is highly mechanized using white labour. The lower part of the Rio Grande Valley is becoming an active competitor of Florida and California in the citrus fruit and winter truck-crop industries. It grows more grapefruit than California and is trying to establish a market in the southern states since it is too far away from the north-east. The coastal margin is ill-drained, but immediately inland partial clearance of forest has permitted the cultivation of bananas, rice and sugar cane. Farming is very primitive, the hoe and digging stick or wooden plough still being used, and water is carried from stream to the fields in pots. Northern Yucatan is renowned for the commercial production of henequen,

FIG. 191. *Aerial view of New Orleans.*

a fibre used in rope and twine manufacture. Drainage is perhaps more important than irrigation to the expansion of the agricultural area of southern Louisiana and Mississippi, but farther west, water supplies from wells and streams supplement the small rainfall.

Minerals. The mineral wealth of the Gulf coastlands lies in petroleum, natural gas, sulphur and salt. The oilfield extends eastwards from Texas into Louisiana behind the coastal marshes and lagoons, and even on the swamp-lands themselves. Drilling is now taking place from island platforms offshore. Texas produces a third of the nation's oil and Louisiana takes second place. The petroleum is well suited to the production of fuel oils and is being used for industrial purposes along the coast. One of the largest refineries is at Baton Rouge and gas is being fed into Florida and Georgia by pipeline. Beaumont, Houston and San Antonio are major refining centres in Texas. Mexico also has her share of the Gulf oil but output has declined considerably. There is an extension of the Texan field close to the border, and from Tampico southwards into Tehuantepec, with the Poza Rica field near Tuxpan as the most important. Most of the petroleum is refined at Tampico but gas pipelines link the fields with Mexico City and Monterrey. Unfortunately Tampico is not accessible to large tankers.

These coastlands of the U.S.A. have always been considered agricultural, but now they are approaching a more balanced economy with increasing industrialization, largely because of the increased production of oil. At one time the oil was moved by pipeline to refiners in the north-east, and the manufactured goods brought back to the south. Now the large variety of manufactures required in the area are made in the Gulf ports and cities: synthetic rubber, nylon, fertilizers, insecticides, solvents and plastics, all derivatives of oil, sulphur and salt. The metal industries use imported raw materials from Jamaica, British Guiana and the Ozark Plateau and, while the aluminium industry of Canada is based on hydro-electricity, here natural gas is used as the source of power.

Population

Although urban population has increased rapidly along these coastlands, statistics for the states to which they belong show relatively slower growth. From 1940-50 the population of Texas grew by 20 per cent, Louisiana by 13 per cent and Alabama by 8 per cent while that of Mississippi declined by 0·2 per cent. Between 1950 and 1960 the rates in Texas and Alabama stayed roughly the same but in Louisiana it increased to 21·4 per cent. The change here obviously resulted from the increase of production of natural gas and the improved industrial economy, and this increase is continuing.

In Louisiana, the coastal people and especially those in New Orleans are of Spanish or French descent (Creoles) while in the delta there is a large number of Islanos introduced by the Spanish from the Canary Islands and still traditionally occupied in fishing and trapping. Further north the people are mainly English-speaking. There is apparently little change in the number of negroes along the coast but northwards it is declining through emigration to the northern cities. In Mexico, most of the coastal people are mestizos with more Indian blood than white, and there is a tendency to move to the towns and particularly to Mexico City.

Towns. All the major cities are ports. *New Orleans* is situated on the north-eastern side of the Mississippi delta and is sheltered from southerly hurricanes; it lies on a narrow piece of land between a river meander and Lake Pontchartrain, an old lagoon now serving as an inland harbour since the cutting of a canal. New Orleans has varied exports drawn from an extensive hinterland, its imports coming mainly from the West Indies, the Caribbean Islands and the west coast of South America via the Panama Canal. It is the leading grain port and the chief foreign port of the southern states.

The other ports are *Mobile* in Alabama, connected to the Appalachian coalfields by canal and railway; *Baton Rouge*, the furthest inland deep-water port up the Mississippi; and *Galveston* and *Houston* in Texas, linked together by a ship canal. Houston is now larger than New Orleans and handles more tonnage. Its main trade is in oil and petroleum products but it is now sharing in the coffee, jute, bananas and newsprint traffic. With the change in the position of the focus of cotton production, Houston is the most important cotton port in U.S.A. The chief general port of Mexico is *Vera Cruz*; *Tampico* is another Mexican port on the Gulf Coast.

QUESTIONS

1. Describe farming in the Gulf Coastlands in relation to their relief.
2. Describe and account for the changes now taking place in the utilization of the land and in the distribution of population in the Gulf Coastlands.
3. Comment on the positions of New Orleans and Houston; compare and contrast their importance.

CHAPTER 44

THE WEST INDIES

ALTHOUGH it is probable that the Vikings were the first Europeans to sight the New World, the islands and mainland around the Caribbean Sea were the first regions to be settled by Europeans. The geography of the region appears at first to be remarkably uniform, but although the islands have a common mountain backbone (often volcanic) and lie in the Trade Wind Belt, there has been considerable variation in both economic and political development.

The largest islands, Cuba, Jamaica, Hispaniola and Puerto Rico, form the western end of the curve as the Greater Antilles, while the Lesser Antilles, containing the Leeward and Windward Isles, form the main loop at the eastern end. Cuba and the two parts of Hispaniola are independent. The economy of Puerto Rico is largely dependent on the U.S.A., and the rest of the islands have formed part of British, French and Dutch colonial systems. First Spain, then northern European countries and lastly U.S.A. have had some control in the affairs of the Caribbean. Where the Spanish method of exploitation survived too long, as in Central America, development was retarded. The British form of colonization increased the production of raw materials required at home and thus helped to hasten the development of the islands. Cuba, although nowadays violently anti-American, was for sixty years an American protectorate.

The Federation of the West Indies (which was an attempt to link up the former British colonies) was formed, but inter-island rivalries retarded integration to such an extent that it lasted only four years. The conception of a Federation of the British West Indies had much to commend it because of the ease in the planning of the economy. Air transport and common strategy would perhaps have improved. The refusal of British Guiana (now Guyana) and British Honduras on the mainland and the later withdrawal of Trinidad and Jamaica have prevented this idea coming to fruition.

Relief and Structure

The map on p. 367 shows the West Indies as the Caribbean meeting place of many fold systems: the relief and structure are complex. In the north, the Greater Antilles have belts of granite lying against flat waterless limestone platforms, while the curve of the Lesser Antilles is a zone of volcanic islands; the Bahamas are composed of coral limestone. Barbados is outside

FIG. 192. *Sugar cane, the West Indies' chief crop.*

the curve of the Lesser Antilles and is linked structurally with the Caribbean Coastal Range of mountains through Trinidad. The highest points in the islands are about 7,000 feet but their Atlantic edge is marked by a series of ocean depths to below 30,000 feet.

Climate

The entire area is influenced by the Trade Winds blowing round the Sub-Tropical High Pressure area (Fig. 153) which persist throughout the year. Average temperatures are high, 24° C. (75° F.), and the modifying influence of the sea gives low temperature ranges. Centres of very low atmospheric pressure are often generated over the warm ocean currents and tropical cyclones (or hurricanes) move across the West Indies to the Gulf Coast causing severe damage to crops and buildings. Rainfall is always heavy on the windward slopes of the mountains with autumn the driest season. There are occasional droughts, but not necessarily at the same time over all the islands.The Trade Winds are drying winds and a long drought can be a serious problem to sugar cane farming. The heavy rains on the windward slopes stimulate the growth of a rain forest with tall trees and no undergrowth; where the rainy season is short there is a savanna vegetation of scattered trees and grassland. Between these extremes there may be jungle or scrub thickets.

Population

The West Indies are populated today mainly by people of negro or mulatto (white × negro) origin. Dominica still has a few *Caribs*, the remnants of the cannibalistic aboriginal people of the West Indies. Some of the islands are overpopulated, such as Barbados, which has a population density of 1,400 per square mile, Puerto Rico (683) and Jamaica (366). There have been heavy migrations from Puerto Rico to the U.S.A., especially New York and Chicago, and from the British West Indies to the United Kingdom, presenting special problems in the receiving countries; in spite of this the population of the British West Indies has practically doubled from 2 millions since 1920. There have also been large movements of people within the islands from rural areas to the cities and ports to take up factory and dock work: Havana, for example, has a metropolitan population of 1¼ millions. Whites are relatively few in the islands, and only in Puerto Rico and Cuba do they appear in any numbers. With the abolition of slavery, Chinese and Indian labourers were brought in and Asians now comprise, for example, one third of the population of Trinidad.

Farming

Sugar cane is the main crop of many of the islands and the total production of the whole area continues to grow, although there are fluctuations within individual islands. In St. Lucia and Grenada, production is declining rapidly and in St. Vincent and Dominica it has disappeared from the export trade altogether, home requirements being made up by importing from Trinidad. The industry in St. Lucia has almost closed down because of the shortage of labour owing to emigration and the burning-out of the cane fields for the cultivation of bananas. In St. Kitts, despite the introduction of other industries, the sugar industry continues to dominate the economy although the vulnerability was shown in 1961, when owing to labour troubles cane was not cut and production was considerably lower than it was in 1960.

The West Indies became prosperous when slave labour was introduced into the cane sugar plantations. Large plantations have now been broken up and there are many small holdings, and although sugar is still very valuable and usually forms the main export, other crops have been developed. The following crops are cultivated on most of the islands: sugar, tobacco, coffee, cocoa, sisal, citrus fruits, pineapples, cotton and bananas, though their relative importance varies considerably.

Islands

Cuba is the largest of the islands and is in many ways typical of them. it has high, rugged mountains in the south-east, a low central ridge and extensive lowlands with soils of limestone origin.

At the beginning of the twentieth century Cuba was an economic colony of the U.S.A. and this was the main feature until the revolution. About 80 per cent of the total exports of Cuba was sugar, the remainder being tobacco (mainly as leaves), copper and manganese, and partly-processed nickel.

FIG. 193. *Cuba: propaganda in a Havana brewery, one of the many factories built after the revolution.*

Cuba was essentially a monoculture country with one export commodity and one import market—the U.S.A.—and relied almost entirely on foreign trade. The change in political leadership in 1959 caused the loss of the market for sugar and of supplies of U.S.A. goods. Today, although diversification, especially in agriculture, has taken place, sugar is still the most valuable crop. In 1963, although sugar production fell from 4·8 million metric tons to 3·8 million, this was offset by a rise of 6 per cent in industry as a whole. New markets have been opened, especially in the U.S.S.R. The island has considerable potential, particularly in nickel, and Cuba may well become the site of a valuable metallurgical industry. The cattle industry is also developing rapidly.

Puerto Rico is the smallest island of the Greater Antilles and like Cuba has a fairly large white population. The island has a strong economic link with U.S.A. Its economy is more diverse, but there is the dependence upon sugar production which is so common throughout the islands. Coffee and oranges are grown together on the slopes of the western highlands, and tobacco, grapefruit, pineapples and vegetables are produced with subsistence crops on a share-cropping basis. Both the increased population and soil erosion, as a result of deforestation, necessitated import of food from U.S.A., but as on the Gulf Coastlands, there is an attempt to balance the economy by introducing industries such as the manufacture of rayon, cement and consumer goods and installing a hydro-electric power station, an oil refinery and a chemical plant. The island, however, has become too dependent on U.S.A., for while welfare conditions and roadways have been improved to an extraordinary extent, the U.S.A. tariff policy does not allow complete freedom of trade.

Standards of living are better than elsewhere in the Caribbean and the development of tourism has been rapid.

Jamaica, the largest of the former British West Indies, is only one-tenth the size of Cuba. In structure, climate and population it is similar to Puerto Rico except that the highlands are not volcanic. Strategically, it is more valuable since it lies opposite the Windward Passage and therefore controls a Caribbean trade route, but Puerto Rico has the advantage of being much nearer to its main market than Jamaica. The latter's present value lies in the improved balance of its economy. It has its commercial agriculture of sugar, bananas, citrus, cocoa, coffee and coconuts; it is one of the most important bauxite-producing countries of the world; it has numerous light industries and the tourist industry is growing fast. As usual throughout the islands, many people depend on subsistence crops of maize, rice and vegetables, and on the dry areas, sheep and goats are reared.

Bauxite-mining in the centre of the island behind Port Maria, and the production of alumina have been established for over twenty years and large quantities are exported to Kitimat in British Columbia and to the Gulf ports. Jamaica is the largest world producer of bauxite.

In **Trinidad** the chief mineral is petroleum, obtained from drillings in the Soldado field in the Gulf of Paria and off the east coast. The island refines some of its own oil and part of the Venezuelan supply. Asphalt, after 400 years of exploitation, is still being obtained from the Pitch Lake, which has produced over 7 million tons. New manufacturing industries are using the natural gas and locally refined petroleum for power. There is a diverse population and consequently a mixed economy, with East Indians concentrating on agriculture (sugar, coconuts, citrus fruits and cocoa) and Negroes on the industries. Because of its special products, Trinidad has a larger overseas trade than any other of the West Indian areas in the Commonwealth and it is the only one where the value of exports exceeds that of imports.

Barbados has an extensive cap of coral 300 ft. thick, which gives a dry surface but allows the formation of underground water which can be reached by bores for domestic purposes. Sugar cane is intensively grown and accounts for over 90 per cent of the exports and its easterly position makes it an entrepot port for the southern Caribbean.

Tourism

The immediate future of the West Indies lies partly in the expansion of its tourist industry, for the climate and relative short air journeys from north-east U.S.A. place them in a favourable position to share in the tourist industry with Florida. Pleasure cruises from Britain, first in the banana boats and now in the larger liners (which have lost some of their passenger traffic to the airlines), also help to maintain a flourishing industry. In Barbados a new extension has been made to Seawell Airport runway for the operation of the largest commercial jet aircraft, and in Antigua and Jamaica large new hotels are being built and golf courses laid out. The Bahamas have developed a more purely tourist economy than any other country in the world.

Towns

The only towns of any size are ports. *Havana* is by far the largest and has a well-sheltered harbour. Since it faces Florida, 200 miles away across the Strait of Florida it had until the appearance of the new political regime a share in the tourist trade; it is now partly dependent upon the manufacture of consumer goods, cigars and cigarettes and oil-refining. *Kingston*, the capital of Jamaica, is in a better position and controls the route through the Windward Passage to the Panama Canal. Its harbour, which has been reconstructed so as to dock a number of ships simultaneously, is protected by a long spit called "the Palisados" on which there is an airbase, and the town is sheltered on the landward side by the Blue Mountains. *Bridgetown*, the capital of Barbados, has a strategic position controlling the south-eastern approaches to the Caribbean Sea. *San Juan* is the capital of Puerto Rico with nearly half a million people and has an important airport.

Summary

In contrast with the other regions already considered, the West Indies is a purely tropical area, even with the moderating influence of the sea. The islands produce few of the foods containing the minerals and proteins associated normally with those of temperate lands; grass is poor and animal products therefore absent. Tropical foods are monotonous and because imports are expensive the diet is mainly one of rice and beans; the temperature and humidity are high and not conducive to hard labour. The domination of sugar in the export trade has also allowed only primitive and inefficient methods of growing alternative foods for home consumption. The standard of living is low but is being raised continuously through planned development, e.g. the granting of loans, formation of co-operative organizations for the provision of seed and fertilizers, and for marketing products. It is the desire for better living standards which has caused the large-scale migration away from the area, but with the establishment of new light industries to produce consumer goods and the expansion of the tourist industry there should be less dependence on imports.

QUESTIONS

1. Discuss the validity of the following statement: "While Canada produces the necessities of life, the West Indies produce only luxuries."
2. What factors favour the expansion of the tourist industry in the West Indies?

CHAPTER 45

THE ATLANTIC OUTLOOK

EASTERN U.S.A. shows as many variations in physical, economic and social structure as Western U.S.A., and the journey northwards from Florida takes the traveller from a subtropical to a cool temperate climate, from a predominantly agricultural region with scattered townships to a predominantly industrial region with very large cities and urban agglomerations. It is a journey along a narrowing coastal plain as far as New York. From west to east, however, there are wide differences in the structure and relief across the Appalachians, the Piedmont and the coastal plain, as was shown in Chapter 33.

Relief

The Atlantic Plain consists of young sediments which overlie the same sort of old crystalline rocks that now make up the Piedmont Plateau. It slopes gradually eastwards ending in a marked continental shelf. It is not level, for resistant outcropping limestones have resulted in low scarps with seaward dip slopes. The immediate coastal zone has many marine marshes, dunes and bay bars. The rivers leave the Piedmont by falls which mark the eastern limit of the crystalline structure and the head of river navigation. With the submergence of the land north of Cape Hatteras, the Fall Line comes closer to the coast and reaches it at Trenton and near New York. Here the rivers have estuaries which lead to good natural harbours between the peninsulas, while in the south they have broad flood plains.

New England physically forms part of the Appalachians, separated from the main ranges by the Hudson-Mohawk Gap. Recent glaciation has stripped off the soil on the uplands and laid down moraines on the lowlands, while fluvio-glacial streams have added stony drift and outwash sands. Ice has also been responsible for diversion of drainage and the formation of lakes, rapids and waterfalls as on the Canadian Shield. Submergence resulted in the varied coastal directions and in the rias of Maine, while longshore drift has given bay bars and spits such as Cape Cod.

Climate

Everywhere there is well-distributed rainfall of about 40 in. per year but only in the north is snow experienced. Average temperatures naturally decrease from south to north, largely because of winter temperatures.

Farming

The types of farming vary according to climate, soil, accessibility to markets and tradition. Only in the alluvial valleys of the Appalachian foothills and the Piedmont are the soils really good; elsewhere they are sandy and require heavy manuring, or, as in New England, have to be cleared of stones.

Cotton is grown in Georgia and the two Carolinas, on the inner part of the plain and the lower Piedmont, with maize and beans as the chief subsistence crops. Tobacco is the great money crop of North Carolina and southern Virginia. Spring vegetables are grown for the northern markets. Farther north into Maryland and Pennsylvania, two factors influence the farming character: first, the original settlers from Europe and Britain adapted their traditional methods to the soil and climate of the new lands and were thus mixed farmers; and second, the large coastal cities provided a market for milk, vegetables, eggs and poultry. The light soils of the coastal region are suitable for fruit and vegetables, and the farms are linked by a good road system over comparatively short distances to the seaboard cities. So much is produced that canning and freezing at Camden and Baltimore is a flourishing industry.

In New England, farming has never been easy. The lack of extensive lowlands, the short growing season, the bleak winter and the stony soils must have been discouraging to the early settlers. The only advantage now is the presence of many nearby industrial centres which serve as a market for their milk, poultry, hay, vegetables and fruit. Farming in the Piedmont follows that of the coastal plain closely, but although soils in the valleys are rather more fertile than those near the coast, farming is less intensive and less scientific, and there has been much soil erosion because of the overcultivation of maize without rotation.

Industry

There are four main industrial regions: (1) Central Alabama and the Tennessee Valley, (2) Pennsylvania, (3) the seaboard cities and (4) New England, each with its own particular form of development and modern character.

Central Alabama, with its chief towns of Bessemer and Birmingham, has been the site of an iron and steel industry for eighty years, and has the advantage of haematite iron ore, excellent coking coal and limestone in very close proximity to each other. The chief output is rails for Asia and South America and construction steel.

The most important feature of the area is the all-purpose *Tennessee Valley Scheme*. Before 1933, the valley with its 50 in. rainfall a year was an area of poor agriculture and serious soil erosion through gullying; the river was useless for navigation and subject to damaging floods, not only along its own course

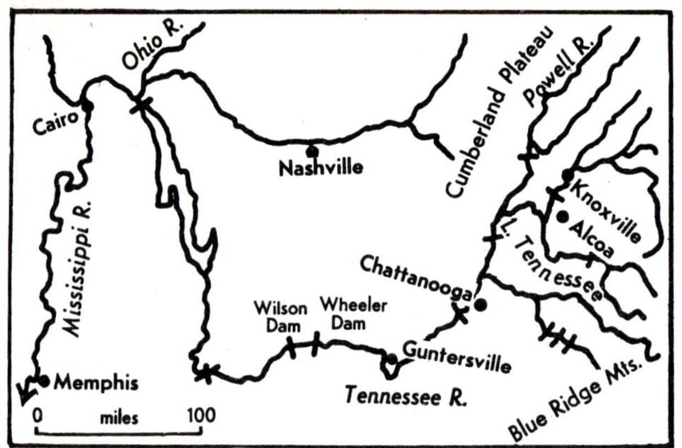

FIG. 194. *Tennessee Valley Project.*

but into the lower Mississippi. The scheme has covered the building of about thirty dams on the main river and its tributaries, with locks and channels, and power plants. This has ensured navigation all year round to the Mid-West and the Gulf Coast, and has eliminated the flood danger. The power supplies have encouraged the establishment of industries such as the manufacture of chemicals for fertilizers, copper smelting, aluminium refining, paper mills and an atomic energy plant near Knoxville. The lakes behind the dams attract tourists. Fertilizers, soil conservation methods, and power supplies have invigorated agriculture. Maize, beans, vegetables and fodder crops are grown in a variety of rotations. The two main towns are Knoxville with its textiles, furniture and metal industries, and Chattanooga which makes iron goods.

Pennsylvania. The greatest manufacturing area is in western Pennsylvania. The coal seams yield good coking, steam and domestic fuel. The coal is used locally in the heavy iron and steel industry, in industry along the shores of the Great Lakes, in the Atlantic seaboard cities, and, via Baltimore, in New England. The movement of coal has been expedited by the many passes and gaps in the highlands. Petroleum is still extracted from the area but natural gas (obtained locally as well as being piped from the south) is more important as a fuel.

The iron and steel industry depends fundamentally on the ores of the shores of Lake Superior, the coking coals of the Appalachians, and the limestones of the Michigan peninsula and islands of Lake Erie. These areas are linked by the Great Lakes waterway and by a system of road, rail and canal transport so that the industry appears in three areas: the southern end of Lake Michigan centred at Chicago and Gary, the Lake Erie shores and the Pittsburg area. Everywhere the industry utilizes modern techniques; plants have their own coke ovens, blast furnaces and steel works and much of the steel is made in electric furnaces. Originally Pittsburgh had its own iron-ore but the cutting of the Soo Canals enabled the haematite ores of the Upper Lakes region to be imported. The ore ports such as Duluth had deep water and

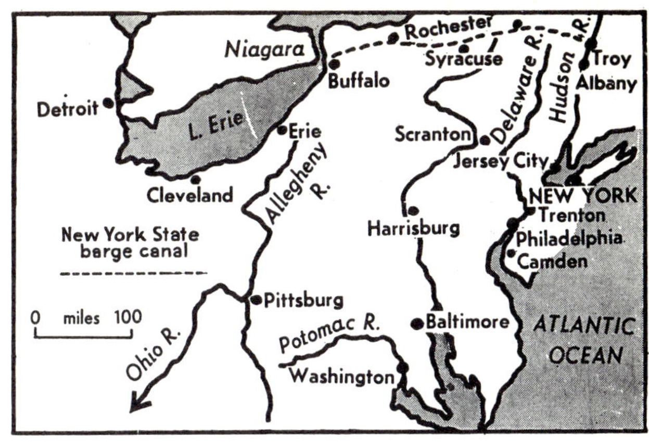

FIG. 195. *Seaboard cities of the North-East.*

were located within very easy reach of the ore. With the declining quantities of this ore, the industry will be maintained with ores from Labrador. Markets for the steel are the engineering and agricultural industries of the Middle-West, the Detroit motor industry and railway and highway construction.

Pittsburg is the regional capital and developed because of its important position controlling the routes to the Potomac and Susquehanna valleys. It became a hub of waterways and then of railways, and, because of its mineral resources, a commercial and manufacturing centre.

Eastern Pennsylvania centred on *Harrisburg* is not as well developed an industrial area as Western Pennsylvania. Its main asset was its own anthracite and the ore from the Cornwall mines east of the Susquehanna.

Seaboard Cities. The industrial activities of New York, Philadelphia and Baltimore rely on hydro-electric power, the gaps in the highlands which form routeways into the interior, and shipping facilities. None of these great cities has mineral resources close at hand.

Hydro-electric power, the ease with which fuel can be obtained from the Appalachian coalfields, and the ability to import all kinds of metals, have made these cities significant metallurgical centres. Philadelphia's early industries of spinning and weaving, iron smelting, and wagon and boat building have been transformed into large-scale textile, steel, locomotive and shipbuilding industries.

The domination of the Atlantic coast settlements by New York did not begin until the Erie Canal, now called the New York State Barge Canal, was opened in 1825, and made New York the only seaport served by an effective canal route. The others had rivers which led by trellised routes into the Appalachians; New York had the Mohawk valley which provided the only easy break through the highlands. Until the canal was built, goods from the interior were too expensive on the coast, as were imported goods offered for sale in the interior. The canal therefore stretched the hinterland of New York to include the Great Lakes, and this commercial advantage served New York well when canals gave way to railways.

FIG. 196. *New York: "Financial District" on the tip of Manhattan Island.*

At the present time, the traffic on the canal is mainly confined to the transport of grain. The Mohawk valley offers an easy route for roads and railways just as it did for the canal but vehicles can travel quicker. Thus the canal has become a "white elephant" but the Eire Canal Zone, an industrial belt about twenty miles wide along the canal, is a long line of cities from New York to Buffalo. The latter, at the junction of the Lakes and Canal routes, has a wide variety of manufactures and handles many different raw materials. Hydro-electricity from the Niagara River serves the mills of Buffalo and Rochester as well as the Canadian shores of Lake Ontario and Erie.

New York stands where the Great Lakes route meets the very important trans-Atlantic route and at this junction are Manhattan and Long Island, left as a result of submergence at the mouth of the Hudson. Its harbour comprises five tidal highways connected to the open ocean by two good channels, deep, well-sheltered and free from ice all the year. Although tidal range is only about 4½ feet, there is sufficient scour to keep the channels silt-free. The small tidal range allows the use of finger piers instead of dock basins as used at London, thus increasing the length of the water frontage. Because of the increasing size of ships, especially oil tankers, channels have been deepened; along the Manhattan side, the channel used by trans-Atlantic liners has been deepened to 45 feet. The port handles almost everything and, like London, makes almost everything, small goods inside the city and the larger goods in the surrounding metropolitan area.

New England, like Switzerland, is one of those regions of the world which has made the most of its natural resources and has overcome most of its disadvantages. The poor soils made farming difficult but encouraged the early settlers to take a share in the Atlantic fishing. Their surplus salt fish and their need for West Indian sugar gave them an interest in the triangular slave trade and led to their own shipbuilding industry—their clippers could outsail the British naval craft until the introduction of iron ships and steam. New England was the first part of the U.S.A. to be settled by a white population which was independent of the European mother-country, and

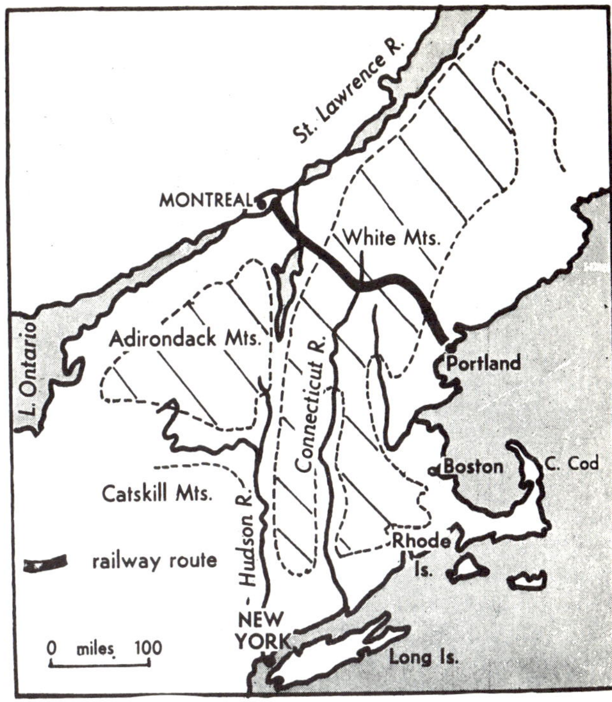

FIG. 197. *Simplified map of New England.*

they had no money crops as were harvested in the south. Instead they made their own clothing, shoes, implements and household needs in their own homes.

Settlements

With the change from cottage to factory industry, the rapids and falls supplied the power and the glacial hollows the water supply. The lowlands of New England have a higher proportion of the population engaged in manufacturing than any other region, in spite of not having any coal, iron or raw materials. The original textile and footwear industries have been retained and the products are for the most part made from a small bulk of raw material but with a high degree of skill.

The concentration of industry is shown by the way in which only 10 per cent of the people in the south and south-east are rural while in Vermont and northern Maine the figure rises to 60 per cent. Here another location of the U.S.A. tourist industry occurs, for scenery, history and tradition combine to attract the people of the industrial zones of the Appalachians and from the seaboard cities.

Boston, the largest city of New England, has a deep, sheltered, tidal harbour formed by the estuaries of two small rivers, the Charles and the Mystic. Like the rest of the east coast ports, it benefits from facing the Old World, but unlike New York, it had no easy access into the interior, until the Hoosac

Tunnel was cut for the Boston-Albany Railway. This came too late to prevent the diversion of New England traffic to New York. Another population feature to note is the northward movement of negroes, so that New York State has now more negroes than any southern state. This region of the U.S.A. is the main one to show a decline in the rate of population increase over the last twenty years. This decline was noticeable in New England and in the coal and steel areas of Pennsylvania and West Virginia, though the latter was the only state to show an actual decline in population total. On the other hand, there are still very high densities of population, with Rhode Island 812 people per square mile, Jersey City 806, Massachusetts 612, Connecticut 517.

QUESTIONS

1. Describe the differences in relief and structure to be found when travelling from west to east across the Appalachian system.
2. Discuss the importance of the iron and steel industry in Pennsylvania; New York's domination of the Atlantic coast settlements; the concentration of New England's factory industries into the south and south-east.
3. With the help of a sketch map, describe the major communication links between the Atlantic seaboard and the Mississippi Basin.

CENTRAL AMERICA: COUNTRIES AND RESOURCES

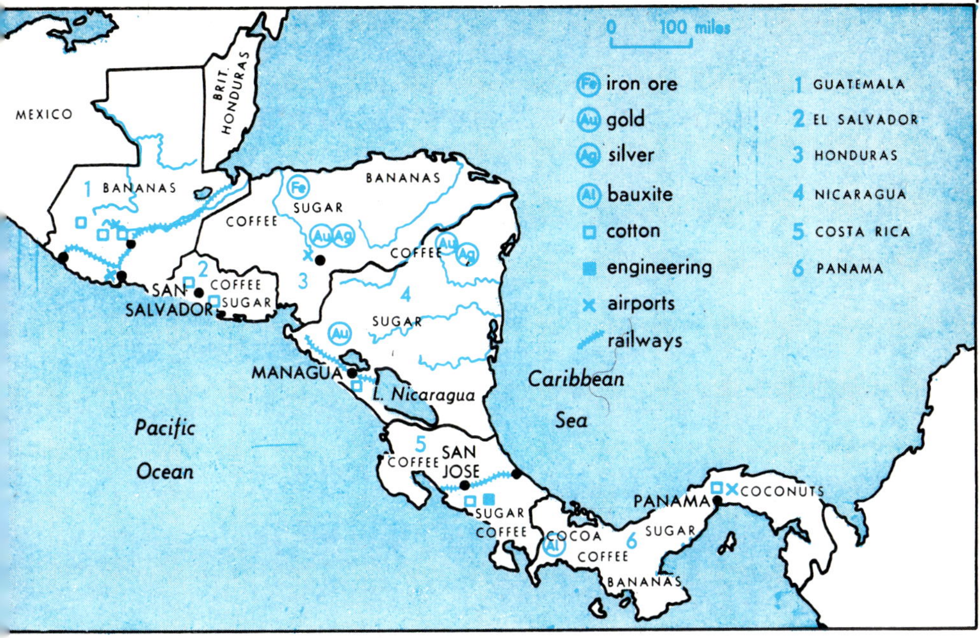

0 100 miles

iron ore
gold
silver
bauxite
cotton
engineering
airports
railways

1 GUATEMALA
2 EL SALVADOR
3 HONDURAS
4 NICARAGUA
5 COSTA RICA
6 PANAMA

MEXICO
BRIT. HONDURAS
BANANAS
COFFEE
SUGAR
SAN SALVADOR
MANAGUA
L. Nicaragua
Pacific Ocean
Caribbean Sea
SAN JOSE
COCOA
PANAMA
COCONUTS

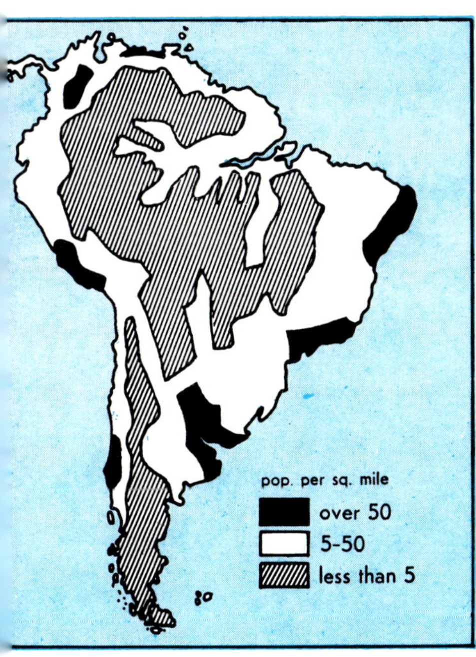

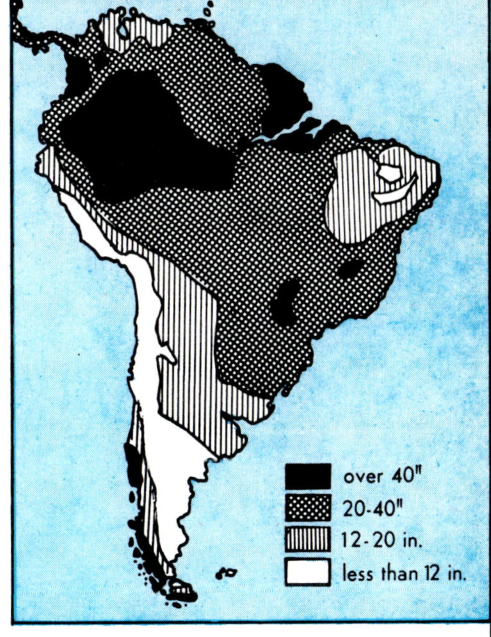

SOUTH AMERICA: POPULATION (left); ANNUAL RAINFALL (right)

AUSTRALIA: DISTRIBUTION OF FARMING, AND TYPES

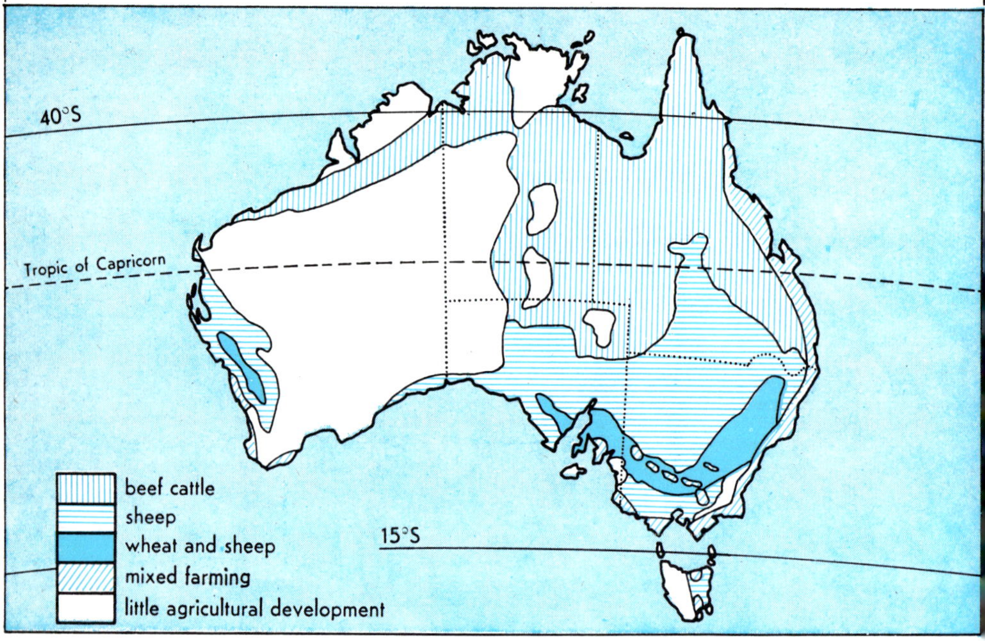

40°S
Tropic of Capricorn
15°S

beef cattle
sheep
wheat and sheep
mixed farming
little agricultural development

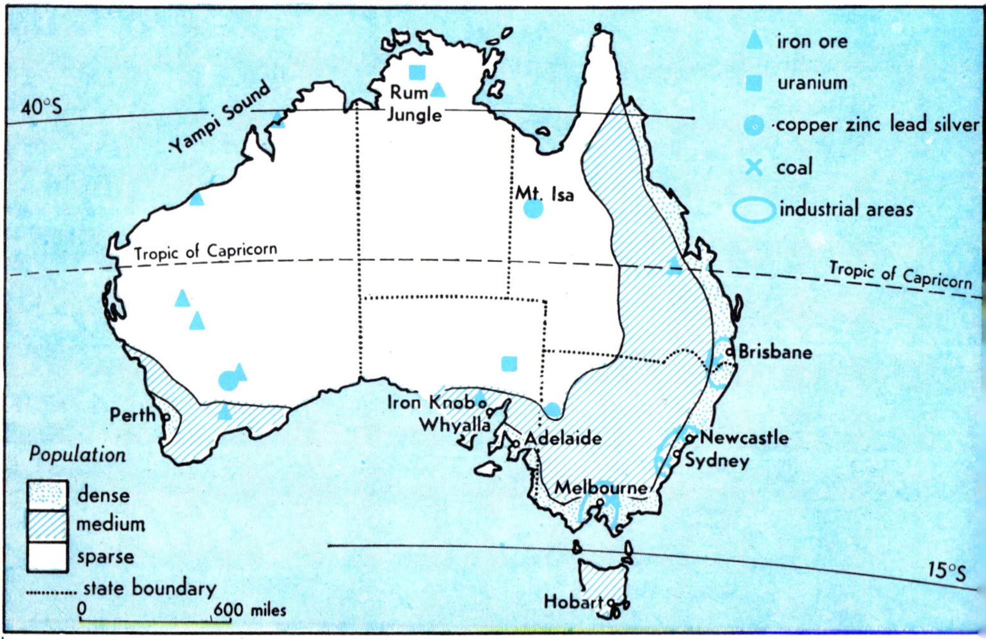

40°S
Yampi Sound
Rum Jungle
Mt. Isa
Tropic of Capricorn
Tropic of Capricorn
Perth
Iron Knob
Whyalla
Adelaide
Brisbane
Newcastle
Sydney
Melbourne
Hobart
15°S

iron ore
uranium
copper zinc lead silver
coal
industrial areas

Population
dense
medium
sparse
state boundary
0 600 miles

AUSTRALIA: POPULATION, MINERALS AND INDUSTRY

SOUTHERN CONTINENTS

CHAPTER 46

SOUTH AMERICA

SOUTH AMERICA lies between latitudes 12° N. and 54° S. of the Equator. Its maximum width occupies equatorial regions and it tapers gradually to the south and more sharply to the north. Most of this triangular-shaped continent lies within the Tropics. See map on page 453.

Relief

Structurally it is very similar to North America, with a long unbroken chain of young fold mountains along the west coast, large areas of lowland in the interior, and the remains of large block mountains near the east coast.

The great western cordillera of the Andean mountains was formed by folding, fracturing and uplift in Tertiary times. Earth movement is still occurring along this chain, as is shown by occasional earthquakes, as happened in Chile in 1961. Uplift of the mountain chain has resulted in mountains over 20,000 feet high, the remnants of extinct volcanoes forming the highest peaks. The almost unbroken chain creates a barrier to east-west communication. There are few passes, and these are sometimes over 11,000 feet, e.g. Uspallata Pass. The Andes can be divided into three regions. The northern cordillera has a distinct series of valleys and ridges, the main valleys following the structural lines of the chain. In the central region the range breaks up into the two main chains enclosing the Bolivian Plateau, a large area of interior drainage over 12,000 feet high. The third region, where these two chains unite once again, forms the southern cordilleran region: here snowfields and glaciers become more numerous as the distance from the Equator increases. South of 45° S. the coastline and the coast ranges have been flooded by the sea and a Norwegian-like fiord coast exists. The proximity of the mountain ranges to the west coast is reflected in the short, swift-flowing rivers that find their way towards the west coast. Their estuaries provide little opportunity for the development of modern harbours. The western coastal plain is very narrow, disappearing altogether in the south.

OCL/GEO 2/2C*

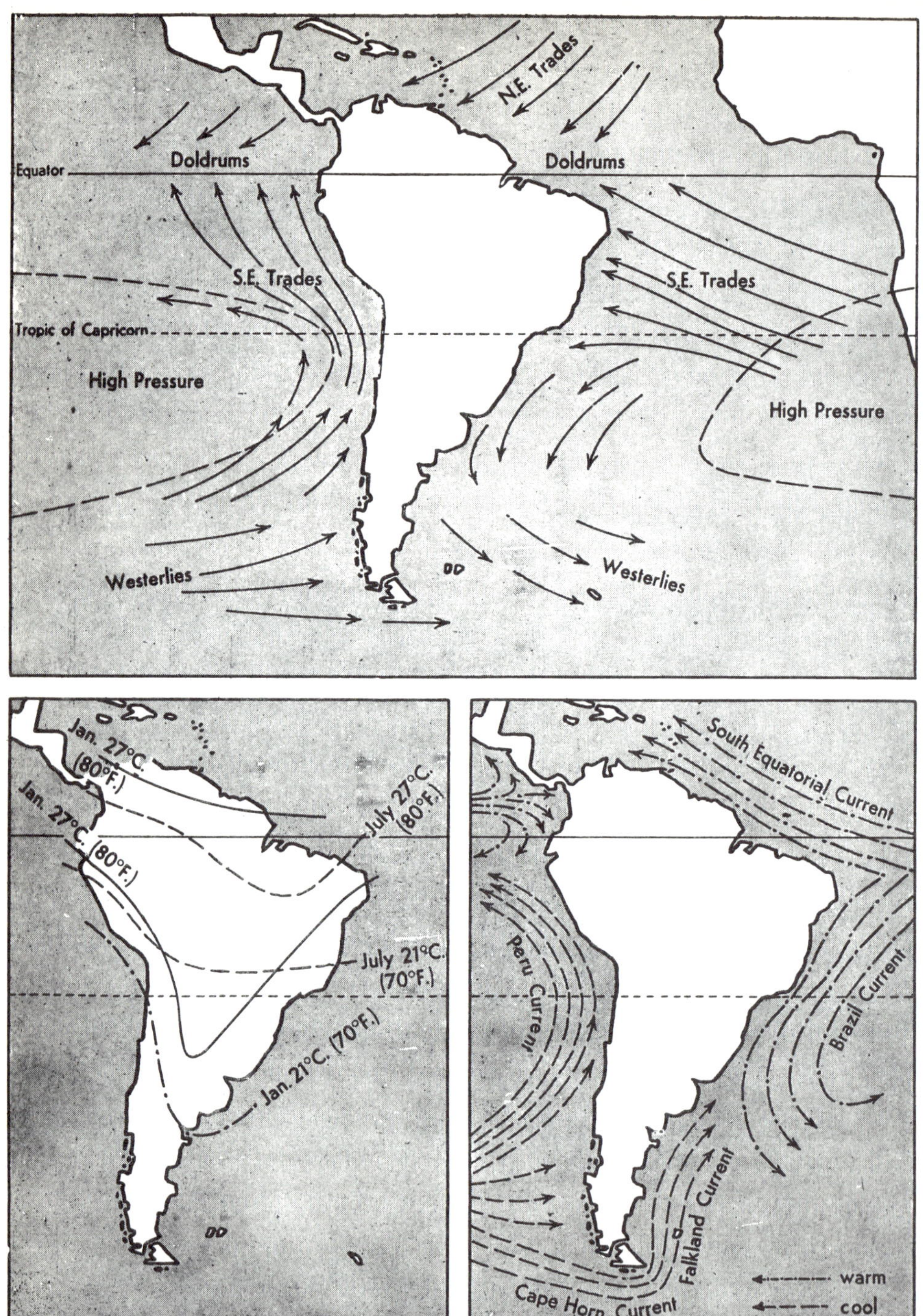

FIG. 198. *General climatic features of South America: air currents (top); January and July isotherms (bottom left); ocean currents (bottom right).*

Near the east coast lie three major highland plateaux—the Guiana Highlands, the Brazilian Highlands and the Patagonian Highlands. These ancient blocks of crystalline rock, with their thin covering of sedimentary deposits, have undergone considerable erosion, which has resulted in deeply dissected river valleys, leaving large flat-topped structures with steep sides called *mesas*. In the Guiana Highlands, the north-eastern border has a larger area of coastal lowland than is usual on the east coast. The Brazilian Highland block is tilted westwards and the main drainage is away from the east coast—with one major exception, the São Francisco River, which breaks through the eastern rim of the plateau to the east coast, where once again the coastal plain is relatively narrow. This location of the highland areas means that the streams drain towards the centre of the continent before uniting into three or four major rivers draining towards the east. Two important results from this are the lack of easy routeways into the continent, retarding its development; and the possibility of greater development of the major waterways as a means of carrying trade into the heart of the continent. The length of these rivers, their low gradient and the relatively few rapids on their lower courses have made them the arteries of the continent for centuries.

The three major river systems of South America occupy the major areas of lowland. The Orinoco, rising in the Guiana Highlands, curves eastwards round the plateau, reaching the Atlantic across flat plains, known as *llanos*. Since the maximum depth of the Orinoco is only 11 feet, it cannot be used by ocean-going vessels; and rapids 600 miles inland completely interrupt any form of navigation. The River Amazon drains an area of some two-and-a-half million square miles, from the Andes (where the upper reaches of the main stream are known as the Marañon) to the Atlantic. Ocean-going vessels can reach Iquitos, which lies 2,000 miles upstream, and smaller vessels can reach the Manseriche Gorge, where the Marañon leaves the Andes. Many of its tributaries are navigable for short distances, until their courses are interrupted by rapids; e.g. Porto Velho, 200 miles up the Madeira River can be reached by vessels up to 6,000 tons. The third lowland region, the basin of the River Plate, is by far the most important because it lies in the warm temperate zone; although smaller than Amazonia, the land drained by the Amazon, it contains nearly four times as many people. The River Plate system consists of three major streams—the Pilcomayo from the Andes, and the Paraguay and Parana rivers that drain the Brazilian plateau.

These three vast areas of lowland, separated by badly drained, ill-defined watersheds, are built of silt brought down by the rivers into an area from which the sea has retreated in relatively recent geological times. The flatness of the area results in tides up to 500 miles up the Amazon, and in annual flooding of extensive areas of land in Amazonia.

Climate

The major climatic features of South America are illustrated, but certain points should be noted. It is the only southern continent to project far into the temperate zone; and, unlike the land areas of the Northern Hemisphere, it tapers towards the Pole. It is only in north-west Argentina that the temperate continental type of climate exists. The Andes with their perpetual Arctic climate at the summit—even at the Equator—not only modify the climate with their altitude, but also form a marked meteorological barrier, causing rain shadows in certain places. Temperatures along the west coast are greatly affected by the cool Humboldt, or Peru, current of the Pacific. Although this current greatly influences the formation of cloud and fog, there is little rain and there are coastal deserts in Chile and Peru. See map on page 453.

Colonization

European colonization followed the journeys of the great Spanish and Portuguese explorers. There were two distinct streams, the effects of which are seen today in language and culture. The Spaniards worked from their base in the West Indies and moved down the Central Andean Plateau, thence to Chile and Argentina; the Portuguese colonized the eastern coast of Brazil. By the Treaty of Tordesillas in 1494, the Pope awarded the lands west of roughly the 48° meridian to Spain and those to the east to Portugal. Portugal ultimately controlled all Brazil, today the largest political unit in South America.

The Portuguese colonization was a very gradual process. Portugal relied on her links with the sea, thus penetration inland was achieved by river valleys. The coastal plain was hot and unhealthy and the densely forested areas impeded rapid advance. Furthermore, the population of Portugal itself was small and at the same time she was trying to establish colonies in other parts of the world. But though slower, their colonization was much more peaceful than that of the Spaniards.

The Spanish conquest was essentially a military one, and followed the old and easy routeways of the Indians along the Andes, avoiding the inhospitable coastal plains and the densely forested areas to the east. Spanish control was established with the defeat of the three most powerful Indian groups—the *Incas* of Peru, the *Chibchas* of Colombia and the *Araucanians* of Chile. After this rapid conquest the Spaniards penetrated much more slowly into the more difficult country of Southern Chile and Argentina. Up to this point, there was a noticeable similarity between the Spanish and the Portuguese patterns of colonization.

The French, Dutch and English did not gain as much territory as they might have done, since they were fully occupied with fighting among themselves. Although their attacks in the seventeenth century succeeded in obtaining some of the better parts of Brazil, they were gradually pushed back to the poorer areas beyond the Amazon, that are now known as the Guianas.

Along the lines of her conquests, Spain rapidly established feudal rule. The situation quickly became that of a land-owning minority, while the majority became landless peasants with no middle class. There followed three centuries of corrupt administration by the Spanish with little development of the territory's vast potential. Spain did not have the resources to develop or fully colonize these territories, and by 1825 European settlement of South America consisted of a few isolated strips across the continent, a situation mainly resulting from the difficulties of travel and communication. In the Portuguese area the pattern of colonization was similar but was achieved with less violence, so that when the break with Portugal came, it was much more gradual and less impetuous.

The first thirty years of the nineteenth century saw a successful revolt by a resentful Indian population against this feudalism imported from Spain and Portugal, a feudalism that gave no real place to the subservient Indians.

The new republics that grew up established their frontiers along the ill-defined boundaries of the old Spanish vice-royalties. This accounts for parts of Colombia, Peru and Ecuador lying to the east of the Andes—these have become "problem" areas, since it is difficult for the Pacific states to develop them. Furthermore, the lack of accurate frontiers has led to boundary disputes, particularly when some valuable mineral or other resource has been discovered along the border area. The overthrow of European power led to the gradual development of a middle-class and the slow growth of trade and industrialization. However, the effect of Spanish occupation, plus the American Monroe Doctrine, which kept South America cut off from the rest of the world, prevented South America making full use of her potential.

People of South America Today

Indians of a great variety of cultural levels—from Stone Age hunters and food-collectors, to members of fairly advanced societies—still form the chief element in large areas of the interior of Brazil and the countries of the Andes. The largest European influence in the population is the Spanish and Portuguese—the Iberian; this is a natural result of three centuries of occupation and the continued use of the Spanish and Portuguese languages and cultures. Probably the descendants of the Iberians, who are most evident in Argentina and Uruguay, do not outnumber the Indians. The Spanish and Portuguese found Indian labour insufficient and imported large numbers of slaves from Africa, their descendants being particularly noticeable in north-east Brazil.

The greater part of the population of South America is a blend of these three elements. All possible combinations exist, the largest single group, about one-third of the population, being the *mestizos*, the Iberian-Indian mixture. The Iberian-Negroid mixture, the *mulattos*, is common in Brazil, while the Indian-Negroid combination, the *zambos*, or *cafusus*, comprises a relatively small unit.

Since these countries gained their independence, there has been a steady emigration of Europeans to the continent, particularly Spanish and Portuguese, who provide the biggest single national group; but Germans, Poles, Swiss, British, French and people from many countries in south-east Europe live there. The twentieth century has brought emigrants from Japan, India, China and countries of the Middle East. They have either settled in the large towns or work on the plantations. This mixing up of races has had the advantage that there is little or no possibility of racial problems (as, for example, between black and white) such as have beset other countries.

The map on page 453 shows the distribution of population. Although relief, climate and vegetation are the principal factors governing distribution, mineral wealth has attracted many people into the less hospitable areas. The differences in political stability, government policies and economic state of the various countries result in differing densities in areas of similar environment, e.g. the Oriente of Peru and Ecuador.

The following table demonstrates the rate of population growth of many South American countries. The U.S.A. is included for comparison.

	1950	**1960**	**1967**
Argentina	15,900,000 (1947)	20,009,000	23,255,000
Brazil	51,900,000	70,968,000	85,655,000
Chile	5,933,000 (1952)	7,340,000	9,137,000
Peru	6,673,000 (1940)	10,365,000 (1961)	12,385,000
Venezuela	5,035,000	7,524,000 (1961)	9,352,000
U.S.A.	151,326,000	179,323,000	199,118,000

North Andean States

The North Andean states, including Colombia, Ecuador, Peru and Bolivia, can be divided into three main regions—the coastal lowlands, the cordilleran area and the Oriente, the areas to the east of the Andes. There are local divergences from this pattern, e.g. Bolivia has no outlet to the sea and Colombia has several additional distinct regions. However we shall first look at the general pattern.

Oriente. This consists of the lowlands of the Amazon and Orinoco headwaters, and is an inhospitable region, isolated from the rest of the country by the great cordilleras of the Andes. Bolivia, Colombia and Ecuador have done little to develop this area leaving it as a region of sparse population, mainly Indian, practising agriculture at subsistence level. Ecuador abandoned all attempts to find oil here in 1950. Peru, however, is developing the area with a

FIG. 199. *Oil-refinery at Talara, a port in north Peru.*

three-pronged attack—agriculture, minerals and communications. Tea, cocoa, coffee and tobacco plantations have been established, particularly in the Pereno valley, and the Ganso Azul petroleum on the Patchitea River is being developed. Till recent years the rivers offered the best routeways, but now new roads and railways are being constructed which are opening up this region.

Andean Cordilleras. For historical and environmental reasons this has long been the major zone of settlement in these countries. Since temperatures become lower with increasing altitude, as one moves higher up the Andes one passes through different zones of climate and vegetation, each with its own individual agricultural pattern.

The economy of the Andean region depends on pastoralism and mineral-workings. Although most of the agriculture is at subsistence level, plantations have developed, particularly where irrigation is possible. Pastoralism includes sheep, llama and alpaca.

The great mineral wealth of the Andes, especially in precious metals and gemstones, attracted the early Spanish settlers. Today, however, industry is centred around other minerals. Cerro de Pasco in Peru is the centre of an area rich in copper, lead, zinc, bismuth and vanadium, while other parts of the same country are noted for cinnabar. Paz del Río has Colombia's only integrated iron and steel plant, while the emerald mines at Muzo are the most productive in the world. Oruro in Bolivia, once important for silver, is the centre of Bolivia's tin-mining.

Coastal Zone. This narrow strip between the Andes and the Pacific Ocean graduates from deserts in Peru in the south to the tropical lowlands in Colombia in the north.

(*a*) *Peruvian Desert.* This is a landscape of coastal oases which stand out against the desert wilderness. Streams from the Andes are utilized for irrigation allowing sugar and cotton to be grown commercially as well as subsistence crops. In the south, vines and tropical fruits are important. The produce of these oases has been a powerful factor in the growth of Lima and its outport Callao.

Petroleum, however, is more valuable than agriculture in this region, Peru being the third largest producer, after Venezuela and Colombia. Of several fields, that near Negritos, which is worked by a Canadian company, is the most profitable.

Although the River Santos is of little use for irrigation, its development for hydro-electric power has greatly influenced the largest national industrial development in Peru. Iron and steel works have been started. Rich iron ore (60 per cent iron content) is obtained from the Marcona field in South Peru (exported via the new port of San Juan). There is also coal in the Santos valley and zinc refining, the manufacture of superphosphates and cement are now important. Chimbote is the main port for this development.

In Ecuador the coastal region gradually changes from the Peruvian deserts in the south to the equatorial *chotos* of Colombia in the north. The main agricultural region is in the lowlands around Guayaquil, where plantation crops such as cocoa, bananas and rice thrive. The relative importance of these is greatly influenced by the fluctuations of world markets.

Ecuador has been developing this region rapidly and the discovery of petroleum on the Santa Elena peninsula has been a great advantage. In the north, shifting agriculture and the working of alluvial gold are still important, as is the growing of Toquilla straw for the manufacture of so-called Panama hats.

In Colombia, the coastal plain is one of the wettest areas of the continent, where dense forests make communications difficult and land utilization is limited to forest clearings.

(*b*) *Magdalena Valley and Caribbean Lowlands.* The valley's main importance lies in that it forms a routeway into this country where the lack of communications is a major economic problem. Many of the settlements are river ports acting as distribution centres for the surrounding areas. Old mining centres have become the focal points in the agricultural landscape where coffee is the main cash crop.

The Caribbean Lowlands are an area of tropical grassland with scrub forest. Thus cattle pastoralism is the main occupation. Arable farming is found round the larger centres.

CHILE

Chile occupies a strip of territory stretching 2,700 miles along the western side of the Andes. Its maximum width is 100 miles. With this latitudinal range it crosses three of the world's major climate belts, which provide a suitable regional division.

Northern Deserts. This is a hot dry region, in parts of which the only rain is a shower once in seven years. However, streams from melted snows on the high Andes enable water to be brought to mining communities, often by pumping and piping. The wealth of the northern deserts lies in their deposits of nitrate of soda, copper and iron ore. The nitrates are a valuable export, being used in the manufacture of chemicals, fertilizers and explosives. The industry suffered greatly from the competition of artificial fertilizers, but improvements in methods of working the deposits has helped to maintain the industry. Iodine is found in association with the nitrates.

FIG. 200. *Chile.*

In the south, there is a gradual transition to semi-desert and in this region there are large deposits of low-grade copper ores at Chuquicamata, Potrerillos and El Salvador. Large-scale methods make the working of this economic. Iron is also important at El Tofo, which supplies both the Chilean iron and steel centre of Huachipato and the U.S.A. mills at Sparrows Point.

Sheep-rearing is also an occupation of the semi-desert area, the flocks and herdsmen moving to the Andes for the summer (transhumance).

Mediterranean Heartland. With its Mediterranean climate, this is the richest agricultural region of Chile, producing all the standard Mediterranean crops, notably vines and wheat. The Central Valley is the most highly developed area and Andean streams provide irrigation water.

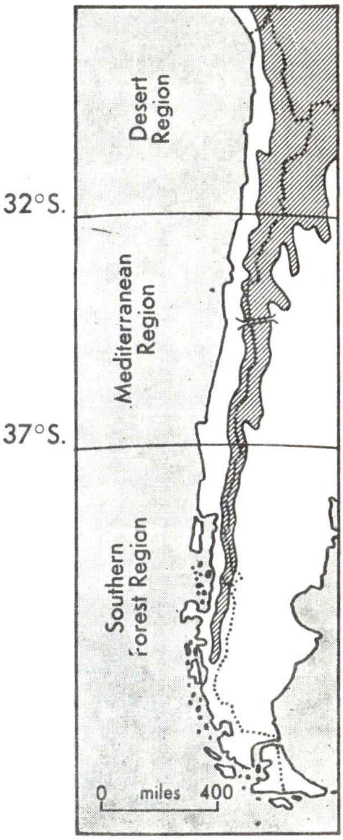

This region has many advantages—agricultural potential, central position between the mineral areas of northern Chile and the forests of the south, good connexions with the coast and the magnificent harbour at Talcahuano. Industry is flourishing using hydro-electric power from the Andes, and supplies of coal are available from Arauco Bay.

Southern Chile. In this region the cool temperate climate prevails and crops such as oats, hay and potatoes are common. However, the colonization has been slow and the area is still a valuable forest region. Some magnetite ores are mined near Corral. Petroleum has been found on Tierra del Fuego.

VENEZUELA

Although bigger than Britain and France together, Venezuela has a population not much larger than that of Greater London. For long it was a backward country, but with large resources of petroleum and iron-ore it is now developing rapidly.

FIG. 201. *Venezuela: Oil derricks on L. Maracaibo.*

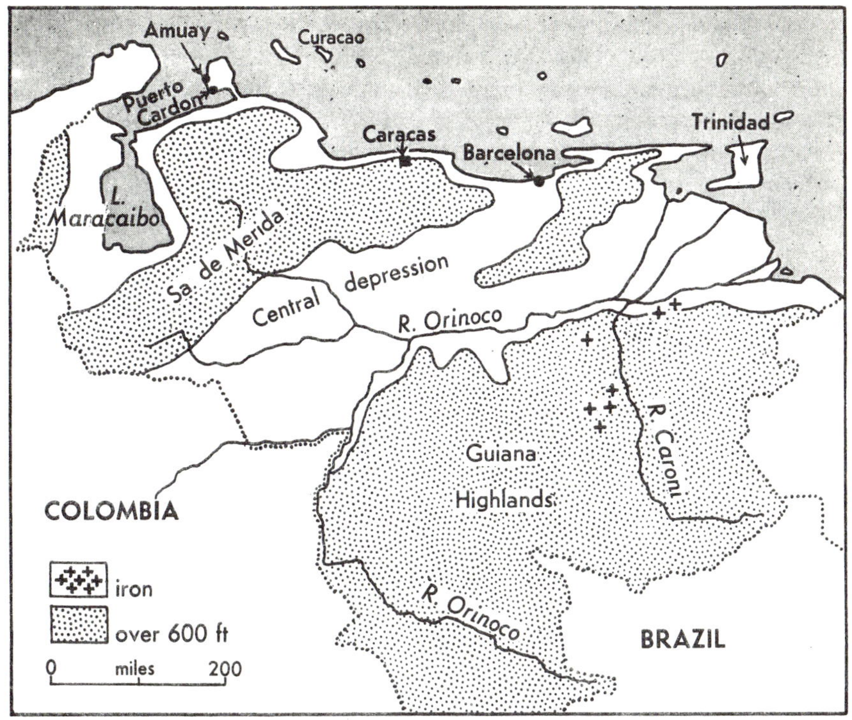

FIG. 202. *Simplified map of Venezuela.*

The major oil-producing area at present is the Lake Maracaibo basin. It is underlain by limestone and sandstone and now there are over 2,500 oil wells, on both the lake and the surrounding shore. A bar at the mouth prevented large tankers entering the lake, but a dyke-protected channel constructed in 1955 has improved the position. Oil, however, is still piped to coastal centres such as Amuay and El Cardón. The islands of Aruba and Curaçao are also important outports.

Agriculture is now very much a secondary occupation in this region. There are a few plantations for sugar, cocoa and coconuts and some shifting agriculture is still practical in the forests, but petroleum is of more economic importance and accounts for the great expansion of the old trading cities.

The Central Highlands were the original core of Venezuelan settlement. Their central depression is still the main agricultural region of Venezuela with its plantations of cocoa, sugar cane, cotton and coffee, and cattle are also important. Agriculture is also significant in the valleys leading down to the coast: because of the agricultural wealth of the area there is a higher density of population here, including *Caracas*, the capital.

The *llanos* of the Orinoco Basin form a vast featureless plain. Forests fringe many of the water courses, but with a climatic régime of wet summers and dry winters, savanna grassland is the predominant vegetation. This then is an area of cattle-rearing. The area is not as highly developed as it might be owing to the long period of Spanish control, lack of communications, disease and the alternate problems of drought and floods. The last mentioned is being overcome by large irrigation schemes. The economy of this area is gradually changing with the exploitation of the Maturín petroleum basin and the discovery in 1951 of one of the largest iron-ore (60 per cent iron content) deposits in the world. The deposits are found in the Caroni tributary valley. U.S.A. companies are developing this field, the ore being shipped to Pennsylvania. This development has established the Orinoco as a routeway.

To the south of the Orinoco, the llanos gradually give place to the Guiana Highlands, a vast undeveloped area with a considerable potential but unlikely to be utilized for some considerable time. Some diamonds are mined in the Upper Caroni valley, and gold at El Callao.

Petroleum is Venezuela's greatest wealth and the programme of "sowing the petroleum" is activating the exploitation of her vast mineral resources, the development of agriculture and the improvement of communications.

GUIANAS

The Guianas contain the last remnants of colonialism on the South American mainland. There are less than 1 million people, of whom 600,000 live in Guyana (formerly British Guiana)—the most developed. French Guiana, the least developed, has only 30,000 inhabitants. Each has two major regions: (1) a flat, featureless coastal strip about fifty miles wide, reaching the sea in a mangrove-fringed coastline; (2) part of the Guiana Highlands.

(1) **Coastal Zone.** In this region the coastal marshes, stretching up to ten miles inland, have been drained to give the only agricultural region. Sugar and rice are the main crops, with oranges a valuable export.

(2) **Guiana Highlands.** The Highlands are clothed in dense forests developed on lateritic soils. Some timber is produced, e.g. greenheart, which being extremely resistant is suitable for such heavy construction as lock gates and ships' hulls. Agriculture is of little importance. Minerals, particularly bauxite, the ore from which aluminium is obtained, are becoming increasingly important. Guyana and Dutch Guiana are now one of the world's leading producers of bauxite. Mining has been developed by North American companies and the ore is taken to Canada and the U.S.A. for processing. Diamonds and gold are also important.

Development of these colonies has been hampered by poor inland communications, inadequate harbour facilities owing to the shallow seas around the coast and the labour shortage since the end of the slave trade.

BRAZIL

Brazil, the giant of South America, is not only the largest country in the southern continent, but only the U.S.S.R., China and Canada exceed it in size. Although still a developing nation, it is already making its impact felt in the Americas. Let us consider it area by area.

Amazonia. The Amazon Lowlands extend over a million square miles and contain less than two million people, the majority of whom live in the eastern part, a quarter of a million living in Belem city alone. The area has a characteristic equatorial climate with twin maxima of temperature and rainfall. Convectional rain occurs almost every day and many parts experience over 100 in. per year. Thus it is a region of high temperatures and high humidity, where vast areas of tropical rain forest called *selvas* grow on poor alluvial soils.

At first sight these luxurious forests would appear to have a considerable timber wealth and, if cleared, would give good agricultural land. However, they consist mainly of hardwoods and one square mile may contain as many as a thousand different species, making them unsuitable for economic exploitation. When cleared, the soils provide poor agricultural land, and the only arable farming is found around the towns and the Belem-Braganca railway. Bananas, cocoa, cotton, maize, manioc and tobacco are grown, while Japanese immigrants cultivate jute and pepper. In some areas cattle-rearing is attempted. Plantation agriculture, notably for rubber, has been tried

FIG. 203. *General view of the coastal town of Bahia.*

but has met with little success, owing to its dependence on fluctuating world prices, distance from markets and labour shortages.

The shifting agriculture of the Indians is still the major method of land utilization in this area. A patch of forest is cleared by burning and cultivated for a few years before the people move on. Agriculture is supplemented by hunting and collecting from the surrounding forests. Brazil nuts, mainly from the Tocantins Basin, are a valuable export.

The area has few important mineral reserves although manganese is now being obtained in Amapá territory. Brazil has high hopes of finding oil along the foothills of the Andes.

Towns such as *Manaos* are located at the confluence of streams and originated as slave-collecting points for the plantations, around Jesuit mission stations, or in association with the rubber boom which collapsed in 1940. This boom led to the construction of the main railways. Many of the towns have declined but others still act as loading centres along the river, the main artery of communication.

Atlantic Coastlands. The Atlantic margins of Brazil, although having many subdivisions, can be considered as a single unit. The north-eastern area is the most fertile with three crops dominating the coastlands—sugar, cotton and cocoa. The wet coastlands with their rich soils made this one of the world's major sugar-producing areas, but owing to competition the area does not supply all Brazil's requirements. The cotton-growing area is in the transition zone between the arable (sugar) coastal lowlands and the drier pastoral interior of the Brazilian plateau. Although cocoa-production is second only to that of the plantations of West Africa, its development is hindered by all the well-known defects of Brazilian agriculture—absentee landlords, lack of attention to cultivation methods, and planting trees without adequate protection. Thus in an area with a high potential, one finds abandoned estates.

The mineral wealth of this area is not great, although the petroleum field at Bahia is the only one known in Brazil at present. Along the coast the evaporation of sea water provides an important source of salt.

The southern part of the Atlantic margins is the most densely peopled of all Brazil and contains over half its total population. The first stage in the opening-up of the area was associated with roaming pastoralists, some of whom found gold and diamonds in Minas Gerais. This brought a great increase in settlement in the early eighteenth century and established *Rio de Janeiro* as the country's capital and major port. This phase had passed by the early nineteenth century, and the immigrants focused their attention on coffee, particularly in the Paraiba valley. With rapidly expanding coffee markets in Europe and North America and the suitability of the well-drained soils, *fazendas* (farms) rapidly spread westward, causing a demand for an increased stream of immigrant labour.

FIG. 204. *Brasilia: the president's palace. The city has been built to a comprehensive plan involving a number of ambitious architectural projects.*

The present tendency is towards industrialization, utilizing the vast resources of hydro-electric power, with Sao Paulo becoming the major industrial centre. The industrial pattern reflects the earlier stages of development. Gold is mined south of Bello Horizonte, and industrial diamonds from Diamantina. The most important mining, however, is the rich, easily worked reserves of iron in the Serra de Espinhaco, and of ferro-alloys, particularly tungsten, in the same area. The agricultural and pastoral wealth of the industrial hinterland also provides the basis for a considerable number of processing and packaging industries, and the gradual evolution of chemical, mechanical, metallurgical and electrical industries is slowly completing the industrial pattern.

In the temperate south of Brazil, south of Sao Paulo, grasslands with some forests predominate. This region is Brazil's main source of home-produced meat.

Central States. The two vast states of Matto Grosso and Goias form one of the least-known areas of all Latin America. Here the characteristic vegetation is the *campos*, which is like the savanna grasslands of Africa but with a much higher density of trees. Isolation, the region's main problem in development, is being overcome by railways and new roads emerging from the industrial areas of the south-east. Brazil is determined to develop her interior as is witnessed by her building of the new capital, **Brasilia,** in the margins of this area. This is still essentially a cattle-ranching area with subsistence agriculture, but in recent years plantation agriculture, particularly coffee growing, has been spreading into the area.

URUGUAY

Uruguay structurally belongs to the Brazilian Highlands, but the landscape is rather that of wide rolling plains. Temperatures range from 10-12° C. (50-55° F.) in winter to 21-27° C. (70-80° F.) in summer, with rain at all seasons of the year. The mean annual rainfall is just over 40 in. Thus grassland is the predominant vegetation, and the country ideally suited to pastoralism. Considerable quantities of pedigree stock have been imported from Britain and both cattle and sheep products are major export items from ports such as *Montevideo*, the capital, and *Fray Bentos*.

The fertile alluvial soils along the Plate estuary encourage arable farming and there is a considerable diversity of crops, including sugar beet, tangerines and vines. Some fishing is carried on and tourism is becoming increasingly important on the beaches east of Montevideo. The Plate estuary is also the main industrial area of the country, processing the produce of agriculture and pastoralism. Uruguay is also trying to develop some new industries using imported raw materials, e.g. chemical and electrical goods.

PARAGUAY

The development of Paraguay has been hindered by many political difficulties—disastrous dictatorships and Indian revolts, and it is one of the most backward countries in South America today. It can be divided into two differing regions—Western and Eastern Paraguay.

Western Paraguay lies between the Paraguay and the Pilcomayo rivers. It is really part of the Gran Chaco and much of the country is covered with forest jungle and swamp. The forests contain quebracho, the tree from which tannin is obtained. This is the country's major export to the U.S.A. Some commercial timber is also obtained. On the grassland areas, some cattle-ranching takes place along with the cultivation of cotton and other crops but lack of communications is restricting development.

In **Eastern Paraguay,** where the rainfall is lower (about 40 in.), there are extensive savanna grasslands. This is part of the western slope of the Brazilian Highlands. Although cattle-ranching is important, the herds are of poor quality and hides are still the major product. Around the towns like Asunción, the capital, there is some cultivation. Tobacco, cotton and oranges are the major crops and some coffee is also grown for export. Lack of coal, oil and political stability have hampered industrial development.

ARGENTINA

Argentina is the largest of the Spanish states and can be conveniently divided into four regions:

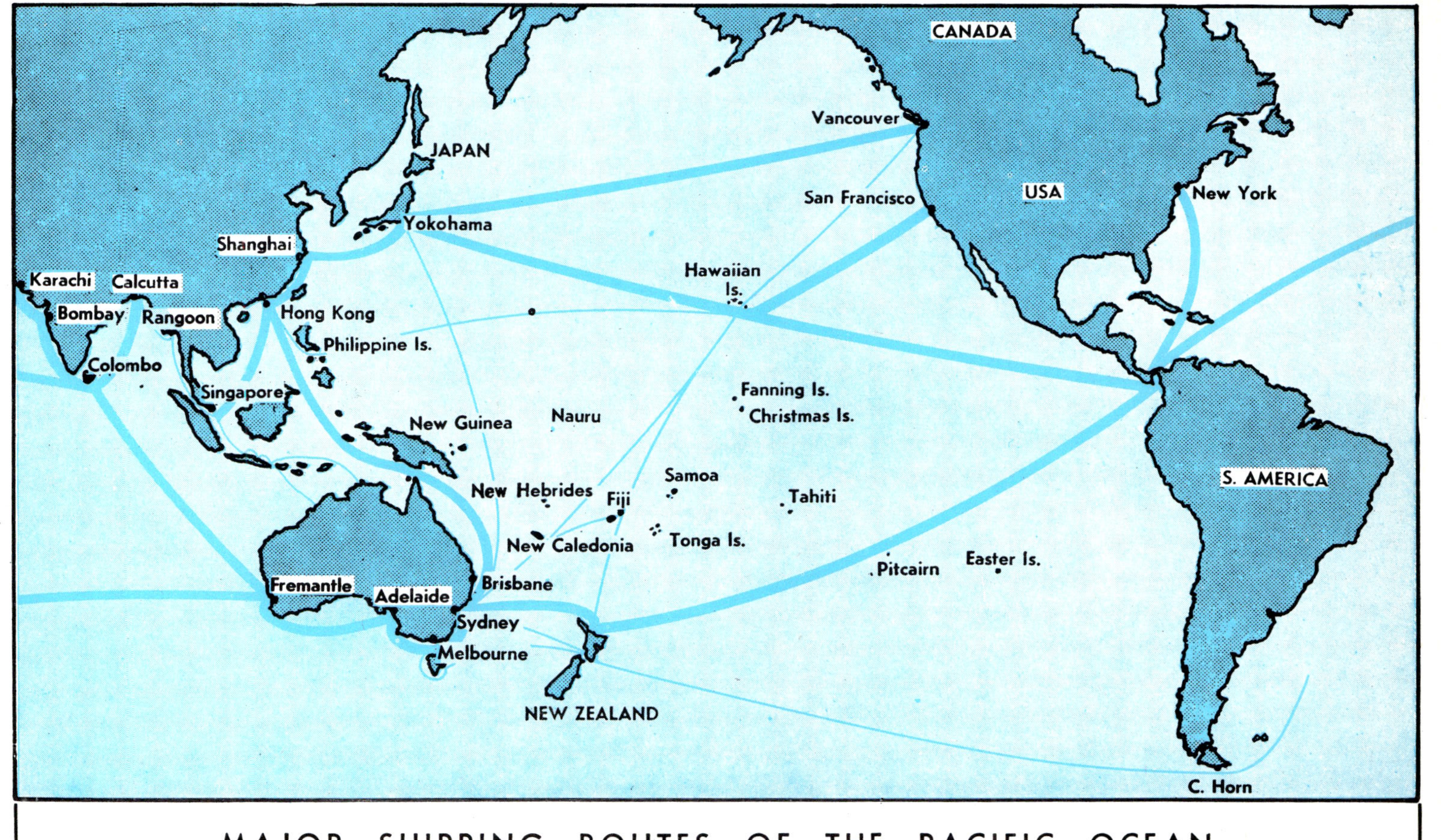

MAJOR SHIPPING ROUTES OF THE PACIFIC OCEAN

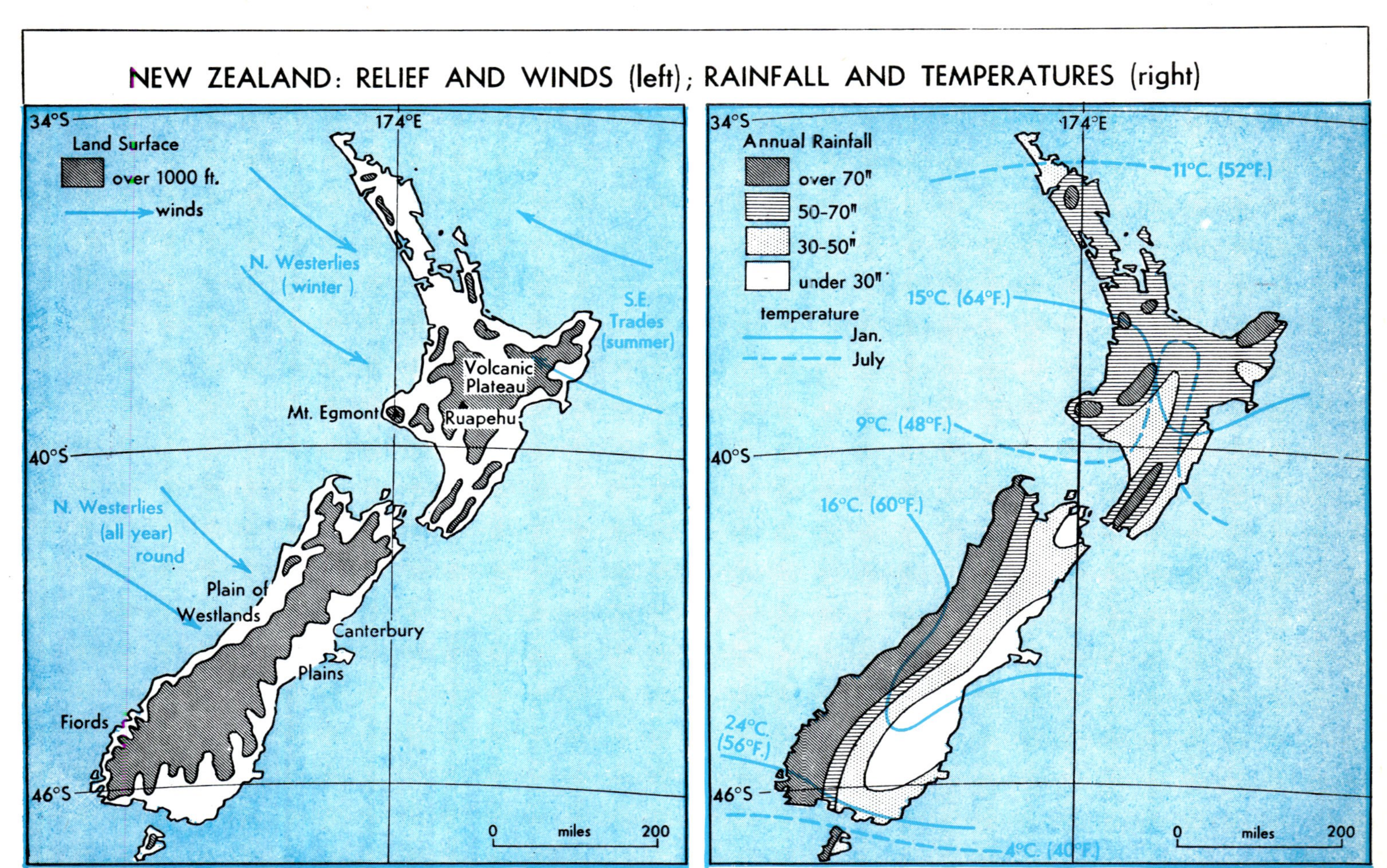
NEW ZEALAND: RELIEF AND WINDS (left); RAINFALL AND TEMPERATURES (right)
34°S
174°E
40°S
46°S
Land Surface
over 1000 ft.
winds
N. Westerlies (winter)
S.E. Trades (summer)
Volcanic Plateau
Mt. Egmont
Ruapehu
N. Westerlies (all year) round
Plain of Westlands
Canterbury
Plains
Fiords
0 miles 200
34°S
174°E
40°S
46°S
Annual Rainfall
over 70"
50-70"
30-50"
under 30"
temperature
Jan.
July
11°C. (52°F.)
15°C. (64°F.)
9°C. (48°F.)
16°C. (60°F.)
24°C. (56°F.)
4°C. (40°F.)
0 miles 200

1. North-West Andes
2. Chaco and North-East
3. Pampa and Entre Rios
4. Patagonia

North-West Andes. This was the area of earliest settlement in Argentina, with first Indians and then Spaniards moving out from the earlier zones of settlement along the Andes. The settlement pattern closely follows the availability of water, upon which agriculture depends. In the far north-west around Tucuman, which experiences about 35 in. of rain a year, sugar is grown, sometimes assisted by irrigation, on large plantations. Farther south where it is more arid, a pastoral economy predominates, mainly involving sheep and goats. The flocks move to the Andes during the summer.

However, where there is ground water, or where Andean streams flow into *playas* (temporary lakes), there is a very different landscape—that of oases so that the area is sometimes referred to as the Garden of the Andes. In these, alfalfa, cotton, wheat, figs, olives and grapes flourish. In fact, this is the major wine-producing region of Argentina. The principal oases are Mendoza, San Juan and San Rafael.

Chaco and North-East. This area is in many ways a continuation of the landscape of the neighbouring areas of Bolivia, Paraguay and the Brazilian Plateau. The principal activity in the Chaco is forest exploitation. The red quebracho tree is used for making railway sleepers, locomotive fuel and fencing posts, while the white quebracho yields tannin. In the grassland areas of both the Chaco and the north-east, cattle-ranching predominates, despite the cattle tick parasite. Maize is grown as fodder, and there are occasional plantations of cotton and tobacco. This, however, is one of the most backward areas of Argentina.

Pampa and Entre Rios. The vast monotonous plains of the Pampa, which now produce 80 per cent of Argentinian exports, were avoided by early settlers. Because of a number of factors their potential began to be realized in the nineteenth century. These factors included:

1. Increasing political stability.
2. Expanding markets—particularly with the growth of industrial areas in Europe and North America.
3. General improvements in agricultural methods.
4. Available capital—notably from Britain with its increasing industrial prosperity.
5. Immigrant labour to develop the area.
6. The development of railways.
7. Introduction of refrigerated ships.

The raising of beef cattle is a very important occupation throughout the whole of the Pampa and Entre Rios, and, where conditions are suitable, there is also the commercial growing of wheat, maize and flax. The sketch map shows the distribution of these crop zones.

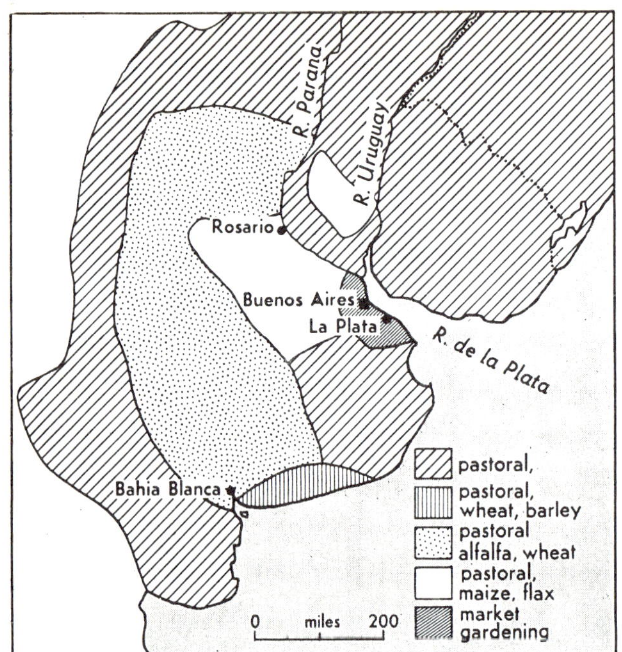

FIG. 205. *Argentina: simplified map showing the main farming zones.*

Buenos Aires, with its population of 7 million including suburbs (the largest city in the Southern Hemisphere), dominates the human geography of the Pampa. There is a market-gardening zone in the fertile loess soils around the Plate estuary. The emergence of the Pampa as a vast agricultural area has led to the development of the associated packing and processing industries in ports such as La Plata. The importance of the region is reflected not only in the density of population but also in the density of the railway network focusing on Buenos Aires.

Patagonia. Much of Patagonia lies in the rain shadow of the Andes, and considerable parts of it experience under 10 in. of rain per annum. There is a gradual transition from the grassland vegetation into the desert landscape of the south. In the nineteenth century, British and Spanish shepherds established themselves there and today Patagonia is a major wool- and mutton-producing area. In the north, where irrigation is possible, some arable farming is carried on.

Argentina's principal oilfield is situated at Comodoro Rivadavia and a pipeline carries natural gas 1,100 miles to Buenos Aires. Some poor lignite (brown coal) is found in the far south-west of the region.

QUESTIONS

1. Make a regional division of Chile and describe each region in turn.
2. Give a brief account of one of the following: Patagonia; the Pampas.
3. Describe the landscape associated with one of the major rivers of South America.

CHAPTER 47

AUSTRALASIA

THE name Australasia refers to the Commonwealth of Australia, New Zealand and the many groups of islands in the Pacific Ocean, often termed Oceania.

AUSTRALIA

Australia can be divided into three major structural units:

1. Western Plateau
2. Eastern Highlands
3. Central Lowlands

The **Western Plateau** is a remnant of the ancient continent of Gondwanaland. In its arrangement of rocks, it is very similar to other remnants such as the Deccan of India and the Brazilian Plateau. Steep slopes separate the tableland from the narrow coastal plain on the west, but beyond these the plateau slopes away to the north and east in a monotonous landscape at about 1,200 feet. Above this, ridges of more resistant rocks stand out as higher ground, such as the Macdonnell and Musgrave ranges. This is an area of sparse vegetation which gradually changes to complete sandy desert in the north-west. Although inland drainage to salt lakes, such as Lake Amadeus, is common on the west coast, some small streams drain to the Indian Ocean. In the south lies the porous limestone area of the Nullarbor Plains—an area with no surface drainage and very little settlement.

The **Eastern Highlands,** with an average height of 2,500 feet, form a barrier-range along the east coast extending into Tasmania in the south. They consist of a number of plateau blocks with important gaps between them, such as the Kilmore Gate and the Cassilis Gate, enabling communication with the interior. Short, swift-flowing streams drain to the east coast, which has been submerged to give ria estuaries, such as Sydney Harbour. Around these rias are areas of lowland.

The 1,000-mile long *Great Barrier Reef*, the largest coral reef in the world, lies 50-150 miles off the eastern coast. The reef has been built up from the limestone remains of tiny creatures, which can live only in shallow, clear water warmer than 20° C. (68° F.). The reef affords some shelter from onshore winds but it is more of a problem than an advantage to shipping, as the waters between it and the coast are shallow and have dangerous tides.

In contrast to the steep slope on the east, the Eastern Highlands slope gently down to the Central Lowlands on the west.

The **Central Lowlands** in their northern part form the Great Artesian Basin. Rain falling in the highland margins percolates through the sandy layers between impervious clays to form underground reservoirs of water. When a well is bored, the water may rise to the surface of its own accord, or it may have to be pumped up. The water is brackish and unsuitable for human use, but its availability is a vital factor to the cattle and sheep farmers of the region.

AUSTRALIA—RELIEF AND CLIMATE

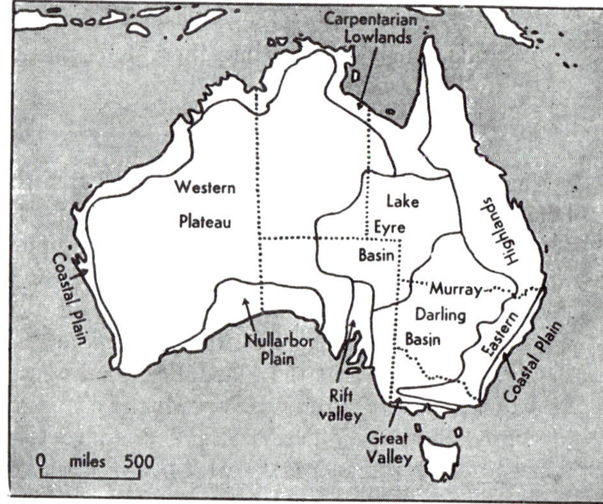

FIG. 206. *Regional division based on relief.*

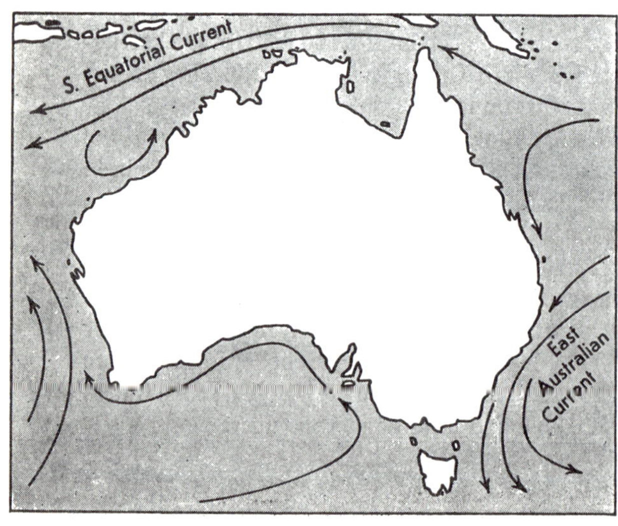

FIG. 207. *Ocean currents affecting Australia.*

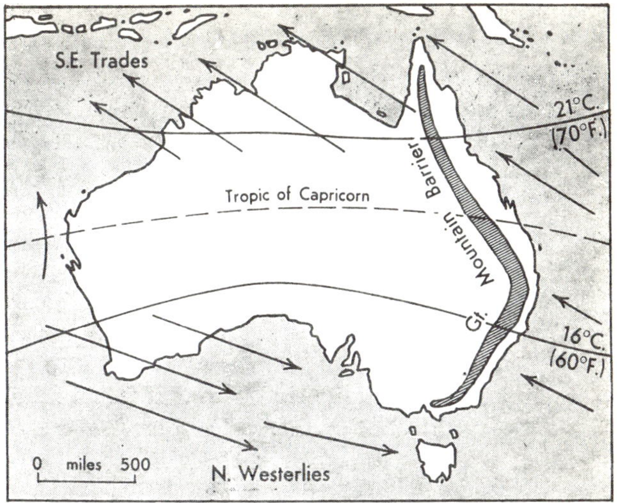

FIG. 208. *Wind systems and isotherms (July).*

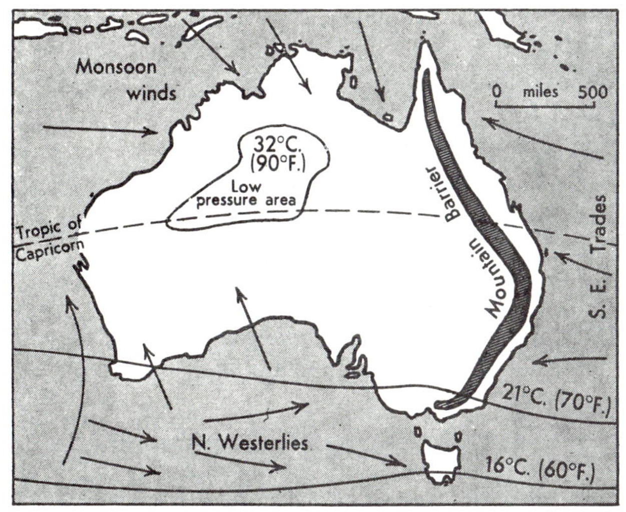

FIG. 209. *Wind systems and isotherms (January).*

FIG. 210. *Annual rainfall distribution.*

The southern portion of the lowlands includes the rift valley region to the west and the Murray-Darling Basin, where rivers fed by springs and snow-melt in the spring bring sufficient water for better pastoral farming than in the north, and in good years allow navigation far inland by small craft. The Lake Eyre basin is a vast area of inland drainage.

The south-western tip of Australia and the area round Adelaide experience a Mediterranean climate which merges, as one moves northwards, into desert conditions. In the north-east, there is a monsoon region, while farther south on the east coast, the climate is warm temperate. Tasmania, in the south, has a climate similar to that found in the British Isles.

Maps of farming, mineral resources and industries appear on page 454.

Western Australia

Western Australia covers one-third of the total area of Australia. The two major agricultural areas are (1) in the north, which experiences monsoonal rain in summer, and (2) the south-western tip, lying within the winter-rain Mediterranean belt. From these two areas, aridity increases towards Central Australia, with the vegetation changing from scrubland to desert.

Northern Districts. With a rainfall of 20-40 in. a year, this area can support cattle-rearing. A freezing and processing plant has been established at *Wyndham*, but the long distances which the cattle had to be driven to reach the plant meant a considerable loss in weight. Slaughterhouses have been built at Glen Roy Station, some 200 miles from Wyndham, and the carcasses are flown to Wyndham. The transport charges are offset by the increased weight of the cattle. Iron-ore is worked on Yampi Sound. Cotton and food-stuffs are grown under irrigation from the Kunanurra dam on the Ord river.

Eastern Deserts. These deserts are inhabited only by a few aborigines.

Western Scrubland Region. Mineral wealth attracted the first settlers to this area with the discovery of gold at *Kalgoorlie* and *Coolgardie.* In the north, in and around the Hamersley Range, immense iron-ore deposits have been found in the 1960s, and new towns, ports, and railways have sprung up. A big oil field on Barrow Island came into production in 1967.

South-West Region. Rainfall decreases inland from a maximum of 40 in. a year on the coast and there are a series of agricultural zones related to rainfall. The area with the heaviest rainfall is well wooded especially with hardwoods, e.g. eucalyptus. Fruit growing and mixed farming are important occupations. In the next belt, with a rainfall of 20-12 in., wheat becomes the predominant product and tomatoes are particularly notable around Geraldton. Cattle and sheep are important throughout the south-west.

Coal around Collie, iron-ore at Koolyanobbing, and bauxite in the Darling Range near Perth are important mineral resources. Nickel has also been discovered by prospecting companies.

Northern Territory

The Northern Territory has only 59,500 inhabitants, of whom a third are full-blooded aboriginals. The effects of climate and vegetation restrict development. Monsoon rains bring as much as 60 in. of rain a year to Darwin, the major centre and airport in the north, but the rainfall decreases to a few inches in the southern part of the state. The raising of beef cattle is the main activity and is centred in three areas—Alice Springs, the Barkly Tablelands and the Victoria River districts. Here also the cattle lose considerable weight in the long walks to the Wyndham meat works, the fattening stations of Queensland or the railheads at Birdum and Alice Springs.

Mineral working in the state is developing rapidly. Manganese and iron ore mines and an alumina plant have recently come into production and there is a major lead-zinc deposit. Copper and gold mining is also expanding, and some oil and natural gas have been found.

South Australia

Some 80 per cent of the state experiences under 10 in. of rain a year. This district is characterized by large tracts of sandy desert and arid landscapes. In the artesian basin area of Lake Eyre, sheep rearing is possible but with low densities of stock per square mile. Thus the main settlements are in the Murray-Darling basin and in the south-east.

The Murray Plains have developed considerably with the help of extensive irrigation. Large areas are devoted to sheep and cattle and there is a marked difference in the type of agriculture to the north and south of the 36th parallel. In the wetter areas to the south, the main crop is oats, while dairy farming is carried on in the south-east. To the north of the 36th parallel, as the climate becomes warmer and rainfall decreases, wheat predominates. North of the Murray River it is too dry for wheat and pastoralism takes over with much lower densities of cattle and sheep than are found farther south.

The Eastern Highlands consist of ancient crystalline rocks with considerable mineral reserves. Small quantities of silver, lead and gold are worked, but the once-important deposits of copper are now almost exhausted.

Between the highlands and Eyre Peninsula, lies the Great Valley, a partly submerged rift valley forming an important corridor and the main outlet for the agricultural resources of the state. *Adelaide* is the main centre, another centre being Port Pirie, which smelts the mineral products from Broken Hill. The northern part of the valley around Lake Torrens is of little value, but the southern portion is fertile. Wheat dominates the agricultural scene as it does in the valleys of the eastern highlands and Eyre Peninsula.

Eyre Peninsula is also developing its pastoral resources, and iron deposits are being worked at Iron Knob and in the Middleback Ranges. Coal deposits are becoming increasingly important at Leigh Creek.

Queensland

Queensland can be divided into four major regions:

1. Western Highlands
2. Western Plains
3. Great Dividing Range
4. Coastal Plain

Western Highlands. The value of this area of old rocks lies in its mineral wealth. The Mount Isa area near Cloncurry, once important for gold and copper, now obtains its main income from lead and zinc production.

Western Plains. In the wetter monsoonal areas to the north, cattle-production predominates, while in the drier south (much of which depends on artesian water) sheep-farming takes over. Supplies of artesian water are not inexhaustible and as the flow has shown signs of diminishing some care is now being taken in its use. Petroleum has been found at Moonie, and the production of bauxite is very important at Weipa on Cape York Peninsula.

Eastern Highlands. These form a considerable barrier to communication with the interior and, apart from their agricultural value, have important mineral deposits—especially of gold, copper and tin. Gold-production has been declining rapidly.

Coastal Plains. This is the most highly developed area of Queensland and there is a close relationship between climate and agriculture. In the warm temperate south, north of Brisbane the capital, maize is the major grain crop. But in the monsoonal area to the north, sugar plantations dominate the landscape. Labour shortages are hindering the development of the sugar industries, since it is difficult to obtain white labour for the sugar fields. The climate also encourages the growth of crops such as cotton, tobacco and tropical fruits. There is a chain of small port towns along the coast, particularly where railways lead into the interior. Open-cast coal is worked at Moura.

New South Wales

In 1788, with the establishment of the settlement at Port Dixon in Sydney harbour, this area became an uncertain outpost of British colonization. Today New South Wales is the foremost primary-producer and industrialized state in Australia. In the eighteenth century, John Macarthur first encouraged British industrialists to see Australia as a source of wool, and it was the wool industry which was the foundation of the present prosperity of the state. The warm, dry regions to the west of the Great Dividing Range are well suited to sheep breeding. Droughts are a problem and overstocking in the early years hampered development. The tendency now is to breed smaller herds, and there have been large-scale water-conservation and irrigation schemes.

One such scheme is the Snowy River Scheme, initiated in 1949, whereby the waters of the Snowy River are diverted into the headwaters of the Murray and Murrumbidgee to provide irrigation water and hydro-electric power. This available power should lead to industrial development. Other major reservoirs are the Hume Weir on the Murray, the Burrinjuck Dam on the Murrumbidgee and the Wyangala Dam on the Lachlan. The last two mentioned also have major hydro-electric power schemes. New South Wales has been slow to develop its large hydro-electric power potential, for its coal supplies produce sufficient industrial fuel.

Although compared to the sheep population, cattle are relatively unimportant, they supply the large local market with beef, while dairying is mainly established in the wetter coastal belt. Arable farming is well developed on the fertile areas of the western slopes, the Central Plains and the Riverina. This is the main wheat-growing area of Australia.

New South Wales is well endowed with minerals and the principal metal ores in the Broken Hill area are silver, lead and zinc.

Major coalfields lie to the north and south of Sydney—the northern around Newcastle. These have stimulated industry, particularly the iron and steel industry at Newcastle and Port Kembla. Most of the iron-ore is imported from other states. Like most other aspects of the Australian economy, the coalmines and iron and steel works suffer from labour shortages. To make full use of her potential, Australia still requires considerable numbers of immigrants and is attracting people from Britain and Europe.

Most of the population of the state live in the industrial areas around Sydney, and efforts are being made to decentralize industry.

Victoria

Physical factors and position have been very influential in the growth of the Victorian economy. The central position between the eastern and western markets of Australia is a distinct advantage in a country where long distances are reflected in high transport costs. The Southern Alps act as a partial barrier to movement between Melbourne, the state capital, and the interior.

Wheat, the mainstay of the agricultural pattern, is grown mainly in the Wimmera, Mallee and Northern districts of the state, where the rainfall is at least 15 in. per year. The climate is also suitable throughout the state for fruit- and vegetable-growing, and the canning industry has been established at Shepparton in the Goulburn Valley. More than half of the Australian dried-fruit production comes from the irrigated areas around Mildura.

Although a small state, Victoria ranks with the larger states as a pastoral producer: as early as 1850 the sheep population was over 6 million, and the importance of sheep has continued despite setbacks, as when a series of drought years accur. Wool-production and quality has declined a little in recent years because of the increasing preference for fat lamb production. The location of Victoria has been instrumental in developing a valuable dairying industry.

Mineral wealth greatly influenced the early settlement of the state. The discovery of gold at Bendigo and Ballarat in 1851, triggered off the Victorian gold rush. Some gold is still produced, but the industry is no longer a major employer. Coal has become the most important mineral. Small deposits of black coal at Wonthaggi are of declining importance; and the brown coal-fields of Welshpool, Altona-Balliang, Lal Lal and the Latrobe Valley, all within 100 miles of Melbourne, are more valuable. Open-cast coal is worked at Yallourn, where the thermal electric station supplies more than half the state's requirements. This has been supplemented by a new hydro-electric station in the Kiewa Valley. Natural gas has been found in Bass Strait.

Thus there is adequate power to sustain considerable industry. The main field of industrial development has been in the processing of agricultural produce and the manufacture of agricultural machinery.

Tasmania

Tasmania was discovered in 1642 by Tasman, a Dutchman. The island, much of which consists of a dissected plateau at about 3,500 feet above sea-level, lies in the Westerly Wind belt. Rainfall is high, only falling below 30 in. a year in sheltered places. River valleys cut deep into the highland area and these and the narrow coastal plains are the main zones of activity.

Agriculture, pastoral farming and mining are the chief occupations. Copper, silver and gold are mined on Mount Lyell, while zinc and lead are obtained at Rosebery. Queenstown is the major mining town. Recent exploitation of the forests stimulated the making of paper pulp.

FIG. 211. *Aerial view of Canberra showing the radial pattern of buildings.*

NEW ZEALAND

Since the whole of the Pacific basin is surrounded by a series of ranges of young fold mountains with associated volcanoes, it has been called "The Fiery Ring of the Pacific." New Zealand, whose three main islands, North Island, South Island and Stewart Island, lie 1,200 miles south-east of Australia, is part of this ring and is consequently mountainous. There are eighteen peaks over 10,000 feet and less than a quarter of the land surface is below 600 feet. The main present-day volcanic zone is in the eastern and central parts of North Island where active volcanoes, such as Ruapehu, fringe a volcanic plateau. Big depressions exist in the lava plateau, filled with water, such as Lake Taupo. Geysers and hot springs are numerous.

In South Island the mountain range clings to the west coast leaving the plain of Westland as a narrow coastal lowland. These mountains form a barrier to east-west communication, and are crossed with difficulty by roads and railways. In the south-west the seaward ends of the glaciated valleys have been drowned to form fiords. The Canterbury Plains on the east coast are the largest area of lowland in New Zealand. Although there are many rivers in both islands few are navigable, See map on page 472.

The main features of New Zealand's climate are demonstrated in the sketch maps on page 472. Note the relationship of rainfall to relief and prevailing wind direction. Thus the highland areas form a marked division between the pastoral economies of New Zealand: the wetter areas to the west of the

FIG. 212. *Auckland: note the bulk oil installations in the foreground.*

mountains are primarily cattle regions, while the drier regions to the east are sheep regions. This accounts for the preponderance of sheep farms in South Island, where the great lowland area of the Canterbury Plains runs along the east coast. A similar relationship can be shown for agriculture and forestry, the main forest areas being on the west-facing mountain slopes.

Although New Zealand is roughly the same distance from the Equator as Italy, its summer temperatures are comparable to those of Britain and it does not experience the droughts which handicap so much of Australia. In the north of the country, although grapes can be grown in the open air, they are seldom used for wine. Wheat and oats are the main crops, the latter being mainly used as fodder. With a total population of just over two million, New Zealand is almost self-sufficient in foodstuffs.

Since the rainfall is heavier than one finds in Australia, the pasture is richer. The colony was first settled in 1840 and from the start wool has been a major feature of its economy. Although merino sheep were introduced last century, they do not thrive as well as British breeds. Another indication of the similarity of climate to that of Britain is the successful introduction of British grasses. A tremendous increase in the prosperity of the islands came with the development of refrigeration. The first cargo of frozen meat left New Zealand in 1882 and, as the trade increased, there was a gradual change-over to the dual-purpose cross-breed sheep yielding both good meat and good wool. Refrigeration also saw the commencement of the dairying industry, the districts around Mt. Egmont being particularly important. Pasture is used as economically as possible, one strip being grazed at a time, the damp climate ensuring its recovery before it is again required. The butter and cheese factories are operated on a co-operative ownership basis.

Fruit is also a valuable export and the Nelson district is the largest apple-producing area. Oranges and lemons are grown in the Auckland peninsula, and the sheltered valleys of the Otago Plateau also yield abundant fruit.

Compared to its agricultural wealth, minerals play only a minor part in New Zealand's economy. Gold in Westland, Marlborough and the Coromandel Peninsula attracted many settlers in the gold rushes of the 1860s and at the turn of the century, but production has now declined. Coal occurs in many areas but production is small. The important areas are in the Buller and Gray districts of South Island and in the North Auckland region. Kauri gum found in the latter area is a fossilized resin used in the manufacture of linoleum and varnish, and was formerly of considerable value.

With limited coal deposits, hydro-electric power is of major importance. High relief, heavy rainfall without seasonal droughts, low evaporation rates and lakes which can be used as natural reservoirs, mean that New Zealand has great possibilities in this field. The major centres of development at present are on the Waikiti River in South Island and Waikato River in North Island.

Just over 8 per cent of the population are Maoris, living mainly in North Island. Perhaps surprisingly, nearly half of New Zealand's population are city-dwellers, and Auckland is the largest city with nearly 577,000 inhabitants. The ports have developed industries processing the agricultural products of the country and carry on an international trade.

PACIFIC ISLANDS

There are many small islands scattered over the Pacific Ocean, most of which are coral or volcanic in origin. They can be classified into Melanesia, Micronesia, Polynesia and The Hawaiian Islands.

Melanesia includes the long chain of islands from New Guinea to the Fiji group. Most of these islands differ in origin from the rest of the Pacific Islands in that they were joined to Australia at one period of their history. The Loyalty Islands, however, are coral.

Micronesia, as the name implies, consists of the numerous small islands lying to the north of Melanesia, such as the Caroline, and Gilbert and Ellice groups. Except for the Ladrone Archipelago, which is volcanic, most of these are coral atolls.

Polynesia covers the vast area to the east of Fiji. Some of the islands are volcanic, e.g. Tongan and Samoan groups, while others are coral, e.g. Cook Islands. The Society Islands are volcanic with fringing coral reefs.

The coral islands generally have poor soils and vegetation in contrast to the rich volcanic islands. Most have a tropical climate and produce plantation crops such as bananas, sugar, coconuts and copra. New Caledonia has reserves of nickel, while Nauru is an important phosphate producer.

The Hawaiian Islands form a state of the U.S.A. Many shipping lines connect the islands with the U.S.A., Canada, China and Japan (see map page 471). Honolulu, the largest city and port, has a profitable tourist industry. Sugar and pineapples are the most important crops grown. Farming is highly commercialized and highly mechanized, the main export trade being with the U.S.A.

QUESTIONS

1. Comment briefly on the relative importance of rainfall in different parts of Western Australia.
2. Write a short essay entitled "Sheep Farming in Australia."
3. Compare and contrast the physical geography of North and South Island, New Zealand.

CHAPTER 48

ANTARCTICA

ANTARCTICA is larger than Europe and covers some 5-6 million square miles around the South Pole. In Antarctica the Ice Age still exists, even more so than in Greenland. The land is almost covered by ice, which juts out beyond the land area in the form of shelf ice. In parts the ice is as thick as 15,000 feet and the whole continent is depressed by its weight. It is believed that the ice does not cover one single land mass but at least three separate islands. Around Antarctica the continental shelf is depressed by some 600 fathoms beyond the normal depth for a continental shelf.

The eastern part of Antarctica is a Pre-Cambrian shield area, on which lie horizontally bedded sediments. These are mainly sandstones, the product of a former desert age; others are coal seams, indicating former sub-tropical conditions. The western part of Antarctica dates from Mesozoic and Tertiary times and was formed at the same time as the Andes of South America. Mount Erebus is still an active volcano comparable to those of the Andes. Between eastern and western Antarctica there is a zone of rift valleys and block mountains, the tops of the mountain ranges standing out above the level of the ice.

Climate. It is difficult to generalize about the climate of Antarctica owing to a lack of recording stations. In summer the coastal temperatures rise to freezing point, but the mean annual temperature is between —30° C. (—22° F.) and —50° C. (—58° F.). Temperatures well below —75° C. (—103° F.) have been recorded inland, where temperature is very much modified by altitude. Depressions sometimes cross the continent, causing precipitation in the form of snow. Precipitation decreases inland, being equivalent to 12-20 in. of rain a year at the coast and only 3 in. inland.

In the interior, the accumulation of snow is less than 3 feet per annum, consequently the ice is not deepened to any great extent. However, there is no melting of ice inland and only small quantities melt at the margins, so that at present there is an almost exact balance between precipitation and melting. This has not always been the case, there being evidence that in former times the ice was much thicker.

This area has no vegetation except on the northern tip of Graham Land, and there are no indigenous animals left on the land. In the surrounding seas, however, life is prolific, and seals and penguins breed on some sections of the coast.

Human Activity in Antarctica

Whaling is carried on in the seas surrounding Antarctica; each whaling expedition has a large factory ship, up to about 50,000 tons in size, where the whales are cut up and processed. Each factory ship is served by some twelve "catchers," fast boats capable of sailing at 16 knots and carrying guns that fire a harpoon fitted with an explosive into the whale's body. Helicopters and sounding devices are used to help to locate the whales. The trade has much contracted in recent years, and only the Soviet Union, Norway and Japan take part in it. International agreements have been drawn up in an attempt to limit the number of whales caught in any one season, and to avoid what happened in the seventeenth century when seals were almost exterminated by the hunters.

Scientific stations which study all aspects of the continent have been established. Many of these stations were established in the International Geophysical Year of 1957-58. At them work glaciologists, geologists, surveyors, explorers and, perhaps most important of all, meteorologists. The latter require a considerable amount of further information about the climate and weather of Antarctica, since these play a very important part in the climate and weather of much of the Southern Hemisphere. This information will be essential if flying over the South Pole develops in the same way as the North Pole routes have developed. This is not likely to happen in the very near future owing to the lower density of population in the southern continents. Such flying routes would be long: e.g. Buenos Aires to Sydney is 8,000 miles, Sydney to Cape Town is 7,500 miles. There would be little chance of survival for passengers if planes were forced down near the South Pole, so a safety and meteorological base would be essential. McMurdo Sound, centre of recent exploration work, could probably provide such a centre.

Political Geography. As there were no native people in Antarctica, there has been little real political activity there until recently. In 1907 Britain claimed the Falkland Islands Dependencies Sector, and later Norway and Australia laid claim to other sectors. Britain also claimed the Ross Sea Dependency Sector, but handed her claim on to New Zealand. Britain claimed the Falkland Islands Dependencies Sector on grounds of actual occupation. Other areas were claimed on the grounds of having been explored by the countries concerned. After the Second World War, Argentina and Chile claimed parts of the Falkland Islands Dependencies Sector, but these claims were not pressed. One sector is unclaimed but the U.S.A. is interested in it. The U.S.A. does not recognize any claims and rather feels that this area should be under international sovereignty. Thus in Antarctica we see some of the problems of nationalism versus internationalism.

Why are the various countries interested in this ice-bound continent? The Andes, of which the western mountains of Antarctica are a continuation,

FIG. 213. *The Russian scientific base at Mirny.*

are rich in minerals and it is possible that high mineral concentrations exist too in Antarctica. With improving methods of mining, it is possible that these mineral concentrations could be profitably worked, as those in Greenland have been.

This area is considered to be of great strategic value, but even in this age of nuclear weapons this importance is overrated. The strait between Antarctica and South America may in time, however, increase in importance as a seaway.

QUESTION

1. Assess the importance of Antarctica in the world today.

REVISION SUMMARY—NORTH AMERICA

THE CONTINENT AS A WHOLE

A. GENERAL *Page*

Early voyages: Continent first sighted by Vikings who encouraged the journeys of Columbus, Vespucci and Cabot. 361

B. POSITION AND CLIMATE

Continent is one of contrasts from tundra to desert and from high ranges to plains; large area in temperate latitudes has rich resources and is favourable to European settlement. 361

C. RELIEF 362

Seven relief regions covering three highland areas and three lowland areas on the mainland, and the highland of the West Indies.

(1) **Canadian Shield:** Rich ore deposits as a result of volcanic action and intrusive igneous action. 362

(2) **Eastern Highlands,** with their Caledonian trend, have resulted from the complicated folding (in Armorican times) of a peneplain, and a subsequent series of alternating uplifts and depressions through to Tertiary times; they include the Appalachian coalfields and the Fall Line. 364

(3) **Western Cordillera:** Produced by more recent Alpine folding; three ranges with intermont plains and plateaux, and narrow coastal plains. Complicated river courses originated from local movements and volcanic action, while the whole area has been subject to intense glaciation. 365

(4) **Central Lowlands:** The largest area of plains; consist of horizontally-bedded Cretaceous and Tertiary sedimentary rocks, forming a broad belt from north to south. 365

(5) **Atlantic Coastal Plain:** Scarpland appearance, narrow in the north where subsidence has produced an indented coast; the plain widens south of Cape Hatteras as a result of emergence and there are open river valleys, coastal marshes and sandy reefs. 366

(6) **Gulf Coast:** Follows the coast of the Gulf of Mexico; Mississippi delta. 366

(7) **West Indies:** The remnants of a fold system accompanied by volcanic action which separate the Gulf from the Caribbean. 367

Page

CLIMATE AND NATURAL VEGETATION OF NORTH AMERICA

A. CLIMATE

(1) **Latitude** and the **size** of the continent are reflected in: 368

- (i) the variations of temperature from Arctic to subtropical;
- (ii) the vast area below freezing point in winter; and
- (iii) the large bends in the isotherms.

(2) Seven different types of **air mass** affect the continent. 368

(3) **Influence of ocean currents** 370

- (i) The warming-up of the westerlies on the coast of British Columbia and of the winds on the Florida coast.
- (ii) The fogs which occur along the coast of California and over the Grand Banks of Newfoundland.

(4) **The Chinook** of the eastern slopes of the Western Cordillera is air drawn in by depressions in the Central Lowland and warmed under compression in descent. 371

B. CLIMATIC REGIONS

The continent may be divided into ten climatic regions. The latitudinal spread is shown by regions 1 (arctic coastlands), 3, 5 and 8 (cool temperate), 4 (warm temperate), and 10 (tropical). Regions 2, 3 and 7 are characterized by wide temperature ranges because of the immense size of the continent. The influence of the oceans and their currents is shown in regions 5, 6 and 8 and wide variations in temperature and rainfall between the mountain slopes and valley floors are seen in regions 5 and 9. 371

C. NATURAL VEGETATION

Although most of the natural vegetation has been cleared, a covering of tundra and forest, semi-desert and swamp, still remains in places. 372

(1) **Forests** include the rich stands of conifers of the western mountain slopes, the northern taiga, the hardwood deciduous forests of the east and interior, and the evergreen woodlands of the south-east. 372

(2) **Grasslands** occur in the interior (the true prairie) and the east. 372

(3) **Tundra** is found in the north. 372

(4) **Deserts and Semi-deserts**—occur on the intermontane plateaux and the Lower Californian Peninsula. 373

(5) **Swamp** conditions are found in some wetter sub-tropical areas. 373

(6) **Tropical** conditions: Mexico and West Indies have vegetation ranging from grasslands to equatorial rain forest according to variations in altitude and rainfall. 373

Page

DISTRIBUTION OF POPULATION

A. GENERAL

Most of the people live in the south-eastern part. Low densities are found in the extreme north, where average temperatures are very low and soils are thin, and in the west where, because of the highland and intermontane semi-deserts, occupations are limited to pastoral and irrigation farming. The trading and industrial centres in the agricultural zones of Canada and the Middle West, and the intensively irrigated farmland of the west coast, are areas of high local densities within moderate average density. 374

B. CANADA

(1) Development is taking place on the margins of the old habitable regions and several points should be noted: 374

(i) The greatest density is in the industrial heart, in the St. Lawrence Lowlands and the Lakes Peninsula.

(ii) People in the Maritime Provinces live mainly along the coast.

(iii) There is an even distribution in the sparsely-populated agricultural area of the southern prairies.

(iv) There are scattered rural centres in British Columbia and the Canadian Shield.

(v) The lowest density (under five per square mile) is in the highlands of Alberta.

C. U.S.A.

All the advantages of timber, fuel and power resources, fertile soils, healthy climate and excellent harbour sites have enabled the north-east to become the most densely populated part. The farmland area of the centre and south is evenly populated with some industrial patches of higher density. The lowest densities are again on the west except where there are commercial and mining centres. The character is essentially urban with an annual decrease of 2 per cent in the farming population. 375

D. MEXICO

The south is more densely populated than the north where mining and pastoral farming in the drier regions provide the only occupations. In the east, oil and gas deposits have stimulated the growth of large industrial centres. In the West Indies population distribution is very uneven. 375

E. POPULATION GROWTH 375

Page

THE ST. LAWRENCE APPROACH

A. POSITION AND RELIEF

The region covers the long depression of the course of the St. 380
Lawrence and spreads farther westwards to include the Great Lakes region. The Frontenac Axis separates the two parts and causes the Lachine Rapids. The Maritime Provinces and Newfoundland region is the northward extension of the Appalachians, while along the course of the St. Lawrence, the highland comes close to the river with steep and unbroken slopes which are not amenable to settlement. Winter ice is a further disadvantage. The Lakes Peninsula region is crossed by the abrupt Niagara Escarpment which is cut by the Niagara River to give the Falls. The Great Lakes area occupies a pre-glacial river basin draining to the Atlantic, with the warp hollows which were filled with meltwater at the close of the Ice Age. Glacial retreat was irregular, overflow channels were cut and marine sediments laid down.

B. REGIONS

	Relief, structure, minerals, etc.	Climate	Occupations	Towns	Page
(i) Maritime Provinces and Newfoundland	Continuation of Appalachians; coalfields; excellent soils of Nova Scotia; iron ore of Newfoundland; cover of glacial drift; broken coast.	Temp. range 17° — 22° C. (30° to 40° F.); rainfall over 40 in. all seasons; 130-150 days of frost.	Fishing on Grand Banks; pastoral farming in Prince Edward Island; saw mills; steel.	*Halifax, St. John, St. John's.*	384
(ii) St. Lawrence Lowlands	Ridges parallel to the river; glacial soils.	Temp. range slightly higher than in (i); rainfall 50 in.	Asbestos; maple syrup; timber; hydro-electricity.	*Quebec, Montreal.*	388
(iii) Upper St. Lawrence	Frontenac Axis; Lachine Rapids; new seaway.	Range over 28° C. (50° F.); rainfall 40 in. with summer max.	Dairying; saw mills.	*Ottawa.*	389
(iv) Great Lakes	Glacial drift; Niagara escarpment, river and falls	Affected by the lakes; warmer winters, lower rainfall, fogs, fewest frost days.	Marketgardening; fruit growing; dairying industry (cars, steel, machinery).	*Toronto.*	389

Page

THE CANADIAN SHIELD

A. GENERAL

The Shield is a peneplain that was at one time covered with ice; this removed soil and left hollows and drift-covered lowlands with irregular drainage. The Clay Plains round Hudson Bay are undulating and covered with boulder clay. 391

B. CLIMATE

(1) **North:** Arctic; summer temperature reaches about 10° C. (50° F.); permafrost; rainfall less than 10 in.; very little snow. 392

(2) **Hudson Bay:** Cold interior: summer temperatures 19° C. (66° F.); long frost period; rainfall 26 in. with summer max. 392

(3) **South:** Similar to that of Hudson Bay, but rainfall is slightly higher (30 in.); depressions responsible for rain and snow. 392

C. NATURAL VEGETATION—tundra or coniferous forests. 393

D. OCCUPATIONS

(1) **Fur-trapping** and **trading.** 393

(2) Extensive **mineral wealth, timber** and **power potential** has encouraged development. 393

(3) **Farming**—very restricted. 393

Farming	Mining	Power
(1) Clay Belt: fertile lake muds with mixed farming, but forestry more important. (2) S. Ontario: fodder crops, dairying and vegetables in land cleared of forests and drained.	(1) Western Edge: gold, uranium. *Yellowknife, Port Radium* and *Uranium City.* (2) Southern Edge: copper, nickel, cobalt, gold, *Flin Flon, Sheraton, Sudbury, Kirkland Lake.* (3) Labrador: haematite at *Schefferville.*	(1) Abitibi Canyon, Saguenay and St. Maurice valley for papermaking, and aluminium industry at Arvida. (2) *Otto Holden* on the Ottawa R. for timber and pulp. (3) *Bersimis* in Quebec.

E. GROWTH OF SETTLEMENTS 395

F. THE TUNDRA 396

THE CENTRAL LOWLANDS

A. GENERAL

The region stretches from the Arctic Sea to the Gulf of Mexico and comprises all kinds of alluvial material resting on sedimentary rocks. Drainage is to (i) the Arctic by the Mackenzie; (ii) Hudson Bay by the Saskatchewan; (iii) the Atlantic via the Great Lakes and the St. Lawrence; and (iv) the Gulf of Mexico by the Mississippi River. It also stretches through (i) the pioneer belt of the 397

Page

Mackenzie Basin with its slow population growth; (ii) the undulat- 397
ing plains of the Canadian Prairies with its particular farming development and the more recent development of its coal, oil and natural gas deposits; and (iii) the successive crop belts of the U.S.A. with the same mineral value. There is both latitudinal and longitudinal variation in soil and climate over the Canadian Prairies, but in the U.S.A., corresponding changes are more longitudinal. Southwards, the belts merge into each other with no well-defined boundaries, and the growing seasons become gradually longer southwards.

B. REGIONS

(1) The **Mackenzie Lowlands.** 397

(2) The **Prairies** of southern Canada and continuing into northern U.S.A., often referred to as the Spring Wheat Belt. 399

(3) The **Corn Belt.** 403

(4) The **Cotton Belt.** 406

Physical	Climate	Occupations, etc.	Page
(1) River irregular in flow and navigability; delta is tundra; valleys forested.	Sub-Arctic; short growing season.	Forest clearings: roots, oats, dairying; fur trapping, but little lumbering. *Inuvik* is the new town; pioneer belt; slow population growth.	397
(2) Undulating plains; longitudinalescarpments divide them into terraces. Dark brown soils. Temperate grasslands.	Spring and summer max. of rainfall (12-20 in.). Winter temperature below —18°C. (0° F.), summer, above 15° C. (60° F.). 100 days free from frost.	Spring wheat belt; but increasing mixed farming in the Parklands. Ranching in drier areas. Development of coal, oil and natural gas deposits in Alberta. Increasing industrialization. *Edmonton* now larger than *Winnipeg*. Others *Calgary*, *Regina*, *Minneapolis*, *St. Paul*.	399
(3) Undulating plain on Mississippi drainage; brown glacial soils.	Summer temperature 24° C. (75° F.) with warm nights. 30-40 in. rain with slight summer max. 110-150 days free from frost.	Maize fed to cattle and pigs. Winter wheat, fodder crops, tobacco (rotation and artificial fertilizers). Interior coal and oil fields and pipelines. *Chicago* is economic capital and industrial centre. *St. Louis* a river port.	404
(4) Mississippi flood plain; alluvial; black soils. Old outcrop of Ozark—Ouachita Hills.	Warmer than (3) [8-27° C. (47-80° F.)]. Summer max. of rain, 50 in. per annum, decreasing westwards. 200 days free from frost.	Cotton, tobacco, maize. Introduction of new crops, mechanization, and manufacturing industries. Alabama coalfield; oil and gas. Two settlement patterns: (*a*) on the Mississippi, e.g. *Memphis*; (*b*) edge of High Plains, e.g. *Dallas*.	406

Page

THE HIGH PLAINS AND DUST BOWL

A. GENERAL

The Central Lowlands rise across the "Break of Plains" (the zone between 98° W. and 100° W. with less than 15 in. of rain per annum) to the High Plains, a series of plateaux with edges dissected by streams draining to the Mississippi and the Gulf of Mexico. The wetter and better soils of the east stimulate the growth of tall grass prairie; the more alkaline soils of the west have a short grass prairie cover with a lack of timber everywhere. The Indian buffalo plains provided simple crossing to the Pacific and became recognized as ranching country; there are few minerals and few towns. 408

B. FARMING 408

West	East
Spring maximum of rain, 12 in. Drier brown alkaline soils. Short grass prairie. Extensive beef-cattle ranching feeding on alfalfa, sugar beet for markets outside the region. Sheep-rearing (transhumance in summer).	Spring maximum of rain, 15 in. Better, wetter, dark soils. Long grass prairie. Wheat, sugar beet, alfalfa, cotton. Dry farming. Irrigation from the Colorado-Big Thompson schemes. *Denver*, *Pueblo* as industrial centres.

C. THE DUST BOWL

This area, extending through Kansas, Oklahoma and Texas, is a well-marked region largely because of the effects of soil erosion; irrigation and flood control schemes have been introduced. 415

ACROSS THE WESTERN CORDILLERA

A. GENERAL

The area has a complex structure, it forms a barrier to westward movement and shows a great diversity of climate. 416

B. REGIONS

(1) **Mountain chains** of the Rocky Mts. on the east; mining and smelting (e.g. at *Trail* in British Columbia), lumbering and ranching; dairying in the open woodlands; sheep rearing under transhumance conditions; towns such as *Calgary*, *Denver*, 418

Page

Pueblo, *Monterrey* are situated on routes through the mountains from the High Plains. 419

(2) The **Central Basin and Range Country,** an area of mountains, plateaux and basins with the Rocky Mountain Trench. Rainfall under 10 in. per annum; desert in the south-west; mainly convectional rain with summer maximum; large-temperature ranges and sunshine totals; agriculture using irrigation or dry farming; fur trapping (Yukon), orchard fruits (British Columbia), wheat (Palouse); irrigation schemes of **Grand Coulee,** the **Great Basin** and **Colorado.** 419

(3) The **Western Ranges;** coastal mountains and inner ranges with intervening longitudinal valleys; wet western slopes, timber covered; winter snows; latitudinal variation in temperatures. 422

ALASKA TO OREGON

A. GENERAL

The Pacific region has a longitudinal coastline, mountain chains, islands and peninsulas, and experiences cool temperate climate. Important area for fish, timber and dairying, but lack of level land generally restricts crop cultivation. However, there are two significant farming areas—the Willamette and Kent Valleys and the dyked shorelands of the Puget Sound. 424

B. FARMING

The chief features are the meat industry in Alaska, dairying in British Columbia and in the Puget Sound region; Kent Valley—vegetables; Willamette Valley—grain, grass and fruit. 426

C. DEVELOPMENT

Deposits of coal (Fairbanks, Vancouver Island), petroleum and natural gas (Anchorage) stimulated the growth of industries and the region is one of large urban growth, e.g. *Vancouver* has three-quarters of the population of British Columbia. Communication by sea is very important with *Vancouver*, *Seattle* and *Portland* as commercial ports as well as regional capitals. 427

CALIFORNIA AND FLORIDA

Both California and Florida have shown rapid increases in population, industrialization and agricultural development. They differ considerably in relief, climate and water supply. Most of the Californian population live in industrial areas, especially in Los Angeles and San Francisco. Florida has a fairly uniform distribution of population. 430

	California	Florida	Page
Relief	Central Valley, a trough with a deep alluvial cover. Rift valley of Golden Gate. Longitudinal coastline.	Low-lying peninsula. Coast of sand bars and lagoons. Sink holes through subsidence in limestone, and swamplands (e.g. **Everglades**). Porous surface rocks.	430, 433
Climate	Mild winters, much sunshine. Depressions bring high rainfall to Coast Range. Winds offshore in the south. Cool Californian current produces coastal fogs, which lower temperatures.	High average temperatures, 13° C. (55° F.) in winter to 27° C. (80° F.) in summer. Onshore air blows over the warm Gulf Stream all seasons. High annual rainfall of 50 in. High winter sunshine figure.	430, 433
Occupations	Crops mainly from the Central Valley; cotton, barley, vegetables, sugar beet, fruit and rice. Irrigation **(Central Valley Project)** and long growing season. Variety of fruits. Citrus industry. Gold mining and petroleum refineries. Modern consumer industries at *Los Angeles* and *San Francisco.* Machinery, aircraft, fruit canning, food processing.	Irrigation necessary because of porous soils, but use of underground water. Citrus groves, tourist industry, truck farming, sugar cane, tung oil, beef cattle. Phosphate mining. Manufacturing of consumer goods and processing of foods.	431, 434

THE GULF COASTLANDS

A. RELIEF

This region has a coast of lagoons, swamps and sand spits broken by the Mississippi delta, and inner low-lying plains of marine deposits. 436

B. CLIMATE—warm winters, hot summers; rainfall varies. 436

C. OCCUPATIONS

(1) **Agriculture** is important in the west where the black soils of the drained swamplands are used for the growing of rice and sugar cane. Methods vary from the mechanized cultivation of rice in Louisiana using negro labour, cultivation of cotton and vegetables in Texas, again mechanized by using white labour and irrigation, to the primitive farming of the Mexican coast. 437

(2) **Mineral production** is also important (petroleum, gas, sulphur and salt), and with agriculture has produced a more balanced economy. There is a corresponding increase in urban population. The chief towns are ports, e.g. *New Orleans*, *Mobile*, *Galveston* and *Houston*. 438

Page

THE WEST INDIES

A. GENERAL

Although fairly uniform in structure and climate, the separate islands of the West Indies have developed differently as a result of the many different outside influences—British, French, Dutch, Spanish and American. Population distribution is uneven and the only towns of any size are ports. Many people have emigrated from the islands to the U.S.A. and U.K., presenting special problems in the receiving countries. Subtropical and tropical crops are important everywhere, but minerals such as bauxite, petroleum and iron ore are to be found only in specific islands. Tourism is increasing in value. 440

B. THE ISLANDS

(1) **Greater Antilles**—Cuba, Jamaica, Hispaniola and Puerto Rico. 440
(2) **Lesser Antilles**—Leeward and Windward Islands. 440

Cuba is the largest island. Valuable products include sugar, tobacco, metals—especially manganese and copper. The island has considerable potential. 442

Puerto Rico: Smallest of the Greater Antilles, and has a strong economic link with the U.S.A. Most important products are sugar, coffee, oranges, tobacco, grapefruit, pineapples and vegetables. Manufacturing industries have been introduced, e.g. rayon, cement and consumer goods. 443

Jamaica: Largest of the West Indies Commonwealth countries. Important commercial crops are sugar, bananas, citrus fruits, coffee and coconuts. The island is the largest producer of bauxite. 444

Trinidad: Chief mineral is petroleum. 444

THE ATLANTIC OUTLOOK

A. RELIEF

Main features: 446

(1) Variation of structure and relief of the Appalachian region.
(2) Young sediments of the Atlantic Plain south of New York with seaward dip slopes from low escarpments.
(3) Fall Line from the Piedmont.
(4) Effects of recent glaciation in New England.

B. CLIMATE

Rainfall is well distributed. Temperatures vary according to latitude. 446

	Page
C. FARMING	
(1) The many types of farming reflect the varied relief and the accessibility to markets.	447
(2) Cotton, maize and beans, tobacco and spring vegetables in the south; mixed farming and market gardening in Maryland and Pennsylvania; milk, poultry, hay and fruit in New England.	447
D. REGIONS	
There are four main industrial regions:	447
(1) **Central Alabama** and the **Tennessee Valley** with the scheme of drainage, flood control, irrigation and hydro-electric power production; iron and steel industry, metallurgy, chemicals, paper, agriculture.	447
(2) **Pennsylvania** is the greatest and depends on its coal, the Great Lakes waterway and the passes and gaps in the highlands. *Pittsburgh* and *Harrisburg* are respectively the regional capitals of Western and Eastern Pennsylvania.	448
(3) The **Seaboard Cities** of *New York*, *Philadelphia* and *Baltimore* have the advantages of hydro-electricity of the Fall Line, the shipping facilities and the gaps in the Appalachians, particularly in the case of New York, where the Great Lakes route meets the transatlantic route.	449
(4) **New England,** where the industries have remained because of inertia, for there is no coal, no iron ore and no raw materials.	450
E. DISTRIBUTION OF POPULATION	
(1) Population is densest in industrial areas.	451
(2) Rate of population increase has declined over the last twenty years, especially in New England, and in the coal and steel areas of Pennsylvania and West Virginia.	451

REVISION SUMMARY—SOUTHERN CONTINENTS

SOUTH AMERICA

A. RELIEF *Page*

(1) The **Andean cordillera** on the west coast rises to over 20,000 feet and forms a major barrier to communications. It may be divided into three major regions: 455

(i) A series of ridges and valleys in the north;

(ii) A central region where two main chains enclose the *Bolivian Plateau*; and

(iii) The southern region where there is a single chain.

South of 45° S. the narrow coastal plain disappears altogether and is replaced by a fjord coastline.

(2) **Highland plateaux** form the major structural features of the east coast—the *Guiana Highlands, Brazilian Plateau* and *Patagonian Highlands.* 457

(3) Three major rivers drain the main areas of lowland on the continent. The Orinoco system drains the **llanos.** The Amazon drains the area of equatorial rain forest known as the **selvas,** and ships are able to travel far inland. The most important is the third, the River Plate, as it drains a warm, temperate lowland. 457

B. CLIMATE—see maps. 456

C. COLONIZATION 458

D. REGIONS

(1) **The North Andean states:** There are three contrasting regions: 460

(i) The area to the east of the Andes which is more like the neighbouring areas of Brazil and Paraguay. Communications with the area are being improved and it is hoped to develop plantations. The sparse population mainly consists of Indians practising subsistence agriculture.

(ii) The Andean cordilleras, which for centuries have been the main zone of settlement. A wide range of minerals of economic importance exists, agriculture is being developed on a larger scale with plantations, and pastoralism on the Andean slopes. 461

(iii) The coastal zone ranges from hot deserts in south Peru to tropical rain forests in Colombia. Although oil is the most valuable product of the desert areas, tropical crops are 462

Page

grown where Andean streams provide water for irrigation. The tropical rain forests of *Colombia* are a poor area. Cattle rearing is significant in the warm temperate grasslands along the *Caribbean shores*. 462

(2) **Chile:** Three major regions can be distinguished: 463

(i) The northern desert region, where the economic wealth lies in the mineral deposits of nitrates, copper and iron.

(ii) The central and most heavily populated region, which has a Mediterranean climate and the typical agriculture associated with it. Industrialization is increasing here.

(iii) A cool temperate southern region, where forests predominate; agriculture includes hay, oats and potatoes.

(3) **Venezuela** 464

(i) Although Venezuela is a large country with a very small population, the discovery of vast resources of petroleum and iron and ore have put it in the forefront of world economics and encouraged considerable development.

(ii) The discovery of oil has brought significant changes in the settlement pattern and agriculture has become of secondary importance.

(iii) Although the interior of the country is largely underdeveloped, the revenue from oil will enable further exploitation of Venezuela's vast potential.

(4) **The Guianas** 466

(i) These contain the last remnants of colonialism on the South American mainland.

(ii) In the highlands, minerals are of considerable importance with Guyana and Dutch Guiana being leading world producers of bauxite.

(iii) The coastal region is the main agricultural region, with sugar, rice and oranges as the principal crops.

(5) **Brazil:** Although the largest of the South American republics, Brazil has vast sparsely populated areas with little development, e.g. the Amazon lowlands with their dense tropical rain forest (**selvas**). Indian subsistence agriculture is still the main form of land utilization. Modern methods of cultivation are only found near settlements which have developed as trading centres along the river and the railway. The underdevelopment in the **campos** of the interior states is gradually being overcome. Although plantations are being developed, cattle ranching is still the mainstay of the economy. The most highly 467

Page

developed area of Brazil is the Atlantic coastlands, particularly in the southern part where there is a high population density. Hydro-electric power has encouraged extensive industrialization. Agriculture, particularly pastoral farming, is still very important, as also is coffee growing in the north of the coastlands. 468

(6) **Uruguay:** The rolling plains of Uruguay with their fertile soils and warm temperate climate have stimulated extensive arable farming and cattle rearing. 470

(7) **Paraguay:** There is little industrialization, and agriculture is backward. 470

(8) **Argentina** 470

(i) In the rain shadow of the Andes, the settlement pattern of north-west Argentina is very much controlled by the availability of water.

(ii) Pastoralism is common, but in the north a variety of crops, particularly vines, are grown.

(iii) Patagonia, also in the rain shadow of the Andes, has a cool temperate climate.

(iv) Although sheep grazing has long been the main occupation of the area, the region now also supplies most of Argentina's oil.

(v) The raising of beef cattle and its associated industries are the principal occupations of the *Pampa* and have encouraged the growth of the great cities around the Plate estuary.

(vi) Wheat, maize and alfalfa are grown in the fertile crescent from Bahia Blanca to Rosario.

(vii) The **Chaco** of north-east Argentina is more similar to Paraguay than the rest of Argentina.

AUSTRALASIA

AUSTRALIA

A. REGIONS

Australia can be divided into three regions:

(1) The **Western Plateau,** with a very sparse vegetation which becomes sandy desert in the north. There are many areas of inland drainage. 475

(2) The **Eastern Highlands** form a barrier range along the east coast, but important gaps allow communication with the interior. The Great Barrier Reef lies offshore. 475

Page

(3) The **Central Lowlands,** which mainly occupy the Great Artesian Basin. 476

B. REGIONS:

(1) **Western Australia** 478

(i) A large part of the state is desert and scrublands, where minerals such as iron, gold, asbestos, copper and tin are worked.

(ii) In the northern districts, which have monsoonal rain in winter, cattle ranching is the main occupation.

(iii) The most densely populated area in Western Australia is the south-western tip, which has a Mediterranean climate, and as rainfall gradually decreases inland, a distinct sequence of agricultural belts can be distinguished.

(2) **Northern Territory** has a very small population, and although the exploitation of minerals is increasing, the raising of beef cattle is of more economic value. 479

(3) **South Australia** 479

(i) Pastoral farming is the principal occupation in the Murray-Darling Basin, where great use has been made of irrigation.

(ii) Higher rainfall in the south encourages the growing of wheat.

(iii) There is a long tradition of mineral-working, and silver and lead are still mined.

(iv) Iron and coal are becoming increasingly important.

(4) **Queensland** may be divided into four regions: 480

(i) **Western Highlands**—mineral working.

(ii) **Western Plains**—cattle production in the monsoonal north. Sheep in the drier south where artesian supplies are the main source of water. Petroleum and bauxite.

(iii) **Eastern Highlands**—a barrier to communications, and has considerable mineral wealth.

(iv) **The Coastal Plains**—there is a strong influence of climate on the type of agriculture, varying from maize-growing in the warm temperate south to sugar-growing in the tropical north. Towns are strung along the coast.

(5) **New South Wales** 480

(i) Major irrigation and hydro-electric schemes, such as the Snowy River project.

Page

(ii) The warm, dry regions to the west of the Great Dividing Range are ideal for sheep rearing; major coalfields on the east coast have encouraged industrialization. 480

(iii) The problem of labour shortage is being gradually overcome by immigrant Europeans.

(6) **Victoria.** Wool production has shown a slight decline in recent years. The processing of agricultural produce is an important industry; coal mining has replaced gold as the major extractive industry. 481

NEW ZEALAND

The "Fiery Ring" of the Pacific is seen in the volcanoes and hot springs in eastern and central North Island. Young fold mountains form a spinal chain in South Island. There is only a very narrow coastal plain in the west, and the main agricultural areas of South Island are on the *Canterbury Plains* to the east. The wetter western districts are associated with cattle-rearing and forestry on the mountain slopes, while on the drier Canterbury Plains sheep are reared. Wool and lamb production are of great economic value. New Zealand is also self-sufficient in food supply. Fruit is another important source of revenue. 483

Hydro-electric power has been extensively developed. Minerals are no longer particularly important in the New Zealand economy and the processing and packing of agricultural products is the main industry.

PACIFIC ISLANDS

There are many small islands scattered over the Pacific Ocean, most of which are coral or volcanic in origin. The coral islands are generally infertile contrasting with the rich volcanic islands. Many of the islands, especially the Hawaiian Islands, lie on important trade routes. 485

ANTARCTICA

Antarctica is an almost completely ice-covered continent, where the ice is often many thousands of feet thick. The area has a polar climate and there is little vegetation except on the northern tip of Graham Land. Whaling and sealing are of considerable economic value. The other "settlement" in the Antarctic is for scientific reasons. Many stations have been established, especially during the *International Geophysical Year*. 486

It is thought that Antarctica may have considerable wealth, as the geological structures of South America are continued into the continent, but for a long time to come exploitation may be too expensive and too difficult.

THE EXAMINATION

THE aim in answering examination questions in Geography is to demonstrate not only that you know the facts but that you are also capable of setting them down in an organized fashion so that they are *relevant* to the question. This involves reading through each question very carefully. What does the question demand from you? No question will try to find out all you know about an area, industry, or relief feature. Rather it will be asking you to select from your knowledge to bring forward the facts which must then be so put together as to be *relevant*. It will also mean demonstrating the *relationships* which exist between one area and another, or two parts of the same area, or different groups of phenomena. Mere lists of facts poured out on the paper are unlikely to get you pass marks.

Some questions may ask you to "compare and contrast." In this case do not simply write two separate descriptions one after the other—this is not answering the question. Tabulating the information in vertical columns side by side may be a useful way of establishing the differences and comparisons for revision purposes but the examination answer should read consecutively, e.g. "The Mediterranean climate has rainfall in winter whereas in areas with a tropical monsoon climate rain is experienced in the summer. Because of this the vegetation in the Mediterranean lands is parched and brown in the summer whereas in areas with a tropical monsoon climate it is live and green . . . etc."

A geographical description of an area involves much more than listing the facts. Think in terms of the natural environment and man's response to it.

In almost every question the answer can be improved by a sketch map or diagrams. You will find it difficult to draw sketch maps unless you have practised them beforehand. It is useful to do your revision in terms of sketch maps, which give the idea of distributions and the relationship of different phenomena to one other. A sketch map can save much writing but must be referred to in the answer it illustrates.

Always provide a brief introduction setting out the salient points of your answer; this gives the reader an idea of what is coming. Finally bring the answer together with a conclusion.

In the suggestions on pages 523-5 some questions are answered more fully than others. They present one way in which the question could be answered. For other questions some of the problems posed by the type of question are indicated without detailed frameworks.

SPECIMEN EXAMINATION PAPER

Two hours are allowed. Answer any four questions. Sketch maps and diagrams should be drawn whenever they serve to illustrate an answer.

1. Draw a sketch map showing the location of the coalfields of the British Isles. For any one coalfield describe the development and characteristics of its manufacturing industries.
2. Describe and account for the farming landscape in two of the following: (*a*) Cheshire; (*b*) North-East Scotland; (*c*) East Anglia.
3. Select *one* upland region of the British Isles and account for the size and distribution of the population.
4. Write a geographical account of *either* Norway *or* Denmark.
5. Choosing *either* the Rhine *or* the Danube, describe and account for the importance of the river to the countries through which it flows.
6. Describe the major areas of hydro-electricity production in Europe and indicate the uses to which the power is put.
7. Compare and contrast the occupations of the inhabitants of New England and California.
8. What is meant by the term "monsoon"? Describe the effects of the monsoon climate on agricultural activities in the Indo-Gangetic lowland.
9. Write geographical notes on two of the following: (*a*) Beef cattle rearing in Queensland; (*b*) The agriculture of the Canterbury Plains; (*c*) The extraction of minerals in Australia.
10. Describe and account for population distribution in South Africa.

OUTLINE ANSWERS TO TEXT QUESTIONS

INTRODUCTORY

1. (*a*) **Location** of Britain in relation to (i) Western Europe; (ii) Rest of World (sketch map to show location in N.W. Europe).
   (*b*) Long **development** of natural resources from prehistoric times; (i) possibilities for agriculture with export of wool, hides, tallow, etc.; (ii) **Mineral wealth** (sketch map to show coal and iron-ore fields); (iii) power resources (e.g. water).
   (*c*) Advantage of early development of **technology** to use natural resources.
   (*d*) **Trade links** established by exploration and development of overseas possessions.
   (*e*) Culminating in **export trade** with developed and developing countries—possibility of importing food to supplement home grown supplies—agricultural specialisation, intensive agriculture use of location to maximum advantage; providing services to the World, e.g. banking, insurance, etc.
2. (*a*) Establish **distribution of population** before Industrial Revolution (i) in areas of good agriculture—mainly in lowland Britain with maximum concentrations of rural population and many market towns; (ii) in areas of poorer agriculture but with low densities of population, e.g. upland areas; (iii) seaports and fishing ports; (iv) in London.
   (*b*) **Changes in population** distribution since Industrial Revolution—rural to urban: (i) to areas with water power, e.g. Pennine Valleys for textiles; (ii) to coastal coalfields, especially where iron-ore available for heavy industrial development; (iii) to inland coalfields.
   (*c*) Most recent changes involve movement of population to **conurbations,** including London, based on modern industries, services and distribution and transportation centres.

CHAPTER 1

1. **Upland Britain:** Mainly in north and west above 600 ft., uplifted blocks of old resistant, mainly crystalline rocks with thin, often acid soils: steep slopes and high plateaux, with marked erosional effects of glaciation; swift rivers, rapids, waterfalls; cliff coasts with sandy bays or rias or sea lochs (fjords); heath, moorland vegetation and bog, except on valley floors. Rigorous climate with heavy precipitation, short growing season, high incidence of frost.
   Lowland Britain: Mainly in south and east; below about 600 ft.; vales and lowlands with broad river valleys but with occasional steep slopes (e.g. escarpments); wider rivers with lower gradients; estuaries and low coasts; younger sedimentary rocks with cover of glacial deposits (north of the Thames-Severn line); few areas of non-cultivated ground; soils generally better and deeper than in uplands; climate less severe and longer growing season, lesser incidence of frost, lower rainfall.

2. Occurrences north of Thames-Severn line: differences between uplands and lowlands: provide examples of different types indicating modifications of earlier land forms.

   Uplands—mainly erosional forms due to action of ice and meltwater with some depositional forms in valleys.

   Lowlands—mainly depositional forms but with local erosion by ice and especially by meltwater.

   Erosional forms—modification of cross and long sections of former river valleys to give truncated spurs and valley floor hollows (now lake-filled); tendency towards a U-shaped cross section, corries, aretes, drowned glaciated valleys (sea lochs); bare rock surfaces; meltwater channels.

   Depositional forms: (*a*) ground moraine, i.e. boulder clay (e.g. drumlins); (*b*) meltwater deposits, i.e. sand and gravel.

CHAPTER 2

1. **Temperature** higher than in surrounding rural areas; "heat island effect"; variations in temperature between centre and suburbs due to atmospheric pollution; higher winter temperatures, especially at night.

   Effect of **air pollution** on frequency of fogs—changes as a result of Clean Air Act. Variations within the city boundary resulting from different types of land-use, e.g. parks and built-up areas.
2. **Tundra climate** with arctic vegetation followed by Boreal period with birch as dominant tree in the early stages but, with warming up of climate, hazel and pine with some oak and alder became the characteristic vegetation. In Atlantic times, the "climatic optimum"—warmer and wetter—led to more widespread mixed oak and alder with much development of peat on the wetter western uplands. Deterioration of climate as summers became cooler and wetter with alder, beech and oak forests to give the general characteristics of the climate since the beginning of the Iron Age (about 9000 B.C.).
3. Variations in **temperature** and **precipitation** from season to season and year to year as a result of the unpredictable passage of different air masses. Give examples, using weather associated with different air masses. Unpredictable periods of **drought** and **heavy rainfall.** Severity of **frost**, and frost frequency variable from year to year; but farming in lowland valleys may be affected by frost more than on the marginal slopes. Drought conditions may not only cause lower yields but also blowing away of top soil (e.g., example from Fens).

CHAPTER 3

1. Important to draw a sketch map to show (*a*) distribution of coal fields; (*b*) main *areas* of production of hydro-electric power; (*c*) North Sea gas fields; (*d*) *main* nuclear power stations. Summarise **energy supplies** available (coal, electricity, oil and gas, water) and indicate their location using division between upland and lowlands as guide. Draw attention to the sketch map in your answer.
2. Basically the result of unequal distribution of **water supplies** compared with the distribution of **population;** surface water mainly in wetter north and west uplands while the population is mainly concentrated in the south and east lowlands. In the latter, underground supplies from aquifiers in the sedimentary rocks now being exhausted. Increases in water demands by people and industry.

Draw attention to cost of transporting water and high costs of desalination. (Sketch map to show wetter and drier areas).

CHAPTER 4

1. Note the phrase "in part" implying that factors other than climate, relief and soils are important. Draw sketch maps to show (*a*) upland and lowland areas; (*b*) wetter west and drier east; (*c*) decreasing length of growing season from south to north. A useful cross-section would show **changes in land-use** from low to high ground. Use examples from north and south, east and west, lowland and upland, bringing in references to slope, altitude, soils, etc., and the **farming** types of the different areas.
2. Outline the difficulties for **agriculture** and livestock rearing in the uplands, e.g. climate, relief soils, remoteness from markets, etc. Show how alternative method of land-use, e.g. **forestry**, may provide employment and a "crop" which can give a return under such difficult environments. **Fishing** (provide some details) offers another opportunity for a living where the land itself cannot support a large population. Use sketch maps to show distributions and draw examples from e.g. the Scottish Highlands and Islands.
3. Draw attention to (*a*) changes in **arable farming**—increased yields of all main crops with applications of fertilizers and pesticides; (*b*) improvements in **grassland management**; (*c*) increases in **livestock production.** Mention some of the reasons for these changes, e.g. application of **science and technology**, government **subsidies,** etc., **mechanization**, new systems of **marketing** (e.g. poultry) and transport. Provide examples from a wide variety of regions.

CHAPTER 5

1. This can be answered, in part, by drawing sketch maps to show how the **boundaries** between, say, counties fail to coincide with relief region boundaries, e.g. Yorkshire and Lancashire which contain both lowland and upland. For the city regions show, for instance, how the influence of **London** spreads over a number of counties and physical regions.
2. Draw a sketch map to show location of interior **coalfields** of England, main railway (use a recent atlas) and motorway systems. Emphasise development of **industries,** especially light, science-based and assembly industries and the build-up of population. Using the sketch maps show how the centrality of location favours marketing and distribution of consumer and export products. Give some attention to quality of agriculture in the Midlands.

CHAPTER 6

1. A good answer to this question will incorporate sketch maps of (*a*) the **site** of London and (*b*) its **location** in Southern England with reference to the western seaboard of Europe.

   Points to include are characteristics of the site in relation to the Thames, growth of the city from early times to the modern urban sprawl with influence beyond the built-up area of the city and its suburbs. Mention local and through-route transport problems, industrial development, airport problems, etc.

2. A sketch map to show the distribution of the **chalk lands** and the **clay vales** will be of assistance here. Give the main characteristics of the scarps and dip-slopes, water gaps, dry valleys, etc., using examples given in the text—pay special attention to the **Weald.** For the soils, distinguish between chalk land and clay vale soils, linking the argument with the presence or lack of surface water.

CHAPTER 7

1. Incorporate a sketch map to show (*a*) high and low ground; (*b*) main rivers; (*c*) distribution of coal and salt fields. In your answer show how the local natural resources have been developed to give **textile** and **chemical** industries; conclude by indicating the problems which the older industries introduced and describe the new industries which are being developed.
2. A good answer will incorporate points such as **raw materials** (wool) from the Pennine Upland, availability of **water** and waterpower, a short account of the early **textile** industries and a summary of the modern divisions between the various branches of the textile trades.

CHAPTER 8

1. There are several examples to choose from but select that which provides the maximum opportunity to discuss the **development** in order to explain the characteristics of **industry.** North-east England or South Wales would be very suitable; a sketch map to show location of the coalfield(s), estuaries, ports, is essential. Show how heavy industry of the past is being changed to incorporate modern science-based industries. Indicate the population problems of such areas.
2. A sketch map to show the heavy industrial areas is advisable, e.g. Central Scotland, especially Lanarkshire and Clydeside, North-east England, South Wales. Outline the characteristics of **heavy industries** and emphasise dependence of such areas on one or two industries. Outline the **difficulties** of their economies and show how government intervention has been necessary to provide new forms of employment—Special Areas, economic planning regions, etc.
3. Your answer should draw attention to the **differences** in the resource base of the three main physical regions of Scotland. Show how Central Scotland has the advantages of low relief, gentle slopes and areas of glacial deposition for **agriculture** compared with the Highlands and the Southern Uplands. A sketch map to show the distribution of **coalfields** in Scotland and the firths of the Clyde and Forth will help to develop the argument that the natural resources of the Central Scotland area are superior to other areas. Make some reference to the **industries** which have been developed.

CHAPTER 9

1. A question such as this demands an *orderly* description, which does not simply retail the facts, but attempts to link them together. An attempt should be made to show by sketch cross-sections how land-use changes with slope and altitude from valley floors to mountain summits. A good answer will show that there is competition in land-use where conditions of relief and soil allow, for instance, farming, forestry, recreational activities and water storage. Emphasise the role of the estates in determining what the land-use of a particular area will be.

2. Both **coastal** and **deep-sea fishing** based on north-east Scotland, favourably sited near North Sea and Icelandic fishing grounds. In the nearer waters **seine-netters** catch **cod** and **haddock,** and **herring** in summer and early autumn. Towns engaged include Buckie, Macduff, Peterhead, Fraserburgh, Stonehaven, Aberdeen, all over-dependent on the fishing industry. Shellfish also caught. **Trawlers** operate from Aberdeen. **Ancillary industries** include shipbuilding, marine engineering, fish-processing, box-making.

CHAPTER 10

1. Summarise the main characteristics of the Welsh massif in terms of its **land-use** and **economic** activities; indicate those which are of special significance to the lowlands, e.g. cattle-rearing, water supply, recreational areas.
2. (*a*) An orderly description is essential: deal first with **cereals,** etc., indicating the factors that make this a favourite cereal-growing area; then deal with **livestock.** Finally, draw the two together with reference to **mixed farming,** cattle-feeding. (*b*) This answer requires more than a description of the Fens; it must answer the question *why?* This involves a comparison with other lowland areas which have a similar situation, e.g. Somerset Levels or the carse-lands of the Forth; the aim is to show that in scale and in intensity of effort, the **reclamation** of the Fenland is quite unique in the British Isles.

CHAPTER 11

1. The **textile industry** was originally based on locally grown flax (now insufficient, so supplies imported); local water suitable for retting of flax stems. The industry was advanced by settlement of Huguenot weavers in early eighteenth century, and later, by the position of Ulster opposite Ayrshire and Cumberland coalfields. The industry has declined in recent years and the area is one of heavy unemployment. Fine linen goods still produced, but there is increasing use of man-made fibres and cotton. **Belfast** is centre of the industry; **other towns** involved include: Londonderry, Portadown, Ballymena, Larne.
2. **Belfast** much more a centre of industry than is Dublin, and Northern Ireland as a whole is more industrialized than the Republic. However, many new industries have been established in the south. Belfast's chief industries are shipbuilding and textiles (both depressed). There is also aircraft manufacture, light engineering, electrical engineering, milling, distilling and food and tobacco industries. **Dublin** principally involved in exporting dairy products, live cattle, potatoes. Brewing and distilling are important industries and provide major exports. Imports include wheat, fertilizers, tropical products, etc. Belfast imports coal, iron and other minerals, food, flax and tobacco. Dublin and Belfast are both important passenger ports dealing with the mainland. Belfast deals mainly with Clydeside, Heysham, Merseyside: Dublin with Holyhead and Liverpool.

CHAPTER 12

1. Without extremes of climate and weather: **influence of sea.** No insuperable physical obstacles to internal communication. Very long coastline—seaborne communications (small area penetrated by tongues of sea water). Most densely-populated continent.
2. Oldest: **Caledonian,** associated with Scotland and Norway. Middle period: **Hercynian,** associated with central Germany, central France, Iberia. Most recent: **Tertiary** Alpine, associated with the Pyrenees, Alps, Carpathians.

3. **North European Plain** good example of glaciated lowland; Finnish lowlands show glacial deposition (poor drainage, many lakes and swamps). **Alps** best example of glaciated mountains: sharply sculpted, U-shaped valleys, hanging valleys, etc. Glaciers still at work. Coast of Norway heavily glaciated; fiords.

CHAPTER 13

1. The relatively warm water of **North Atlantic Drift** makes climate of north-west European coastal belt warmer in winter than might be expected from its northerly latitude. Seas ice-free.
2. Moving west to east winters progressively more intense and summers hotter. In west rain more evenly spread, with **cyclonic rainfall** (from **depressions).** In east much rain from summer thunderstorms **(thermal).** East therefore more **continental** in climatic type.
3. Temperature decreases with altitude, but in winter air may drain into valleys making them colder than adjoining slopes. Permanent snow at very high altitudes. **Aspect:** south-facing slopes sunnier and warmer than northern slopes. West-facing slopes tend to be wetter than east-facing (which cannot gather moist Atlantic air).

CHAPTER 14

1. (*a*) Man has cleared forests, drained swamps, canalized rivers, cultivated land (introducing many non-European crops). (*b*) **Tundra:** wet moorland type of vegetation; found in far northern latitudes or, farther south, on upland surfaces. Usually devoid of trees, large areas of bog. **Loess:** fine yellowish dust probably blown from edge of retreating Quaternary ice-sheets. On it form fertile dark soils—e.g. **chernozem.** Very suitable for grain crops. **Rendzina:** dark reddish soil, rubbery when wet, particularly evident on limestones in eastern Poland; good for grain. **Podsols:** acid soil poor in humus. Tends to be very wet. Occurs in northerly latitudes where climate also unsuitable for farming.

CHAPTER 15

1. Europeans are of **very mixed** racial origins. Tall, long-headed, blond in north; shorter, darker people with either round or long heads around Mediterranean; medium build, brunette colouring, round to broad heads in Central Europe. **Germanic** languages spoken by about third Europe's population—English, German, Dutch, Danish, Swedish, etc. **Slavonic** languages spoken by another third—Russian, Polish, Czech, Serbo-Croat, etc. **Romance** languages make up another quarter—French, Italian, Spanish, Portuguese, Romanian. Remainder made up of Ural-Altaic, Asiatic languages (Finnish, Estonian) and Celtic and Basque. Emigration during nineteenth century. Effects of increased birthrate and decreased deathrate. Urbanization of populations.
2. First lines appeared in 1830s, but most major trunk lines built 1840-1900; secondary routes built until 1914; at this later period some trunk lines still being built in Eastern Europe. Since 1914 railways built have usually been short (except some in U.S.S.R.) and for specific reasons (e.g. Norwegian ore-carrying line to Narvik). Trunk lines tend to reflect pre-1914 political situation in Europe—particularly with regard to former Austro-Hungarian territory.

CHAPTER 16

1. (*a*) **Norway** has milder winters than Sweden because of influence of North Atlantic Drift. Mostly agricultural country. Fishing important. Also large merchant fleet. Fiords. Importance of hydro-electric power. Ice-free port of

Narvik terminal for iron-ore carrying railway from Sweden. **Sweden** largest and most industrialized Scandinavian country. Importance of iron ore (Kiruna); forest industries. High-quality manufactures. Hydro-electric power. **Denmark** mainly agricultural—particularly dairy products. Population dense by Scandinavian standards. Scientific agricultural methods. Imports electricity. Importance of domestic goods as exports. Shipbuilding also important.

2. **East German industry state-controlled** (Communist system), **West Germany private enterprise.** Both great industrial powers, particularly West—advantages in natural resources (particularly **coal**) in Saar, Ruhr and Aachen coalfields. East Germany has lignite and potassium salts. Investment, political situation and large population stimulated development of manufacturing industries in West; greater concentration on **utility goods** in East. **Heavy industry** in both. **Chemicals** very important East; **automobile industry** and manufacturing in West.

CHAPTER 17

1. Plains of European Russia (west of Urals). East of Urals plains of Asiatic Russia. Central Siberia dissected ancient plateau. Northern Siberia still rising. Southern Siberia built of great tilted blocks of country; signs of recent volcanic activity and many recently formed troughs (Lake Baykal).

CHAPTER 18

1. A **short hot summer,** with dust and thunderstorms; brief autumn; **long very cold winter** with heavy snow (in Siberia lack of adequate **snow cover**—because of the high winds sweeping vast open plains—leads to **permafrost**). Spring very wet as snow and frost thaws; also some rain and snow showers in spring.
2. Vast undulating grassland; few trees. In European Russia and Siberia all best steppe **cultivated.** Black soils developed over fertile **loess.** Parts of steppe badly gullied. Very cold in winter.

CHAPTER 19

1. Population of 225 million divided into over 160 national groups (three-quarters of population Slavs). Great Russian official language. Fifteen largest national groups form independent Soviet Socialist Republics (hence "Union of"). Legislation has to be passed by both Soviet of the Union (elected by universal suffrage) and Soviet of Nationalities—made up of twenty-five Deputies from the fifteen Republics.

CHAPTER 20

1. **Main fields** (three-quarters of reserves) now between **Urals** and **Volga.** Another group of fields is around the flanks of the Caucasus—notably near **Groznyy** and **Baku.** These have been exploited for longer than above.
2. **Three-quarters** of Soviet output comes from industrial belts of the **Ukraine** and **Urals.** A tenth comes from western Siberia, and slightly less than this from the Moscow region.
3. Organized into **state farms** and **collective farms**; no privately owned farms. State is sole customer for produce of state farms and collectives—with the exception of the peasants' own smallholdings on collectives. Mixed farming predominates in European Russia. Dairying, rye and flax in northern forest areas; also increasing area being sown with newly developed varieties of wheat that ripen in very short growing season. Maize in Byelorussia. Sugar beet and potatoes in

southern wooded steppe. Wheat on steppes, particularly in Ukraine. Sunflowers in southern farming belt. In wooded steppe and steppe many animals stall-fed from crop waste. Vines in Caucasia and Crimea. Market gardening in vicinity of big towns.

CHAPTER 21

1. Vast distances, severe climate; marshy ground, wide rivers, permafrost east of Yenisay. Rivers frozen in winter and flow north-south, whereas direction of arterial movement in modern Russia mainly west-east.

CHAPTER 22

1. **Hindrances** to the early explorers were mainly the lack of good harbours owing to the smooth coastline. On the rivers the waterfalls and rapids, combined with the irregular seasonal flow, made penetration of the interior by river difficult.
2. The heaviest **rainfall** is found in a broad belt athwart the Equator, where constant heat gives rise to convectional currents which cause regular heavy downpours. This rainfall belt moves northwards and southwards in response to the movement of the overhead sun, so that to the north and south of the Equatorial zone are areas of heavy summer rains—the savannas. Still farther north and south are areas where rainfall is very low or absent—the hot deserts. These are regions of high pressure and outward blowing winds, so that moist air cannot penetrate. The extreme north and south are regions of winter rain and summer drought—the "Mediterranean" areas.
3. **Rift Valley:** an area of the earth's surface where the rocks have sunk between parallel faults. Hence the valley is usually bounded by continuous lines of mountains.

   Potential Water Power: water power which is capable of being used but has not yet been utilized, e.g. a waterfall.

   Trade Winds: regular winds blowing in tropical latitudes. There is a belt north of the Equator (the north-east Trades) and another to the south of the Equator (the south-east Trades).

   Convectional Rain: rain caused by ascending currents of moist air. It is common in the tropics, and in temperate latitudes during thunderstorms.

   Monsoon: a seasonal reversal of wind such as occurs in India, West Africa and north Australia.

   Pastoralism: a mode of life where the mainstay of the economy is the herd of sheep, cattle or other animals. In Africa the word commonly implies a nomadic existence.

CHAPTER 23

1. **Cape Town** is in an area of Mediterranean climate, where the rainfall is moderate in amount and comes chiefly in winter. The vegetation is adapted to hot dry summers and mild winters, and consists largely of shrubs with trees on the higher ground. Beyond the mountains, rainfall declines northwards, and vegetation changes to grassland and then semi-desert scrub. In the Cape, the chief crops are wheat, fruit and vegetables, the rest of the section northwards to the Kalahari produces practically no crops.

   A similar sequence is found in the journey from Durban westwards. From the rainy east coast, with its luxuriant subtropical vegetation and abundant crops of sugar cane, bananas, maize and citrus fruit, we pass over the Drakensbergs once again into the lower rainfall of the grass veld where maize is the chief crop and westwards to the desert.

2. **South Africa** became the most industrialized country in Africa because it was settled by large groups of Europeans with an advanced culture who throve in the excellent climates of the region. Moreover, South Africa is rich in minerals such as gold and diamonds which provided the capital for economic development, and in iron ore and coal and many other raw materials for an industrialized economy.

CHAPTER 24

1. The African farm is small, and takes the form of a number of scattered strips. Farming is unscientific, and yields are low. Only a few traditional crops are grown. The white settler's farm is very large, well organized, compact and grows cash crops such as tobacco, maize, fruit, sugar and cotton. Much capital is employed and the farmer's standard of living is high.
2. **Minerals in Zambia.** The chief mineral product is copper, and there is a large export of this metal. **Kariba Dam:** this large dam on the Zambezi was completed in 1960. Large amounts of electricity are produced, and it is hoped to develop a fishing industry in the lake.

   Shifting Agriculture. This is the traditional form of land use over much of Africa. Patches of ground are cleared of forest or scrub, and the debris burned. Crops are grown in the soil thus fertilized, and cropping continues for a number of years. Once the soil is exhausted the process is repeated on another patch of ground.
3. Relatively low density of population in the Congo Basin may be accounted for in terms of the following: unfavourable climate; areas of dense forest; low standard of development of the population; high deathrate (caused by diseases); no great mineral wealth except in extreme south; poor communications. Export potential chiefly tropical plantation crops not developing high population density.

CHAPTER 25

1. The figures show a close correspondence between relief and temperature in the case of the three stations given, the highest (Nairobi) being 8° C. (14° F.) cooler in the hot season and 9° C. (16° F.) cooler in cool season than the lowest station. This is characteristic of temperature figures all over East Africa, and in fact over the whole world.
2. Land use in interior Tanganyika is chiefly governed by the rainfall and incidence of the tsetse fly.

CHAPTER 26

1. The vegetation changes between Lagos and Timbuktu from dense rain forest of Equatorial type, through savanna, with long grass and scattered trees, then savanna with scantier vegetation, until eventually the desert is reached. Corresponding changes in climate occur. At Lagos, heavy rain throughout the year gives way to summer rain in the savanna and little or no rain farther north. Similarly, temperature changes from constant heat at Lagos to a distinct summer-winter rhythm in the north.
2. The railway network could be completed by interior links to join the railways which run coastward. The handicaps include the necessity to import nearly all equipment; and damage caused by termites.

3. **Cocoa:** grown chiefly by peasant farmers, not in large commercial plantations. The small cocoa trees grow in shelter of other larger trees, and produce large pods which are cut open to extract beans. These are fermented, dried, bagged and exported, chiefly to European countries and North America.

Groundnuts: grown mainly in savanna areas in sandy soil. Harvested and bagged chiefly for export to western countries where they are pressed to extract oil used for margarine and soap.

Cotton is grown in drier areas of north and centre, sometimes with irrigation, e.g. on the Niger. Lint and seed picked and separated. Lint pressed into bales for export, seed also exported and pressed for oil.

CHAPTER 27

1. **The Climate of Algiers:** Typically Mediterranean, with mild winters and hot summers and rainfall chiefly in winter. Inland rainfall decreases and summer temperatures increase.

Relief of the Maghrib: a series of high Alpine ranges arranged in echelon, with intervening plateaux and lowlands. In Algeria, the Plateau of the Shotts occupies much of the interior and there is a discontinuous coastal range. In Morocco there is a coastal plateau.

Sahara Desert: world's largest hot desert. Variety of surface features—sand, clay, stones, rock, dunes. High mountains in centre. Low rainfall (under 10 inches) everywhere. Scanty population except at occasional oases and in Nile valley. Recent large oil and gas discoveries.

Minerals of North Africa: oil and natural gas fields in Algeria and Libya. Phosphates and iron ore in Maghrib, together with lead, zinc, manganese and copper.

Irrigation in Egypt: most important type now perennial, i.e. all year. Water from reservoirs, created by large dams. Older types of irrigation by shaduf, Archimedean screw, wells, etc. Egypt absolutely dependent on irrigated food supplies, hence the importance of the Aswan High Dam.

Manufactures in Algeria: chiefly concentrated in large towns—Oran, Algiers, Constantine. Main products: glass, textiles, soap, leather, woollens, superphosphates and wine.

Gezira Scheme: large irrigated area watered from a reservoir on the Blue Nile. Most important crop is cotton.

Lake Victoria experiences an equatorial climatic regime, with constant high temperature and heavy rainfall in every month. Here the vegetation is dense forest and savanna, i.e. grassland with scattered trees. Downstream climate changes to savanna, i.e. hot rainy summers and mild dry winters. In the Sudan, climate and vegetation become desert, with practically no rainfall or vegetation, and this continues to mouth of the river. Below Aswan, river flows through green belt of cultivated and irrigated land.

The Egyptian birthrate is high and recently modern medicine has lowered the deathrate. The area has been a centre of population from the earliest times because it is exceptionally fertile (Nile silt). It is intensively cultivated and irrigated so that food crops and cotton can be grown all year.

CHAPTER 28

1. **Land bridge** between Asia and Africa. **Sea routes** Europe/Far East with reference to Suez. Spread of religions and ideas from here, e.g. Islam; irrigation; agriculture; cities. Some mention of the role of **oil**, but in context.

2. The **desert** with the reasons for its occurrence (subtropical high pressures; dry, land winds). Expansion of the topic just mentioned to describe desert landforms but with emphasis on the restrictions they impose on movement and cultivation. Lack of adjacent wet highlands to supply rivers for **irrigation** (contrast Egypt); development of this to explain the limited groundwater resources. **Mineral resources** limited except for some oil in the east.
3. The **mode of occurrence** of oil; leading to the location of oil-bearing zones. Review of the **countries** with oil resources. The pattern of **trade** and **transport** of oil, with special reference to pipelines; canal dues; political conflicts. The **economic significance** to the countries concerned: its potential role in developing the backward and poorer regions.

CHAPTER 29

1. Draw a simple map. Begin with the wet Bengal delta, with **jute,** two **rice crops** in the region of prolonged monsoon rains and high temperatures, plus silt renewed by annual flooding. Moving westwards, explain the decreasing amounts of rainfall and length of wet season; the increasing need for irrigation; the cooler winters and associated changes in crops. **Overlap zone** of rice (summer), wheat (winter); sugar cane; well irrigation, then canals. **Wheat/cotton zone** of Punjab associated with canals and winter rain. Lower Indus; entirely dependent on irrigation.
2. Three main types: canal, well, tank. **Canals:** (*a*) ancient (e.g. Cauvery), (*b*) modern, e.g. Punjab and many peninsular rivers; the advantages and disadvantages; inundation and perennial. **Wells:** Location in relation to water table; old, shallow wells compared with modern tube wells. **Tanks:** Location mainly on non-porous peninsular rocks; only prolong growing season; contrast with dams which store from year to year.
3. Draw a map to show the **relative positions** and describe briefly. Compare the **sites,** e.g. *Calcutta* on the difficult delta compared with Bombay on series of islands; *Karachi* west of the delta; *Colombo* on a sandy coastline, requiring breakwaters and lighters. Discuss the **functions** viz. *Calcutta, Karachi* tapping agricultural hinterland initially; *Bombay* related to the cotton region; *Colombo* on world shipping routes and exports tea and rubber.

CHAPTER 30

1. Describe **location** of the regions; i.e. on Monsoon deltas. The **clearing of jungle** in Irrawaddy and Mekong deltas in the nineteenth century in response to the food requirements of plantation workers. Thailand different in being already settled. The **markets** for rice; Malaya, Ceylon, Indonesia, Japan, and why these countries import rice.
2. The basic **volcanic soil.** Efforts made by the Dutch to produce export crops; their introduction of intensive methods, and their extension of irrigation.

CHAPTER 31

1. Draw a simple map. **Climate:** wet south, dry north; warm south, cold north (winter). **Vegetation:** forested south; grassland, bare hills of north. **Relief:** hilly south, plateaux and alluvial plains and shallow seas of north. **Crops:** rice in south, wheat/millets of north. Industrial north, rural predominance even greater in south.
2. State importance of **size**; population; agricultural predominance; industrial resources. Approaches to **planning**; Five-Year Plan; emphasis on development

of heavy industry and then the expansion of consumer industries; railway construction and developments in communications; attempts to increase food production.

CHAPTER 32

1. Population pressure in relation to cultivable land. Relative lack of resources following defeat in 1945. Dependence on trade; loss of mainland markets, lack of developed markets in S.E. Asia; competition with western countries.

CHAPTER 33

1. British Columbia or Newfoundland (coastal areas) and the Canadian Shield or the Great Lakes' shores (inland areas) would provide suitable examples. Characteristic features of glaciation are present especially the fiords of British Columbia and Newfoundland; similar to the Shield with bare rocks, rock basins, morainic surfaces, distribution of erratics, diversions of drainage. In general these glaciated areas are not amenable to development; fiords often provide suitable harbours; the Shield Lakes used for water storage for hydro-electric power.

CHAPTER 34

1. (i) **British** type southwards to 43° N., (ii) **Mediterranean** type of the Californian coast, (iii) **desert** of South California. Factors which govern these climatic types: Western Cordillera behind the coastlands; presence of high and low pressure centres of the eastern Pacific with their associated wind systems; the Californian current.
2. (*a*) Chief influence is the lowering of temperatures owing to distance from the sea. Note Montreal is not frozen up because of the Labrador current.
   (*b*) **Mississippi Basin** receives considerable summer rainfall from winds from the Gulf of Mexico and the amount decreases from south to north; it also receives rain from depressions moving west to east and north-east so that the amount of rainfall increases from west to east over the Basin.
   In both (*a*) and (*b*) relevant climatic figures should be quoted.
3. Chief areas are in the intermontane plateaux of the Western Cordillera.
   Temperate: Snake River Plateau and the Great Basin. **Tropical and Subtropical:** Lower Colorado River basin and north-west Mexico, the peninsula of Lower California, north-east Mexico and across the Rio Grande into Texas. For explanation of distribution see pages 369-70.

CHAPTER 35

1. Compare and contrast: influence of position, climate, resources. In particular note labour demand created by industry.
2. **Eskimo** and **Aztec** (see pages 378-9). Treatment for the second part of this question will be similar to that of Question 1.

CHAPTER 36

1. Because of rugged relief, inclement climate, thin soils, fishing was originally the only occupation. Since the 1920s a number of new industries have been developed such as lumbering, pulp making and mining. Development has been stimulated by abundant supplies of hydro-electric power.

2. **Quebec:** situated where the St. Lawrence narrows; overlooked by high plateau. **Montreal** is an island site, hill feature. Montreal is at the limit of navigation and is a far greater port. Similarity in amount of trade, but Montreal more important because of the traffic of the prairies and Great Lakes. Montreal has therefore developed as a manufacturing centre. French influence on both.
3. Both in the south of Canada, where the population is greater than in the north; close to the Great Lakes (transport); valuable mineral deposits and sources of power. Traditional importance.

CHAPTER 37

1. (i) The greatest benefit will come to the parts of the Shield lying close to the Great Lakes for transport, timber and minerals, e.g. the industrial expansion of Toronto.
   (ii) The exploitation of Labrador **iron ore** and the transport of the ore to the industrial zones of the interior.
2. First locate the mining areas. Difficulties owing to climate, isolation and the production of food. Progress by development of roads and air transport and use of power for smelting of ores, expansion of dairying.

CHAPTER 38

1. Basin occupies part of the Plains, Shield and Western Cordillera. Answer should include where the river rises, lakes through which it flows, delta and the climate of the area. Development has been slow because of sub-Arctic climate, poor coast; commercial farming restricted to Peace River district; difficulties of transport; disadvantage of building on ground which experiences permafrost. High potential values of timber, ores and hydro-electric power.
2. In Canada, particularly in west. Saskatchewan and Alberta, early farming was subsistence, then extensive wheat farming with ranching in the drier Alberta; methods of dry farming and use of the short grass prairie. In U.S.A. the early settlements towards the Mississippi were established in forest clearings and in a humid climate, but beyond the Mississippi land agriculture was difficult until the introduction of modern farm machinery. Thus here ranching preceded agriculture.
3. Separate into **Spring Wheat** and **Winter Wheat** zones. Advantages of soil, climate (with evaporation rate as an important factor) and relief. Routes via Great Lakes, Hudson Bay, Vancouver and the Panama Canal, and some by the Mississippi.

CHAPTER 39

1. Contrast rainfall totals and temperatures and note seasonal distribution of rainfall. Differences caused by height, latitude, effect of chinook and of influence of tropical maritime air from the Gulf of Mexico.
2. Bookwork: see page 415.

CHAPTER 40

1. Use of river valleys, passes and cols across the watersheds, e.g. trails across Canadian Rockies through the three main passes, and the Fraser Thompson and Skeena valleys; the Oregon Trail over South Pass in Wyoming, and through the valleys of the Green, Snake and Columbia Rivers; California, Spanish-Santa Fé, Gila River trails in the south of the U.S.A. and all mainly starting from Kansas

City. Present system still uses passes from junctions on the east of the Rockies (atlases should be consulted and sketch map drawn).

2. The physical background can be shown best by sketch map showing the rivers, the surrounding highlands and the positions of the dams. "Economic importance" entails flood control, hydro-electric power generation as well as irrigation; the associated improvement of land for agriculture.
3. Say where the area is and name the specialized farming.

Staked Plains: level, sink holes, marginal rainfall, tufty grass, oil pumps, fences; drought-resistant wheat, improvements in cattle rearing, winter feed (hay) for cattle, increases in cotton production.

Okanagan: fruit on old lake shores and deltas; long, hot sunny summers; irrigation.

Salt Lake Basin: cereals (dry farming), sugar beet, alfalfa, vine under irrigation.

Coachilla Valley: dates; truck farming; cotton by irrigation from the Salton Sink by means of Coachilla Branch Canals. Palouse and the Indian reservations.

CHAPTER 41

1. Strategic position; ice-free harbours; resources include gold, copper, fish, fur and oil; potential in increasing timber production, and coal and water power; many potential grazing lands; improved communications essential to development.
2. Vancouver: long shoreline, ice-free inlet, extensive dock area; railway terminus and airport; trade with Asia and expanding manufacturing industry. Prince Rupert too far from populous areas; landlocked, island harbour at the mouth of the Skeena; fishing as the sole occupation.

CHAPTER 42

1. **Fruit-growing:** California experiences a Mediterranean climate over most of the area, with a long growing season; no frost in the south; covers a wide latitudinal area, therefore large variety of fruit; the alluvial soils of the Valley; co-operative marketing methods; need for high-quality products; abundant labour. Florida similar, with warm influence from the sea and onshore winds, equable moist climate; therefore a winter harvest of fruit for special markets in eastern U.S.A.

   Irrigation: see pages 431 and 434.
2. Extreme differences in relief and climate; similarities in agriculture and tourism; California has the oil, forestry, engineering and manufacturing industries while Florida has phosphate mining and a number of other industries are being developed. Greater importance of irrigation in California.

CHAPTER 43

1. **Relief:** the coast of sand spits, lagoons and swamps, and the warm, humid subtropical climate, the marine and river muds. **Crops** include sugar cane, rice and vegetables, tung oil. The area is too wet for cotton growing.
2. Increasing importance of oil, salt and sulphur. Expansion of manufacturing industries especially chemicals. Metal industries using bauxite, iron and zinc and especially the steel industry at Houston. Hence the population is becoming more urban with people migrating from the poorer inland areas.
3. **New Orleans:** river and a sea port, linked by rail to the Appalachian and Mississippi cities; focus for trade from the West Indies and from Panama and has a wide range of manufacturing industries based on the local raw materials.

Houston: larger than New Orleans; less variety of industries most of which are based on oil from Texas-Louisiana. Chief industries include meat packing, and the manufacturing of cotton, chemicals and steel.

CHAPTER 44

1. Canada has the products of a temperate area—grain, dairy produce and fruit—while West Indies has tropical products—fruit, sugar, rum, cocoa. Canada produces the luxury furs while Jamaica produces bauxite and Trinidad produces oil.
2. Chief advantages are warm waters and sandy beaches, fishing as a sport, equable climate, the proximity of the U.S.A. and Western Europe.

CHAPTER 45

1. **Main relief regions:** Cumberland Plateau of gently warped Carboniferous and more recent sedimentary rocks; Ridge and Valley region of highly folded and faulted sedimentary rocks varying from Cambrian to Carboniferous and built up by Hercynian earth movements; Great Appalachian Valley; Blue Ridge Mountains of the older Appalachians of crystalline rocks uplifted in Caledonian times; Piedmont Plateau of the same system which formed the coastal plain of an old continent, uplifted and marked at its edge by the Fall Line. These regions are best shown by sketch map and cross section.
2. (i) Points mentioned should include coking, and steam coals, early iron ore mining, importance of the Ohio Valley and the Great Lakes.
   (ii) Value as a port site (position on islands at a river mouth and hence its numerous waterfronts); at the terminus of the easiest route across the Appalachians; opposite Europe.
   (iii) The forested and mountainous interior impeded movements westwards; the south-east was the first area to be settled; here the first domestic industries were established (based on local raw materials), later expansion.
3. (i) Hudson-Mohawk Gap to the Great Lakes; (ii) New York and Philadelphia via the Delaware and Ohio Rivers; (iii) from Chesapeake Bay, route across the Appalachians is difficult and passengers now usually travel by air to Nashville (Tennessee); (iv) navigability of the Tennessee River has been improved by the T.V.A.; (v) the main rail route goes round the southern end of the Appalachians via Atlanta.

CHAPTER 46

1. Regional division can be made on basis of climatic regions. Sketch map showing latitudinal divisions and basis thereof.
   Northern Deserts: dry area; snow melting from Andes provides valuable water supply; some sheep farming in semi-arid areas (transhumance to Andes); main value of area is mineral wealth—nitrates, copper, iron.
   Central Chile: Mediterranean climate—best agricultural area of Chile; usual Mediterranean crops; Andes provide irrigation water and hydro-electric power; rapid industrialization; coal; most densely settled area.
   Southern Chile: cool temperate climate, similar to north-west Europe; oats and potatoes basis of agriculture but still largely a forest region; some minerals.
2. (*a*) **Patagonia**
   Position, relief and **climate.** Stress rain shadow effect of Andes giving almost desert conditions. **Agriculture**—some arable farming in the north where irrigation possible, but most of Patagonia devoted to pastoral farming. Sheep industry began by British and Spanish shepherds. Now major mutton and wool producing

area of Argentina. Marketing of produce—importance of markets in North America and Western Europe. **Value of technological advances,** e.g. refrigerated ships, etc. **Mineral wealth:** Argentina's principal oilfield at Comodoro Rivadavia; also natural gas piped to *Buenos Aires*; some lignite.

(*b*) **Pampas**

Sketch maps essential, drawing attention to cattle areas, the fertile crescent and the crop zones therein. Indication of isohyets particularly valuable. The monotonous landscape of the Pampas and the climatic factors. Describe crop growing of **fertile crescent.** Account for the rise of the cattle farming industry of the Pampas; use made of the entire animal; processing, packing and export centres.

3. Amazon drains one of the largest basins in the world, flowing eastwards from the Andes. Although the basin covers over two million square miles and ocean-going vessels can reach *Iquitos* 2,000 miles inland while many of the tributaries are also navigable, the whole region has a small population. This may be partly explained by the **equatorial climate** of the region: high daily convectional rainfall (over 100 inches per annum); high temperatures; high humidity all the year round. Dense tropical rain forest—**selvas.**

   Densely luxuriant forest landscape. Uneconomic to exploit forestry. If cleared—area has poor soils. Some tropical agriculture only around towns. Main method of land utilization is still shifting cultivation of Indians. Cattle rearing and plantation agriculture attempted. Minerals—growth of towns at collecting points, e.g. *Manaos.*

CHAPTER 47

1. Sketch maps showing isohyets essential. **Eastern desert area:** little economic potential. **North-western part:** 20-40 inches rainfall; cattle; problems of the industry. **Western scrubland region:** again unsuitable for agriculture but valuable mineral wealth has stimulated development. **South-west region:** distinct zoning of agriculture owing to rainfall.
2. Sketch map showing main sheep-farming areas of Australia and also showing isohyets. Discuss the relationship between sheep farming and water supply. Draw particular attention to rainfall, rivers and irrigation, artesian wells. Types of sheep and size of farms. Examine the marketing of sheep products—wool and mutton. Processing centres and export routes.
3. Sketch map showing relief. Mention mountainous nature of the country and how it fits into the general pattern of the Pacific. Describe parallel ranges of **Southern Alps** near west coast of South Island. Otago dissected plateau—influence of **glaciation—fiords.** Contrast lower mountains on east coast of North Island. Main areas of present-day **volcanicity** in North Island.

   Compare areas of lowland on both islands drawing particular attention to **Canterbury Plains.** Comment on the coastlines.

   Examine the drainage patterns on both islands noticing the respective value of the rivers to man and bringing attention to contrasts within each island, e.g. between east and west coasts of South Island.

CHAPTER 48

1. Description of the ice-covered continent—stressing the remote possibilities of settlement. Continent may have some **mineral wealth.** Some value if **transpolar air routes** developed—unlikely at present owing to: (*a*) great distances between southern hemisphere cities, and (*b*) low population density of southern hemisphere. Strategic and scientific value.

ANSWERS TO QUESTIONS IN THE SPECIMEN EXAMINATION PAPER

1. An important coalfield in the British Isles occurs in North Staffordshire. It is situated on the west side of the southern Pennines where the rivers, which unite to form the upper Trent, have cut a series of valleys trending roughly north and south in a low plateau of coal-bearing rocks. Based on this coalfield are the Potteries with towns situated mainly in the valleys, along which also run the canals and railways. Iron works were also developed in this area but more industries are required to give a greater variety of employment.

The coalfield has rich seams which, according to locality, can be worked for long-flame coals for firing pottery and tile kilns, for coking and gas coals. The relief of ridges and valleys made it possible to mine the coal by both adits or drifts and deep shafts. Along with the coal there are deposits of clay-band and black-band ironstone, which together became the basis of the iron industry. In addition, clays suitable for coarse earthenware or for the making of firebricks and building bricks were also available. The manufacture of pottery seems to

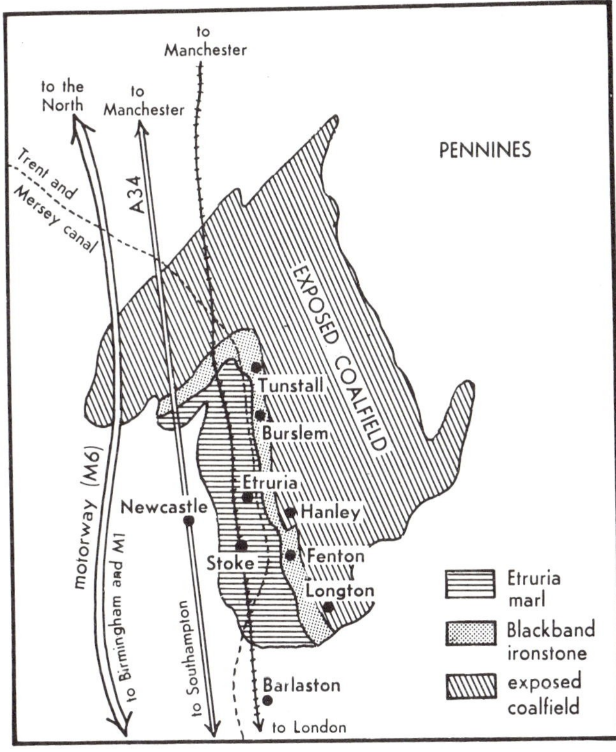

have begun as a cottage industry in the fourteenth century. To improve the quality of the pottery the area began to import, using river and packhorse, clay and flint from Dorset and Devon. This gave a whiter ware, superior to the coarse red earthenware pots produced from local clays. The most important changes took place in the late eighteenth century when Wedgwood introduced china clay from Cornwall and made it easier to import by encouraging the construction of a canal from the Mersey to the Trent. On the banks of this canal at Etruria, he established the first large factory for high-quality chinaware. Tunstall, Burslem and Hanley continued making coarse earthenware glazed for sanitary purposes with salt from Cheshire. The sketch map shows how towns are situated in relation to the coal and marl. The latter is used for engineering bricks, flooring and roofing tiles.

In recent years electric kilns have tended to replace the older coal-fired kilns and have reduced the smoke haze over the Potteries.

Iron smelting began in areas where charcoal was available but later developed at several places on the coalfield: often near the canals. Most of these iron works closed down with the changeover to steel, although Shelton is now a large iron and steel plant, using ore from the East Midlands and abroad, while Goldendale survives as an iron foundry using local coking coal because of its specialization in high-quality foundry iron.

The concentration on coal mining, pottery and iron works has recently been changed. New industries such as the manufacture of tyres, wire and cables give additional employment, but the landscape is still dominated by many characteristics of the older industries. Perhaps the situation of the Potteries in relation to the main trunk road and railway routes between London and the north-west will make it a main centre for future distributive industries.

2. **Description:**
   (1) Relief and land forms, character of glaciation, e.g. erosion or deposition, soil types, climate.
   (2) Hedgerows, stone dykes or fences.
   (3) Size and shape of fields.
   (4) Character of farm houses and farm buildings, orchards, etc.
   (5) Effect of agricultural type on above, e.g.:
      (i) dairying: permanent and rotation grass, hay barns, silage towers, etc.;
      (ii) beef cattle: grass and arable for fodder crops;
      (iii) large-scale cereal farming: wheatfields of East Anglia.

3. **Choose area,** e.g. Lake District, Wales, Southern Uplands, Scottish Highlands.
   Introduction: Relief, climate, soils, remoteness and difficulty of communications restrict agriculture and industries. Areas offer little employment and consequently usually sparsely populated. Population tends to be concentrated in the valleys. Tendency towards depopulation.
   Relief: Controls of altitude, effect of glaciation—importance of valleys—problems of agriculture in glaciated valleys.
   Climate: Control of altitude, higher precipitation; frequency of frosts—restrictions on agriculture.
   Soils: Effects of glaciation—thin, poor soils, leaching, podsols, etc.
   Occupations: Mainly pastoralism; sheep, some cattle, small amount of arable; forestry; quarrying; mining; basically dispersed rural population mainly in valleys. Few large towns.

4. (1) Draw **sketch maps** to show the major features of relief, climate, occupations

and major settlements.

(2) In the answer include small sketches and diagrams, e.g. Norwegian fiords and possible areas of settlement, etc.

(3) A good geographical account should stress man's relationships between man with his environment. It is useful to think in terms of (*a*) physical environment; (*b*) man's response.

(4) Requires a good introduction and conclusion.

5. (1) Draw a **sketch map** to show the river in relation to the major relief features and national boundaries.

(2) Importance of the river in (*a*) highland, (*b*) lowland, areas.

(3) Importance in relation to (*a*) communications, (*b*) transport and raw materials and manufactured articles, (*c*) hydro-electricity development, (*d*) floods and agriculture, (*e*) town locations, e.g. confluence and gap towns, town sites, (*f*) international boundaries.

6. Requires a **sketch map** to show areas of production and, if possible, larger scale diagrams of individual H.E.P. projects. Then group the areas by type of power station, and detail the uses of the electric power for industry, communications, and domestic.

7. This is a question which requires great care in presentation. It must *not* be a list of the occupations of the New Englanders followed by the occupations of the Californians. The answer should be based on the geographical differences between the two areas and man's response to these differences. Some of the differences depend on physical controls, relief, climate, soils, etc.; some depend on geological structure; others on climate. At the same time the variations in historical development are also significant. Sketch maps and diagrams would be very valuable.

8. (1) Define the term "monsoon" and give a brief account of the climate characteristics.

(2) Draw a sketch map to show (*a*) the Indo-Gangetic lowland with the major relief areas on the margins, and its rivers; (*b*) variations in rainfall (decreasing up the Ganges valley from the Bay of Bengal).

(3) Discuss the strong seasonal rhythm of the monsoon and its influence on agriculture, e.g. rice planting and harvesting, etc.

(4) Emphasize agricultural problems of irrigation and problems of population in relation to monsoon reliability, etc.

9. "Geographical" refers to the relation of the activity to the natural environment including the problems to be faced and the solutions which have been reached to establish the activity. Emphasize not only local relationships, e.g. water problems in Queensland, but also broader relationships, e.g. problem of markets and communications.

10. The best way to describe the distribution of population is by a sketch map using say "dense," "medium" and "sparse" categories. Try to bring out the reasons for the variations in population density in terms of the physical environment and the way it has influenced the growth of population by offering opportunities for agriculture, mining or industry, etc. Discuss briefly the racial problems.

INDEX

ACKNOWLEDGEMENTS

The publishers wish to thank the following for the use of photographs on the pages indicated: Aerofilms and Aero Pictorial Ltd., 61, 66, 97, 363. Australian News and Information Bureau, 482. British Iron and Steel Federation, 83. British Petroleum Ltd., 8 (centre), 40. Camera Press Ltd., 443. J. Allan Cash, 21, 35, 43, 109, 126, 149, 245, 246, 263, 328. European Community Information Service, 167, 169, 176, 178. Fairchild Aerial Survey, Inc., 366. FIAT, 193. Gas Council, 278. I.C.I. Fibres Ltd., 53. Interfoto, 190. Keystone Press Agency Ltd., 152, 194, 206, 209, 441. John Laing and Son Ltd., 59. LKAB, 171. London Brick Company, 46. Marples Ridgeway Ltd., 8 (top). National Film Board of Canada, 364, 384, 385, 389, 394, 395, 400, 401, 425. Royal Netherlands Embassy, 199. High Commissioner of New Zealand, 483. Novosti Press Agency, 200, 202, 207, 211, 212, 215, 217 (top), 219, 226, 227, 228, 229, 232, 233, 234, 236. Paul Popper Ltd., 33, 140, 144, 155, 183, 217 (bottom), 286, 291, 293, 299, 305, 317, 318, 319, 320, 329, 333, 433, 461, 464, 467, 469, 488. Pictorial Press Ltd., 142, 147, 151, 222. Portuguese State Information and Tourist Office, 173, 198. P.A.-Reuter, 103, 111, 416. U.A.R. Tourist and Information Centre, 274. U.K. Atomic Energy Authority, 8 (bottom). United Press International, 450. U.S. Information Service, 376, 377, 414, 420, 432, 437, 438. Yan, 170. Zambian Information Services, 258